Introduction to Inorganic Chemistry

Chemistry 121

Spencer L. Seager | Michael R. Slabaugh

CENGAGE
Learning·

Australia • Brazil • Japan • Korea • Mexico • Singapore • Spain • United Kingdom • United States

CENGAGE
Learning·

**Introduction to Inorganic Chemistry
Chemistry 121**

Chemistry for Today: General, Organic, and Biochemistry, Eighth Edition
Spencer L. Seager | Michael R. Slabaugh

© 2014, 2011 Cengage Learning. All rights reserved.

Library of Congress Control Number: 2012946128

Senior Project Development Manager:
Linda deStefano

Market Development Manager:
Heather Kramer

Senior Production/Manufacturing Manager:
Donna M. Brown

Production Editorial Manager:
Kim Fry

Sr. Rights Acquisition Account Manager:
Todd Osborne

For product information and technology assistance, contact us at
Cengage Learning Customer & Sales Support, 1-800-354-9706
For permission to use material from this text or product,
submit all requests online at **cengage.com/permissions**
Further permissions questions can be emailed to
permissionrequest@cengage.com

This book contains select works from existing Cengage Learning resources and
was produced by Cengage Learning Custom Solutions for collegiate use. As such,
those adopting and/or contributing to this work are responsible for editorial
content accuracy, continuity and completeness.

Compilation © 2013 Cengage Learning
ISBN-13: 978-1-285-91521-0

ISBN-10: 1-285-91521-6

Cengage Learning
5191 Natorp Boulevard
Mason, Ohio 45040
USA
Cengage Learning is a leading provider of customized learning solutions with
office locations around the globe, including Singapore, the United Kingdom,
Australia, Mexico, Brazil, and Japan. Locate your local office at:
international.cengage.com/region.

Cengage Learning products are represented in Canada by Nelson Education, Ltd.
For your lifelong learning solutions, visit **www.cengage.com/custom.**
Visit our corporate website at **www.cengage.com.**

Printed in the United States of America

Brief Contents

Chemistry for Today

General, Organic, and Biochemistry

1 Matter, Measurements, and Calculations

Case Study

Purpose: These scenarios introduce you, the student, to diverse situations that a health-care professional might encounter. The purpose of the case studies is to stimulate inquiry; for that reason, we've placed them at the beginning of each chapter of the book. Vocabulary and scenarios may be unfamiliar to you, but our intention is to stimulate questions in you, to pique your curiosity. Medicine has long been described as an art as well as a science. The questions raised by these case studies rarely have a single correct answer. With the knowledge that you gain from this text, and your future training, acceptable answers to the questions raised in our scenarios will become apparent.

Disclaimer: Some of the case studies are based on real-life situations. In such cases, names have been changed to protect the individual's anonymity. Any resemblance to a particular person is purely coincidental. Models are used in all photos illustrating the cases. No photos of actual people experiencing particular medical scenarios are ever used in this text.

Case Study: Mary was excited to visit the obstetrician. At 38 weeks of gestation, Mary was tired of being pregnant! Mary's doctor said that she would consider inducing labor if the fetus was ready. The ultrasound technician took a measurement of the fetus' femur in millimeters from the computer screen. It was consistent with 39 weeks of gestation. She also measured the amount of amniotic fluid in the spaces between the fetus' body and the interior of the womb. The combination of the femur length and the smaller-than-normal amount of amniotic fluid influenced the doctor's decision to approve an induction that afternoon. At birth, the baby weighed 5 lbs 13 oz, and her development was consistent with 35–36 weeks gestation—clearly, this fetus was not ready to be born.

How important are measurements to health professionals in their decision making? What other factors might have the physician considered?

Follow-up to this Case Study appears at the end the chapter before the Concept Summary.

Learning Objectives

When you have completed your study of this chapter, you should be able to:

1 Explain what matter is. (**Section 1.1**)

2 Explain the difference between the terms *physical* and *chemical* as applied to the properties of matter and changes in matter. (**Section 1.2**)

3 Describe matter in terms of the accepted scientific model. (**Section 1.3**)

4 On the basis of observation or information given to you, classify matter into the correct category of each of the following pairs: heterogeneous or homogeneous, solution or pure substance, and element or compound. (**Section 1.4**)

5 Recognize the use of measurement units in everyday activities. (**Section 1.5**)

6 Recognize units of the metric system, and convert measurements done using the metric system into related units. (**Section 1.6**)

7 Express numbers using scientific notation, and do calculations with numbers expressed in scientific notation. (**Section 1.7**)

8 Express the results of measurements and calculations using the correct number of significant figures. (**Section 1.8**)

9 Use the factor-unit method to solve numerical problems. (**Section 1.9**)

10 Do calculations involving percentages. (**Section 1.10**)

11 Do calculations involving densities. (**Section 1.11**)

C hemistry is often described as the scientific study of matter. In a way, almost any study is a study of matter, because matter is the substance of everything. Chemists, however, are especially interested in matter; they study it and attempt to understand it from nearly every possible point of view.

The chemical nature of all matter makes an understanding of chemistry useful and necessary for individuals who are studying in a wide variety of areas, including the health sciences, the natural sciences, home economics, education, environmental science, and law enforcement.

Matter comes in many shapes, sizes, and colors that are interesting to look at and describe. Early chemists did little more than describe what they observed, and their chemistry was a descriptive science that was severely limited in scope. It became a much more useful science when chemists began to make quantitative measurements, do calculations, and incorporate the results into their descriptions. Some fundamental ideas about matter are presented in this chapter, along with some ideas about quantitative measurement, the scientific measurement system, and calculations.

1.1 What Is Matter?

Learning Objective

1. Explain what matter is.

Definitions are useful in all areas of knowledge; they provide a common vocabulary for both presentations to students and discussions between professionals. You will be expected to learn a number of definitions as you study chemistry, and the first one is a definition of *matter.* Earlier, we said that matter is the substance of everything. That isn't very scientific, even though we think we know what it means. If you stop reading for a moment and look around, you will see a number of objects that might include people, potted plants, walls, furniture, books, windows, and a TV set or radio. The objects you see have at least two things in common: Each one has mass, and each one occupies space. These two common characteristics provide the basis for the scientific definition of matter. **Matter** is anything that has mass and occupies space. You probably understand what is meant by an object occupying space, especially if you have tried to occupy the same space as some other object. The resulting physical bruises leave a lasting mental impression.

You might not understand the meaning of the term *mass* quite as well, but it can also be illustrated "painfully." Imagine walking into a very dimly lit room and being able to just barely see two large objects of equal size on the floor. You know that one is a bowling ball and the other is an inflated plastic ball, but you can't visually identify which is which. However, a hard kick delivered to either object easily allows you to identify each one. The bowling ball resists being moved much more strongly than does the inflated ball. Resistance to movement depends on the amount of matter in an object, and **mass** is an actual measurement of the amount of matter present.

The term *weight* is probably more familiar to you than *mass,* but the two are related. All objects are attracted to each other by gravity, and the greater their mass, the stronger the attraction between them. The **weight** of an object on Earth is a measurement of the gravitational force pulling the object toward Earth. An object with twice the mass of a second object is attracted with twice the force, and therefore has twice the weight of the second object. The mass of an object is constant no matter where it is located (even if it is in a weightless condition in outer space). However, the weight of an object depends on the strength of the gravitational attraction to which it is subjected. For example, a rock that weighs 16 pounds on Earth would weigh about 2.7 pounds on the moon because the gravitational attraction is only about one-sixth that of Earth. However, the rock contains the same amount of matter and thus has the same mass whether it is located on Earth or on the moon.

Despite the difference in meaning between mass and weight, the determination of mass is commonly called "weighing." We will follow that practice in this book, but we will use the correct term *mass* when referring to an amount of matter.

matter Anything that has mass and occupies space.

mass A measurement of the amount of matter in an object.

weight A measurement of the gravitational force acting on an object.

A Central Science

Chemistry is often referred to as the "central science" because it serves as a necessary foundation for many other scientific disciplines. Regardless of which scientific field you are interested in, every single substance you will discuss or work with is made up of chemicals. Also, many processes important to those fields will be based on an understanding of chemistry.

As you read this text, you will encounter chapter opening photos dealing with applications of chemistry in the health care professions. Within the chapters, other Chemistry Around Us boxes focus on specific substances that play essential roles in meeting the needs of society.

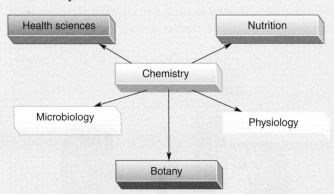

Chemistry is the foundation for many other scientific disciplines.

We also consider chemistry a central science because of its crucial role in responding to the needs of society. We use chemistry to discover new processes, develop new sources of energy, produce new products and materials, provide more food, and ensure better health.

Chemicals are present in everything we can touch, smell, or see. Chemistry is all around us.

1.2 Properties and Changes

Learning Objective

2. Explain the difference between the terms *physical* and *chemical* as applied to the properties of matter and changes in matter.

When you looked at your surroundings earlier, you didn't have much trouble identifying the various things you saw. For example, unless the decorator of your room had unusual tastes, you could easily tell the difference between a TV set and a potted plant by observing such characteristics as shape, color, and size. Our ability to identify objects or materials and discriminate between them depends on such characteristics. Scientists prefer to use the term *property* instead of characteristic, and they classify properties into two categories, physical and chemical.

Physical properties are those that can be observed or measured without changing or trying to change the composition of the matter in question—no original substances are destroyed, and no new substances appear. For example, you can observe the color or measure the size of a sheet of paper without attempting to change the paper into anything else. Color and size are physical properties of the paper. **Chemical properties** are the properties matter demonstrates when attempts are made to change it into other kinds of matter. For example, a sheet of paper can be burned; in the process, the paper is changed into new substances. On the other hand, attempts to burn a piece of glass under similar conditions fail. The ability of paper to burn is a chemical property, as is the inability of glass to burn.

physical properties Properties of matter that can be observed or measured without trying to change the composition of the matter being studied.

chemical properties Properties matter demonstrates when attempts are made to change it into new substances.

Cosmetics: Complex Mixtures and Complex Regulations

The federal Food, Drug, and Cosmetic (FD&C) Act defines a cosmetic as anything applied directly to the human body for cleansing, beautifying, promoting attractiveness, or altering the appearance without affecting the body's structure or functions. According to this definition, mixtures as diverse as a modern roll-on deodorant and henna, a colored plant extract used in ancient times as well as today to dye hair, are classified as cosmetics. However, it is interesting to note that according to the FD&C Act, soap is not legally considered to be a cosmetic.

The sale of cosmetics in the United States is regulated by the federal Food and Drug Administration (FDA), but the regulatory requirements applied to the sale of cosmetics are not nearly as stringent as those applied to other FDA-regulated products. With the exception of color additives and a few prohibited substances, cosmetics manufacturers may use any ingredient or raw material in their products and market the products without obtaining FDA approval. The regulation that provides consumers with the greatest amount of information about the chemical composition of cosmetics comes not from the FDA, but from the Fair Packaging and Labeling Act. This act requires that every cosmetic product must be labeled with a list of all ingredients in order of decreasing quantity. For example, many skin-care products contain more water than any other ingredient, so water is listed first.

Any cosmetic product that is also designed to treat or prevent disease, or otherwise affect the structure or functions of the human body, is regulated as both a drug and a cosmetic, and must meet the labeling requirements for both. Some well-known examples of this type of product are dandruff shampoos, fluoride toothpastes, and antiperspirants/deodorants. A good way to tell if you are buying a cosmetic that is also regulated as a drug is to see if the first item on the ingredient label is listed as an "active ingredient." Regulations require that the active ingredient be identified and listed first, followed by the cosmetic ingredients in order of decreasing amounts.

Many different types of products are classified as cosmetics. Each one must have a list of ingredients on the label.

---

physical changes Changes matter undergoes without changing composition.

chemical changes Changes matter undergoes that involve changes in composition.

You can easily change the size of a sheet of paper by cutting off a piece. The paper sheet is not converted into any new substance by this change, but it is simply made smaller. **Physical changes** can be carried out without changing the composition of a substance. However, there is no way you can burn a sheet of paper without changing it into new substances. Thus, the change that occurs when paper burns is called a **chemical change.** ❱ Figure 1.1 shows an example of a chemical change, the burning of magnesium metal. The bright light produced by this chemical change led to the use of magnesium in the flash powder used in early photography. Magnesium is still used in fireworks to produce a brilliant white light.

❱ Example 1.1

Classify each of the following changes as physical or chemical: (a) a match is burned; (b) iron is melted; (c) limestone is crushed; (d) limestone is heated, producing lime and carbon dioxide; (e) an antacid seltzer tablet is dissolved in water; and (f) a rubber band is stretched.

Solution
Changes b, c, and f are physical changes because no composition changes occurred and no new substances were formed.

The others are chemical changes because new substances were formed. A match is burned—combustion gases are given off, and matchstick wood is converted to ashes. Limestone is heated—lime and carbon dioxide are the new substances. A seltzer tablet is dissolved in water—the fizzing that results is evidence that at least one new material (a gas) is produced.

▶ **LEARNING CHECK 1.1** Classify each of the following changes as physical or chemical, and, in the cases of chemical change, describe one observation or test that indicates new substances have been formed: (a) milk sours, (b) a wet handkerchief dries, (c) fruit ripens, (d) a stick of dynamite explodes, (e) air is compressed into a steel container, and (f) water boils.

1 A strip of magnesium metal.

2 After being ignited with a flame, the magnesium burns with a blinding white light.

3 The white ash of magnesium oxide from the burning of several magnesium strips.

Figure 1.1 A chemical change occurs when magnesium metal burns.

Among the most common physical changes are changes in state, such as the melting of solids to form liquids, the sublimation of solids to form gases, or the evaporation of liquids to form gases. These changes take place when heat is added to or removed from matter, as represented in ▶ Figure 1.2. We will discuss changes in state in more detail in Chapter 6.

Figure 1.2 Examples of physical change.

A Solid iodine becomes gaseous iodine when heated.

B Liquid benzene becomes solid benzene when cooled.

1.3 A Model of Matter

Learning Objective

3. Describe matter in terms of the accepted scientific model.

scientific models Explanations for observed behavior in nature.

Model building is a common activity of scientists, but the results in many cases would not look appropriate on a fireplace mantle. **Scientific models** are explanations for observed behavior. Some, such as the well-known representation of the solar system, can easily be depicted in a physical way. Others are so abstract that they can be represented only by mathematical equations.

Our present understanding of the nature of matter is a model that has been developed and refined over many years. Based on careful observations and measurements of the properties of matter, the model is still being modified as more is learned. In this book, we will concern ourselves with only some very basic concepts of this model, but even these basic ideas will provide a powerful tool for understanding the behavior of matter.

The study of the behavior of gases such as air, oxygen, and carbon dioxide by some of the earliest scientists led to a number of important ideas about matter. The volume of a gas kept at a constant temperature was found to change with pressure. An increase in pressure caused the gas volume to decrease, whereas a decrease in pressure permitted the gas volume to increase. It was also discovered that the volume of a gas maintained at constant pressure increased as the gas temperature was increased. Gases were also found to have mass and to mix rapidly with one another when brought together.

A simple model for matter was developed that explained these gaseous properties, as well as many properties of solids and liquids. Some details of the model are discussed in Chapter 6, but one conclusion is important to us now. All matter is made up of particles that are too small to see (see ▶ Figure 1.3). The early framers of this model called the small particles *molecules*. It is now known that molecules are the constituent particles of many, but not all, substances. In this chapter, we will limit our discussion to substances made up of molecules. Substances that are not made of molecules are discussed in Sections 4.3 and 4.11.

The results of some simple experiments will help us formally define the term *molecule*. Suppose you have a container filled with oxygen gas and you perform a number of experiments with it. You find that a glowing splinter of wood bursts into flames when placed in the gas. A piece of moist iron rusts much faster in the oxygen than it does in air. A mouse or other animal can safely breathe the gas.

Figure 1.3 A hang glider soars far above the ground. How does this feat confirm that air is matter?

Now suppose you divide another sample of oxygen the same size as the first into two smaller samples. The results of similar experiments done with these samples would be the same as before. Continued subdivision of an oxygen sample into smaller and smaller samples does not change the ability of the oxygen in the samples to behave just like the oxygen in the original sample. We conclude that the physical division of a sample of oxygen gas into smaller and smaller samples does not change the oxygen into anything else—it is still oxygen. Is there a limit to such divisions? What is the smallest sample of oxygen that will behave like the larger sample? We hope you have concluded that the smallest sample must be a single molecule. Although its very small size would make a one-molecule sample difficult to handle, it would nevertheless behave just as a larger sample would—it could be stored in a container, it would make wood burn rapidly, it would rust iron, and it could be breathed safely by a mouse.

molecule The smallest particle of a pure substance that has the properties of that substance and is capable of a stable independent existence. Alternatively, a molecule is the limit of physical subdivision for a pure substance.

We are now ready to formally define the term *molecule*. A **molecule** is the smallest particle of a pure substance that has the properties of that substance and is capable of a stable independent existence. Alternatively, a molecule is defined as the limit of physical subdivision for a pure substance.

In less formal terms, these definitions indicate that a sample of pure substance—such as oxygen, carbon monoxide, or carbon dioxide—can be physically separated into smaller and smaller samples only until there is a single molecule. Any further separation cannot be done physically, but if it were done (chemically), the resulting sample would no longer have the same properties as the larger samples.

The idea that it might be possible to chemically separate a molecule into smaller particles grew out of continued study and experimentation by early scientists. In modern terminology, the smaller particles that make up molecules are called atoms. John Dalton (1766–1844) is generally credited with developing the first atomic theory containing ideas that are still used today. The main points of his theory, which he proposed in 1808, can be summarized in the following five statements:

1. All matter is made up of tiny particles called atoms.
2. Substances called elements are made up of atoms that are all identical.
3. Substances called compounds are combinations of atoms of two or more elements.
4. Every molecule of a specific compound always contains the same number of atoms of each kind of element found in the compound.
5. In chemical reactions, atoms are rearranged, separated, or combined, but are never created nor destroyed.

Early scientists used graphic symbols such as circles and squares to represent the few different atoms that were known at the time. Instead of different shapes, we will use representations such as those in ▶ Figure 1.4 for oxygen, carbon monoxide, and carbon dioxide molecules.

Figure 1.4 Symbolic representations of molecules.

The three pure substances just mentioned illustrate three types of molecules found in matter. Oxygen molecules consist of two oxygen atoms, and are called **diatomic molecules** to indicate that fact. Molecules such as oxygen that contain only one kind of atom are also called **homoatomic molecules** to indicate that the atoms are all of the same kind. Carbon monoxide molecules also contain two atoms and therefore are diatomic molecules. However, in this case the atoms are not identical, a fact indicated by the term **heteroatomic molecule.** Carbon dioxide molecules consist of three atoms that are not all identical, so carbon dioxide molecules are described by the terms **triatomic** and heteroatomic. The words *diatomic* and *triatomic* are commonly used to indicate two- or three-atom molecules, but the word **polyatomic** is usually used to describe molecules that contain more than three atoms.

diatomic molecules Molecules that contain two atoms.

homoatomic molecules Molecules that contain only one kind of atom.

heteroatomic molecules Molecules that contain two or more kinds of atoms.

triatomic molecules Molecules that contain three atoms.

polyatomic molecules Molecules that contain more than three atoms.

▶ Example 1.2

Use the terms *diatomic, triatomic, polyatomic, homoatomic,* or *heteroatomic* to classify the following molecules correctly:

Solution

A. Polyatomic and heteroatomic (more than three atoms, and the atoms are not all identical)

B. Polyatomic and homoatomic (more than three atoms, and the atoms are identical)

C. Diatomic and homoatomic (two atoms, and the atoms are identical)

D. Triatomic and heteroatomic (three atoms, and the atoms are not identical)

E. Diatomic and heteroatomic (two atoms, and the atoms are not identical)

▶ **LEARNING CHECK 1.2** Use the terms *diatomic, triatomic, polyatomic, homoatomic,* or *heteroatomic* to classify the following molecules correctly:

a. Water molecules have been found to contain two hydrogen atoms and one oxygen atom.

b. Molecules of ozone contain three oxygen atoms.

c. Natural gas is made up primarily of methane molecules, which contain one atom of carbon and four atoms of hydrogen.

The subdivision of molecules into smaller particles is a chemical change. How far can such subdivisions of molecules go? You are probably a step ahead of us and have guessed that the answer is atoms. In fact, this provides us with a definition of atoms. An **atom** is the limit of chemical subdivision. In less formal terms, atoms are the smallest particles of matter that can be produced as a result of chemical changes. However, all chemical changes do not necessarily break molecules into atoms. In some cases, chemical changes might just divide a large molecule into two or more smaller molecules. Also, as we will see later, some chemical changes form larger molecules from smaller ones. The important point is that only chemical changes will produce a division of molecules, and the smallest particles of matter that can possibly be produced by such a division are called atoms.

1.4 Classifying Matter

Learning Objective

4. On the basis of observation or information given to you, classify matter into the correct category of each of the following pairs: heterogeneous or homogeneous, solution or pure substance, and element or compound.

Unknown substances are often analyzed to determine their compositions. An analyst, upon receiving a sample to analyze, will always ask an important question: Is the sample a pure substance or a mixture? Any sample of matter must be one or the other. Pure water and sugar are both pure substances, but you can create a mixture by stirring together some sugar and pure water.

What is the difference between a pure substance and a mixture? Two differences are that a **pure substance** has a constant composition and a fixed set of physical and chemical properties. Pure water, for example, always contains the same proportions of hydrogen and oxygen and freezes at a specific temperature. A **mixture** of sugar and water, however, can vary in composition, and the properties will be different for the different compositions. For example, a glass of sugar water could contain a few crystals of sugar or several spoonfuls. Properties such as the sweetness and freezing point would vary depending on the amount of sugar present in the mixture.

Another difference is that a pure substance cannot be physically separated into simpler substances, whereas a mixture can theoretically be separated into its components. For example, if we heat a sugar-and-water mixture, the water evaporates, and the sugar remains. We say mixtures can theoretically be separated because some separations are very difficult to achieve. ▶ Figure 1.5 summarizes these ideas.

Pure substances, and mixtures such as sugar water, are examples of **homogeneous matter**—matter that has a uniform appearance and the same properties throughout.

atom The limit of chemical subdivision for matter.

pure substance Matter that has a constant composition and fixed properties.

mixture A physical blend of matter that can theoretically be physically separated into two or more components.

homogeneous matter Matter that has the same properties throughout the sample.

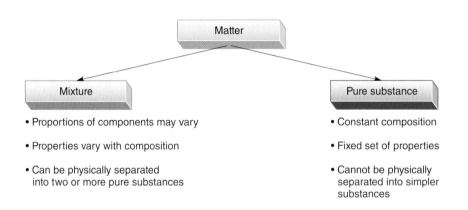

Figure 1.5 Mixtures and pure substances.

Matter	
Mixture	**Pure substance**
• Proportions of components may vary	• Constant composition
• Properties vary with composition	• Fixed set of properties
• Can be physically separated into two or more pure substances	• Cannot be physically separated into simpler substances

Homogeneous mixtures such as sugar water are called **solutions** (see ❱ Figure 1.6). Mixtures in which the properties and appearance are not uniform throughout the sample are examples of **heterogeneous matter.** The mixture of rock salt and sand that is spread on snowy roads during the winter is an example.

Commonly, the word *solution* is used to describe homogeneous liquid mixtures such as sugar water, but solutions of gases and solids also exist. The air around us is a gaseous solution, containing primarily nitrogen and oxygen. The alloys of some metals are solid solutions. For example, small amounts of copper are often added to the gold used in making jewelry. The resulting solid solution is harder than gold and has greater resistance to wear.

solutions Homogeneous mixtures of two or more pure substances.

heterogeneous matter Matter with properties that are not the same throughout the sample.

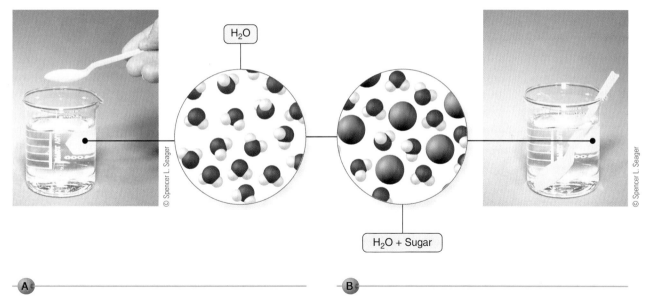

Figure 1.6 Sugar and water Ⓐ form a solution when mixed Ⓑ.

Most matter is found in nature in the form of heterogeneous mixtures. The properties of such mixtures depend on the location from which samples are taken. In some cases, the heterogeneity is obvious. In a slice of tomato, for example, the parts representing the skin, juice, seeds, and pulp can be easily seen and identified because they look different. Thus, at least one property (e.g., color or texture) is different for the different parts. However, a sample of clean sand from a seashore must be inspected very closely before slight differences in appearance can be seen for different grains. At this point, you might be thinking that even the solutions described earlier as homogeneous mixtures would appear to be heterogeneous if they were looked at closely enough. We could differentiate between sugar and water molecules if sugar solutions were observed under sufficient magnification. We will generally limit ourselves to differences normally visible when we classify matter as heterogeneous on the basis of appearance.

Earlier, we looked at three examples of pure substances—oxygen, carbon monoxide, and carbon dioxide—and found that the molecules of these substances are of different types. Oxygen molecules are diatomic and homoatomic, carbon monoxide molecules are diatomic and heteroatomic, and carbon dioxide molecules are triatomic and heteroatomic. Many pure substances have been found to consist of either homoatomic or heteroatomic molecules—a characteristic that permits them to be classified into one of two categories. Pure substances made up of homoatomic molecules are called **elements,** and those made up of heteroatomic molecules are called **compounds.** Thus, oxygen is an element, whereas carbon monoxide and carbon dioxide are compounds.

It is useful to note a fact here that is discussed in more detail later in Section 4.11. The smallest particles of some elements and compounds are individual atoms rather than molecules. However, in elements of this type, the individual atoms are all of the same kind, whereas in compounds, two or more kinds of atoms are involved. Thus, the classification of a pure substance as an element or a compound is based on the fact that only one kind of atom is found in elements and two or more kinds are found in compounds. In both cases, the atoms may be present individually or in the form of homoatomic molecules (elements) or heteroatomic molecules (compounds). Some common household materials are pure substances (elements or compounds), such as aluminum foil, baking soda, and table salt.

> ▶ **LEARNING CHECK 1.3** Classify the molecules represented below as those of an element or a compound:

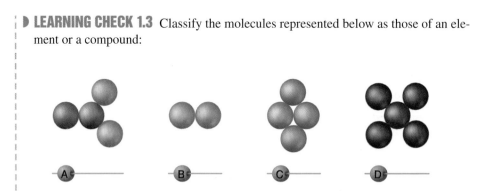

The characteristics of the molecules of elements and compounds lead us to some conclusions about their chemical behavior. Elements cannot be chemically subdivided into simpler pure substances, but compounds can. Because elements contain only one kind of atom and the atom is the limit of chemical subdivision, there is no chemical way to break an element into any simpler pure substance—the simplest pure substance is an element. On the other hand, because the molecules of compounds contain more than one kind of atom, breaking such molecules into simpler pure substances is possible. For example, a molecule of table sugar can be chemically changed into two simpler molecules (which are also sugars) or into atoms or molecules of the elements carbon, hydrogen, and oxygen. Thus, compounds can be chemically subdivided into simpler compounds or elements. ▶ Figure 1.7 summarizes these ideas, and ▶ Figure 1.8 illustrates a classification scheme for matter based on the ideas we have discussed.

element A pure substance consisting of only one kind of atom in the form of homoatomic molecules or individual atoms.

compound A pure substance consisting of two or more kinds of atoms in the form of heteroatomic molecules or individual atoms.

Figure 1.7 Elements and compounds.

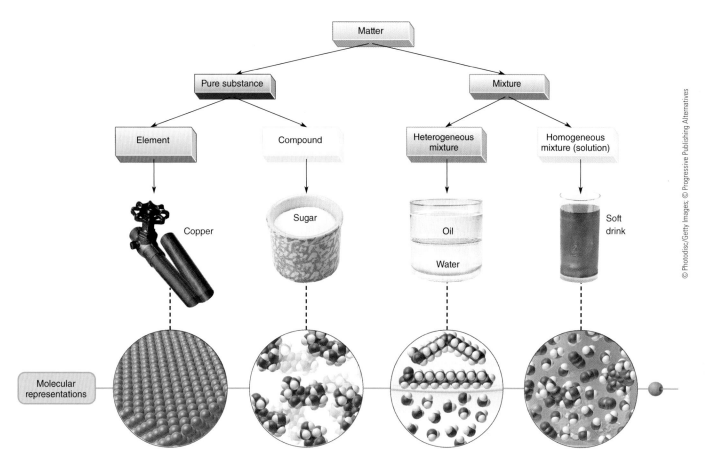

Figure 1.8 A classification scheme for matter.

▶ Example 1.3

When sulfur, an element, is heated in air, it combines with oxygen to form sulfur dioxide. Classify sulfur dioxide as an element or a compound.

Solution

Because sulfur and oxygen are both elements and they combine to form sulfur dioxide, the molecules of sulfur dioxide must contain atoms of both sulfur and oxygen. Thus, sulfur dioxide is a compound because its molecules are heteroatomic.

▶ **LEARNING CHECK 1.4** Suppose an element and a compound combine to form only one product. Classify the product as an element or a compound.

1.5 Measurement Units

Learning Objective

5. Recognize the use of measurement units in everyday activities.

Matter can be classified and some physical or chemical properties can be observed without making any measurements. However, the use of quantitative measurements and calculations greatly expands our ability to understand the chemical nature of the world around us. A measurement consists of two parts, a number and an identifying unit. A number expressed without a unit is generally useless, especially in scientific work. We constantly make and express measurements in our daily lives. We measure the gallons of gasoline put into our cars, the time it takes to drive a certain distance, and the temperature on a hot or cold day. In some of our daily measurements, the units might be implied or understood. For example, if someone said the temperature outside was 39, you would probably assume this was 39 degrees Fahrenheit if you lived in the United States, but in most other parts of the world, it would be 39 degrees Celsius. Such confusion is avoided by expressing both the number and the unit of a measurement.

All measurements are based on units agreed on by those making and using the measurements. When a measurement is made in terms of an agreed-on unit, the result is expressed as some multiple of that unit. For example, when you purchase 10 pounds of potatoes, you are buying a quantity of potatoes equal to 10 times the standard quantity called 1 pound. Similarly, 3 feet of string is a length of string 3 times as long as the standard length that has been agreed on and called 1 foot.

The earliest units used for measurements were based on the dimensions of the human body. For example, the foot was the length of some important person's foot, and the biblical cubit was the length along the forearm from the elbow to the tip of the middle finger. One problem with such units is obvious; the size of the units changed when the person on whom they were based changed because of death, change in political power, and so on.

As science became more quantitative, scientists found that the lack of standard units became more and more of a problem. A standard system of units was developed in France about the time of the French Revolution and was soon adopted by scientists throughout the world. This system, called the *metric system,* has since been adopted and is used by almost all nations of the world. The United States adopted the system but has not yet put it into widespread use.

In an attempt to further standardize scientific measurements, an international agreement in 1960 established certain basic metric units, and units derived from them, as preferred units to be used in scientific measurements. Measurement units in this system are known as SI units after the French Système International d'Unités. SI units have not yet been totally put into widespread use. Many scientists continue to express certain quantities, such as volume, in non-SI units. The metric system in this book is generally based on accepted SI units but also includes a few of the commonly used non-SI units.

1.6 The Metric System

Learning Objective

6. Recognize units of the metric system, and convert measurements done using the metric system into related units.

The metric system has a number of advantages compared with other measurement systems. One of the most useful of these advantages is that the metric system is a decimal system in which larger and smaller units of a quantity are related by factors of 10. See ▶ Table 1.1 for a comparison between the metric and English units of length—a meter is slightly longer than a yard. Notice in Table 1.1 that the units of length in the metric system are related by multiplying a specific number of times by 10—remember, $100 = 10 \times 10$ and $1000 = 10 \times 10 \times 10$. The relationships between the units of the English system show no such pattern.

TABLE 1.1 Metric and English Units of Length

	Base Unit	Larger Unit	Smaller Unit
Metric	1 meter	1 kilometer = 1000 meters	10 decimeters = 1 meter 100 centimeters = 1 meter 1000 millimeters = 1 meter
English	1 yard	1 mile = 1760 yards	3 feet = 1 yard 36 inches = 1 yard

basic unit of measurement A specific unit from which other units for the same quantity are obtained by multiplication or division.

The relationships between units of the metric system that are larger or smaller than a **basic** (defined) **unit** are indicated by prefixes attached to the name of the basic unit. Thus, 1 kilometer (km) is a unit of length that is 1000 times longer than the basic unit of 1 meter (m), and a millimeter (mm) is only $\frac{1}{1000}$ the length of 1 m. Some commonly used prefixes are given in ▶ Table 1.2.

TABLE 1.2 Common Prefixes of the Metric System

Prefix[a]	Abbreviation	Relationship to Basic Unit	Exponential Relationship to Basic Unit[b]
mega-	M	$1,000,000 \times$ basic unit	$10^6 \times$ basic unit
kilo-	k	$1000 \times$ basic unit	$10^3 \times$ basic unit
deci-	d	$1/10 \times$ basic unit	$10^{-1} \times$ basic unit
centi-	c	$1/100 \times$ basic unit	$10^{-2} \times$ basic unit
milli-	m	$1/1000 \times$ basic unit	$10^{-3} \times$ basic unit
micro-	μ	$1/1,000,000 \times$ basic unit	$10^{-6} \times$ basic unit
nano-	n	$1/1,000,000,000 \times$ basic unit	$10^{-9} \times$ basic unit
pico-	p	$1/1,000,000,000,000 \times$ basic unit	$10^{-12} \times$ basic unit

[a]The prefixes in boldface (heavy) type are the most common ones. [b]The use of exponents to express large and small numbers is discussed in Section 1.7.

Area and volume are examples of **derived units** of measurement; they are obtained or derived from the basic unit of length:

$$\text{area} = (\text{length})(\text{length}) = (\text{length})^2$$

$$\text{volume} = (\text{length})(\text{length})(\text{length}) = (\text{length})^3$$

The unit used to express an area depends on the unit of length used.

derived unit of measurement A unit obtained by multiplication or division of one or more basic units.

▶ Example 1.4

Calculate the area of a rectangle that has sides of 1.5 and 2.0 m. Express the answer in units of square meters and square centimeters.

Solution

$$\text{area} = (\text{length})(\text{length})$$

In terms of meters, area = $(1.5 \text{ m})(2.0 \text{ m}) = 3.0 \text{ m}^2$. Note that m² represents meter squared, or square meters. In terms of centimeters, area = $(150 \text{ cm})(200 \text{ cm}) = 30,000 \text{ cm}^2$. The lengths expressed in centimeters were obtained by remembering that 1 m = 100 cm.

▶ **LEARNING CHECK 1.5** The area of a circle is given by the formula $A = \pi r^2$, where r is the radius and $\pi = 3.14$. Calculate the area of a circle that has a radius of 3.5 cm.

The unit used to express volume also depends on the unit of length used in the calculation. Thus, a volume could have such units as cubic meters (m³), cubic decimeters (dm³), or cubic centimeters (cm³). The abbreviation cc is also used to represent cubic centimeters, especially in medical work. The liter (L), a non-SI unit of volume, has been used as a basic unit of volume by chemists for many years (see ▶ Figure 1.9). For all practical purposes, 1 L and 1 dm³ are equal volumes. This also means that 1 milliliter (mL) is equal to 1 cm³. Most laboratory glassware is calibrated in liters or milliliters.

▶ Example 1.5

A circular Petri dish with vertical sides has a radius of 7.50 cm. You want to fill the dish with a liquid medium to a depth of 2.50 cm. What volume of medium in milliliters and liters will be required?

Figure 1.9 A liter is slightly larger than a quart.

© Spencer L. Seager

Solution

The volume of medium required will equal the area of the circular dish (in square centimeters, cm^2) multiplied by the liquid depth (in centimeters, cm). Note that the unit of this product will be cubic centimeters (cm^3). According to Learning Check 1.5, the area of a circle is equal to πr^2, where $\pi = 3.14$. Thus, the liquid volume will be

$$V = (3.14)(7.50 \text{ cm})^2(2.50 \text{ cm}) = 442 \text{ cm}^3$$

Because $1 \text{ cm}^3 = 1 \text{ mL}$, the volume equals 442 mL. Also, because $1 \text{ L} = 1000 \text{ mL}$, the volume can be converted to liters:

$$(442 \text{ mL})\left(\frac{1 \text{ L}}{1000 \text{ mL}}\right) = 0.442 \text{ L}$$

Notice that the milliliter units canceled in the calculation. This conversion to liters is an example of the factor-unit method of problem solving, which is discussed in Section 1.9.

▶ **LEARNING CHECK 1.6** A rectangular aquarium has sides with lengths of 30.0 cm and 20.0 cm, and a height of 15.0 cm. Calculate the volume of the aquarium, and express the answer in milliliters and liters.

The basic unit of mass in the metric system is 1 kilogram (kg), which is equal to about 2.2 pounds in the English system. A kilogram is too large to be conveniently used in some applications, so it is subdivided into smaller units. Two of these smaller units that are often used in chemistry are the gram (g) and milligram (mg) (see ▶ Figure 1.10). The prefixes *kilo-* (k) and *milli-* (m) indicate the following relationships between these units:

$$1 \text{ kg} = 1000 \text{ g}$$
$$1 \text{ g} = 1000 \text{ mg}$$
$$1 \text{ kg} = 1,000,000 \text{ mg}$$

Figure 1.10 Metric masses of some common items as found in a 0.4-g paper clip, 3.0-g razor blade, 3.1-g penny, and 4.7-g nickel.

▶ Example 1.6

All measurements in international track and field events are made using the metric system. Javelins thrown by female competitors must have a mass of no less than 600 g. Express this mass in kilograms and milligrams.

Solution

Because $1 \text{ kg} = 1000 \text{ g}$, 600 g can be converted to kilograms as follows:

$$600 \text{ g} \times \frac{1 \text{ kg}}{1000 \text{ g}} = 0.600 \text{ kg}$$

Also, because 1 g = 1000 mg,

$$600 \text{ g} \times \frac{1000 \text{ mg}}{1 \text{ g}} = 600{,}000 \text{ mg}$$

Once again, the units of the original quantity (600 g) were canceled, and the desired units were generated by this application of the factor-unit method (see Section 1.9).

▶ **LEARNING CHECK 1.7** The javelin thrown by male competitors in track and field meets must have a minimum mass of 0.800 kg. A javelin is weighed and found to have a mass of 0.819 kg. Express the mass of the weighed javelin in grams.

Temperature is difficult to define but easy for most of us to measure—we just read a thermometer. However, thermometers can have temperature scales that represent different units. For example, a temperature of 293 would probably be considered quite high until it was pointed out that it is just room temperature as measured using the Kelvin temperature scale. Temperatures on this scale are given in kelvins, K. (Notice that the abbreviation is K, not °K.)

The Celsius scale (formerly known as the centigrade scale) is the temperature scale used in most scientific work. On this scale, water freezes at 0°C and boils at 100°C under normal atmospheric pressure. A Celsius degree (division) is the same size as a kelvin of the Kelvin scale, but the two scales have different zero points. ▶ Figure 1.11 compares the two scientific temperature scales and the familiar Fahrenheit scale. There are 100 Celsius degrees (divisions) between the freezing point (0°C) and the boiling point (100°C) of water. On the Fahrenheit scale, these same two temperatures are 180 degrees (divisions) apart (the freezing point is 32°F and the boiling point is 212°F). Readings on these two scales are related by the following equations:

$$°C = \frac{5}{9} \left(°F - 32°\right) \tag{1.1}$$

$$°F = \frac{9}{5} \left(°C\right) + 32° \tag{1.2}$$

As mentioned, the difference between the Kelvin and Celsius scales is simply the zero point; consequently, readings on the two scales are related as follows:

$$°C = K - 273 \tag{1.3}$$

$$K = °C + 273 \tag{1.4}$$

Notice that Equation 1.2 can be obtained by solving Equation 1.1 for Fahrenheit degrees, and Equation 1.4 can be obtained by solving Equation 1.3 for kelvins. Thus, you need to remember only Equations 1.1 and 1.3, rather than all four.

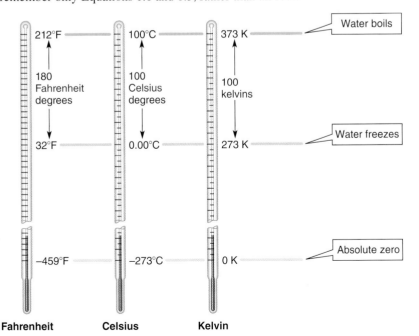

Figure 1.11 Fahrenheit, Celsius, and Kelvin temperature scales. The lowest temperature possible is absolute zero, 0 K.

▶ Example 1.7

A temperature reading of 77°F is measured with a Fahrenheit thermometer. What reading would this temperature give if a Celsius thermometer were used?

Solution
The change is from a Fahrenheit reading to a Celsius reading, so Equation 1.1 is used:

$$°C = \frac{5}{9}\left(°F - 32°\right) = \frac{5}{9}\left(77° - 32°\right) = \frac{5}{9}\left(45°\right) = 25°$$

Thus, the reading on a Celsius thermometer would be 25°C.

▶ **LEARNING CHECK 1.8** What Kelvin thermometer reading would correspond to the 77°F reading described in Example 1.7?

The last units discussed at this point are derived units of energy. Other units will be introduced later in the book as they are needed. The metric system unit of energy is a joule (J), pronounced "jewl." A joule is quite small, as shown by the fact that a 50-watt light bulb uses 50 J of energy every second. A typical household in the United States uses several billion joules of electrical energy in a month.

The calorie (cal), a slightly larger unit of energy, is sometimes used by chemists. One calorie is the amount of heat energy required to increase the temperature of 1 g of water by 1°C. The calorie and joule are related as follows:

$$1 \text{ cal} = 4.184 \text{ J}$$

The nutritional calorie of the weight watcher is actually 1000 scientific calories, or 1 kcal. It is represented by writing *calorie* with a capital C (Calorie, abbreviated Cal). ▶ Table 1.3 contains a list of the commonly used metric units, their relationship to basic units, and their relationship to English units.

TABLE 1.3 Commonly Used Metric Units

Quantity	Metric Unit	Relationship to Metric Basic Unit	Relationship to English Unit
Length	meter (m)	Basic unit	1 m = 1.094 yd
	centimeter (cm)	100 cm = 1 m	1 cm = 0.394 in.
	millimeter (mm)	1000 mm = 1 m	1 mm = 0.0394 in.
	kilometer (km)	1 km = 1000 m	1 km = 0.621 mi
Volume	cubic decimeter (dm^3)	Basic unit	1 dm^3 = 1.057 qt
	cubic centimeter (cm^3 or cc)	1000 cm^3 = 1 dm^3	1 cm^3 = 0.0338 fl oz
	liter (L)	1 L = 1 dm^3	1 L = 1.057 qt
	milliliter (mL)[a]	1000 mL = 1 dm^3	1 mL = 0.0338 fl oz
Mass	gram (g)	1000 g = 1 kg	1 g = 0.035 oz
	milligram (mg)	1,000,000 mg = 1 kg	1 mg = 0.015 grain
	kilogram (kg)	Basic unit	1 kg = 2.20 lb
Temperature	degree Celsius (°C)	1°C = 1 K	1°C = 1.80°F
	kelvin (K)	Basic unit	1 K = 1.80°F
Energy	calorie (cal)	1 cal = 4.184 J	1 cal = 0.00397 BTU[b]
	kilocalorie (kcal)	1 kcal = 4184 J	1 kcal = 3.97 BTU
	joule (J)	Basic unit	1 J = 0.000949 BTU
Time	second (s)	Basic unit	Same unit used

Note: [a]1 mL = 1 cm^3. [b]A BTU (British thermal unit) is the amount of heat required to increase the temperature of 1 pound of water 1°F.

Effects of Temperature on Body Function

The human body has the ability to remain at a relatively constant temperature even when the surrounding temperature increases or decreases. Because of this characteristic, we humans are classified as warm-blooded. In reality, our body temperature varies over a significant range depending on the time of day and the temperature of our surroundings. "Normal" body temperature is considered to be 37.0°C when measured orally. However, this "normal" value can fluctuate between a low of 36.1°C for an individual just waking up in the morning to a value as high as 37.2°C just before bedtime in the late evening.

In addition to this regular variation, our body temperature fluctuates in response to extremes in the temperature of our surroundings. In extremely hot environments, the capacity of our perspiration-based cooling mechanism can be overtaxed, and our body temperature will increase. Body temperatures more than 3.5°C above normal begin to interfere with bodily functions. Body temperatures above 41.1°C can cause convulsions and can result in permanent brain damage, especially in children.

Hypothermia occurs when the body's internal heat generation is not sufficient to balance the heat lost to very cold surroundings. As a result, the body temperature decreases, and at 28.5°C the afflicted person appears pale and might have an irregular heartbeat. Unconsciousness usually results if the body temperature gets lower than 26.7°C. At these low temperatures respiration also slows and becomes shallow, resulting in a decrease in the delivery of oxygen to body tissues.

Thus, the body temperature of we warm-blooded humans is really not constant. Even though we have built in heating and cooling systems, their capacity to maintain a constant, normal body temperature is limited.

The effects of body temperature on body function using the Celsius scale.

1.7 Large and Small Numbers

Learning Objective

7. Express numbers using scientific notation, and do calculations with numbers expressed in scientific notation.

Numbers are used in all measurements and calculations. Many numbers are readily understood and represented because of common experience with them. A price of 10 dollars, a height of 7 feet, a weight of 165 pounds, and a time of 40 seconds are examples of such numbers. But how do we handle numbers like the diameter of a hydrogen atom (about one hundred-millionth of a centimeter) or the distance light travels in 1 year—a light-year (about 6 trillion miles)? These numbers are so small and large, respectively, that they defy understanding in terms of relationships to familiar distances. Even if we can't totally relate to them, it is important in scientific work to be able to conveniently represent and work with such numbers.

Scientific notation provides a method for conveniently representing any number including those that are very large or very small. In **scientific notation,** numbers are represented as the product of a nonexponential term and an exponential term in the general form $M \times 10^n$. The nonexponential term M is a number between 1 and 10 (but not equal to 10) written with a decimal to the right of the first nonzero digit in the number. This position of the decimal is called the **standard position.** The exponential term is a 10 raised to a whole number exponent n that may be positive or negative. The value of n is the number of places the decimal must be moved from the standard position in M to be at the original position in the number when the number is written normally without using scientific notation. If n is positive, the original decimal position is to the right of the standard position. If n is negative, the original decimal position is to the left of the standard position.

scientific notation A way of representing numbers consisting of a product between a nonexponential number and 10 raised to a whole-number exponent that may be positive or negative.

standard position for a decimal In scientific notation, the position to the right of the first nonzero digit in the nonexponential number.

▶ Example 1.8

The following numbers are written using scientific notation. Write them without using scientific notation.

a. 3.72×10^5 **b.** 8.513×10^{-7}

Solution

a. The exponent 5 indicates that the original position of the decimal is located 5 places to the right of the standard position. Zeros are added to accommodate this change:

$$3.72 \times 10^5 = 372,000. = 372,000$$

standard   original
position   position

b. The exponent -7 indicates that the original position of the decimal is 7 places to the left of the standard position. Again, zeros are added as needed:

$$8.513 \times 10^{-7} = .0000008513$$

original   standard
position   position

▶ Example 1.9

Write the following numbers using scientific notation:

a. 8725.6   **b.** 0.000729

Solution

a. The standard decimal position is between the 8 and 7: 8.7256. However, the original position of the decimal is 3 places to the right of the standard position. Therefore, the exponent must be $+3$:

$$8725.6 = 8.7256 \times 10^3$$

b. The standard decimal position is between the 7 and 2: 7.29. However, the original position is 4 places to the left of standard. Therefore, the exponent must be -4:

$$0.000729 = 7.29 \times 10^{-4}$$

The zeros to the left of the 7 are dropped because they are not significant figures (see Section 1.8); they only locate the decimal in the nonscientific notation and are not needed in the scientific notation.

▶ **LEARNING CHECK 1.9** Some of the following numbers are written using scientific notation, and some are not. In each case, rewrite the number using the notation in which it is not written.

a. 5.88×10^2 **c.** 3.915×10^{-4} **e.** 36.77
b. 0.000439   **d.** 9870   **f.** 0.102

▶ Example 1.10

Determine which of the following numbers are written correctly using scientific notation. For those that are not, rewrite them correctly.

a. 001.5×10^{-3} **b.** 28.0×10^2 **c.** 0.35×10^4

Solution

a. Incorrect; the zeros to the left are not needed. The correct answer is 1.5×10^{-3}.

b. Incorrect; the decimal is not in the standard position. Move the decimal 1 position to the left and increase the exponent by 1 to give the correct answer of 2.80×10^{3}.

c. Incorrect; the decimal is not in the standard position. Move the decimal 1 position to the right and decrease the exponent by 1 to give the correct answer of 3.5×10^{3}.

▶ **LEARNING CHECK 1.10** Determine which of the following numbers are written correctly using scientific notation. For those that are not, rewrite them correctly.

a. 62.5×10^{4} **b.** 0.0098 **c.** 0.0041×10^{-3} **d.** 7.85×10^{2}

The multiplication and division of numbers written in scientific notation can be done quite simply by using some characteristics of exponentials. Consider the following multiplication:

$$(a \times 10^{y})(b \times 10^{z})$$

The multiplication is done in two steps. First, the nonexponential terms a and b are multiplied in the usual way. The exponential terms 10^{y} and 10^{z} are multiplied by adding the exponents y and z and using the resulting sum as a new exponent of 10. Thus, we can write

$$(a \times 10^{y})(b \times 10^{z}) = (a \times b)(10^{y+z})$$

Division is done similarly. The nonexponential terms are divided in the usual way, and the exponential terms are divided by subtracting the exponent of the bottom term from that of the top term. The final answer is then written as a product of the resulting nonexponential and exponential terms:

$$\frac{a \times 10^{y}}{b \times 10^{z}} = \left(\frac{a}{b}\right)(10^{y-z})$$

Multiplication and division calculations involving scientific notation are easily done using a hand calculator. ▶ Table 1.4 gives the steps, the typical calculator procedures (buttons to press), and typical calculator readout or display for the division of 7.2×10^{-3} by 1.2×10^{4}.

TABLE 1.4 Using a Calculator for Scientific Notation Calculations

Step	Procedure	Calculator Display
1. Enter 7.2	Press buttons 7, ., 2	7.2
2. Enter 10^{-3}	Press button that activates exponential mode (EE, Exp, etc.)	7.2^{00}
	Press 3	7.2^{03}
	Press change-sign button (±, etc.)	7.2^{-03}
3. Divide	Press divide button (÷)	7.2^{-03}
4. Enter 1.2	Press buttons 1, ., 2	1.2
5. Enter 10^{4}	Press button that activates exponential mode (EE, Exp, etc.)	1.2^{00}
	Press 4	1.2^{04}
6. Obtain answer	Press equals button (=)	$6.^{-07}$

▶ Example 1.11

Do the following operations:

a. $(3.5 \times 10^{4})(2.0 \times 10^{2})$ **c.** $(4.6 \times 10^{-7})(5.0 \times 10^{3})$

b. $\dfrac{3.8 \times 10^{5}}{1.9 \times 10^{2}}$ **d.** $\dfrac{1.2 \times 10^{3}}{3.0 \times 10^{-2}}$

Solution

a. $(3.5 \times 10^4)(2.0 \times 10^2) = (3.5 \times 2.0)(10^4 \times 10^2)$
$$= (7.0)(10^{4+2}) = 7.0 \times 10^6$$

b. $\dfrac{3.8 \times 10^5}{1.9 \times 10^2} = \dfrac{3.8}{1.9} \times \dfrac{10^5}{10^2} = (2.0)(10^{5-2}) = 2.0 \times 10^3$

c. $(4.6 \times 10^{-7})(5.0 \times 10^3) = (4.6 \times 5.0)(10^{-7} \times 10^3)$
$$= (23)(10^{-7+3}) = 23 \times 10^{-4}$$

To get the decimal into the standard position, move it 1 place to the left. This changes the exponent from -4 to -3, so the final result is 2.3×10^{-3}. (This number in decimal form, 0.0023, can be written correctly as either 23×10^{-4} or 2.3×10^{-3}, but scientific notation requires that the decimal point be to the right of the first nonzero number.)

d. $\dfrac{1.2 \times 10^3}{3.0 \times 10^{-2}} = \dfrac{1.2}{3.0} \times \dfrac{10^3}{10^{-2}} = (0.40)(10^{3-(-2)}) = 0.40 \times 10^5$

Adjust the decimal to the standard position and get 4.0×10^4. If these examples were done using a calculator, the displayed answers would normally be given in correct scientific notation.

> ▶ **LEARNING CHECK 1.11** Perform the following operations, and express the result in correct scientific notation:
>
> **a.** $(2.4 \times 10^3)(1.5 \times 10^4)$ **c.** $\dfrac{6.3 \times 10^5}{2.1 \times 10^3}$
>
> **b.** $(3.5 \times 10^2)(2.0 \times 10^{-3})$ **d.** $\dfrac{4.4 \times 10^{-2}}{8.8 \times 10^{-3}}$

The diameter of a hydrogen atom, mentioned earlier as one hundred-millionth of a centimeter, is written in scientific notation as 1.0×10^{-8} cm. Similarly, 1 light-year of 6 trillion miles is 6.0×10^{12} mi.

1.8 Significant Figures

Learning Objective

8. Express the results of measurements and calculations using the correct number of significant figures.

Every measurement contains an uncertainty that is characteristic of the device used to make the measurement. These uncertainties are represented by the numbers used to record the measurement. Consider the following small square:

In (A), the length of one side of the square is measured with a ruler divided into centimeters. It is easy to see that the length is greater than 1 cm, but not quite 2 cm. The length is recorded by writing the number that is known with certainty to be correct (the 1) and writing an estimate for the uncertain number. The result is 1.9 cm, where the .9 is the estimate. In (B), the ruler is divided into tenths of centimeters. It is easy to see that the length is at least 1.8 cm, but not quite 1.9 cm. Once again the certain numbers (1.8) are written, and an estimate is made for the uncertain part. The result is 1.86 cm. When measurements are recorded this way, the numbers representing the certain measurement plus one number representing the estimate are called **significant figures.** Thus, the first measurement of 1.9 cm contains two significant figures, and the second measurement of 1.86 cm contains three significant figures.

significant figures The numbers in a measurement that represent the certainty of the measurement, plus one number representing an estimate.

The maximum number of significant figures possible in a measurement is determined by the design of the measuring device and cannot be changed by expressing the measurement in different units. The 1.8-cm length determined earlier can also be represented in terms of meters and millimeters as follows:

$$1.8 \text{ cm} = 0.018 \text{ m} = 18 \text{ mm}$$

In this form, it appears that the length expressed as 0.018 m contains four significant figures, but this is impossible; a measurement made with a device doesn't become more certain simply by changing the unit used to express the number. Thus, the zeros are not significant figures; their only function is to locate the correct position for the decimal. Zeros located to the left of nonzero numbers, such as the two zeros in 0.018 cm, are never considered to be significant. Thus, 12.5 mg, 0.0125 g, and 0.0000125 kg all represent the same measured mass, and all contain three significant figures.

Zeros located between nonzero numbers or trailing zeros located at the end of numbers will be considered significant. Thus, 2050 μL, 2.050 mL, and 0.002050 L all represent the same volume measurement, and all contain four significant figures.

The rule that specifies counting trailing zeros as significant is generally followed by scientists, but some quantities are expressed with trailing zeros that are not significant. For example, suppose you read in a newspaper that the population of a city is 1,250,000 people. Should the four trailing zeros be considered significant? If they are, it means that the population is known with certainty to the nearest 10 people and that the measurement has an uncertainty of only plus or minus 1 person. A more reasonable conclusion is that the census is correct to the nearest 1000 people. This could be represented as 1250 thousand or, using scientific notation, 1.250×10^6. In either of these representations, only four significant figures are used. In scientific notation, the correct number of significant figures is used in the nonexponential term, and the location of the decimal is determined by the exponent. In this book, large numbers will always be represented by scientific notation instead of using nonsignificant trailing zeros. However, you are likely to encounter nonsignificant trailing zeros in other reading materials. The rules for determining the significance of zeros are summarized as follows:

1. Zeros not preceded by nonzero numbers are not significant figures. These zeros are sometimes called *leading zeros.*
2. Zeros located between nonzero numbers are significant figures. These zeros are sometimes called *buried* or *confined zeros.*
3. Zeros located at the end of a number are significant figures. These zeros are sometimes called *trailing zeros.*

▶ Example 1.12

Determine the number of significant figures in each of the following measurements, and use scientific notation to express each measurement using the correct number of significant figures:

a. 24.6°C   **b.** 0.036 g   **c.** 15.0 mL   **d.** 0.0020 m

Solution

a. All the numbers are significant: three significant figures, 2.46×10^1 °C.

b. The leading zeros are not significant: two significant figures, 3.6×10^{-2} g.

c. The trailing zero is significant: three significant figures, 1.50×10^1 mL.

d. The leading zeros are not significant, but the trailing zero is: two significant figures, 2.0×10^{-3} m.

▶ **LEARNING CHECK 1.12** Determine the number of significant figures in each of the following measurements:

a. 250 mg   **c.** 0.0108 kg   **e.** 0.001 mm
b. 18.05 mL   **d.** 37°C   **f.** 101.0 K

▶ **LEARNING CHECK 1.13** Use scientific notation to express each of the following measurements using the correct number of significant figures:

a. 101 m   **c.** 0.00230 kg   **e.** 21.65 mL
b. 1200 g   **d.** 1296°C   **f.** 0.015 km

Most measurements that are made do not stand as final answers. Instead, they are usually used to make calculations involving multiplication, division, addition, or subtraction. The answer obtained from such a calculation cannot have more certainty than the least certain measurement used in the calculation. It should be written to reflect an uncertainty equal to that of the most uncertain measurement. This is accomplished by the following rules:

1. The answer obtained by multiplication or division must contain the same number of significant figures as the quantity with the fewest significant figures used in the calculation.
2. The answer obtained by addition or subtraction must contain the same number of places to the right of the decimal as the quantity in the calculation with the fewest number of places to the right of the decimal.

To follow these rules, it is often necessary to reduce the number of significant figures by rounding answers. The following are rules for rounding:

1. If the first of the nonsignificant figures to be dropped from an answer is 5 or greater, all the nonsignificant figures are dropped, and the last significant figure is increased by 1.
2. If the first of the nonsignificant figures to be dropped from an answer is less than 5, all nonsignificant figures are dropped, and the last significant figure is left unchanged.

Remember, if you use a calculator, it will often express answers with too few or too many figures (see ▶ Figure 1.12). It will be up to you to determine the proper number of significant figures to use and to round the calculator answer correctly.

Figure 1.12 Calculators usually display the sum of 4.362 and 2.638 as 7 (too few figures), and the product of 0.67 and 10.14 as 6.7938 (too many figures).

Example 1.13

Do the following calculations, and round the answers to the correct number of significant figures:

a. $(4.95)(12.10)$ **b.** $\dfrac{3.0954}{0.0085}$ **c.** $\dfrac{(9.15)(0.9100)}{3.117}$ **d.** $\dfrac{320}{4.00}$

Solution

All calculations are done with a hand calculator, and the calculator answer is written first. Appropriate rounding is done to get the final answer.

a. Calculator answer: 59.895

The number 4.95 has three significant figures, and 12.10 has four. Thus, the answer must have three significant figures:

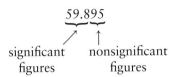

The first of the nonsignificant figures to be dropped is 9, so after both are dropped, the last significant figure is increased by 1. The final answer containing three significant figures is 59.9.

b. Calculator answer: 364.16471

The number 3.0954 has five significant figures, and 0.0085 has two. Thus, the answer must have only two:

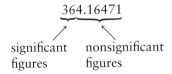

The first of the nonsignificant figures to be dropped is 4, so the last significant figure remains unchanged after the nonsignificant figures are dropped. The correct answer then is 360. If this is written as 360, it will contain three significant figures. However, the answer can be written with the proper number of significant figures by using scientific notation. The final correct answer containing two significant figures is 3.6×10^2.

c. Calculator answer: 2.6713186

The number 9.15 has three significant figures, 0.9100 has four, and 3.117 has four. Thus, the answer must have only three:

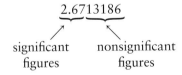

Appropriate rounding gives 2.67 as the correct answer.

d. Calculator answer: 80

Both the numbers 320 and 4.00 contain three significant figures, so the answer should also have three. However, modern calculators usually do not display trailing zeros that follow the decimal. So, the appropriate number of trailing zeros must be added. The correct answer should be expressed as 80.0, which contains three significant figures.

▶ **LEARNING CHECK 1.14** Do the following calculations, and round the answers to the correct number of significant figures:

a. $(0.0019)(21.39)$ **b.** $\dfrac{8.321}{4.1}$ **c.** $\dfrac{(0.0911)(3.22)}{1.379}$

▶ Example 1.14

Do the following additions and subtractions, and write the answers with the correct number of significant figures:

a. $1.9 + 18.65$ **b.** $15.00 - 8.0$ **c.** $1500 + 10.9 + 0.005$ **d.** $5.1196 - 5.02$

Solution

In each case, the numbers are arranged vertically with the decimals in a vertical line. The answer is then rounded so it contains the same number of places to the right of the decimal as the smallest number of places in the quantities added or subtracted.

a.
$$\begin{array}{r} 1.9 \\ 18.65 \\ \hline 20.55 \end{array}$$

The answer must be expressed with one place to the right of the decimal to match the one place in 1.9.

Correct answer: 20.6 (Why was the final 5 increased to 6?)

b.
$$\begin{array}{r} 15.00 \\ -8.0 \\ \hline 7.0 \end{array}$$

The answer must be expressed using one place to the right of the decimal to match the 8.0. A typical calculator answer would probably be 7, requiring that a zero be added to the right of the decimal to provide the correct number of significant figures.

c.
$$\begin{array}{r} 1500 \\ 10.9 \\ 0.005 \\ \hline 1510.905 \end{array}$$

The answer must be expressed with no places to the right of the decimal to match the 1500.

Correct answer: 1511 (Why was the final 0 of the answer increased to 1?)

d.
$$\begin{array}{r} 5.1196 \\ -5.02 \\ \hline 0.0996 \end{array}$$

The answer must be expressed with two places to the right of the decimal to match the 5.02.

Correct answer: 0.10 (What rounding rule was followed?)

Notice that the answer to (a) has more significant figures than the least significant number used in the calculation. The answer to (d), on the other hand, has fewer significant figures than either number used in the calculation. This happened because the rule for dealing with addition and subtraction focuses on the number of figures located to the right of the decimal and is not concerned with the figures to the left of the decimal. Thus, the number of significant figures in the answer is sometimes increased as a result of addition and decreased as a result of subtraction. You should be aware of this and not be confused when it happens.

▶ **LEARNING CHECK 1.15** Do the following additions and subtractions, and round the answers to the correct number of significant figures:

a. $8.01 + 3.2$ **c.** $4.33 - 3.12$
b. $3000 + 20.3 + 0.009$ **d.** $6.023 - 2.42$

exact numbers Numbers that have no uncertainty; numbers from defined relationships, counting numbers, and numbers that are part of simple fractions.

Some numbers used in calculations are **exact numbers** that have no uncertainty associated with them and are considered to contain an unlimited number of significant trailing zeros. Such numbers are not used when the appropriate number of significant figures is determined for calculated answers. In other words, exact numbers do not limit the number of significant figures in calculated answers. One kind of exact number is a number used as part of a defined relationship between quantities. For example, 1 m contains exactly 100 cm:

$$1 \text{ m} = 100 \text{ cm}$$

Thus, the numbers 1 and 100 are exact. A second kind of exact number is a counting number obtained by counting individual objects. A dozen eggs contains exactly 12 eggs, not 11.8 and not 12.3. The 12 is an exact counting number. A third kind of exact number is one that is part of a simple fraction such as $\frac{1}{4}$, $\frac{2}{3}$, or the $\frac{5}{9}$ used in Equation 1.1 to convert Fahrenheit temperature readings into Celsius readings.

1.9 Using Units in Calculations

Learning Objective

9. Use the factor-unit method to solve numerical problems.

Some beginning chemistry students are concerned about not being able to solve numerical chemistry problems. They may say, "I can work the problems, I just can't set them up." What they are really saying is "I can do the arithmetic once the numbers are properly arranged, but I can't do the arranging."

This section presents a method for arranging numbers that will work for most of the numerical problems you will encounter in this course. This method has a number of names, including the factor-unit method, the factor-label method, and dimensional analysis. We will call it the factor-unit method. It is a systematic approach to solving numerical problems and consists of the following steps:

Step 1. Write down the known or given quantity. Include both the numerical value and units of the quantity.

Step 2. Leave some working space and set the known quantity equal to the units of the unknown quantity.

Step 3. Multiply the known quantity by one or more factors, such that the units of the factor cancel the units of the known quantity and generate the units of the unknown quantity. These **factors** are fractions derived from numerical relationships between quantities. These relationships can be definitions or experimentally measured quantities. For example, the defined relationship 1 m = 100 cm provides the following two factors:

factors used in the factor-unit method Fractions obtained from numerical relationships between quantities.

$$\frac{1 \text{ m}}{100 \text{ cm}} \quad \text{and} \quad \frac{100 \text{ cm}}{1 \text{ m}}$$

Step 4. After you get the desired units, do the necessary arithmetic to produce the final answer.

▶ Example 1.15

Use the factor-unit method and numerical relationships from Table 1.3 to calculate the number of yards in 100 m.

Solution

The known quantity is 100 m, and the unit of the unknown quantity is yard (yd).

Step 1. 100 m

Step 2. 100 m $= $ yd

Step 3. 100 m̶ $\times \dfrac{1.094 \text{ yd}}{1 \text{ m̶}} = $ yd

The factor $\dfrac{1.094 \text{ yd}}{1 \text{ m}}$ came from the numerical relationship 1 m = 1.094 yd found in Table 1.3.

Step 4. $\dfrac{(100)(1.094)\text{yd}}{1} = 109.4$ yd

This answer should be rounded to 109 yd, an answer that contains three significant figures, just as 100 m does. The 1 m in the factor is an exact number used as part of a defined relationship, so it doesn't influence the number of significant figures in the answer.

Example 1.16

A laboratory technician uses a micropipet to measure a 50-μL (50-microliter) sample of blood serum for analysis. Express the sample volume in liters (L).

Solution

The known quantity is 50 μL, and the unit of the unknown quantity is liters.

Step 1. $50 \ \mu L$

Step 2. $50 \ \mu L \qquad\qquad\qquad\qquad = L$

Step 3. $50 \ \mu L \times \dfrac{1 \times 10^{-6} \ L}{1 \ \mu L} = L$

The factor $\dfrac{1 \times 10^{-6} \ L}{1 \ \mu L}$ came from the numerical relationship $1 \ \mu L = 1 \times 10^{-6} \ L$ described in Table 1.2.

Step 4. $\dfrac{(50)(1 \times 10^{-6} L)}{1} = 50 \times 1 = 50$

The answer is expressed using the same number of significant figures as the 50 μL because 1 and 1×10^{-6} are exact numbers by definition.

▶ **LEARNING CHECK 1.16** Creatinine is a substance found in the blood. An analysis of a blood serum sample detected 1.1 mg of creatinine. Express this amount in grams by using the factor-unit method. Remember, the prefix milli means $\frac{1}{1000}$, so 1 g = 1000 mg.

Example 1.17

One of the fastest-moving nerve impulses in the body travels at a speed of 400 feet per second (ft/s). What is the speed in miles per hour?

Solution

The known quantity is 400 ft/s, and the unit of the unknown quantity is miles per hour (mi/h).

Step 1. $\dfrac{400 \ ft}{s}$

Step 2. $\dfrac{400 \ ft}{s} \qquad\qquad\qquad\qquad\qquad = \dfrac{mi}{h}$

Step 3. $\left(\dfrac{400 \ ft}{s}\right)\left(\dfrac{1 \ mi}{5280 \ ft}\right)\left(\dfrac{60 \ s}{1 \ min}\right)\left(\dfrac{60 \ min}{1 \ h}\right) = \dfrac{mi}{h}$

The factors came from the following well-known numerical relationships: 1 mi = 5280 ft, 1 min = 60 s, 1 h = 60 min. All numbers in these factors are exact numbers based on definitions.

Step 4. $\dfrac{(400)(1)(60)(60) \ mi}{(5280)(1)(1) \ h} = 272.7 \ \dfrac{mi}{h}$

Rounding to three significant figures, the same as in 400 ft/s, gives 273 mi/h.

▶ **LEARNING CHECK 1.17** A world-class sprinter can run 100 m in 10.0 s. This corresponds to a speed of 10.0 m/s. Convert this speed to miles per hour. Use information from Table 1.3.

1.10 Calculating Percentages

Learning Objective

10. Do calculations involving percentages.

The word *percent* literally means per one hundred. It is the number of specific items in a group of 100 such items. Since items are seldom found in groups of exactly 100, we usually have to calculate the number of specific items that would be in the group if it did contain exactly 100 items. This number is the percentage, and the calculation follows:

$$\text{percent} = \frac{\text{number of specific items}}{\text{total items in the group}} \times 100 \tag{1.5}$$

$$\% = \frac{\text{part}}{\text{total}} \times 100 \tag{1.6}$$

In Equation 1.6, the word *part* is used to represent the number of specific items included in the total.

▶ Example 1.18

A college has 4517 female and 3227 male students enrolled. What percentage of the student body is female?

Solution

The total student body consists of 7744 people, of which 4517 are female.

$$\% \text{ female} = \frac{\text{number of females}}{\text{total number of students}} \times 100$$

$$\% \text{ female} = \frac{4517}{7744} \times 100 = 58.33$$

Rearrangement of Equation 1.6 gives another useful percent relationship. Because $\% = \text{part/total} \times 100$,

$$\text{part} = \frac{(\%)(\text{total})}{100} \tag{1.7}$$

According to Equation 1.7, the number of specific items corresponding to a percentage can be calculated by multiplying the percentage and total, then dividing the product by 100.

▶ Example 1.19

The human body is approximately 70% water by mass. What is the mass of water in a 170-pound (lb) person?

Solution

In this problem, what we would classify as the part is the mass of water in a person who weighs a total of 170 lb. Substitution into Equation 1.7 gives

$$\text{mass of water} = \frac{(\%)(\text{total})}{100} = \frac{(70)(170 \text{ lb})}{100} = 119 \text{ lb}$$

We should round our answer to only two significant figures to match the two in the 70%. Do this by using scientific notation; the answer is 1.2×10^2 lb.

▶ **LEARNING CHECK 1.18**

a. A student is saving money to buy a computer that will cost a total of $1200. The student has saved $988. What percentage of the purchase price has been saved?

b. In a chemistry class of 83 students, 90.4% voted not to have the final exam. How many students wanted to take the exam?

Study Skills 1.1

Help with Calculations

Many students feel uneasy about working chemistry problems that involve the use of mathematics. The uneasiness is often increased if the problem to be solved is a story problem. One tip that will help you solve such problems in this textbook is to remember that almost all of these problems are one of two types: those for which a specific formula applies and those where the factor-unit method is used. When you do homework or take quizzes or examinations and encounter a math-type problem, your first task should be to decide which type of problem it is, formula or factor-unit.

In this chapter, formula-type problems were those that dealt with percentage calculations (see Examples 1.18 and 1.19) and the conversion from one temperature scale to another (see Example 1.7). If you decide a problem is a formula type, Step 1 in solving it is to write down the formula that applies. This important step makes it easier to do the next step because a formula is like a road map that tells you how to proceed from one point to another. Step 2 is to identify the given quantity and put its number and units into the formula. This will leave only the answer missing in most formula-type problems. You can then obtain the answer by doing the appropriate calculations.

In this textbook, the factor-unit method discussed in Section 1.9 is used for most problems that require mathematical calculations. This method simplifies problem solving and should be mastered so it can be used where it applies. The beauty of this method is that it mimics your natural, everyday way of solving problems. This real-life method usually involves identifying where you are, where you want to go, and how to get there. The factor-unit method follows the same pattern: Step 1, identify the given number and its units; Step 2, write down the unit of the desired answer; Step 3, put in factors that will convert the units of the given quantity into the units of the desired answer.

Thus, we see that working story problems is not as difficult a task as it might first appear. A key is to see through all the words and find what is given (number and units). Then look for what is wanted by focusing on key words or phrases like "how much," "what is," and "calculate." Finally, use one of the two methods, formula or factor-unit, to solve the problem.

The steps to follow in solving both types of problems are summarized in the following flow charts.

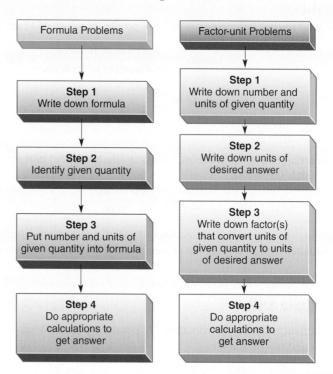

1.11 Density

Learning Objective

11. Do calculations involving densities.

A discussion of the density of matter will conclude the topics of this chapter. Density is a physical property of matter, so it can be measured without changing the composition of the sample of matter under investigation. **Density** is the number obtained by dividing the mass of a sample of matter by the volume of the same sample.

$$\text{density} = \frac{\text{mass}}{\text{volume}} \quad (1.8)$$

$$d = \frac{m}{V} \quad (1.9)$$

density The number given when the mass of a sample of a substance is divided by the volume of the same sample.

We see from these equations that once a numerical value has been obtained for the density, two factors are available that relate mass and volume, and these may be used to solve problems.

▶ Example 1.20

The density of iron metal has been determined to be 7.2 g/cm^3.

a. Use the density value to calculate the mass of an iron sample that has a volume of 35.0 cm^3.
b. Use the density value to calculate the volume occupied by 138 g of iron.

Solution

The value of the density tells us that one cubic centimeter (cm^3) of iron has a mass of 7.2 g. This may be written as 7.2 g = 1.0 cm^3, using two significant figures for the volume. This relationship gives two factors that can be used to solve our problems:

$$\frac{7.2 \text{ g}}{1.0 \text{ cm}^3} \quad \text{and} \quad \frac{1.0 \text{ cm}^3}{7.2 \text{ g}}$$

a. The sample volume is 35.0 cm^3, and we wish to use a factor to convert this to grams. The first factor given above will work.

$$35.0 \text{ cm}^3 \times \frac{7.2 \text{ g}}{1.0 \text{ cm}^3} = 252 \text{ g (calculator answer)}$$

$$= 2.5 \times 10^2 \text{ g (properly rounded answer)}$$

b. The sample mass is 138 g, and we wish to convert this to cubic centimeters (cm^3). The second factor given above will work.

$$138 \text{ g} \times \frac{1.0 \text{ cm}^3}{7.2 \text{ g}} = 19.17 \text{ cm}^3 \text{ (calculator answer)}$$

$$= 19 \text{ cm}^3 \text{ (properly rounded answer)}$$

▶ **LEARNING CHECK 1.19** Aluminum metal has a density of 2.7 g/cm^3.

a. Calculate the mass of an aluminum sample with a volume of 60.0 cm^3.
b. Calculate the volume of an aluminum sample that has a mass of 98.5 g.

For some substances, density is rather easily determined experimentally by direct measurement. The mass of a sample is obtained by weighing the sample. The sample volume can be calculated if the sample is a regular solid such as a cube. If the sample is a liquid or an irregular solid, the volume can be measured by using volumetric apparatus such as those shown in ▶ Figure 1.13. Densities of solids are often given in units of g/cm^3, and those of liquids in units of g/mL because of the different ways the volumes are determined. However, according to Table 1.3, 1 cm^3 = 1 mL, so the numerical value is the same regardless of which of the two volume units is used.

Figure 1.13 Glassware for measuring volumes of liquids. Clockwise from top center: buret, graduated cylinder, syringe, pipet, and volumetric flask.

© Jeffrey M. Seager

Health Information on the Web

The World Wide Web contains more than 90,000 sites that are related to health issues or health problems or that sell health products. Many of these sites are very good resources for individuals concerned about such topics. Unfortunately however, there are also sites run by scam artists who are simply interested in trying to make money at the expense of gullible or uneducated web surfers. The information provided by these sites is not only useless but might also be dangerous. There have been reports of sites run by individuals claiming to have a "miracle cure" for serious diseases. The sites encourage individuals to stop taking their prescription medication and instead buy the new miracle product. Another characteristic of a site to be avoided is one that claims to have a physician who will diagnose or treat you without requiring you to have a proper examination and consultation. As a general rule, you should use common sense. Another good idea is to find websites that are already linked with organizations you are familiar with or recognize as being legitimate. The following sites are very helpful and contain accurate and complete information.

Healthfinder: *www.healthfinder.gov*
USDA's Food & Nutrition Research Briefs: *www.nal.usda.gov/fnic/usda/fnrb*
National Cancer Institute: *www.cancer.gov*
Centers for Disease Control and Prevention: *www.cdc.gov*
U.S. Department of Health & Human Services: *www.ahrq.gov/data/infoqual.htm*
U.S. National Library of Medicine: *www.nlm.nih.gov/medlineplus/*
New England Journal of Medicine: *www.nejm.org*

Journal of the American Medical Association: *www.ama-assn.org/public/journals/jama/*
American Cancer Society: *www.cancer.org*
Harvard University: *www.health.harvard.edu*
Mayo Clinic: *www.mayoclinic.com*
American Academy of Family Physicians: *www.familydoctor.org*
Johns Hopkins Bayview Medical Center: *www.hopkinsbayview.org/healthylinks/*
Medical Library Association: *www.mlanet.org/resources/userguide.html*
NOAH: New York Online Access to Health: *www.noah-health.org/*

An enormous amount of health information is available through home computers and the internet.

© iStockphoto.com/Daniel Laflor

▶ Example 1.21

a. A hypodermic syringe was used to deliver 5.0 cc (cm^3) of alcohol into an empty container that had a mass of 25.12 g when empty. The container with the alcohol sample weighed 29.08 g. Calculate the density of the alcohol.

b. A cube of copper metal measures 2.00 cm on each edge and weighs 71.36 g. What is the density of the copper sample?

c. According to its owner, a chain necklace is made of pure gold. In order to check this, the chain was weighed and found to have a mass of 19.21 g. The chain was then put into a graduated cylinder that contained 20.8 mL of water. After the chain was put into the cylinder, the water level rose to 21.9 mL. The density of pure gold was looked up in a handbook and found to be 19.2 g/cm^3. Is the chain made of pure gold?

Solution

a. According to Table 1.3, 1 cc (or cm^3) = 1 mL, so the volume is 5.0 mL. The sample mass is equal to the difference between the mass of the container with the sample inside and the mass of the empty container:

$$m = 29.08 \text{ g} - 25.12 \text{ g} = 3.96 \text{ g}$$

Unless otherwise noted all art on this page © Cengage Learning 2014

The density of the sample is equal to the sample mass divided by the sample volume:

$$d = \frac{m}{V} = \frac{3.96 \text{ g}}{5.0 \text{ mL}} = 0.79 \text{ g/mL (rounded value)}$$

b. The volume of a cube is equal to the product of the three sides:

$$V = (2.00 \text{ cm})(2.00 \text{ cm})(2.00 \text{ cm}) = 8.00 \text{ cm}^3$$

The density is equal to the mass divided by the volume:

$$d = \frac{m}{V} = \frac{71.36 \text{ g}}{8.00 \text{ cm}^3} = 8.92 \text{ g/cm}^3 \text{ (rounded value)}$$

c. The volume of the chain is equal to the difference in the water level in the cylinder with and without the chain present:

$$V = 21.9 \text{ mL} - 20.8 \text{ mL} = 1.1 \text{ mL} = 1.1 \text{ cm}^3$$

The density is equal to the mass divided by the volume:

$$d = \frac{m}{V} = \frac{19.21 \text{ g}}{1.1 \text{ cm}^3} = 17 \text{ g/cm}^3 \text{ (rounded value)}$$

The experimentally determined density is less than the density of pure gold, so the chain is not made of pure gold.

▶ LEARNING CHECK 1.20

a. A pipet was used to put a 10.00-mL sample of a liquid into an empty container. The empty container weighed 51.22 g, and the container with the liquid sample weighed 64.93 g. Calculate the density of the liquid in g/mL.

b. A box of small irregular pieces of metal was found in a storage room. It was known that the metal was either nickel or chromium. A 35.66-g sample of the metal was weighed and put into a graduated cylinder that contained 21.2 mL of water (▶ Figure 1.14). The water level after the metal was added was 25.2 mL (Figure 1.14, right). Was the metal nickel (density = 8.9 g/cm³) or chromium (density = 7.2 g/cm³)?

Figure 1.14 Measuring the volume of irregular metal pieces.

DR. MELINA B. JAMPOLIS, M.D.

Dr. Melina B. Jampolis, M.D. is a Los Angeles, California, Physician Nutrition Specialist—a specialty practiced by only a few hundred physicians in the United States. She focuses exclusively on nutrition for weight loss and disease prevention and treatment.

Ask an Expert 1.1

Q: Does food density matter when you're trying to lose weight?

A: When it comes to losing weight, it seems like there is a new fad diet out every month. Low fat, low carb, high protein, vegan, vegetarian low carb—the choices are endless. What really works for weight loss? The answer varies among individuals and likely depends on a combination of genes, food preferences, and lifestyle. But one of the most important concepts for achieving permanent weight loss is the energy density of the food you consume. Energy density refers to the number of calories per gram in a given type of food. Eating foods that have a lower energy density allows you to eat larger, more satisfying, portions while consuming fewer calories. Research has shown that this is important both mentally and physically when it comes to weight loss. Beginning at about the age of 3 or 4, we begin to eat in response to visual cues in addition to physiological cues such as feeling full. Studies show that when presented with larger portions, most people eat significantly more yet do not report feeling more full. By lowering the energy density of the foods you consume, your portions will be visually more substantial and will also physically take up more space in your stomach.

Energy density depends on three factors: water, fiber, and fat. Water has the strongest relationship to energy density. Foods high in water content like fruits, vegetables, soups, hot cereal, and low-fat or fat-free dairy have a lower energy density because water contributes to the weight and size of a serving but not the calories. Dried fruit and dry snacks like pretzels and crackers are higher in energy density, even if they are fat free, because they don't contain water.

High-fiber foods like whole grains, beans, and, again, fruits and vegetables, have a lower energy density because fiber contributes to the weight of the food but is not well absorbed and digested by the body, and so it contributes far fewer calories. Fat, on the other hand, is very dense calorically. It contains 9 calories per gram compared to approximately 4 calories per gram for both protein and carbohydrates. So foods that are high in fat, like oils, nuts, cheese, salad dressing, high-fat meat, cream-based sauces, and desserts have a higher energy density. Oil is particularly energy dense because it is not only high in fat, it also lacks fiber and contains no water.

Although there are no standard categories for energy density, one of the leading researchers in the field, Dr. Barbara Rolls, suggests the following:

Very low energy density = 0.0–0.6 kcal/g
Low energy density = 0.6–1.5 kcal/g

Moderate energy density = 1.5–4 kcal/g
High energy density = >4 kcal/g

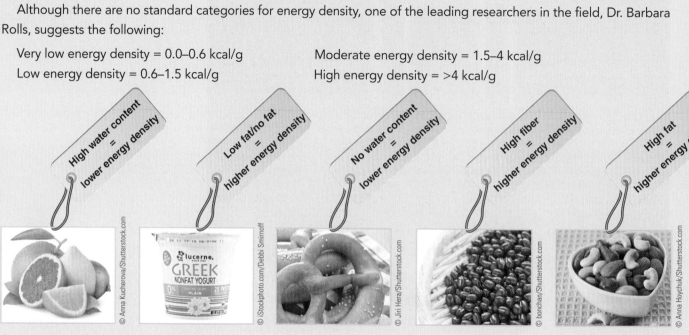

High water content = lower energy density

Low fat/no fat = higher energy density

No water content = lower energy density

High fiber = higher energy density

High fat = higher energy den

By eating more foods that have lower energy density, restricting portions of foods that have high energy density, and modifying recipes to decrease the energy density (substituting apple sauce for oil in baking, combining chopped mushrooms with ground meat before grilling, making soups instead of casseroles), you can lose or maintain your weight with far less effort and more success than following the "diet of the month."

Case Study Follow-up

Examples of the importance of measurement abound. Overdose has resulted when caregivers have failed to consider an accurate weight for a patient or when grams have been confused with milligrams in administering medication. Liver damage has occurred in children who were given infant-strength pain-reliever, which is more concentrated, at a volume that would be correct for a children's strength dosage (ISMP, 2002).

In this case study, the physician decided to induce labor based on fetal femur length and possible oligohydramnios (insufficient amniotic fluid). She could have also considered how much effacement (thinning and shortening of the cervix) and dilation (opening of the cervix) had occurred—other signs of how ready a woman is to give birth. This physician may also have made a better choice if she had encouraged the mother to wait patiently until her due date. Impatience is never a good reason for an induction of labor. Compelling reasons for medical induction include pregnancy that exceeds the due date by more than two weeks, infection, early separation of the placenta, and ruptured membranes without contractions (Mayo, 2011).

Concept Summary

What Is Matter?

Matter, the substance of everything, is defined as anything that has mass and occupies space. Mass is a measurement of the amount of matter present in an object. Weight is a measure of the gravitational force pulling on an object.

Objective 1, Exercise 1.2

Properties and Changes

Chemical properties cannot be determined without attempting to change one kind of matter into another. Physical properties can be determined without attempting such composition changes. Any change in matter that is accompanied by a composition change is a chemical change. Physical changes take place without the occurrence of any composition changes.

Objective 2, Exercises 1.8 a & b and 1.10 b & c

A Model of Matter

Scientific models are explanations for observed behavior. The results of many observations led scientists to a model for matter in which all matter is composed of tiny particles. In many substances, these particles are called molecules, and they represent the smallest piece of such substances that is capable of a stable existence. Molecules, in turn, are made up of atoms, which represent the limit of chemical subdivision for matter. The terms *diatomic, triatomic, polyatomic, homoatomic,* and *heteroatomic* are commonly used to describe the atomic composition of molecules.

Objective 3, Exercise 1.12

Classifying Matter

It often simplifies things to classify the items being studied. Some useful categories into which matter can be classified are heterogeneous, homogeneous, solution, pure substance, element, and compound. All matter is either heterogeneous or homogeneous. Heterogeneous matter is a mixture in which the properties and appearance are not uniform. Homogeneous matter is either a mixture of two or more pure substances (and the mixture is called a solution), or it is a pure substance. If it is a pure substance, it is either an element (containing atoms of only one kind) or a compound (containing two or more kinds of atoms).

Objective 4, Exercises 1.18, 1.22, and 1.24

Measurement Units

All measurements are based on standard units that have been agreed on and adopted. The earliest measurements were based on human body dimensions, but the changeable nature of such basic units made the adoption of a worldwide standard desirable.

Objective 5, Exercise 1.28

The Metric System

The metric system of measurement is used by most scientists worldwide and all major nations except the United States. It is a decimal system in which larger and smaller units of a quantity are related by factors of 10. Prefixes are used to designate relationships between the basic unit and larger or smaller units of a quantity.

Objective 6, Exercises 1.30 and 1.40

Large and Small Numbers

Because of difficulties in working with very large or very small numbers in calculations, a system of scientific notation has been devised to represent such numbers. In scientific notation, numbers are represented as products of a nonexponential number and 10 raised to some power. The nonexponential number is always written with the decimal in the standard position (to the right of the first nonzero digit in the number). Numbers written in scientific notation can be manipulated in calculations by following a few rules.

Objective 7, Exercises 1.48 and 1.60

Significant Figures

In measured quantities, the significant figures are the numbers representing the part of the measurement that is certain, plus one number representing an estimate. The maximum number of significant figures possible in a measurement is determined by the design of the measuring device. The results of calculations made using numbers from measurements can be expressed with the proper number of significant figures by following simple rules.

Objective 8, Exercises 1.64 and 1.66

Using Units in Calculations

The factor-unit method for doing calculations is based on a specific set of steps. One crucial step involves the use of factors that are obtained from fixed numerical relationships between quantities. The units of the factor must always cancel the units of the known quantity and generate the units of the unknown or desired quantity.

Objective 9, Exercise 1.82

Calculating Percentages

The word *percent* means per one hundred, and a percentage is literally the number of specific items contained in a group of 100 items. Few items always occur in groups of exactly 100, so a calculation can be done that determines how many specific items would be in a group if the group actually did contain exactly 100 items.

Objective 10, Exercise 1.92

Density

The density of a substance is the number obtained by dividing the mass of a sample by the volume of the same sample. Measured values of density provide two factors that can be used with the factor-unit method to calculate the mass of a substance if the volume is known, or the volume if the mass is known.

Objective 11, Exercise 1.98

Key Terms and Concepts

Atom (1.3)
Basic unit of measurement (1.6)
Chemical changes (1.2)
Chemical properties (1.2)
Compound (1.4)
Density (1.11)
Derived unit of measurement (1.6)
Diatomic molecules (1.3)
Element (1.4)
Exact numbers (1.8)

Factors used in the factor-unit method (1.9)
Heteroatomic molecules (1.3)
Heterogeneous matter (1.4)
Homoatomic molecules (1.3)
Homogeneous matter (1.4)
Mass (1.1)
Matter (1.1)
Mixture (1.4)
Molecule (1.3)
Physical changes (1.2)

Physical properties (1.2)
Polyatomic molecules (1.3)
Pure substance (1.4)
Scientific models (1.3)
Scientific notation (1.7)
Significant figures (1.8)
Solutions (1.4)
Standard position for a decimal (1.7)
Triatomic molecules (1.3)
Weight (1.1)

Key Equations

1.	Conversion of temperature readings from one scale to another (Section 1.6)	$^\circ C = \dfrac{5}{9}\left(^\circ F - 32^\circ\right)$	Equation 1.1
		$^\circ F = \dfrac{9}{5}\left(^\circ C\right) + 32^\circ$	Equation 1.2
		$^\circ C = K - 273$	Equation 1.3
		$K = {}^\circ C + 273$	Equation 1.4
2.	Calculation of percentage (Section 1.10)	$\text{percent} = \dfrac{\text{number of specific items}}{\text{total items in the group}} \times 100$	Equation 1.5
		$\% = \dfrac{\text{part}}{\text{total}} \times 100$	Equation 1.6
3.	Calculation of number of items representing a specific percentage of a total (Section 1.10)	$\text{part} = \dfrac{(\%)(\text{total})}{100}$	Equation 1.7
4.	Calculation of density from mass and volume data (Section 1.11)	$d = \dfrac{m}{V}$	Equation 1.9

Exercises

Even-numbered exercises are answered in Appendix B.

Blue-numbered exercises are more challenging.

What Is Matter? (Section 1.1)

1.1 A heavy steel ball is suspended by a thin wire. The ball is hit from the side with a hammer but hardly moves. Describe what you think would happen if this identical experiment were carried out on the moon.

1.2 Explain how the following are related to each other: matter, mass, and weight.

1.3 Tell how you would try to prove to a doubter that air is matter.

1.4 Which of the following do you think is likely to change the most when done on Earth and then on the moon? Carefully explain your reasoning.

   a. The distance you can throw a bowling ball.

   b. The distance you can roll a bowling ball on a flat, smooth surface.

1.5 The attractive force of gravity for objects near Earth's surface increases as you move toward Earth's center. Suppose you are transported from a deep mine to the top of a tall mountain.

   a. How would your mass be changed by the move?

   b. How would your weight be changed by the move?

1.6 Earth's rotation causes it to bulge at the equator. How would the weights of people of equal mass differ when one was determined at the equator and one at the North Pole? (See Exercise 1.5.)

Properties and Changes (Section 1.2)

1.7 Classify each of the following as a physical or chemical change, and give at least one observation, fact, or reason to support your answer.

   a. An inflated balloon gets larger when heated

   b. A baking cake burns

   c. A bottle of milk sours

   d. A stretched rubber band breaks

1.8 Classify each of the following as a physical or chemical change, and give at least one observation, fact, or reason to support your answer.

  a. A stick is broken into two pieces

  b. A candle burns

  c. Rock salt is crushed by a hammer

  d. Tree leaves change color in autumn

1.9 Classify each of the following properties as physical or chemical. Explain your reasoning in each case.

  a. Iron melts at 1535°C

  b. Alcohol is very flammable

  c. The metal used in artificial hip-joint implants is not corroded by body fluids

  d. A 1-in. cube of aluminum weighs less than a 1-in. cube of lead

  e. An antacid tablet neutralizes stomach acid

1.10 Classify each of the following properties as physical or chemical. Explain your reasoning in each case.

  a. Mercury metal is a liquid at room temperature

  b. Sodium metal reacts vigorously with water

  c. Water freezes at 0°C

  d. Gold does not rust

  e. Chlorophyll molecules are green in color

A Model of Matter (Section 1.3)

1.11 A sample of liquid alcohol is frozen to a solid, then allowed to melt back to a liquid. Have the alcohol molecules been changed by the process? Explain your answer.

1.12 Succinic acid, a white solid that melts at 182°C, is heated gently, and a gas is given off. After the gas evolution stops, a white solid remains that melts at a temperature different from 182°C.

  a. Have the succinic acid molecules been changed by the process? Explain your answer.

  b. Is the white solid that remains after heating still succinic acid? Explain your answer.

  c. In terms of the number of atoms contained, how do you think the size of succinic acid molecules compares with the size of the molecules of the white solid produced by this process? Explain your answer.

  d. Classify molecules of succinic acid by using the term *homoatomic* or *heteroatomic*. Explain your reasoning.

1.13 A sample of solid elemental phosphorus that is deep red in color is burned. While the phosphorus is burning, a white smoke is produced that is actually a finely divided solid that is collected.

  a. Have the molecules of phosphorus been changed by the process of burning? Explain your answer.

  b. Is the collected white solid a different substance from the phosphorus? Explain your answer.

  c. In terms of the number of atoms contained, how do you think the size of the molecules of the white solid compares with the size of the molecules of phosphorus? Explain your answer.

  d. Classify molecules of the collected white solid using the term *homoatomic* or *heteroatomic*. Explain your reasoning.

1.14 Oxygen gas and solid sulfur are both made up of homoatomic molecules. The two react to form a single substance, sulfur dioxide. Use the term *homoatomic* or *heteroatomic* to classify molecules of sulfur dioxide. Explain your reasoning.

1.15 Under appropriate conditions, hydrogen peroxide can be changed to water and oxygen gas. Use the term *homoatomic* or *heteroatomic* to classify molecules of hydrogen peroxide. Explain your reasoning.

1.16 Water can be decomposed to hydrogen gas and oxygen gas by passing electricity through it. Use the term *homoatomic* or *heteroatomic* to classify molecules of water. Explain your reasoning.

1.17 Methane gas, a component of natural gas, is burned in pure oxygen. The only products of the process are water and carbon dioxide. Use the term *homoatomic* or *heteroatomic* to classify molecules of methane. Explain your reasoning.

Classifying Matter (Section 1.4)

1.18 Classify each pure substance represented below by a capital letter as an element or a compound. Indicate when such a classification cannot be made, and explain why

  a. Substance A is composed of heteroatomic molecules

  b. Substance D is composed of homoatomic molecules

  c. Substance E is changed into substances G and J when it is heated

1.19 Classify each pure substance represented below by a capital letter as an element or a compound. Indicate when such a classification cannot be made, and explain why.

  a. Two elements when mixed combine to form only substance L

  b. An element and a compound when mixed form substances M and Q

  c. Substance X is not changed by heating it

1.20 Consider the following experiments, and answer the questions pertaining to classification:

  a. A pure substance R is heated, cooled, put under pressure, and exposed to light but does not change into anything else. What can be said about classifying R as an element or a compound? Explain your reasoning.

  b. Upon heating, solid pure substance T gives off a gas and leaves another solid behind. What can be said about classifying T as an element or compound? Explain your reasoning.

  c. What can be said about classifying the solid left in part b as an element or compound? Explain your reasoning.

1.21 Early scientists incorrectly classified calcium oxide (lime) as an element for a number of years. Discuss one or more reasons why you think they might have done this.

1.22 Classify each of the following as homogeneous or heterogeneous:

  a. a pure gold chain

  b. liquid eyedrops

  c. chunky peanut butter

  d. a slice of watermelon

  e. cooking oil

  f. Italian salad dressing

  g. window glass

1.23 Classify each of the following as homogeneous or heterogeneous:

    a. muddy flood water

    b. gelatin dessert

    c. normal urine

    d. smog-filled air

    e. an apple

    f. mouthwash

    g. petroleum jelly

1.24 Classify as pure substance or solution each of the materials of Exercise 1.22 that you classified as homogeneous.

1.25 Classify as pure substance or solution each of the materials of Exercise 1.23 that you classified as homogeneous.

Measurement Units (Section 1.5)

1.26 Briefly discuss why a system of measurement units is an important part of our modern society.

1.27 In the distant past, 1 in. was defined as the length resulting from laying a specific number of grain kernels such as corn in a row. Discuss the disadvantages of such a system.

1.28 Identify at least eight different measuring devices used in your everyday activities (home, work, entertainment, etc.), and for each device, indicate the units used to express the resulting measurement.

The Metric System (Section 1.6)

1.29 Which of the following are expressed in metric units:

    a. The amount of antibiotic in a capsule: 125 mg

    b. The amount of gasoline in a tank: 15 gal

    c. The diameter of a contact lens: 6 mm

    d. The amount of heat used to warm some water: 50 BTU

    e. The distance of a foot race: 100 m

    f. The diameter of a circular table: 6 ft

1.30 Which of the following are expressed in metric units?

    a. Normal body temperature: 37°C

    b. The amount of soft drink in a bottle: 2 L

    c. The height of a ceiling in a room: 8.0 ft

    d. The amount of aspirin in a tablet: 81 mg

    e. The volume of a cooking pot: 4 qt

    f. The time for a short race to be won: 10.2 s

1.31 Referring to Table 1.3, suggest an appropriate metric system unit for each nonmetric unit in Exercise 1.29.

1.32 Referring to Table 1.3, suggest an appropriate metric system unit for each nonmetric unit in Exercise 1.30.

1.33 Referring only to Table 1.2, answer the following questions:

    a. A computer has 12 megabytes of memory storage. How many bytes of storage is this?

    b. A 10-km race is 6.2 mi long. How many meters long is it?

    c. A chemical balance can detect a mass as small as 0.1 mg. What is this detection limit in grams?

    d. A micrometer is a device used to measure small lengths. If it lives up to its name, what is the smallest metric length that could be measured using a micrometer?

1.34 Referring only to Table 1.2, answer the following questions:

    a. Devices are available that allow liquid volumes as small as one microliter (mL) to be measured. How many microliters would be contained in 1.00 liter?

    b. Electrical power is often measured in kilowatts. How many watts would equal 75 kilowatts?

    c. Ultrasound is sound of such high frequency that it cannot be heard. The frequency is measured in hertz (vibrations per second). How many hertz correspond to 15 megahertz?

    d. A chlorine atom has a diameter of 200 picometers. How many meters is this diameter?

1.35 One inch is approximately equal to 2.54 cm. Express this length in millimeters and meters.

1.36 Cookbooks are going metric. In such books, 1 cup is equal to 240 mL. Express 1 cup in terms of liters and cubic centimeters.

1.37 The so-called metric mile race is 1500 m long. What is its length in miles?

1.38 The shotput used by female track and field athletes has a mass of 4.0 kg. What would be the weight of such a shotput in pounds?

1.39 Referring to Table 1.3, answer the following questions:

    a. Which is larger, a liter or a quart?

    b. How many milliliters are in a 12.0-fl-oz soft drink?

    c. Which is larger, a BTU or a kilocalorie?

1.40 Referring to Table 1.3, answer the following questions:

    a. Approximately how many inches longer is a meter stick than a yardstick?

    b. A temperature increases by 65°C. How many kelvins would this increase be?

    c. You have a 5-lb bag of sugar. Approximately how many kilograms of sugar do you have?

1.41 Do the following, using appropriate values from Table 1.3:

    a. Calculate the area in square meters of a circular skating rink that has a 12.5-m radius. For a circle, the area (A) is related to the radius (r) by $A = \pi r^2$, where $\pi = 3.14$.

    b. Calculate the floor area and volume of a rectangular room that is 5.0 m long, 2.8 m wide, and 2.1 m high. Express your answers in square meters and cubic meters (meters cubed).

    c. A model sailboat has a triangular sail that is 25 cm high (h) and has a base (b) of 15 cm. Calculate the area (A) of the sail in square centimeters. $A = \dfrac{(b)(h)}{2}$ for a triangle.

1.42 Using appropriate values from Table 1.3, answer the following questions:

    a. One kilogram of water has a volume of 1.0 dm^3. What is the mass of 1.0 cm^3 of water?

    b. One quart is 32 fl oz. How many fluid ounces are contained in a 2.0-L bottle of soft drink?

    c. Approximately how many milligrams of aspirin are contained in a 5-grain tablet?

1.43 The weather report says the temperature is 23°F. What is this temperature on the Celsius scale? On the Kelvin scale?

Even-numbered exercises answered in Appendix B.

1.44 Recall from Chemistry Around Us 1.3 that a normal body temperature might be as low as 36.1°C in the morning and as high as 37.2°C at bedtime. What are these temperatures on the Fahrenheit scale?

1.45 One pound of body fat releases approximately 4500 kcal of energy when it is metabolized. How many joules of energy is this? How many BTUs?

Large and Small Numbers (Section 1.7)

1.46 Which of the following numbers are written using scientific notation correctly? For those that are not, explain what is wrong.

  a. 02.7×10^{-3}

  b. 4.1×10^{2}

  c. 71.9×10^{-6}

  d. 10^{3}

  e. $.0405 \times 10^{-2}$

  f. 0.119

1.47 Which of the following numbers are written using scientific notation correctly? For those that are not, explain what is wrong.

  a. 3.6×10^{-5}

  b. 3.9^{-2}

  c. 295×10^{3}

  d. 0.05×10^{-3}

  e. 10^{-4}

  f. 13.1×10^{6}

1.48 Write each of the following numbers using scientific notation:

  a. 14 thousand

  b. 365

  c. 0.00204

  d. 461.8

  e. 0.00100

  d. 9.11 hundred

1.49 Write each of the following numbers using scientific notation:

  a. two thousand

  b. 5162

  c. 0.00493

  d. 91.88

  e. 2.7 million

  f. 500

1.50 The speed of light is about 186 thousand mi/s, or 1100 million km/h. Write both numbers using scientific notation.

1.51 A sheet of paper is 0.0106 cm, or 0.0042 in., thick. Write both numbers using scientific notation.

1.52 A single copper atom has a mass of 1.05×10^{-22} g. Write this number in a decimal form without using scientific notation.

1.53 In 2.0 g of hydrogen gas, there are approximately 6.02×10^{23} hydrogen molecules. Write this number without using scientific notation.

1.54 Do the following multiplications, and express each answer using scientific notation:

  a. $(8.2 \times 10^{-3})(1.1 \times 10^{-2})$

  b. $(2.7 \times 10^{2})(5.1 \times 10^{4})$

  c. $(3.3 \times 10^{-4})(2.3 \times 10^{2})$

  d. $(9.2 \times 10^{-4})(2.1 \times 10^{4})$

  e. $(4.3 \times 10^{6})(6.1 \times 10^{5})$

1.55 Do the following multiplications, and express each answer using scientific notation:

  a. $(2.2 \times 10^{2})(3.1 \times 10^{-3})$

  b. $(4.9 \times 10^{-3})(1.8 \times 10^{-2})$

  c. $(9.1 \times 10^{-2})(1.2 \times 10^{1})$

  d. $(8.3 \times 10^{4})(4.1 \times 10^{3})$

  e. $(7.3 \times 10^{3})(6.6 \times 10^{-7})$

1.56 Express each of the following numbers using scientific notation, then carry out the multiplication. Express each answer using scientific notation.

  a. $(144)(0.0876)$

  b. $(751)(106)$

  c. $(0.0422)(0.00119)$

  d. $(128,000)(0.0000316)$

1.57 Express each of the following numbers using scientific notation, then carry out the multiplication. Express each answer using scientific notation.

  a. $(835)(0.00245)$

  b. $(300)(245)$

  c. $(68.3)(421)$

  d. $(32.9)(0.115)$

1.58 Do the following divisions, and express each answer using scientific notation:

  a. $\dfrac{3.1 \times 10^{-3}}{1.2 \times 10^{2}}$

  b. $\dfrac{7.9 \times 10^{4}}{3.6 \times 10^{2}}$

  c. $\dfrac{4.7 \times 10^{-1}}{7.4 \times 10^{2}}$

  d. $\dfrac{0.00229}{3.16}$

  e. $\dfrac{119}{3.8 \times 10^{3}}$

1.59 Do the following divisions, and express each answer using scientific notation:

  a. $\dfrac{154}{2.82}$

  b. $\dfrac{7.6 \times 10^{2}}{5.5 \times 10^{1}}$

  c. $\dfrac{9.1 \times 10^{-5}}{3.4 \times 10^{-2}}$

d. $\dfrac{7.6 \times 10^{3}}{3.8 \times 10^{-4}}$

e. $\dfrac{3.8 \times 10^{-3}}{4.7 \times 10^{4}}$

1.60 Do the following calculations, and express each answer using scientific notation:

a. $\dfrac{(5.3)(0.22)}{(6.1)(1.1)}$

b. $\dfrac{(3.8 \times 10^{-4})(1.7 \times 10^{-2})}{6.3 \times 10^{3}}$

c. $\dfrac{4.8 \times 10^{6}}{(7.4 \times 10^{3})(2.5 \times 10^{-4})}$

d. $\dfrac{5.6}{(0.022)(109)}$

e. $\dfrac{(4.6 \times 10^{-3})(2.3 \times 10^{2})}{(7.4 \times 10^{-4})(9.4 \times 10^{-5})}$

1.61 Do the following calculations, and express each answer using scientific notation:

a. $\dfrac{(8.1 \times 10^{-4})(1.2 \times 10^{5})}{4.4 \times 10^{-3}}$

b. $\dfrac{(0.0022)(4.2)}{(3.1)(0.088)}$

c. $\dfrac{3.6 \times 10^{3}}{(7.2 \times 10^{2})(5.6 \times 10^{-5})}$

d. $\dfrac{(9.4 \times 10^{1})}{(1.2 \times 10^{3})(7.5 \times 10^{2})}$

e. $\dfrac{(1.3 \times 10^{-3})(2.4 \times 10^{-2})}{(6.0 \times 10^{-6})(5.7 \times 10^{4})}$

Significant Figures (Section 1.8)

1.62 Indicate to what decimal position readings should be estimated and recorded (nearest 0.1, .01, etc.) for measurements made with the following devices:

a. A ruler with a smallest scale marking of 0.1 cm

b. A measuring telescope with a smallest scale marking of 0.1 mm

c. A protractor with a smallest scale marking of 1°

d. A tire pressure gauge with a smallest scale marking of 1 lb/in²·

1.63 Indicate to what decimal position readings should be estimated and recorded (nearest 0.1, .01, etc.) for measurements made with the following devices:

a. A buret with a smallest scale marking of 0.1 mL

b. A graduated cylinder with a smallest scale marking of 1 mL

c. A thermometer with a smallest scale marking of 0.1°C

d. A barometer with a smallest scale marking of 1 torr

1.64 Write the following measured quantities as you would record them, using the correct number of significant figures based on the device used to make the measurement:

a. Exactly 6 mL of water measured with a graduated cylinder that has a smallest scale marking of 0.1 mL.

b. A temperature that appears to be exactly 37 degrees using a thermometer with a smallest scale marking of 1°C.

c. A time of exactly nine seconds measured with a stopwatch that has a smallest scale marking of 0.1 second.

d. Fifteen and one-half degrees measured with a protractor that has 1-degree scale markings.

1.65 Write the following measured quantities as you would record them, using the correct number of significant figures based on the device used to make the measurements.

a. A length of two and one-half centimeters measured with a measuring telescope with a smallest scale marking of 0.1 mm.

b. An initial reading of exactly 0 for a buret with a smallest scale marking of 0.1 mL.

c. A length of four and one-half centimeters measured with a ruler that has a smallest scale marking of 0.1 cm.

d. An atmospheric pressure of exactly 690 torr measured with a barometer that has a smallest scale marking of 1 torr.

1.66 In each of the following, identify the measured numbers and exact numbers. Do the indicated calculation, and write your answer using the correct number of significant figures.

a. A bag of potatoes is found to weigh 5.06 lb. The bag contains 16 potatoes. Calculate the weight of an average potato.

b. The foul-shooting percentages for the five starting players of a women's basketball team are 71.2%, 66.9%, 74.1%, 80.9%, and 63.6%. What is the average shooting percentage of the five players?

1.67 In each of the following, identify the measured numbers and exact numbers. Do the indicated calculations, and write your answer using the correct number of significant figures.

a. An individual has a job of counting the number of people who enter a store between 1 P.M. and 2 P.M. each day for 5 days. The counts were 19, 24, 17, 31, and 40. What was the average number of people entering the store per day for the 5-day period?

b. The starting five members of a women's basketball team have the following heights: 6′9″, 5′8″, 5′6″, 5′1″, and 4′11″. What is the average height of the starting five?

1.68 Determine the number of significant figures in each of the following:

a. 0.0400

b. 309

c. 4.006

d. 4.4×10^{-3}

e. 1.002

f. 255.02

1.69 Determine the number of significant figures in each of the following:

a. 0.010

b. 14.84

c. 3.11×10^{3}

d. 157.2

e. 10.05

f. 162.15

1.70 Do the following calculations and use the correct number of significant figures in your answers. Assume all numbers are the results of measurements.

a. $(3.71)(1.4)$

b. $(0.0851)(1.2262)$

c. $\dfrac{(0.1432)(2.81)}{(0.7762)}$

d. $(3.3 \times 10^4)(3.09 \times 10^{-3})$

e. $\dfrac{(760)(2.00)}{6.02 \times 10^{20}}$

1.71 Do the following calculations and use the correct number of significant figures in your answers. Assume all numbers are the results of measurements.

a. $(1.21)(3.2)$

b. $(6.02 \times 10^{23})(0.220)$

c. $\dfrac{(0.023)(1.1 \times 10^{-3})}{100}$

d. $\dfrac{(365)(7.000)}{60}$

e. $\dfrac{(810)(3.1)}{8.632 \times 10^{-1}}$

1.72 Do the following calculations and use the correct number of significant figures in your answers. Assume all numbers are the results of measurements.

a. $0.208 + 4.9 + 1.11$

b. $228 + 0.999 + 1.02$

c. $8.543 - 7.954$

d. $(3.2 \times 10^{-2}) + (5.5 \times 10^{-1})$
(HINT: Write in decimal form first, then add.)

e. $336.86 - 309.11$

f. $21.66 - 0.02387$

1.73 Do the following calculations and use the correct number of significant figures in your answers. Assume all numbers are the results of measurements.

a. $2.1 + 5.07 + 0.119$

b. $0.051 + 8.11 + 0.02$

c. $4.337 - 3.211$

d. $(2.93 \times 10^{-1}) + (6.2 \times 10^{-2})$
(HINT: Write in decimal form first, then add.)

e. $471.19 - 365.09$

f. $17.76 - 0.0479$

1.74 Do the following calculations and use the correct number of significant figures in your answers. Assume all numbers are the results of measurements. In calculations involving both addition/subtraction and multiplication/division, it is usually better to do additions/subtractions first.

a. $\dfrac{(0.0267 + 0.0019)(4.626)}{28.7794}$

b. $\dfrac{212.6 - 21.88}{86.37}$

c. $\dfrac{27.99 - 18.07}{4.63 - 0.88}$

d. $\dfrac{18.87}{2.46} - \dfrac{18.07}{0.88}$
(HINT: Do divisions first, then subtract.)

e. $\dfrac{(8.46 - 2.09)(0.51 + 0.22)}{(3.74 + 0.07)(0.16 + 0.2)}$

f. $\dfrac{12.06 - 11.84}{0.271}$

1.75 Do the following calculations and use the correct number of significant figures in your answers. Assume all numbers are the results of measurements. In calculations involving both addition/subtraction and multiplication/division, it is usually better to do additions/subtractions first.

a. $\dfrac{132.15 - 32.16}{87.55}$

b. $\dfrac{(0.0844 + 0.1021)(7.174)}{19.1101}$

c. $\dfrac{(2.78 - 0.68)(0.42 + 0.4)}{(1.058 + 0.06)(0.22 + 0.2)}$

d. $\dfrac{27.65 - 21.71}{4.97 - 0.36}$

e. $\dfrac{12.47}{6.97} - \dfrac{203.4}{201.8}$
(HINT: Do divisions first, then subtract.)

f. $\dfrac{19.37 - 18.49}{0.822}$

1.76 The following measurements were obtained for the length and width of a series of rectangles. Each measurement was made using a ruler with a smallest scale marking of 0.1 cm.

Black rectangle: $l = 12.00$ cm, $w = 10.40$ cm
Red rectangle: $l = 20.20$ cm, $w = 2.42$ cm
Green rectangle: $l = 3.18$ cm, $w = 2.55$ cm
Orange rectangle: $l = 13.22$ cm, $w = 0.68$ cm

a. Calculate the area (length $\times$ width) and perimeter (sum of all four sides) for each rectangle and express your results in square centimeters and centimeters, respectively, and give the correct number of significant figures in the result.

b. Change all measured values to meters and then calculate the area and perimeter of each rectangle. Express your answers in square meters and meters, respectively, and give the correct number of significant figures.

c. Does changing the units used change the number of significant figures in the answers?

Using Units in Calculations (Section 1.9)

1.77 Determine a single factor derived from Table 1.3 that could be used as a multiplier to make each of the following conversions:

a. 3.4 lb to kilograms

b. 3.0 yd to meters

c. 1.5 oz to grams

d. 40 cm to inches

Even-numbered exercises answered in Appendix B.
42   Blue-numbered exercises are more challenging.

Unless otherwise noted all art on this page © Cengage Learning 2014

1.78 Determine a single factor from Table 1.3 that could be used as a multiplier to make each of the following conversions:

   a. 20 mg to grains

   b. 350 mL to fl oz

   c. 4 qt to liters

   d. 5 yd to meters

1.79 Obtain a factor from Table 1.3 and calculate the number of liters in 1.00 gal (4 qt) by using the factor-unit method of calculation.

1.80 A marathon race is about 26 miles. Obtain a factor from Table 1.3 and use the factor-unit method to calculate the distance of a marathon in kilometers.

1.81 A metric cookbook calls for a baking temperature of 200°C. Your oven settings are in degrees Fahrenheit. What Fahrenheit setting should you use?

1.82 A metric cookbook calls for 250 mL of milk. Your measuring cup is in English units. About how many cups of milk should you use? (NOTE: You will need two factors, one from Table 1.3 and one from the fact that 1 cup = 8 fl oz.)

1.83 An Olympic competitor threw the discus 64.70 m. What is this distance in feet?

1.84 You have a 40-lb baggage limit for a transatlantic flight. When your baggage is put on the scale, you think you are within the limits because it reads 18.0. But then you realize that weight is in kilograms. Do a calculation to determine whether your baggage is overweight.

1.85 You need 3.00 lb of meat that sells for $3.41/lb (i.e., 1 lb = $3.41). Use this price to determine a factor to calculate the cost of the meat you need using the factor-unit method.

1.86 During a glucose tolerance test, the serum glucose concentration of a patient was found to be 131 mg/dL. Convert the concentration to grams per liter.

Calculating Percentages (Section 1.10)

1.87 Retirement age is 65 years in many companies. What percentage of the way from birth to retirement is a 45-year-old person?

1.88 A salesperson made a sale of $467.80 and received a commission of $25.73. What percent commission was paid?

1.89 After drying, 140 lb of grapes yields 28 lb of raisins. What percentage of the grapes' mass was lost during the drying process?

1.90 The recommended daily intake of thiamin is 1.4 mg for a male adult. Suppose such a person takes in only 1.0 mg/day. What percentage of the recommended intake is he receiving?

1.91 The recommended daily caloric intake for a 20-year-old woman is 2000. How many Calories should her breakfast contain if she wants it to be 45% of her recommended daily total?

1.92 Immunoglobulin antibodies occur in five forms. A sample of serum is analyzed with the following results. Calculate the percentage of total immunoglobulin represented by each type.

Type:	IgG	IgA	IgM	IgD	IgE
Amount (mg):	987.1	213.3	99.7	14.4	0.1

Density (Section 1.11)

1.93 Calculate the density of the following materials for which the mass and volume of samples have been measured. Express the density of liquids in g/mL, the density of solids in g/cm^3, and the density of gases in g/L.

   a. 250 mL of liquid mercury metal (Hg) has a mass of 3400 g.

   b. 500 mL of concentrated liquid sulfuric acid (H_2SO_4) has a mass of 925 g.

   c. 5.00 L of oxygen gas has a mass of 7.15 g.

   d. A 200-cm^3 block of magnesium metal (Mg) has a mass of 350 g.

1.94 Calculate the density of the following materials for which the mass and volume of samples have been measured. Express the density of liquids in g/mL, the density of solids in g/cm^3, and the density of gases in g/L.

   a. A 50.0-mL sample of liquid acetone has a mass of 39.6 g.

   b. A 1.00-cup (236-mL) sample of homogenized milk has a mass of 243 g.

   c. 20.0 L of dry carbon dioxide gas (CO_2) has a mass of 39.54 g.

   d. A 25.0-cm^3 block of nickel metal (Ni) has a mass of 222.5 g.

1.95 Calculate the volume and density of a rectangular block of metal with edges of 5.50 cm, 12.0 cm, and 4.00 cm. The block weighs 929.5 g.

1.96 Calculate the volume and density of a cube of lead metal (Pb) that has a mass of 718.3 g and has edges that measure 3.98 cm.

1.97 The volume of an irregularly shaped solid can be determined by immersing the solid in a liquid and measuring the volume of liquid displaced. Find the volume and density of the following:

   a. An irregular piece of the mineral quartz is found to weigh 12.4 g. It is then placed into a graduated cylinder that contains some water. The quartz does not float. The water in the cylinder was at a level of 25.2 mL before the quartz was added and at 29.9 mL afterward.

   b. The volume of a sample of lead shot is determined using a graduated cylinder, as in part (a). The cylinder readings are 16.3 mL before the shot is added and 21.7 mL after. The sample of shot weighs 61.0 g.

   c. A sample of coarse rock salt is found to have a mass of 11.7 g. The volume of the sample is determined by the graduated-cylinder method described in (a), but kerosene is substituted for water because the salt will not dissolve in kerosene. The cylinder readings are 20.7 mL before adding the salt and 26.1 mL after.

1.98 The density of ether is 0.736 g/mL. What is the volume in mL of 280 g of ether?

1.99 Calculate the mass in grams of 125 mL of acetone (d = 0.791 g/mL).

Additional Exercises

1.100 Do the following metric system conversions by changing only the power of 10. For example, convert 2.5 L to mL: 2.5 L = 2.5×10^3 mL

   a. Convert 4.5 km to mm

   b. Convert 6.0×10^6 mg to g

   c. Convert 9.86×10^{15} m to km

   d. Convert 1.91×10^{-4} kg to mg

   e. Convert 5.0 ng to mg

Even-numbered exercises answered in Appendix B.

1.101 A single water molecule has a mass of 2.99×10^{-23} g. Each molecule contains two hydrogen atoms that together make up 11.2% of the mass of the water molecule. What is the mass in grams of a single hydrogen atom?

1.102 It has been found that fat makes up 14% of a 170-lb person's body weight. One pound of body fat provides 4500 kcal of energy when it is metabolized. The person requires 2000 kcal of energy per day to survive. Assume all the energy needed for survival comes from the metabolism of body fat, and calculate the number of days the person could survive without eating before depleting the entire body fat reserve.

1.103 Cooking oil has a density of 0.812 g/mL. What is the mass in grams of 1.00 quart of cooking oil? Use Table 1.3 for any necessary factors.

1.104 A 175-lb patient is to undergo surgery and will be given an anesthetic intravenously. The safe dosage of anesthetic is 12 mg/kg of body weight. Determine the maximum dose of anesthetic in mg that should be used.

1.105 At 4.0°C, pure water has a density of 1.00 g/mL. At 60.0°C, the density is 0.98 g/mL. Calculate the volume in mL of 1.00 g of water at each temperature, and then calculate the percentage increase in volume that occurs as water is heated from 4.0°C to 60.0°C.

Allied Health Exam Connection

The following questions are from these sources:

1. *Nursing School Entrance Exam* © 2005, Learning Express, LLC.

2. *McGraw-Hill's Nursing School Entrance Exams* by Thomas A. Evangelist, Tamara B. Orr and Judy Unrein © 2009, The McGraw-Hill Companies, Inc.

3. *NSEE Nursing School Entrance Exams,* 3rd Edition © 2009, Kaplan Publishing.

4. *Cliffs Test Prep: Nursing School Entrance Exams* by Fred N. Grayson © 2004, Wiley Publishing, Inc.

5. *Peterson's Master the Nursing School and Allied Health Entrance Exams,* 18th Edition by Marion F. Gooding © 2008, Peterson's, a Nelnet Company.

1.106 Which of the following properties is considered a physical property?

  a. flammability

  b. boiling point

  c. reactivity

  d. osmolarity

1.107 Which of the following depicts a chemical process?

  a. Helium is combined with neon

  b. Iron forms rust

  c. Water causes soil erosion

  d. Ice melts

1.108 Which of the following is a mixture?

  a. sodium chloride

  b. rice and beans

  c. magnesium sulfate

  d. water

1.109 If it is 90°F, approximately what temperature is it on the Celsius scale?

  a. 18°C

  b. 32°C

  c. 58°C

  d. 104°C

1.110 If the temperature is 25°C, what is the temperature in °F?

  a. 25°F

  b. 298°F

  c. 0°F

  d. 77°F

1.111 The correct formula for converting Fahrenheit to Celsius is given by: $°C = \dfrac{5}{9}(°F - 32°)$. Convert 72°F into temperature in Celsius.

  a. 72°C

  b. 40°C

  c. 25°C

  d. 22.2°C

1.112 In degrees Kelvin, the freezing pointing of water is:

  a. −273°

  b. 0°

  c. 100°

  d. 273°

1.113 The number of degrees on the Fahrenheit thermometer between the freezing point and the boiling point of water is:

  a. 100 degrees

  b. 180 degrees

  c. 212 degrees

  d. 273 degrees

1.114 A calorie is a form of:

  a. light

  b. heat

  c. darkness

  d. sound

1.115 How many millimeters are there in one centimeter?

  a. 10,000

  b. 1,000

  c. 100

  d. 10

1.116 Convert 4.50×10^2 nm into _____ pm.

  a. 4.50×10^2 pm

  b. 4.50×10^{-2} pm

  c. 4.50×10^{11} pm

  d. 4.50×10^5 pm

1.117 One millimeter contains how many μm?

  a. 10

  b. 100

c. 1,000

d. 10,000

1.118 Convert 4.50×10^2 nm into ___ m.

    a. 4.50×10^2 m

    b. 4.50×10^{11} m

    c. 4.50×10^{-7} m

    d. 4.50×10^8 m

1.119 The quantity 6,180 meters can be rewritten as:

    a. 6.180×10^3 meter.

    b. 6,180 kilometer.

    c. $6,180 \times 10^3$ meter.

    d. 180×10^3 meter.

1.120 The number 1,000,000 is what power of 10?

    a. 10^{-6}

    b. 10^6

    c. 1^6

    d. 0.000001

1.121 What exponent or power of ten would you use to express how many meters are in a kilometer?

    a. 10^5

    b. 10^3

    c. 10^4

    d. 10^2

1.122 Express 0.05620 in exponential notation.

    a. 0.057×10^{-3}

    b. 57×10^{-3}

    c. 563×10^{-4}

    d. 5.62×10^{-2}

1.123 Write the correct answer (correct number of significant figures) for the following calculation: $(27 + 93) \times 5.1558$

    a. 618.697

    b. 618.7

    c. 619

    d. 618.6970

1.124 The oxidation of 1 gram of CHO (carbohydrate) produces 4 calories. How much CHO must be oxidized in the body to produce 36 calories?

    a. 4 grams

    b. 7 grams

    c. 9 grams

    d. 12 grams

1.125 The percentage of oxygen by weight in $Al_2(SO_4)_3$ (atomic weights: Al = 27, S = 32, O = 16) is approximately:

    a. 19

    b. 21

    c. 56

    d. 92

1.126 A 10-percent solution of glucose will contain:

    a. 1 gram of glucose per 1,000 milliliters of solution

    b. 1 gram of glucose per 100 milliliters of solution

    c. 1 gram of glucose per 10 microliters of solution

    d. 10 grams of glucose per 100 milliliters of solution

1.127 The density of gold (Au) is 19.3g/cm^3 and that of iron (Fe) is 7.9g/cm^3. A comparison of the volumes (V) of 50 gram samples of each metal would show that:

    a. $V_{Au} = V_{Fe}$

    b. $V_{Au} < V_{Fe}$

    c. $V_{Au} > V_{Fe}$

    d. There is no predictable relationship between volumes.

Chemistry for Thought

1.128 The following pairs of substances represent heterogeneous mixtures. For each pair, describe the steps you would follow to separate the components and collect them.

    a. wood sawdust and sand

    b. sugar and sand

    c. iron filings and sand

    d. sand soaked with oil

1.129 Explain why a bathroom mirror becomes foggy when someone takes a hot shower. Classify any changes that occur as physical or chemical.

1.130 A 20-year-old student was weighed and found to have a mass of 44.5 kg. She converted this to pounds and got an answer of 20.2 lb. Describe the mistake she probably made in doing the calculation.

1.131 Liquid mercury metal freezes to a solid at a temperature of $-38.9°C$. Suppose you want to measure a temperature that is at least as low as $-45°C$. Can you use a mercury thermometer? If not, propose a way to make the measurement.

1.132 Answer the question contained in Figure 1.3. How does hang gliding confirm that air is an example of matter?

1.133 Show how the factor-unit method can be used to prepare an oatmeal breakfast for 27 guests at a family reunion. The directions on the oatmeal box say that 1 cup of dry oatmeal makes 3 servings.

1.134 A chemist is brought a small solid figurine. The owner wants to know if it is made of silver but doesn't want it damaged during the analysis. The chemist decides to determine the density, knowing that silver has a density of 10.5 g/mL. The figurine is put into a graduated cylinder that contains 32.6 mL of water. The reading while the figurine is in the water is 60.1 mL. The mass of the figurine is 240.8 g. Is the figurine made of silver? Explain your reasoning.

1.135 Refer to Chemistry Around Us 1.2, then check the labels on your toiletries and see if you can identify one product that is regulated as both a drug and a cosmetic.

1.136 Refer to Figure 1.6, then use the model of matter described in Section 1.3 to propose an explanation for the following observation. When two teaspoons of sugar are dissolved in a small glass of water, the volume of the resulting solution is not significantly larger than the original volume of the water.

Even-numbered exercises answered in Appendix B.

© Gregor Schuster/Getty Images

Case Study

Raju, a 15-year-old high-school student, was small for his age. Perhaps he took after his mother, who was quite petite. In the middle of his school year, Raju began having severe headaches, feeling them mostly behind his eyes. Over-the-counter remedies didn't seem to help much. Raju's mother had suffered from migraine headaches for some time. His doctor was inclined to diagnose migraine disorder, but Raju's symptoms didn't seem to completely support that diagnosis. Motivated by intuition, the doctor ordered an MRI scan. An MRI scan produces clear images of soft tissues such as the brain. Raju's scan revealed a golf ball–sized tumor impinging on the pituitary gland, located close to the hypothalamus, at the base of the brain. The tumor was likely the cause of Raju's growth issues as well as the headaches, because the pituitary gland regulates hormones related to growth. Fortunately, Raju's tumor was operable. He is now leading a normal life and annual MRI tests have shown no recurrence of the tumor.

How does an MRI scan work? What element in the body does an MRI scan detect? How important is it for health professionals to consider unlikely explanations for seemingly straightforward symptoms?

Follow-up to this Case Study appears at the end the chapter before the Concept Summary.

Learning Objectives

When you have completed your study of this chapter, you should be able to:

1 Use symbols for chemical elements to write formulas for chemical compounds. **(Section 2.1)**

2 Identify the characteristics of protons, neutrons, and electrons. **(Section 2.2)**

3 Use the concepts of atomic number and mass number to determine the number of subatomic particles in isotopes and to write correct symbols for isotopes. **(Section 2.3)**

4 Use atomic weights of the elements to calculate molecular weights of compounds. **(Section 2.4)**

5 Use isotope percent abundances and masses to calculate atomic weights of elements. **(Section 2.5)**

6 Use the mole concept to obtain relationships between number of moles, number of grams, and number of atoms for elements, and use those relationships to obtain factors for use in factor-unit calculations. **(Section 2.6)**

7 Use the mole concept and molecular formulas to obtain relationships between number of moles, number of grams, and number of atoms or molecules for compounds, and use those relationships to obtain factors for use in factor-unit calculations. **(Section 2.7)**

We introduced some fundamental ideas about matter, atoms, molecules, measurements, and calculations in Chapter 1. In this chapter, these ideas are applied, the mole is defined, and the quantitative nature of chemistry becomes more apparent. A system of symbols is introduced that simplifies the way atoms and molecules are represented.

2.1 Symbols and Formulas

Learning Objective

1. Use symbols for chemical elements to write formulas for chemical compounds.

In Chapter 1, we defined elements as homogeneous pure substances made up of identical atoms. At least 115 different elements are known to exist. This leads to the conclusion that a minimum of 115 different kinds of atoms exist. Eighty-eight of the elements are naturally occurring and therefore are found in Earth's crust, oceans, or atmosphere. The others are synthetic elements produced in the laboratory.

Each element can be characterized and identified by its unique set of physical and chemical properties, but it would be very cumbersome to list all these properties each time a specific element was discussed. For this reason, each element has been assigned a unique name and symbol. Many elements have been named by their discoverer, and as a result the names are varied. Some are based on elemental properties, others come from the names of famous scientists or places, while others are derived from the names of astronomical bodies or mythological characters.

elemental symbol A symbol assigned to an element based on the name of the element, consisting of one capital letter or a capital letter followed by a lowercase letter.

An **elemental symbol** is based on the element's name and consists of a single capital letter or a capital letter followed by a lowercase letter. The symbols for 11 elements are based on the element's name in Latin or German. An elemental symbol is sometimes used to represent an element in a general way or to represent a single atom of an element. ▶ Table 2.1 lists the elements whose names and symbols have been agreed on.

TABLE 2.1 The Chemical Elements and Their Symbols

Ac	actinium	Dy	dysprosium	Mg	magnesium	Rn	radon
Ag	silver *(argentum)*[a]	Er	erbium	Mn	manganese	Ru	ruthenium
Al	aluminum	Es	einsteinium	Mo	molybdenum	S	sulfur
Am	americium	Eu	europium	Mt	meitnerium	Sb	antimony *(stibium)*[a]
Ar	argon	F	flourine	N	nitrogen	Sc	scandium
As	arsenic	Fe	iron *(ferrum)*[a]	Na	sodium *(natrium)*[a]	Se	selenium
At	astatine	Fl	flerovium	Nb	niobium	Sg	seaborgium
Au	gold *(aurum)*[a]	Fm	fermium	Nd	neodymium	Si	silicon
B	boron	Fr	francium	Ne	neon	Sm	samarium
Ba	barium	Ga	gallium	Ni	nickel	Sn	tin *(stannum)*[a]
Be	beryllium	Gd	gadolinium	No	nobelium	Sr	strontium
Bh	bohrium	Ge	germanium	Np	neptunium	Ta	tantalum
Bi	bismuth	H	hydrogen	O	oxygen	Tb	terbium
Bk	berkelium	He	helium	Os	osmium	Tc	technetium
Br	bromine	Hf	hafnium	P	phosphorus	Te	tellurium
C	carbon	Hg	mercury *(hydrargyrum)*[a]	Pa	protactinium	Th	thorium
Ca	calcium	Ho	holmium	Pb	lead *(plumbum)*[a]	Ti	titanium
Cd	cadmium	Hs	hassium	Pd	palladium	Tl	thallium
Ce	cerium	I	iodine	Pm	promethium	Tm	thulium
Cf	californium	In	indium	Po	polonium	U	uranium
Cl	chlorine	Ir	iridium	Pr	praseodymium	V	vanadium
Cm	curium	K	potassium *(kalium)*[a]	Pt	platinum	W	tungsten *(wolfram)*[a]
Cn	copernicium	Kr	krypton	Pu	plutonium	Xe	xenon
Co	cobalt	La	lanthanum	Ra	radium	Y	yttrium
Cr	chromium	Li	lithium	Rb	rubidium	Yb	ytterbium
Cs	cesium	Lr	lawrencium	Re	rhenium	Zn	zinc
Cu	copper *(cuprum)*[a]	Lu	lutetium	Rf	rutherfordium	Zr	zirconium
Db	dubnium	Lv	livermorium	Rg	roentgenium		
Ds	darmstadtium	Md	mendelevium	Rh	rhodium		

[a]Elements with symbols not derived from their English names.

Compounds are pure substances made up of two or more different kinds of atoms. The atoms found in compounds are the same ones found in elements. Thus, the symbols used to represent elements can be combined and used to represent compounds. A molecular compound is depicted by a **compound formula,** in which each atom is represented by an appropriate elemental symbol. When more than one atom of an element is present in a molecule, a subscript is used to indicate the number.

compound formula A representation of the molecule of a compound, consisting of the symbols of the atoms found in the molecule. Atoms present in numbers greater than 1 have the number indicated by a subscript.

Notice the similarity between this practice and the molecular representations in Figure 1.4. The carbon dioxide molecules in Figure 1.4 are represented by the formula CO_2, where C represents an atom of carbon and O an atom of oxygen. The subscript 2 on the oxygen indicates that the molecule contains two atoms of oxygen. Notice that the single carbon atom in the molecule is not indicated by a subscript 1. Subscript 1 is never used in molecular formulas; it is understood. Formulas for molecular compounds are sometimes used to represent a compound in a general way, or to represent a single molecule of a compound. See ▶ Table 2.2 and ▶ Figure 2.1 for other examples of compound formulas.

TABLE 2.2 Examples of Compound Formulas

Compound Name	Molecular Representation	Molecular Formula
Methane		CH_4
Water		H_2O
Carbon monoxide		CO
Hydrogen peroxide		H_2O_2

Figure 2.1 Iron pyrite is a mineral that contains iron and sulfur atoms in a 1:2 ratio, respectively. The golden crystals, called "fool's gold" by experienced miners, have caused (temporary) excitement for many novice prospectors. What is the formula for iron pyrite?

© Charles D Winters/Getty Images

▶ Example 2.1

Write formulas for the following compounds:

a. Nitrogen dioxide: one nitrogen (N) atom and two oxygen (O) atoms
b. Sulfuric acid: two hydrogen (H) atoms, one sulfur (S) atom, and four oxygen (O) atoms

Solution

a. The symbols for the atoms are obtained from Table 2.1. The single N atom will not have a subscript because ones are understood and never written. The two O atoms will be represented by writing a subscript 2. The molecular formula is NO_2.
b. Using similar reasoning, the H atom will have a subscript 2, the S atom will have no subscript, and the O atom will have a subscript 4. Therefore, the molecular formula is H_2SO_4.

▶ **LEARNING CHECK 2.1** Write molecular formulas for the following compounds:

a. Phosphoric acid: three hydrogen (H) atoms, one phosphorus (P) atom, and four oxygen (O) atoms
b. Sulfur trioxide: one sulfur (S) atom and three oxygen (O) atoms
c. Glucose: six carbon (C) atoms, twelve hydrogen (H) atoms, and six oxygen (O) atoms

2.2 Inside the Atom

Learning Objective

2. Identify the characteristics of protons, neutrons, and electrons.

An atom has been defined as the limit of chemical subdivision for matter. On the basis of the characteristics of atoms that have been discussed, you probably have a general (and correct) idea that atoms can be considered to be the units from which matter is made. However, the question of how atoms interact to form matter has not yet been addressed. This interesting topic is discussed in Chapters 3 and 4, but a bit more must be learned about atoms first.

Extensive experimental evidence collected since the middle of the 19th century indicates that atoms are made up of many smaller particles. More than 100 of these subatomic particles have been discovered, and the search for more continues. As yet there is no single theory that can explain all observations involving subatomic particles, but three fundamental particles are included in all current theories: the proton, the neutron, and the electron. Most chemical behavior of matter can be explained in terms of a few of the well-known characteristics of these particles. These important characteristics are mass, electrical charge, and location in atoms. They are summarized in ❭ Table 2.3. The atomic mass unit (u) shown in the table is discussed in Section 2.4.

TABLE 2.3 Characteristics of the Fundamental Subatomic Particles

| Particle | Common Symbols | Characteristics | | | |
		Charge ($\pm$)	Mass (g)	Mass (u)	Location
Electron	e^-	$1-$	9.07×10^{-28}	1/1836	Outside nucleus
Proton	p, p^+, H^+	$1+$	1.67×10^{-24}	1	Inside nucleus
Neutron	n	0	1.67×10^{-24}	1	Inside nucleus

nucleus The central core of atoms that contains protons, neutrons, and most of the mass of atoms.

Figure 2.2 Electrons move rapidly around a massive nucleus. This figure is not drawn to scale. For a nucleus of the size shown, the closest electrons would be at least 80 m away.

In Chapter 1, atoms were described as being very tiny particles. The masses of single atoms of the elements are now known to fall within the range of 1.67×10^{-24} g for the least massive to 5.00×10^{-22} g for the most massive. Thus, it is not surprising to find that the particles that make up atoms have very small masses as well. Even though atomic masses are very small, the mass information given in Table 2.3 indicates that most of the mass of an atom comes from the protons and neutrons it contains. The protons and neutrons are tightly bound together to form the central portion of an atom called the **nucleus.** Because protons each have a +1 electrical charge, and neutrons have no charge, the nucleus of an atom has a positive electrical charge that is equal to the number of protons it contains.

Electrons are negatively charged particles located outside the nucleus of an atom. Protons and electrons carry equal but opposite electrical charges, so a neutral atom that has no electrical charge must have the same number of protons in its nucleus as it has electrons outside the nucleus. The electrons of an atom are thought to move very rapidly throughout a relatively large volume of space surrounding the small but very heavy nucleus (see ❭ Figure 2.2).

Even though subatomic particles exist, the atom itself is the particle of primary interest in chemistry because subatomic particles do not lead an independent existence for any appreciable length of time. The only way they gain long-term stability is by combining with other particles to form an atom.

Looking at Atoms

For many years, scientists believed in the existence of atoms based primarily on the successful use of the atomic theory to explain the observed properties of matter. Direct visual evidence of atoms, however, eluded scientists until 1970, when an electron microscope, which uses electron beams instead of light, obtained images that showed the locations of large, heavy atoms of thorium and uranium.

In 1981, the scanning tunneling microscope (STM) was introduced. A different type of microscope, the STM does not have lenses and a light source like the microscopes with which you are familiar. Instead, an STM uses a very thin, needlelike probe that moves horizontally over the surface of a sample. A high-voltage electrical current passes between the sharp tip of the probe and the sample under investigation. The magnitude of the current passing between the sample and the probe depends on the distance between the tip of the probe and the surface of the sample. As the probe passes over the atoms making up the sample surface, the STM keeps the magnitude of the current constant by moving the probe up and down to maintain a very tiny but constant separation distance between the atoms and the probe tip. It moves up when it goes over a protruding atom and moves down when it is above a space between atoms. A computer detects and processes these tiny up-and-down motions and converts them into a greatly enlarged, three-dimensional image. Because the motion resulted from the atomic bumps on the surface, the image shows the location of the bump-producing atoms on the surface. The inventors of the STM, Gerd Benning, Heinrich Rohrer, and Ernst Ruska, shared the 1986 Nobel Prize in physics for their accomplishment.

A view from a scanning tunneling microscope (STM) shows the locations of atoms protruding from a short strand of a DNA double helix.

2.3 Isotopes

Learning Objective

3. Use the concepts of atomic number and mass number to determine the number of subatomic particles in isotopes and to write correct symbols for isotopes.

Atoms of elements have no electrical charge and so must contain identical numbers of positive protons and negative electrons. However, because neutrons have no electrical charge, their numbers in an atom do not have to be the same as the numbers of protons or electrons. The number of protons in the nucleus of an atom is given by the **atomic number** for the atom. Atomic numbers are represented by the symbol Z. All atoms of a specific element must have the same atomic number. The atomic numbers for each element are the numbers above the elemental symbols of the periodic table inside the front cover of this book. Remember, this is also the number of electrons in the atoms of each element.

The possibility of having different numbers of neutrons combined with one given number of protons to form nuclei leads to some interesting results. For example, three different kinds of hydrogen atoms are known to exist. Each kind of atom contains one proton and one electron, so all have an atomic number of 1. However, the nuclei of the different kinds of hydrogen atoms contain different numbers of neutrons. The most common kind has no neutrons in the nucleus, the next most common has one, and the least common kind has two. The sum of the number of protons and the number of neutrons in a nucleus is called the **mass number** and is represented by the symbol A. Thus, the three kinds of hydrogen atoms all have atomic numbers of 1 and mass numbers of 1, 2, and 3, respectively. Atoms that have the same atomic number but different mass numbers are called **isotopes.** Most elements are made up of mixtures of two or more isotopes. When it is important to

atomic number of an atom A number equal to the number of protons in the nucleus of an atom. Symbolically it is represented by Z.

mass number of an atom A number equal to the sum of the number of protons and neutrons in the nucleus of an atom. Symbolically it is represented by A.

isotopes Atoms that have the same atomic number but different mass numbers. That is, they are atoms of the same element that contain different numbers of neutrons in their nuclei.

distinguish between them, the following notation is used for each isotope: $^A_Z E$, where E is the symbol for the element.

The three isotopes of hydrogen are represented as follows using this notation: ^1_1H, ^2_1H, and ^3_1H. When these symbols are not convenient to use, as in written or spoken references to the isotopes, the elemental name followed by the mass number is used. Thus, the three hydrogen isotopes are hydrogen-1, hydrogen-2, and hydrogen-3. These three isotopes have specific names: protium ($A = 1$), deuterium ($A = 2$), and tritium ($A = 3$).

▶ Example 2.2

Use the periodic table inside the front cover to answer the following questions about isotopes:

a. What are the mass number, atomic number, and isotope symbol ($^A_Z E$) for an atom that contains 7 protons and 8 neutrons?

b. How many neutrons are contained in an atom of nickel-60?

c. How many protons and how many neutrons are contained in an atom with a mass number of 26 and the symbol Mg?

Solution

a. The mass number, A, equals the sum of the number of protons and the number of neutrons: $A = 7 + 8 = 15$. The atomic number, Z, equals the number of protons: $Z = 7$. According to the periodic table, the element with an atomic number of 7 is nitrogen, with the symbol N. The isotope symbol is $^{15}_7N$.

b. According to the periodic table, nickel has the symbol Ni, and an atomic number, Z, of 28. The mass number, 60, is equal to the sum of the number of protons and the number of neutrons. The number of protons is equal to the atomic number, 28. Therefore, the number of neutrons is $60 - 28 = 32$. The atom contains 32 neutrons.

c. According to the periodic table, the element with the symbol Mg is magnesium, which has an atomic number of 12. Therefore, the atom contains 12 protons. Since A, the number of protons plus neutrons is equal to 26, the number of neutrons is $26 - 12$, or 14. The atom contains 14 neutrons.

▶ LEARNING CHECK 2.2

Use the periodic table inside the front cover to answer the following questions about isotopes:

a. What are the atomic number, mass number, and isotope symbol for an atom that contains 4 protons and 5 neutrons?

b. How many neutrons are contained in an atom of chlorine-37?

c. How many protons and how many neutrons are contained in an atom with a mass number of 28 and the symbol Si?

2.4 Relative Masses of Atoms and Molecules

Learning Objective

4. Use atomic weights of the elements to calculate molecular weights of compounds.

Because of their extremely small size, it is very inconvenient to use the actual masses of atoms when the atoms are being characterized or when quantitative calculations are done. In fact, the earliest chemists had no way of determining the actual masses of atoms. For this reason, a system was devised that utilized relative or comparative masses for the atoms. These relative masses are the numbers that are given beneath the symbol and name for each element in the periodic table inside the front cover.

Relative masses provide a simple way of comparing the masses of atoms. For example, the mass of neon atoms, Ne, from the periodic table is given as 20.18. Similarly, the mass

Nonprescription Medicines

The important role of pharmacy in the modern practice of medicine is well known. In a simplified but familiar scenario, a sick patient consults a physician, who diagnoses the ailment and writes a prescription for medicine. The patient takes the prescription to a pharmacist, who prepares and packages the medication, which the patient takes according to the directions on the container. However, it is estimated that nearly 40% of the common, everyday health problems of people in the United States are treated without the aid of either a physician or a pharmacist. This is possible because of the availability of an estimated thousand medicinal products that may be purchased by consumers without a prescription. These nonprescription medications are often called over-the-counter (OTC) drugs or medicines. They are available not only in pharmacies, but also in such places as supermarkets and convenience stores. The range of health problems that can be treated with OTC medicines has grown significantly in recent years as a result of action by the FDA that transferred more than 600 effective prescription-only medicines to nonprescription (OTC) status. However, the widespread availability and use of these products raised significant questions about how best to provide consumers with the adequate directions for proper use that are required by law. In an attempt to improve upon the nonstandardized labeling practices used for OTC medicines, the FDA in 1997 proposed new OTC medicine labeling rules:

All wording must be in "plain English."

A large, easy-to-read type must be used.

3. Labels must follow a consistent design style.
4. Information must be given in a standardized order.
5. Standardized headings and subheadings must be used.

As is often true, progress of this type does not come without a price tag. The FDA estimated it would cost about $14 million to comply with the new labeling rules. However, the Nonprescription Drug Manufacturers Association estimated implementation costs at $155 million for compliance within the required two years and more than $400 million if packaging changes were required to accommodate the new labeling format. The actual costs have not yet been determined, but consumers will certainly pay a significant part of these costs through increased prices in OTC products.

The active ingredients in all these products were originally available only by prescription.

of calcium atoms, Ca, is given as 40.08. These numbers simply indicate that calcium atoms have a mass that is about twice the mass of neon atoms. The exact relationship between the two masses is calculated as follows, using the correct number of significant figures:

$$\frac{\text{Ca atom mass}}{\text{Ne atom mass}} = \frac{40.08}{20.18} = 1.986$$

In a similar way, we arrive at the conclusion that helium atoms are about four times as massive as hydrogen atoms:

$$\frac{\text{He atom mass}}{\text{H atom mass}} = \frac{4.003}{1.008} = 3.971$$

In each case, we have been able to determine the relationship between the masses of the atoms without using the actual masses.

Modern instruments called mass spectrometers allow the actual masses of individual atoms to be measured. These measured masses show the same relationships to each other as do the relative masses:

$$\frac{\text{Ca atom mass}}{\text{Ne atom mass}} = \frac{6.655 \times 10^{-23} \text{ g}}{3.351 \times 10^{-23} \text{ g}} = 1.986$$

$$\frac{\text{He atom mass}}{\text{H atom mass}} = \frac{6.647 \times 10^{-24} \text{ g}}{1.674 \times 10^{-24} \text{ g}} = 3.971$$

Are You at Risk for Osteoporosis?

Osteoporosis, the abnormal thinning of bones that often accompanies aging, may lead to bone fractures, disability, and even death. While women are most susceptible, this serious condition also affects men but usually at a more advanced age than women. A number of significant risk factors have been identified, including the following:

1. A poor diet, especially one low in calcium
2. Advanced age
3. The onset of menopause (or having had ovaries removed)
4. A sedentary lifestyle
5. Smoking
6. A family history of osteoporosis or hip fracture
7. Heavy drinking
8. The long-term use of certain steroid drugs
9. Vitamin D deficiency

Ninety-nine percent of the calcium found in the body is located in the skeleton and teeth, so it is not surprising that the behavior of this metallic element in the body plays a central role in a number of these risk factors for osteoporosis. For example, it is known that in later life the body's ability to absorb calcium from food in the small intestine decreases (factor 2) and such absorption requires the presence of vitamin D (factor 9).

It is well recognized that the best insurance against developing osteoporosis in later life is to build and strengthen as much bone as possible during the first 25 to 35 years of life. Two essential components of this process are eating a healthful diet containing adequate amounts of calcium and vitamin D, and following a healthful lifestyle that includes regular weight-bearing exercise such as walking, jogging, weight lifting, stair climbing, or physical labor. Dietary calcium supplements provide an additional way to enhance calcium intake, especially for individuals who are at risk to develop osteoporosis.

A bone density test provides an effective way to diagnose the presence or extent of osteoporosis in an individual. Such tests measure the absorption of X-rays by bones, and are not invasive or uncomfortable. Bone density tests were not routinely recommended in the past, but some health and wellness organizations now suggest that women who are at high risk should have such a test by the age of 50, and all women over age 65 should be routinely tested.

An active lifestyle is an essential component of building and strengthening bones.

Thus, we see that the relative masses used in the periodic table give the same results as the actual masses when the masses of atoms are compared with one another.

Actual atomic masses are given in mass units such as grams, but the relative values used in the periodic table are given in units referred to as **atomic mass units.** The symbol u will be used to represent atomic mass units throughout the text. The actual mass represented by a single atomic mass unit is $\frac{1}{12}$ the mass of a single carbon-12 atom or 1.661×10^{-24} g.

The relative masses of the elements as given in the periodic table are referred to as atomic masses or atomic weights. We will use the term **atomic weights** in this book. In those cases where the naturally occurring element exists in the form of a mixture of isotopes, the recorded atomic weight is the average value for the naturally occurring isotope mixture. This idea is discussed in Section 2.5.

A convenient way to visualize the concept of relative masses, and to compare the relative masses of atoms, involves the use of the simple child's toy called a see-saw or teeter-totter shown below.

When the masses on both sides of the central pivot are equal, the see-saw will be in balance. When the masses on each side are different, the see-saw will be out of balance, and the side with the greatest mass will go down. This characteristic lends itself to comparing the masses of objects to each other without knowing their actual values in mass units such as grams.

atomic mass unit (u) A unit used to express the relative masses of atoms. One u is equal to $\frac{1}{12}$ the mass of an atom of carbon-12.

atomic weight The mass of an average atom of an element expressed in atomic mass units.

▶ Example 2.3

Round the atomic weights of the periodic table to the nearest whole number, and answer the following questions:

a. How many calcium atoms, Ca, would have to be put on one side of a see-saw to balance one atom of bromine, Br, that was placed on the other side?
b. How many helium atoms, He, would balance one oxygen atom on a see-saw?
c. How many hydrogen atoms, H, would balance one carbon atom, C, on a see-saw?

Solution

a. Calcium atoms have a periodic table atomic weight of 40.08u, which rounds to 40u. Bromine atoms have a periodic table atomic weight of 79.90u, which rounds to 80u. Thus, two calcium atoms would be required to balance one bromine atom as indicated below:

b. Helium atoms have a periodic table atomic weight of 4.003u, which rounds to 4u. Oxygen atoms have a periodic table atomic weight of 16.00u, which rounds to 16u. Thus, four helium atoms will balance one oxygen atom:

c. Hydrogen atoms have a periodic table atomic weight of 1.008u, which rounds to 1u. Carbon atoms have a periodic table atomic weight of 12.01u, which rounds to 12u. Twelve hydrogen atoms will balance one carbon atom:

▶ **LEARNING CHECK 2.3** Round atomic weights of the periodic table to the nearest whole number, and answer the following questions using the see-saw balance idea:

a. How many nitrogen atoms, N, will balance one atom of iron, Fe?
b. How many carbon atoms, C, will balance six helium atoms, He?
c. How many calcium atoms, Ca, will balance two argon atoms, Ar?

Molecules are made up of atoms, so the relative mass of a molecule can be calculated by adding together the atomic weights of the atoms that make up the molecule. Relative molecular masses calculated in this way are called **molecular weights** and are also given in atomic mass units.

molecular weight The relative mass of a molecule expressed in atomic mass units and calculated by adding together the atomic weights of the atoms in the molecule.

▶ Example 2.4

Use atomic weights from the periodic table inside the front cover to determine the molecular weight of urea, CH_4N_2O, the chemical form in which much nitrogenous body waste is excreted in the urine.

Solution

According to the formula given, a urea molecule contains one carbon atom, C, two nitrogen atoms, N, four hydrogen atoms, H, and one oxygen atom, O. The molecular weight is calculated as follows:

$$MW = 1(\text{at. wt. C}) + 2(\text{at. wt. N}) + 4(\text{at. wt. H}) + 1(\text{at. wt. O})$$

$$MW = 1(12.01 \text{ u}) + 2(14.01 \text{ u}) + 4(1.008 \text{ u}) + 1(16.00 \text{ u}) = 60.062 \text{ u}$$

Rounded to four significant figures, the correct answer is 60.06 u.

2.5 Isotopes and Atomic Weights

Learning Objective

5. Use isotope percent abundances and masses to calculate atomic weights of elements.

The atomic weights discussed in Section 2.4 were defined as the relative masses of atoms of the elements. It would have been more correct to define them as the relative masses of average atoms of the elements. Why include the idea of an average atom? Remember, the mass number of an isotope is the sum of the number of protons and neutrons in the nucleus of atoms of the isotope. Also, protons and neutrons both have masses of 1 u (see Table 2.3). The masses of electrons are quite small, so the atomic weights of isotopes are very nearly equal to their mass numbers. For example, all naturally occurring phosphorus is made up of atoms containing 15 protons and 16 neutrons. The mass number of this isotope is 31, and its symbol is $^{31}_{15}P$. The relative mass (atomic weight) of atoms of this isotope is 30.97 u, the same as the atomic weight of the element given in the periodic table. The atomic weight of the isotope and the listed atomic weight of the element are the same because all atoms contained in the element are identical and have the same relative mass.

Naturally occurring chlorine is different. It is a mixture of two isotopes, $^{35}_{17}Cl$ and $^{37}_{17}Cl$. Chlorine-35 has a mass number of 35 and a relative mass of 34.97 u. Chlorine-37 has a mass number of 37 and a relative mass of 36.97 u. Note that the mass numbers and relative masses (atomic weights) of the atoms of an isotope are essentially identical. However, naturally occurring chlorine is a mixture containing both isotopes. Thus, the determined relative mass of chlorine atoms will be the average relative mass of the atoms found in the mixture (see ▶ Figure 2.3).

The average mass of each particle in a group of particles is simply the total mass of the group divided by the number of particles in the group. This calculation requires that the total number of particles be known. However, the percentage of each isotope in a mixture of atoms is easier to determine than the actual number of atoms of each isotope present. The percentages can be used to calculate average masses. Remember that percent means per 100, so we use an imaginary sample of an element containing 100 atoms in the calculation. On this basis, the number of atoms of each isotope in the 100-atom sample will be the percentage of that isotope in the sample. The mass contributed to the sample by each isotope will be the product of the number of atoms of the isotope (the percentage of the isotope) and the mass of the isotope. The total mass of the sample will be the sum of the masses contributed by each isotope. This total mass divided by 100, the number of atoms in our imaginary sample, gives the mass of an average atom, which is the atomic weight of the element.

Figure 2.3 The pieces of fruit in a bowl are somewhat like atoms of the isotopes of an element. Each piece of fruit may have the same color, taste, and texture, but it is unlikely that any two have *exactly* the same mass. The 12 oranges in the bowl weigh a total of 2.36×10^3 g. What is the average mass of each orange in the bowl?

© Mark Stout Photography/Shutterstock.com

▶ **Example 2.5**

Calculate the atomic weight of chlorine, given that the naturally occurring element consists of 75.53% chlorine-35 (mass = 34.97 u) and 24.47% chlorine-37 (mass = 36.97 u).

Solution

$$\text{atomic weight} = \frac{(\% \text{ chlorine-35})(\text{mass chlorine-35}) + (\% \text{ chlorine-37})(\text{mass chlorine-37})}{100}$$

$$= \frac{(75.53)(34.97 \text{ u}) + (24.47)(36.97 \text{ u})}{100}$$

$$= \frac{2641.28\ \text{u} + 904.66\ \text{u}}{100} = \frac{3545.94\ \text{u}}{100} = 35.4594\ \text{u}$$

$$= 35.46\ \text{u}\ (\text{rounded value})$$

This result is slightly different from the periodic table atomic weight value of 35.45 because of slight errors introduced in rounding the isotope masses to four significant figures.

▶ LEARNING CHECK 2.5

a. Naturally occurring fluorine consists of a single isotope, fluorine-19, with a mass of 19.00 u (using four significant figures). Determine the atomic weight of fluorine, and compare your answer with the value given in the periodic table.

b. Naturally occurring magnesium has three isotopes: magnesium-24, magnesium-25, and magnesium-26. Their relative masses and percent abundances are, respectively, 23.99 u (78.70%), 24.99 u (10.13%), and 25.98 u (11.17%). Determine the atomic weight of magnesium, and compare it with the value given in the periodic table.

2.6 Avogadro's Number: The Mole

Learning Objective

6. Use the mole concept to obtain relationships between number of moles, number of grams, and number of atoms for elements, and use those relationships to obtain factors for use in factor-unit calculations.

The atomic weights of the elements given in the periodic table have far more uses than simply comparing the masses of the atoms of various elements. However, these uses are not apparent until some additional ideas are developed. According to the periodic table, the atomic weight of magnesium, Mg, is 24.31 u, and the atomic weight of carbon, C, is 12.01 u. As we have seen, this means that an average magnesium atom has about twice the mass of an average carbon atom. Modern instruments allow us to determine that the actual mass of an average magnesium atom is 4.037×10^{-23} g, and the actual mass of an average carbon atom is 1.994×10^{-23} g.

Even though the actual masses of magnesium and carbon atoms are extremely small, it should still be possible to collect together enough atoms of either element to give a sample of any desired mass. Suppose we wish to collect together enough atoms of each element to give a sample with a mass in grams equal to the atomic weight of each element. That is, we want to collect enough magnesium atoms to give a sample with a total mass of 24.31 g, and enough carbon atoms to give a sample with a total mass of 12.01 g (see ▶ Figure 2.4). How many atoms of each element will be required?

We know the mass of one atom of each element, and this fact will provide us with a factor that will allow us to convert the desired sample mass into the number of atoms required. The given mass of one atom of magnesium can be written as

$$1\ \text{Mg atom} = 4.037 \times 10^{-23}\ \text{g Mg}$$

This known relationship provides two factors that can be used to solve factor-unit problems:

$$\frac{1\ \text{Mg atom}}{4.037 \times 10^{-23}\ \text{g Mg}} \quad \text{and} \quad \frac{4.037 \times 10^{-23}\ \text{g Mg}}{1\ \text{Mg atom}}$$

Our task is to convert the desired sample mass of 24.31 g Mg into the number of Mg atoms in the sample. The first factor will cancel the units "g Mg" and will generate the units "Mg atoms."

$$(24.31\ \cancel{\text{g Mg}}) \times \frac{1\ \text{Mg atom}}{4.037 \times 10^{-23}\ \cancel{\text{g Mg}}} = 6.022 \times 10^{23}\ \text{Mg atoms}$$

Weight 24.31 g

Ⓐ

Magnesium

© Jeffrey M. Seager

© Jeffrey M. Seager

Carbon

Ⓑ

Weight 12.01 g

Figure 2.4 Samples of magnesium and carbon with masses of 24.31 g and 12.01 g, respectively.

In a similar way, the number of C atoms needed to produce a sample with a mass of 12.01 g is calculated:

$$(12.01 \text{ g } \cancel{C}) \times \frac{1 \text{ C atom}}{1.994 \times 10^{-23} \text{ g } \cancel{C}} = 6.022 \times 10^{23} \text{ C atoms}$$

The result that the same number of atoms is required for each sample might seem surprising, but it is a consequence of the sample sizes we wanted to produce. If we collected just one atom of each element, the ratio of the mass of Mg to the mass of C would be equal to the atomic weight of Mg divided by the atomic weight of C:

$$\frac{\text{Mg mass}}{\text{C mass}} = \frac{4.037 \times 10^{-23} \text{ g}}{1.944 \times 10^{-23} \text{ g}} = 2.024 = \frac{\text{At. wt. Mg}}{\text{At. wt. C}} = \frac{24.31 \text{ u}}{12.01 \text{ u}}$$

This result reflects the fact that magnesium atoms have about twice the mass of carbon atoms. If we collected samples of 100 atoms of each element, the ratio of the mass of the Mg sample to the mass of the C sample would still be 2.024 because each sample would have a mass 100 times greater than the mass of a single atom:

$$\frac{\text{Mg mass}}{\text{C mass}} = \frac{(100)(4.037 \times 10^{-23} \text{ g})}{(100)(1.994 \times 10^{-23} \text{ g})} = 2.024$$

It follows that if we collected samples containing 6.022×10^{23} atoms of each element, the ratio of the mass of the Mg sample to the mass of the C sample would still be 2.024 because each sample would have a mass that would be 6.022×10^{23} times greater than the mass of a single atom:

$$\frac{\text{Mg mass}}{\text{C mass}} = \frac{(6.022 \times 10^{23})(4.037 \times 10^{-23} \text{ g})}{(6.022 \times 10^{23})(1.994 \times 10^{-23} \text{ g})} = 2.024$$

These results lead to the following conclusion: Any samples of magnesium and carbon that have mass ratios equal to 2.024 will contain the same number of atoms.

▶ Example 2.6

Show that samples of magnesium and carbon with masses of 9.663 g and 4.774 g, respectively, have a mass ratio of 2.024 and contain the same number of atoms.

Solution
The mass ratio is obtained by dividing the mass of the magnesium sample by the mass of the carbon sample:

$$\frac{\text{Mg mass}}{\text{C mass}} = \frac{9.663 \text{ g}}{4.774 \text{ g}} = 2.024$$

The number of atoms in each is calculated using the factors obtained earlier from the mass of one atom of each element:

$$(9.663 \text{ g } \cancel{Mg}) \times \frac{1 \text{ Mg atom}}{4.037 \times 10^{-23} \text{ g } \cancel{Mg}} = 2.394 \times 10^{23} \text{ Mg atoms}$$

$$(4.774 \text{ g } \cancel{C}) \times \frac{1 \text{ C atom}}{1.994 \times 10^{-23} \text{ g } \cancel{C}} = 2.394 \times 10^{23} \text{ C atoms}$$

▶ **LEARNING CHECK 2.6** Show that samples of magnesium and carbon with masses of 13.66 g and 6.748 g, respectively, have a mass ratio of 2.024 and contain the same number of atoms.

The preceding results for magnesium and carbon may be generalized for all the elements of the periodic table as follows. Any samples of two elements that have a mass ratio equal to the ratio of their atomic weights will contain identical numbers of atoms. In addition, we have seen that if the number of grams of sample is equal numerically to the atomic weight of an element, the number of atoms in the sample is equal to 6.022×10^{23}. ▶ Figure 2.5 illustrates these ideas for particles that are familiar to most of us.

Figure 2.5 An average jelly bean has a mass that is 1.60 times the mass of an average dry bean. Each jar contains the same number of beans. The total mass of jelly beans is 472 g. What is the total mass of the dry beans?

With modern equipment, it is possible to determine the number of atoms in any size sample of an element. However, before this was possible, the practice of focusing on samples with a mass in grams equal to the atomic weights of the elements became well established. It continues today. We have learned that the number of atoms in such samples is 6.022×10^{23}.

We saw in Section 2.4 that molecular weights, the relative masses of molecules, are calculated by adding the atomic weights of the atoms contained in the molecules. The resulting molecular weights are expressed in atomic mass units, just as are the atomic weights of the atoms. The same ideas we used to discuss the actual and relative masses of atoms can be applied to molecules. We will not go through the details but simply state the primary conclusion: A sample of compound with a mass in grams equal to the molecular weight of the compound contains 6.022×10^{23} molecules of the compound.

The number 6.022×10^{23} is called *Avogadro's number* in honor of Amadeo Avogadro (1776–1856), an Italian scientist who made important contributions to the concept of atomic weights. As we have seen, this number represents the number of atoms or molecules in a specific sample of an element or compound. Because of its importance in calculations, the number of particles represented by Avogadro's number is given a specific name; it is called a **mole,** abbreviated mol.

It is sometimes helpful to remember that the word *mole* represents a specific number, just as the word *dozen* represents 12, regardless of the objects being counted. Thus, 6.022×10^{23} people would be called 1 mol of people, just as 12 people would be called 1 dozen people. The immensity of Avogadro's number is illustrated by the results of a few calculations based on 1 mol of people. One mol of people would be enough to populate about 1×10^{14} Earths at today's level. That is 100 trillion Earths. Or, put another way, the present population of Earth is $1 \times 10^{-12}\%$ (0.000000000001%) of 1 mol.

In the development of these ideas to this point, we have used four significant figures for atomic weights, molecular weights, and Avogadro's number to minimize the introduction of rounding errors. However, in calculations throughout the remainder of the book, three significant figures will generally be sufficient and will be used.

In a strict sense, 1 mol is a specific number of particles. However, in chemistry it is customary to follow the useful practice of also letting 1 mol stand for the mass of a sample of element or compound that contains Avogadro's number of particles. Thus, the application of the mole concept to sulfur (at. wt. = 32.1 u) gives the following relationships:

$$1 \text{ mol S atoms} = 6.02 \times 10^{23} \text{ S atoms} = 32.1 \text{ g S}$$

When written individually, these three relationships can be used to generate six factors for use in factor-unit calculations involving sulfur:

$$1 \text{ mol S atoms} = 6.02 \times 10^{23} \text{ S atoms}$$

$$6.02 \times 10^{23} \text{ S atoms} = 32.1 \text{ g S}$$

$$1 \text{ mol S atoms} = 32.1 \text{ g S}$$

mole The number of particles (atoms or molecules) contained in a sample of element or compound with a mass in grams equal to the atomic or molecular weight, respectively. Numerically, 1 mol is equal to 6.022×10^{23} particles.

▶ Example 2.7

Determine the following using the factor-unit method of calculation and factors obtained from the preceding three relationships given for sulfur (S):

a. The mass in grams of 1.35 mol of S
b. The number of moles of S atoms in 98.6 g of S
c. The number of S atoms in 98.6 g of S
d. The mass in grams of one atom of S

Solution

a. The known quantity is 1.35 mol of S, and the unit of the unknown quantity is grams of S. The factor comes from the relationship 1 mol S atoms = 32.1 g S.

$$(1.35 \text{ mol S atoms})\left(\frac{32.1 \text{ g S}}{1 \text{ mol S atoms}}\right) = 43.3 \text{ g S}$$

b. The known quantity is 98.6 g of S, and the unit of the unknown quantity is moles of S atoms. The factor comes from the same relationship used in (a).

$$(98.6 \text{ g S})\left(\frac{1 \text{ mol S atoms}}{32.1 \text{ g S}}\right) = 3.07 \text{ mol S atoms}$$

c. The known quantity is, again, 98.6 g of S, and the unit of the unknown is the number of S atoms. The factor comes from the relationship 6.02×10^{23} S atoms = 32.1 g S.

$$(98.6 \text{ g S})\left(\frac{6.02 \times 10^{23} \text{ S atoms}}{32.1 \text{ g S}}\right) = 1.85 \times 10^{24} \text{ S atoms}$$

d. The known quantity is one S atom, and the unit of the unknown is grams of S. The factor comes from the same relationship used in (c), 6.02×10^{23} S atoms = 32.1 g S. Note that the factor is the inverse of the one used in (c) even though both came from the same relationship. Thus, we see that each relationship provides two factors.

$$(1 \text{ S atom})\left(\frac{32.1 \text{ g S}}{6.02 \times 10^{23} \text{ S atoms}}\right) = 5.33 \times 10^{-23} \text{ g S}$$

▶ **LEARNING CHECK 2.7** Calculate the mass of a single oxygen atom in grams. How does the ratio of this mass divided by the mass of a carbon atom given earlier (1.994×10^{-23} g) compare with the ratio of the atomic weights of oxygen and carbon given in the periodic table?

The mole concept can also be applied to particles that are molecules instead of atoms. The compound carbon dioxide consists of molecules that contain one carbon atom, C, and two oxygen atoms, O. The formula for the molecule is CO_2. The molecular weight of the molecule is calculated as shown earlier by adding together the atomic weight of one carbon atom and the atomic weight of two oxygen atoms:

$$MW = 1 \text{ (at. wt. C)} + 2 \text{ (at. wt. O)} = 1 \text{ (12.0 u)} + 2 \text{ (16.0 u)} = 44.0 \text{ u}$$

Thus, the relative mass of a CO_2 molecule is 44.0 u, and application of the mole concept to CO_2 molecules gives the following relationships:

$$1 \text{ mol } CO_2 \text{ molecules} = 6.02 \times 10^{23} \text{ } CO_2 \text{ molecules} = 44.0 \text{ g } CO_2$$

When written individually, these three relationships can be used to generate six factors for use in factor-unit calculations involving CO_2.

$$1 \text{ mol } CO_2 \text{ molecules} = 6.02 \times 10^{23} \, CO_2 \text{ molecules}$$

$$6.02 \times 10^{23} \, CO_2 \text{ molecules} = 44.0 \text{ g } CO_2$$

$$1 \text{ mol } CO_2 \text{ molecules} = 44.0 \text{ g } CO_2$$

▶ Example 2.8

Determine the following using the factor-unit method of calculation and factors obtained from the preceding three relationships given for carbon dioxide, CO_2:

a. The mass in grams of 1.62 mol of CO_2
b. The number of moles of CO_2 molecules in 63.9 g of CO_2
c. The number of CO_2 molecules in 63.9 g of CO_2
d. The mass in grams of one molecule of CO_2

Solution

a. The known quantity is 1.62 mol of CO_2, and the unit of the unknown quantity is g CO_2. The factor comes from the relationship 1 mol CO_2 molecules = 44.0 g CO_2.

$$(1.62 \text{ mol } CO_2 \text{ molecules})\left(\frac{44.0 \text{ g } CO_2}{1 \text{ mol } CO_2 \text{ molecules}}\right) = 71.3 \text{ g } CO_2$$

b. The known quantity is 63.9 g of CO_2, and the unit of the unknown quantity is mol of CO_2 molecules. The factor comes from the same relationship used in (a).

$$(63.9 \text{ g } CO_2)\left(\frac{1 \text{ mol } CO_2 \text{ molecules}}{44.0 \text{ g } CO_2}\right) = 1.45 \text{ mol } CO_2 \text{ molecules}$$

c. The known quantity is, again, 63.9 g of CO_2, and the unit of the unknown is the number of CO_2 molecules. The factor comes from the relationship $6.02 \times 10^{23} \, CO_2$ molecules = 44.0 g CO_2.

$$(63.9 \text{ g } CO_2)\left(\frac{6.02 \times 10^{23} \, CO_2 \text{ molecules}}{44.0 \text{ g } CO_2}\right) = 8.74 \times 10^{23} \, CO_2 \text{ molecules}$$

d. The known quantity is one CO_2 molecule, and the unit of the unknown is g CO_2. The factor comes from the same relationship used in (c), but the factor is the inverse of the one used in (c).

$$(1 \, CO_2 \text{ molecule})\left(\frac{44.0 \text{ g } CO_2}{6.02 \times 10^{23} \, CO_2 \text{ molecules}}\right) = 7.31 \times 10^{-23} \text{ g } CO_2$$

▶ **LEARNING CHECK 2.8** Calculate the molecular weight of carbon monoxide, CO, in u, then calculate the mass of a single CO molecule in grams. How does the ratio of this mass divided by the mass of a single CO_2 molecule (from Example 2.8 d) compare to the ratio of the molecular weight of CO in u divided by the molecular weight of CO_2 in u (from Example 2.8)?

2.7 The Mole and Chemical Formulas

Learning Objective

7. Use the mole concept and molecular formulas to obtain relationships between number of moles, number of grams, and number of atoms or molecules for compounds, and use those relationships to obtain factors for use in factor-unit calculations.

According to Section 2.1 and as demonstrated in Example 2.8, the formula for a compound is made up of the symbols for each element present. Subscripts following the elemental symbols indicate the number of each type of atom in the molecule represented. Thus, chemical formulas represent the numerical relationships that exist among the atoms in a compound. Application of the mole concept to the atoms making up the formulas provides additional useful information.

Consider water as an example. The formula H_2O represents a 2:1 ratio of hydrogen atoms to oxygen atoms in a water molecule. Since this ratio is fixed, the following statements can be written:

1. 2 H_2O molecules contain 4 H atoms and 2 O atoms.
2. 10 H_2O molecules contain 20 H atoms and 10 O atoms.
3. 100 H_2O molecules contain 200 H atoms and 100 O atoms.
4. 6.02×10^{23} H_2O molecules contain 12.04×10^{23} H atoms and 6.02×10^{23} O atoms.

Statement 4 is significant because 6.02×10^{23} particles is 1 mol. Thus, Statement 4 can be changed to Statement 5:

5. 1 mol of H_2O molecules contains 2 mol of H atoms and 1 mol of O atoms.

▶ Figure 2.6 contains another example of this concept.

Figure 2.6 Liquid carbon disulfide (CS_2) is composed of carbon (left) and sulfur (right), elements that are solids. How many moles of sulfur atoms would be contained in 1.5 mol of CS_2 molecules?

▶ Example 2.9

How many moles of ears, tails, and legs are contained in 1 mol of normal rabbits?

Solution

This example is nonchemical, but it might help you grasp the relationships that exist between the individual parts of a formula and the formula as a whole. The parts of a rabbit are related to a rabbit just as the parts of a formula are related to the entire formula.

Ears. Each rabbit has two ears and a 2:1 ratio exists between the number of ears and the number of rabbits. Therefore, 1 mol of rabbits contains 2 mol of ears.

Tails. The 1:1 ratio of tails to rabbits leads to the result that 1 mol of rabbits contains 1 mol of tails.

Legs. The 4:1 ratio of legs to rabbits leads to the result that 1 mol of rabbits contains 4 mol of legs.

▶ Example 2.10

How many moles of each type of atom are contained in 1 mol of chloroform ($CHCl_3$)?

Solution

Just as each rabbit has two ears, each chloroform molecule contains one C atom, one H atom, and three Cl atoms. Therefore, 1 mol of $CHCl_3$ contains 1 mol of C atoms, 1 mol of H atoms, and 3 mol of Cl atoms.

▶ **LEARNING CHECK 2.9** How many moles of each type of atom would be contained in 0.50 mol of glucose ($C_6H_{12}O_6$)?

© Jeffrey M. Seager

Problems involving the use of the mole often strike fear into the hearts of beginning chemistry students. The good news is that problems involving the use of moles, atoms, molecules, and grams are made easier by using the factor-unit method. The method focuses your attention on the goal of eliminating the unit of the known, or given, quantity and of generating the unit of the answer, or unknown, quantity. Remember, Step 1 is to write down the number and unit of the given quantity. In Step 2, write down the unit of the answer. In Step 3, multiply the known quantity by a factor whose units will cancel that of the known quantity and will generate the unit of the answer. In Step 4, obtain the answer by doing the required arithmetic using the numbers that were introduced in Steps 1–3.

The ability to write the necessary factors for use in Step 3 is essential if you are to become proficient in solving mole problems using the factor-unit method. The factors come from numerical relationships between quantities that are obtained from definitions, experimental measurements, or combinations of the two. The definition of the mole, coupled with experimentally determined atomic and molecular weights, gives the following numerical relationships:

Atom Y: 1 mol Y atoms = 6.02×10^{23} Y atoms = y g Y
where y is the atomic weight of element Y.

Molecule Z:
1 mol Z molecules = 6.02×10^{23} Z molecules = z g Z

where z is the molecular weight of compound Z. Each of these sets of three related quantities will give six factors that can be used in factor-unit problems. Each factor is simply a ratio between any two of the three related quantities such as

$$\frac{1 \text{ mol } Y \text{ atoms}}{6.02 \times 10^{23} \ Y \text{ atoms}}$$

Because each ratio can be inverted, six different factors result from each set of three quantities. See if you can write the five other factors for atom Y. The steps given in the flow charts below will help you.

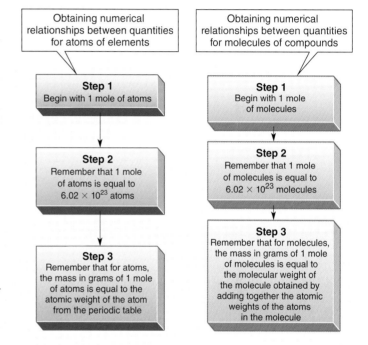

The usefulness of this approach can be increased by remembering and using the mass relationships of the mole concept. Thus, Statement 5 written earlier for water can be changed to Statement 6:

6. 18.0 g of water contains 2.0 g of H and 16.0 g of O.

Or, in a more concise form:

$$18.0 \text{ g } H_2O = 2.0 \text{ g } H + 16.0 \text{ g } O$$

Mass relationships such as these allow percent compositions to be calculated easily.

▶ Example 2.11

Ammonia (NH_3) and ammonium nitrate (NH_4NO_3) are commonly used agricultural fertilizers. Which one of the two contains the higher mass percentage of nitrogen (N)?

Solution

In each case, the mass percentage of N is given by

$$\% \, N = \frac{part}{total} \times 100 = \frac{mass \, of \, N}{mass \, of \, compound} \times 100$$

We will use 1 mol of each compound as a sample because the mass in grams of 1 mol of compound and the mass in grams of N in the 1 mol of compound are readily determined. One mol of NH_3 weighs 17.0 g and contains 1 mol of N atoms, which weighs 14.0 g.

$$\% \, N = \frac{14.0 \, g}{17.0 \, g} \times 100 = 82.4\%$$

Similarly, 1 mol of NH_4NO_3 weighs 80.0 g and contains 2 mol of N atoms, which weigh 28.0 g.

$$\% \, N = \frac{28.0 \, g}{80.0 \, g} \times 100 = 35.0\%$$

▶ **LEARNING CHECK 2.10** Determine the mass percentage of carbon in carbon dioxide (CO_2) and carbon monoxide (CO).

Case Study Follow-up

During an MRI (magnetic resonance imaging) scan, large magnets cause the hydrogen atoms in the body to line up in relation to the magnet. About half point in one direction, and the other half point in the opposite direction, mostly cancelling each other out. A radio-frequency pulse specific to hydrogen is then applied that causes any unmatched atoms to spin in a particular direction at a particular frequency. When the pulse is turned off the spinning atoms emit energy that is detected by the MRI device. A powerful computer uses mathematical formulae to convert this energy information into a meaningful image (Gould, 2010).

Anyone who drives a car knows that a series of complex behaviors can become automatic: A driver may arrive at home with hardly any recollection of the journey there. Health professionals who diagnose common complaints can fall into a routine and automatically diagnose patients with simple conditions when in fact their case is complex or serious. When this occurs, it can result in increased cost to and prolonged suffering for the patient and can delay a timely resolution of the health concern. It can even become life threatening. Busy health professionals have to be diligent in considering a range of possible explanations for the conditions that each individual presents with. Accurate diagnosis depends on focused attention concerning all of a patient's symptoms, family history, behavior, and environment.

Symbols and Formulas

Symbols based on names have been assigned to every element. Most consist of a single capital letter followed by a lowercase letter. A few consist of a single capital letter. Compounds are represented by formulas made up of elemental symbols. The number of atoms of each element in a molecule is shown by subscripts.

Objective 1, Exercise 2.4

Inside the Atom

Atoms are made up of numerous smaller particles of which the most important to chemical studies are the proton, neutron, and electron. Positively charged protons and neutral neutrons have a relative mass of 1 u each and are located in the nuclei of atoms. Negatively charged electrons with a mass of 1/1836 u are located outside the nuclei of atoms.

Objective 2, Exercises 2.10 and 2.12

Isotopes

Most elements in their natural state are made up of more than one kind of atom. These different kinds of atoms of a specific element are called isotopes and differ from one another only in the number of neutrons in their nuclei. A symbol incorporating atomic number, mass number, and elemental symbol is used to represent specific isotopes.

Objective 3, Exercises 2.16 and 2.22

Relative Masses of Atoms and Molecules

Relative masses called atomic weights have been assigned to each element and are tabulated in the periodic table. The units used are atomic mass units, abbreviated u. Relative masses for molecules, called molecular weights, are determined by adding the atomic weights of the atoms making up the molecules.

Objective 4, Exercise 2.32

Isotopes and Atomic Weights

The atomic weights measured for elements are average weights that depend on the percentages and masses of the isotopes in the naturally occurring element. If the isotope percent abundances and isotope masses are known for an element, its atomic weight can be calculated.

Objective 5, Exercise 2.38

Avogadro's Number: The Mole

Avogadro's number of the atoms of an element has a mass in grams equal to the atomic weight of the element. Avogadro's number of molecules has a mass in grams equal to the molecular weight. Avogadro's number of particles is called a mole, abbreviated mol.

Objective 6, Exercises 2.44 a & b and 2.46 a & b

The Mole and Chemical Formulas

The mole concept when applied to molecular formulas gives numerous relationships that yield useful factors for factor-unit calculations.

Objective 7, Exercises 2.50 b and 2.52 b

Key Terms and Concepts

Atomic mass unit (u) (2.4)
Atomic number of an atom (2.3)
Atomic weight (2.4)
Compound formula (2.1)

Elemental symbol (2.1)
Isotopes (2.3)
Mass number of an atom (2.3)
Mole (2.6)

Molecular weight (2.4)
Nucleus (2.2)

Exercises

Even-numbered exercises are answered in Appendix B.

Blue-numbered exercises are more challenging.

You will find it useful to refer to Table 2.1 and the periodic table inside the front cover as you work these exercises.

Symbols and Formulas (Section 2.1)

2.1 Draw a "formula" for each of the following molecules using circular symbols of your choice to represent atoms:

   a. A diatomic molecule of an element

   b. A diatomic molecule of a compound

   c. A triatomic molecule of an element

   d. A molecule of a compound containing one atom of one element and four atoms of another element

2.2 Draw a "formula" for each of the following molecules using circular symbols of your choice to represent atoms:

   a. A triatomic molecule of a compound

   b. A molecule of a compound containing two atoms of one element and two atoms of a second element

   c. A molecule of a compound containing two atoms of one element, one atom of a second element, and four atoms of a third element

   d. A molecule containing two atoms of one element, six atoms of a second element, and one atom of a third element

2.3 Write formulas for the following molecules using elemental symbols from Table 2.1 and subscripts. Compare these formulas with those of Exercise 2.1.

   a. A diatomic molecule of fluorine gas

   b. A diatomic molecule of hydrogen chloride (one hydrogen atom and one chlorine atom)

   c. A triatomic molecule of ozone (a molecular form of the element oxygen)

   d. A molecule of methane (one carbon atom and four hydrogen atoms)

2.4 Write formulas for the following molecules using elemental symbols from Table 2.1 and subscripts. Compare these formulas with those of Exercise 2.2.

   a. A molecule of water (two hydrogen atoms and one oxygen atom)

   b. A molecule of hydrogen peroxide (two hydrogen atoms and two oxygen atoms)

   c. A molecule of sulfuric acid (two hydrogen atoms, one sulfur atom, and four oxygen atoms)

   d. A molecule of ethyl alcohol (two carbon atoms, six hydrogen atoms, and one oxygen atom)

2.5 Determine the number of each type of atom in molecules represented by the following formulas:

   a. ammonia (NH_3)

   b. acetic acid ($C_2H_4O_2$)

   c. boric acid (H_3BO_3)

   d. ethane (C_2H_6)

2.6 Determine the number of each type of atom in molecules represented by the following formulas:

   a. methane (CH_4)

   b. perchloric acid ($HClO_4$)

   c. methylamin (CH_5N)

   d. propane (C_3H_8)

2.7 Tell what is wrong with each of the following molecular formulas and write a correct formula:

   a. H3PO3 (phosphorous acid)

   b. $SICl_4$ (silicon tetrachloride)

   c. SOO (sulfur dioxide)

   d. 2HO (hydrogen peroxide—two hydrogen atoms and two oxygen atoms)

2.8 Tell what is wrong with each of the following formulas and write a correct formula:

   a. HSH (hydrogen sulfide)

   b. $HClO_2$ (chlorous acid)

   c. $2HN_2$ (hydrazine—two hydrogen atoms and four nitrogen atoms)

   d. C2H6 (ethane)

Inside The Atom (Section 2.2)

2.9 Determine the charge and mass (in u) of nuclei made up of the following particles:

   a. 5 protons and 6 neutrons

   b. 10 protons and 10 neutrons

   c. 18 protons and 23 neutrons

   d. 50 protons and 76 neutrons

2.10 Determine the charge and mass (in u) of nuclei made up of the following particles:

   a. 4 protons and 5 neutrons

   b. 9 protons and 10 neutrons

   c. 20 protons and 23 neutrons

   d. 47 protons and 60 neutrons

2.11 Determine the number of electrons that would have to be associated with each nucleus described in Exercise 2.9 to produce a neutral atom.

2.12 Determine the number of electrons that would have to be associated with each nucleus described in Exercise 2.10 to produce a neutral atom.

Isotopes (Section 2.3)

2.13 Determine the number of electrons and protons contained in an atom of the following elements:

   a. sulfur

   b. As

   c. element number 24

2.14 Determine the number of electrons and protons contained in an atom of the following elements:

   a. potassium

   b. Cd

   c. element number 51

2.15 Determine the number of protons, number of neutrons, and number of electrons in atoms of the following isotopes:

   a. $^{25}_{12}Mg$

   b. $^{13}_{6}C$

   c. $^{41}_{19}K$

2.16 Determine the number of protons, number of neutrons, and number of electrons in atoms of the following isotopes:

   a. $^{34}_{16}S$

   b. $^{91}_{40}Zr$

   c. $^{131}_{54}Xe$

2.17 Write symbols like those given in Exercises 2.15 and 2.16 for the following isotopes:

   a. cadmium-110

   b. cobalt-60

   c. uranium-235

2.18 Write symbols like those given in Exercises 2.15 and 2.16 for the following isotopes:

   a. silicon-28

   b. argon-40

   c. strontium-88

2.19 Determine the mass number and atomic number for atoms containing the nuclei described in Exercise 2.9. Write symbols for each atom like those given in Exercises 2.15 and 2.16.

2.20 Determine the mass number and atomic number for neutral atoms containing the nuclei described in Exercise 2.10. Write symbols for each atom like those given in Exercises 2.15 and 2.16.

2.21 Write isotope symbols for neutral atoms with the following characteristics:

   a. Contains 18 electrons and 20 neutrons

   b. A calcium atom with a mass number of 40

   c. An arsenic atom that contains 42 neutrons

2.22 Write isotope symbols for neutral atoms with the following characteristics:

   a. Contains 17 electrons and 20 neutrons

   b. A copper atom with a mass number of 65

   c. A zinc atom that contains 36 neutrons

Relative Masses of Atoms and Molecules (Section 2.4)

2.23 Write the symbols and names for two elements whose average atoms have masses that are within 0.3 u of each other. Don't look beyond element number 83.

2.24 Round atomic weights to the nearest whole number, and determine how many helium atoms would balance one carbon atom on a see-saw.

2.25 Round atomic weights to the nearest whole number, and determine how many lithium atoms would balance two nitrogen atoms on a see-saw.

2.26 What are the symbol and name for an element whose average atoms have a mass that is 77.1% of the mass of an average chromium atom?

2.27 In the first 36 elements, 6 elements have atoms whose average mass is within 0.2 u of being twice the atomic number of the element. Write the symbols and names for these 6 elements.

2.28 What are the symbol and name of the element whose average atoms have a mass very nearly half the mass of an average silicon atom?

2.29 Determine the molecular weights of the following in u:

   a. fluorine (F_2)

   b. carbon disulfide (CS_2)

   c. sulfurous acid (H_2SO_3)

   d. ethyl alcohol (C_2H_6O)

   e. ethane (C_2H_6)

2.30 Determine the molecular weights of the following in u:

   a. sulfur trioxide (SO_3)

   b. glycerin ($C_3H_8O_3$)

   c. sulfuric acid (H_2SO_4)

   d. nitrogen (N_2)

   e. propane (C_3H_8)

2.31 The molecular weight was determined for a gas that is known to be an oxide of nitrogen. The value obtained experimentally was 43.98 u. Which of the following is most likely to be the formula of the gas? NO, N_2O, NO_2.

2.32 A flammable gas is known to contain only carbon and hydrogen. Its molecular weight is determined and found to be 28.05 u. Which of the following is the likely identity of the gas? acetylene (C_2H_2), ethylene (C_2H_4), ethane (C_2H_6).

2.33 Glycine, an amino acid found in proteins, has a molecular weight of 75.07 u and is represented by the formula $C_2H_xNO_2$. What number does x stand for in the formula?

2.34 Serine, an amino acid found in proteins, has a molecular weight of 105.10 u and is represented by the formula $C_yH_7NO_3$. What number does y stand for in the formula?

Isotopes and Atomic Weights (Section 2.5)

2.35 Naturally occurring sodium has a single isotope. Determine the following for the naturally occurring atoms of sodium:

    a. The number of neutrons in the nucleus

    b. The mass (in u) of the nucleus (to three significant figures)

2.36 Naturally occurring aluminum has a single isotope. Determine the following for the naturally occurring atoms of aluminum:

    a. The number of neutrons in the nucleus

    b. The mass (in u) of the nucleus (to three significant figures)

2.37 Calculate the atomic weight of lithium on the basis of the following percent composition and atomic weights of the naturally occurring isotopes. Compare the calculated value with the atomic weight listed for lithium in the periodic table.

$$\text{lithium-6} = 7.42\% \ (6.0151 \text{ u})$$
$$\text{lithium-7} = 92.58\% \ (7.0160 \text{ u})$$

2.38 Calculate the atomic weight of boron on the basis of the following percent composition and atomic weights of the naturally occurring isotopes. Compare the calculated value with the atomic weight listed for boron in the periodic table.

$$\text{boron-10} = 19.78\% \ (10.0129 \text{ u})$$
$$\text{boron-11} = 80.22\% \ (11.0093 \text{ u})$$

2.39 Calculate the atomic weight of silicon on the basis of the following percent composition and atomic weights of the naturally occurring isotopes. Compare the calculated value with the atomic weight listed for silicon in the periodic table.

$$\text{silicon-28} = 92.21\% \ (27.9769 \text{ u})$$
$$\text{silicon-29} = 4.70\% \ (28.9765 \text{ u})$$
$$\text{silicon-30} = 3.09\% \ (29.9738 \text{ u})$$

2.40 Calculate the atomic weight of copper on the basis of the following percent composition and atomic weights of the naturally occurring isotopes. Compare the calculated value with the atomic weight listed for copper in the periodic table.

$$\text{copper-63} = 69.09\% \ (62.9298 \text{ u})$$
$$\text{copper-65} = 30.91\% \ (64.9278 \text{ u})$$

Avogadro's Number: The Mole (Section 2.6)

2.41 Refer to the periodic table and determine how many grams of sulfur contain the same number of atoms as 3.10 grams of phosphorus.

2.42 Refer to the periodic table and determine how many grams of fluorine contain the same number of atoms as 1.60 g of oxygen.

2.43 Write three relationships (equalities) based on the mole concept for each of the following elements:

    a. beryllium

    b. lead

    c. sodium

2.44 Write three relationships (equalities) based on the mole concept for each of the following elements:

    a. silicon

    b. calcium

    c. argon

2.45 Use a factor derived from the relationships written in Exercise 2.43 and the factor-unit method to determine the following:

    a. The number of moles of beryllium atoms in a 25.0-g sample of beryllium

    b. The number of lead atoms in a 1.68-mol sample of lead

    c. The number of sodium atoms in a 120-g sample of sodium

2.46 Use a factor derived from the relationships written in Exercise 2.44 and the factor-unit method to determine the following:

    a. The number of grams of silicon in 1.25 mol of silicon

    b. The mass in grams of one calcium atom

    c. The number of argon atoms in a 20.5-g sample of argon

The Mole and Chemical Formulas (Section 2.7)

2.47 Refer to the periodic table and calculate the molecular weights for the compounds PH_3 and SO_2. Then, determine how many grams of PH_3 contain the same number of molecules as 6.41 g of SO_2.

2.48 Refer to the periodic table and calculate the molecular weights for the compounds BF_3 and H_2S. Then, determine how many grams of BF_3 contain the same number of molecules as 0.34 g of H_2S.

2.49 For each formula given below, write statements equivalent to Statements 1–6 (see Section 2.7):

    a. methane (CH_4)

    b. ammonia (NH_3)

    c. chloroform ($CHCl_3$)

2.50 For each formula given below, write statements equivalent to Statements 1–6 (see Section 2.7):

    a. benzene (C_6H_6)

    b. nitrogen dioxide (NO_2)

    c. hydrogen chloride (HCl)

2.51 Answer the following questions based on information contained in the statements you wrote for Exercise 2.49.

    a. How many moles of hydrogen atoms are contained in 1 mol of CH_4 molecules?

    b. How many grams of nitrogen are contained in 1.00 mol of NH_3?

    c. What is the mass percentage of chlorine in $CHCl_3$?

2.52 Answer the following questions based on information contained in the statements you wrote for Exercise 2.50.

    a. How many moles of hydrogen atoms are contained in 0.75 mol of benzene?

    b. How many oxygen atoms are contained in 0.50 mol of nitrogen dioxide?

    c. What is the mass percentage of chlorine in HCl?

2.53 How many moles of N_2O_5 contain the same number of nitrogen atoms as 3 mol of NO_2?

2.54 How many grams of C_2H_6O contain the same number of oxygen atoms as 0.75 mol of H_2O?

2.55 Determine the mass percentage of nitrogen in NH_3 and N_2H_4.

2.56 Determine the mass percentage of hydrogen in CH_4 and C_2H_6.

2.57 Any of the statements based on a mole of substance (Statements 4–6) can be used to obtain factors for problem solving by the factor-unit method. Write statements equivalent to 4, 5, and 6 for nitrophenol ($C_6H_5NO_3$). Use a single factor obtained from the statements to solve each of the following. A different factor will be needed in each case.

    a. How many grams of nitrogen are contained in 70.0 g of $C_6H_5NO_3$?

    b. How many moles of oxygen atoms are contained in 1.50 mol of $C_6H_5NO_3$?

    c. How many atoms of carbon are contained in 9.00×10^{22} molecules of $C_6H_5NO_3$?

2.58 Any of the statements based on a mole of substance (Statements 4–6) can be used to obtain factors for problem solving by the factor-unit method. Write statements equivalent to 4, 5, and 6 for phosphoric acid (H_3PO_4). Use a single factor obtained from the statements to solve each of the following. A different factor will be needed in each case.

    a. How many grams of hydrogen are contained in 46.8 g of H_3PO_4?

    b. How many moles of oxygen atoms are contained in 1.25 mol of H_3PO_4?

    c. How many atoms of phosphorus are contained in 8.42×10^{21} molecules of H_3PO_4?

2.59 Urea (CH_4N_2O) and ammonium sulfate ($N_2H_8SO_4$) are both used as agricultural fertilizers. Which one contains the higher mass percentage of nitrogen?

2.60 Two iron ores that have been used as sources of iron are magnetite (Fe_3O_4) and hematite (Fe_2O_3). Which one contains the higher mass percentage of iron?

2.61 Both calcite ($CaCO_3$) and dolomite ($CaMgC_2O_6$) are used as dietary calcium supplements. Calculate the mass percentage of calcium in each mineral.

Additional Exercises

2.62 The two major isotopes of uranium metal found in nature are U-235 and U-238. Which isotope has the greater density? Explain your reasoning.

2.63 About one billion (1.0×10^9) peas can fit into a railroad car. What percentage of one mole of peas is this number?

2.64 The mass of a single carbon-12 atom is 1.99×10^{-23} g. What is the mass in grams of a single carbon-14 atom?

2.65 One mole of water molecules, H_2O, has a mass of 18.0 g. What would be the mass in grams of one mole of heavy water molecules, D_2O, where D represents the isotope?

2.66 According to the caption of Figure 2.2, atoms are composed primarily of empty space (99.999% empty). What would happen

to the density of matter (increase or decrease) if the electrons were actually located at the distance from the nucleus shown in the atom drawing of Figure 2.2? Explain your reasoning.

Allied Health Exam Connection

The following questions are from these sources:

1. *Nursing School Entrance Exam* © 2005, Learning Express, LLC.

2. *McGraw-Hill's Nursing School Entrance Exams* by Thomas A. Evangelist, Tamara B. Orr and Judy Unrein © 2009, The McGraw-Hill Companies, Inc.

3. *NSEE Nursing School Entrance Exams,* 3rd Edition © 2009, Kaplan Publishing.

4. *Cliffs Test Prep: Nursing School Entrance Exams* by Fred N. Grayson © 2004, Wiley Publishing, Inc.

5. *Peterson's Master the Nursing School and Allied Health Entrance Exams,* 18th Edition by Marion F. Gooding © 2008, Peterson's, a Nelnet Company.

2.67 The symbol K on the periodic table stands for

    a. potassium

    b. calcium

    c. carbon

    d. phosphorus

2.68 Which one of the following substances is a chemical compound?

    a. blood

    b. water

    c. oxygen

    d. air

2.69 Which of the following is true about compounds?

    a. Compounds are pure substances that are composed of two or more elements in a fixed proportion.

    b. Compounds can be broken down chemically to produce their constituent elements or other compounds.

    c. Both A and B are correct.

    d. Neither A nor B is correct.

2.70 $^{34}_{17}Cl$ has

    a. 17 protons, 17 electrons, and 17 neutrons.

    b. 17 protons, 19 electrons, and 17 neutrons.

    c. 17 protons, 18 electrons, and 17 neutrons.

    d. 34 protons, 34 electrons, and 17 neutrons.

2.71 If two atoms are isotopes, they will

    a. have the same number of protons and neutrons

    b. have the same number of neutrons, but different numbers of protons

    c. have the same number of protons, but different numbers of neutrons

    d. have the same number of neutrons and electrons

2.72 Copper (Cu) has an atomic number of 29 and a mass number of 64. One copper atom, therefore, has how many protons?

a. 27

b. 29

c. 31

d. 35

2.73 Atoms are electrically neutral. This means that an atom will contain

a. more protons than neutrons

b. more electrons than protons

c. an equal number of protons and electrons

d. None of the above

2.74 The negative charged particle found within the atom is the

a. proton.

b. electron.

c. nucleus.

d. neutron.

2.75 Two atoms, L and M, are isotopes. Which of the following properties would they NOT have in common?

a. atomic number

b. atomic weight

c. chemical reactivity

d. the number of protons in the nucleus

2.76 The major portion of an atom's mass consists of:

a. neutrons and protons

b. electrons and protons

c. electrons and neutrons

d. neutrons and positrons

2.77 The mass of an atom is almost entirely contributed by its

a. nucleus

b. protons

c. electrons and protons

d. neutrons

2.78 Which of the following is the chemical symbol for the species that has 16 protons, 17 neutrons, and 18 electrons?

a. $^{33}_{16}S$

b. $^{33}_{17}Cl$

c. $^{35}_{17}Cl$

d. $^{33}_{16}S^{2-}$

2.79 An atom with an atomic number of 58 and an atomic mass of 118 has

a. 58 neutrons

b. 176 neutrons

c. 60 neutrons

d. 116 neutrons

2.80 What is the mass number of an atom with 60 protons, 60 electrons, and 75 neutrons?

a. 120

b. 135

c. 75

d. 195

2.81 Which of the following represents Avogadro's number?

a. 1.66×10^{-24}

b. 1.0×10^{24}

c. 6.022×10^{23}

d. 3.011×10^{23}

2.82 Which of the following has the greatest number of atoms?

a. 1.0 mol N

b. 1.0 g N

c. 1.0 mol NO_2

d. 0.5 mol NH_3

2.83 The formula of carbon dioxide is CO_2. Its molecular weight is 44 amu. A sample of 11 grams of CO_2 contains

a. 1.0 mole of carbon dioxide

b. 1.5 grams of carbon

c. 3.0 grams of carbon

d. 6.0 grams of oxygen

2.84 What is the molar mass of calcium oxide, CaO?

a. 56

b. 28

c. 640

d. 320

2.85 How many grams are contained in 0.200 mol of calcium phosphate, $Ca_3(PO_4)_2$?

a. 6.20

b. 62.0

c. 124

d. 31.0

2.86 How many moles are contained in a 54.0 g sample of Al?

a. 1.0

b. 2.0

c. 0.5

d. 4.0

Chemistry for Thought

2.87 a. Explain how atoms of different elements differ from one another.

b. Explain how atoms of different isotopes of the same element differ from one another.

2.88 The atomic weight of aluminum is 26.98 u, and the atomic weight of nickel is 58.69 u. All aluminum atoms have a mass of

26.98 u, but not a single atom of nickel has a mass of 58.69 u. Explain.

2.89 Answer the question in the caption of Figure 2.3. Would you expect any orange in the bowl to have the exact mass you calculated as an average? Explain.

2.90 Answer the question in the caption of Figure 2.5. Use your answer and the fact that an average jelly bean has a mass of 1.18 g to calculate the number of beans in each jar.

2.91 Answer the question in the caption of Figure 2.6. How many CS_2 molecules would be required to contain 0.25 mol of sulfur atoms?

2.92 In Section 2.4 it was pointed out that an atomic mass unit, u, is equal to $\frac{1}{12}$ the mass of an atom of carbon-12. Suppose one atomic mass unit was redefined as being equal to $\frac{1}{24}$ the mass of a carbon-12 atom. How would this change influence the value of the atomic weight of magnesium?

2.93 How would the change of question 2.92 influence the ratio of the atomic weight of magnesium divided by the atomic weight of hydrogen?

2.94 How would the change of question 2.92 influence the value of Avogadro's number?

Even-numbered exercises answered in Appendix B.
Blue-numbered exercises are more challenging.

3

Electronic Structure and the Periodic Law

Case Study

Maggie was 18 months old. She was very fair, but she was a redhead, so her pale skin didn't seem unusual. Maggie's mother, Suzanne, was a "chocoholic," especially for very dark chocolate. Suzanne was given some 80%-cocoa dark chocolate from Germany. Maggie incessantly asked for a bite of the chocolate. Suzanne gave her a tiny taste, thinking that she surely wouldn't like it. It wasn't even sweet. Maggie loved it and launched into a full-fledged tantrum to be given more. Suzanne was flabbergasted by her daughter's behavior. Maggie's older sister had just taken a nutrition class and suggested that Maggie might be anemic. Suzanne considered her daughter's suggestion and took Maggie to a pediatrician. Sure enough, she was anemic. Her hemoglobin level was 10.5g/dL. The doctor prescribed 3mg of ferrous sulfate per kilogram of body weight and recommended another hematocrit in three months.

How likely is it that Maggie's chocolate craving was related to the anemia. Was it coincidence? Could this be an example of pica (odd cravings due to anemia)?

Follow-up to this Case Study appears at the end the chapter before the Concept Summary.

Learning Objectives

When you have completed your study of this chapter, you should be able to:

1 Locate elements in the periodic table on the basis of group and period designations. (Section 3.1)

2 Determine the number of electrons in designated atomic orbitals, subshells, or shells. (Section 3.2)

3 Determine the number of valence shell electrons and the electronic structure for atoms, and relate this information to the location of elements in the periodic table. (Section 3.3)

4 Determine the following for elements: the electronic configuration of atoms, the number of unpaired electrons in atoms, and the identity of atoms based on provided electronic configurations. (Section 3.4)

5 Determine the shell and subshell locations of the distinguishing electrons in elements, and based on their location in the periodic table, classify elements into the categories given in Figures 3.10 and 3.12. (Section 3.5)

6 Recognize property trends of elements within the periodic table, and use the trends to predict selected properties of the elements. (Section 3.6)

I n Chapter 1, we defined atoms as particles that represent the limit of chemical subdivision. According to this idea, atoms of a specific element cannot be divided into smaller particles or converted into atoms of another element by any physical or chemical change. Then in Chapter 2, we introduced the idea that atoms are, in fact, made up of particles that are smaller than the atoms themselves. Two of these particles, protons and neutrons, form the nuclei of atoms, whereas electrons are located outside the nuclei.

The idea that atoms are made up of subatomic particles implies that it should be possible to obtain even smaller particles from atoms. Scientists have found that it is possible. During chemical changes, electrons are transferred from one atom to another or are shared between atoms. Some details of these processes are given in Chapter 4, but it is known that they depend on the arrangements of the electrons around the nuclei of atoms. These electronic arrangements are one of the major topics of this chapter.

3.1 The Periodic Law and Table

Learning Objective

1. Locate elements in the periodic table on the basis of group and period designations.

By the early 19th century, detailed studies of the elements known at that time had produced an abundance of chemical information. Scientists looked for order in these facts, with the hope of providing a systematic approach to the study of chemistry. Two scientists independently, and almost simultaneously, made the same important contribution to this end. Julius Lothar Meyer, a German, and Dmitri Mendeleev, a Russian, each produced classification schemes for the elements in 1869. Both schemes were based on the **periodic law,** which in its present form is stated as follows: When all the elements are arranged in order of increasing atomic numbers, elements with similar chemical properties will occur at regular (periodic) intervals.

A convenient way to compactly represent such behavior is to use tables. The arrangement of the elements in a table based on the periodic law is called a *periodic table.* In a modern periodic table, such as the one inside the front cover of this book, elements with similar chemical properties are found in vertical columns called **groups** or **families.**

In this book, the groups are designated in two ways. In the U.S. system, a roman numeral and letter, such as IIA, is used. In the IUPAC system, a simple number such as 2, is used. Both designations appear on the periodic table inside the front cover of the book with the IUPAC number in parentheses. We will generally refer to groups in the text by using both the U.S. system and the IUPAC system. Thus, a reference to the second group would be given as group IIA(2), where we have put the IUPAC number in parentheses. The horizontal rows in the table are called **Periods** and are numbered from top to bottom. Thus, each element belongs to both a period and a group of the periodic table.

periodic law A statement about the behavior of the elements when they are arranged in a specific order. In its present form, it is stated as follows: Elements with similar chemical properties occur at regular (periodic) intervals when the elements are arranged in order of increasing atomic numbers.

group or family of the periodic table A vertical column of elements that have similar chemical properties.

period of the periodic table A horizontal row of elements.

▶ Example 3.1

Identify the group and period to which each of the following belongs:

a. P **b.** Cr **c.** element number 30 **d.** element number 53

Solution

a. Phosphorus (P) is in group VA(15) and period 3.
b. Chromium (Cr) is in group VIB(6) and period 4.
c. The element with atomic number 30 is zinc (Zn), which is found in group IIB(12) and period 4.
d. Element number 53 is iodine (I), found in group VIIA(17) and period 5.

▶ **LEARNING CHECK 3.1** Write the symbol for the element found in the following places of the periodic table:

a. Group IVA(14) and period 4 **b.** Group VIIB(7) and period 6

It should be noted that the periodic table as given inside the front cover appears to violate the practice of arranging the elements according to increasing atomic number. Element 72 follows element 57, and 104 follows 89, whereas elements 58–71 and 90–103 are arranged in two rows at the bottom of the table. Technically, these two rows should be included in the body of the table as shown in ▶ Figure 3.1 (top). To save horizontal space, they are placed in the position shown in Figure 3.1 (bottom). This exception presents no problem, as long as it is understood.

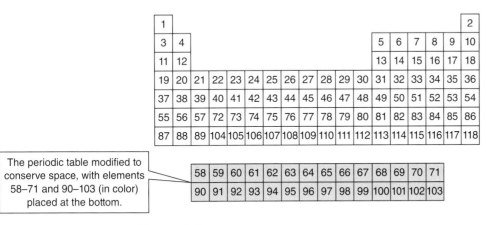

The periodic table with elements 58–71 and 90–103 (area in color) in their proper positions.

The periodic table modified to conserve space, with elements 58–71 and 90–103 (in color) placed at the bottom.

Figure 3.1 Forms of the periodic table.

▶ Example 3.2

a. How many elements are found in period 6 of the periodic table?
b. How many elements are found in group VA(15) of the periodic table?

Solution

a. Period 6 includes elements 55–86, even though elements 58–71 are shown below the main table. Therefore, period 6 contains 32 elements.
b. A count shows group VA(15) contains 6 elements: N, P, As, Sb, Bi, and number 115.

▶ **LEARNING CHECK 3.2** How many elements are found in the following?

a. Period 1 of the periodic table
b. Group IIB(12) of the periodic table

3.2 Electronic Arrangements in Atoms

Learning Objective

2. Determine the number of electrons in designated atomic orbitals, subshells, or shells.

In 1913 Niels Bohr, a young Danish physicist, made an important contribution to our understanding of atomic structure. He was working under the direction of Ernest Rutherford, a British scientist, who had proposed a solar system model for atoms in which negative electrons moved in circular orbits around the positive nucleus, much like the planets move around the sun. Bohr built on this model by proposing that the single electron of a hydrogen atom could occupy orbits only at specific distances from the nucleus, and thus the electron could have only specific energies (see ▶ Figure 3.2). Bohr further proposed that the electron changed orbits only by absorbing or releasing energy. The addition of energy to a hydrogen atom elevated the

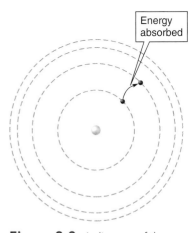

Energy absorbed

Figure 3.2 A diagram of the Bohr hydrogen atom (not drawn to scale; the orbits are actually much larger than the nucleus). The electron is elevated to a higher-energy orbit when energy is absorbed.

DR. MELINA B. JAMPOLIS, M.D.

© Michael Roud Photography

Dr. Melina B. Jampolis, M.D. is a Los Angeles, California, Physician Nutrition Specialist—a specialty practiced by only a few hundred physicians in the United States. She focuses exclusively on nutrition for weight loss and disease prevention and treatment.

Ask an Expert 3.1

Q: Are calcium supplements necessary?

A: Calcium is the most well-known and abundant mineral in the body. It is an essential part of our bones and teeth, but it also plays a role in muscle, nerve, and blood vessel function. Blood levels of calcium, which represent only 1% of total body calcium (the rest is in our bones), are tightly controlled within a narrow range. Unfortunately, if you don't consume adequate amounts of calcium, your body gets what it needs from your bones, which can lead to a decrease in bone density known as osteoporosis. Bone density peaks at about the age of 30, so it is very important to consume adequate amounts of dietary calcium before age 30 to build up bone to optimal levels, and after age 30 it is important to preserve bone as much as possible. Bone undergoes constant turnover, and so it requires a constant supply of calcium. In addition to its role in reducing the risk of osteoporosis, research suggests that adequate dietary calcium may also decrease the risk of colorectal cancer. The recommended daily allowance (U.S. RDA) of calcium for adults (the amount considered adequate to meet the needs of 97–98% of healthy individuals) is shown in the following table:

Age	Men	Women
19–50 years	1000 mg/day	1000 mg/day
51–70 years	1000 mg/day	1200 mg/day
71+ years	1200 mg/day	1200 mg/day

Food sources are the best option for calcium intake, because they provide other important nutrients to support both bone and general health. Dairy products including milk, yogurt, and cheese are the best natural sources. Nondairy foods like Chinese cabbage, kale, broccoli, and fortified products are also good sources. Data from the most recent National Health and Nutrition Survey (NHANES) suggests that females, particularly teenagers and women over the age of 50, are the least likely to get adequate amounts of calcium from food. Also, calcium absorption decreases with age and requires adequate intakes of vitamin D for optimal uptake by the body. According to recent research, a significant percentage of Americans are deficient in Vitamin D, which can be obtained by exposure to sunlight and diet.

Best Food Sources of Calcium

Calcium Supplements

© Morgan Lane Photography/Shutterstock.com; © OSVALDRU/Shutterstock.com; © Gerald Bernard/Shutterstock.com; © GIPhotoStock/Photo Researchers, Inc.

For these reasons, many Americans turn to dietary supplements to meet their daily calcium requirements. There are two major chemical forms of dietary calcium supplements: calcium carbonate and calcium citrate. Calcium carbonate is the least expensive and contains 40% calcium by weight. It requires stomach acid for proper absorption, so it must be taken with meals. It may not be as well absorbed by individuals taking common acid-blocking medications. Calcium citrate is only 21% calcium by weight, but it does not require acid for absorption; research suggests that it may be more effective in postmenopausal women and those who have undergone bariatric (weight loss) surgery. Products like coral calcium, which are often heavily marketed, offer no advantage to justify their much higher cost. Absorption of calcium is best at doses less than 500 mg, so higher doses should be divided and taken at different times of the day.

Is there a downside to calcium supplementation? Some recent studies suggest an increased risk of heart disease, particularly in women who were not previously taking calcium supplements. In addition, calcium supplements (but not calcium from food) might increase the risk of kidney stones, and high levels in men might increase the risk of prostate cancer. The best approach for optimal health is to get most of your calcium from food and supplement appropriately if needed.

electron to a higher-energy orbit located farther from the nucleus. The energy released when an electron dropped from a higher- to a lower-energy orbit appeared as emitted light.

Research into the nature of atoms continued after Bohr's proposal, and in 1926 a revised model of atomic structure was proposed by Erwin Schrödinger, an Austrian physicist, who received the Nobel Prize in physics in 1933 in recognition of this achievement. According to Schrödinger's quantum mechanical model, the precise paths of electrons cannot be determined accurately, as Bohr's model required. It was found that the location and energy of electrons around a nucleus can be specified using three terms: shell, subshell, and orbital. This is somewhat like locating an individual in a city by specifying a street, building, and apartment.

The location of electrons in a **shell** is indicated by assigning a number, n, to the shell and to all the electrons within the shell. The n value of the lowest-energy shell is 1, that of the next higher energy is 2, the next is 3, and so on. Higher n values for a shell correspond to higher energies and greater distances from the nucleus for the electrons of the shell. Electrons in the third shell all have an n value of 3, all have an energy higher than the energies of electrons in shells 1 and 2, and also are located farther from the nucleus than those of shells 1 and 2.

Each shell is made up of **subshells** that are designated by a letter from the group s, p, d, and f. Because all subshells are designated by one of these letters regardless of the shell in which the subshell is found, a combination of both shell number and subshell letter is used to identify subshells clearly. Thus, a p subshell in shell number 2 is referred to as a $2p$ subshell. The number of subshells found in a shell is the same as the value of n for the shell. Thus, shell number 2 ($n = 2$) contains two subshells. The subshells are the $2p$ mentioned earlier and a $2s$. Electrons located in specific subshells are often referred to in terms of the same number and letter as the subshell. For example, we might refer to an atom as having three $2p$ electrons. All electrons within a specific subshell have the same energy.

The description of the location and energy of electrons moving around a nucleus is completed in the quantum mechanical model by specifying an orbital. Each subshell consists of one or more **atomic orbitals,** which are specific volumes of space around nuclei in which electrons move. These atomic orbitals must not be confused with the fixed electron orbits of the original Bohr theory; they are *not* the same. These volumes of space around nuclei have different shapes, depending on the energy of the electrons they contain (see ▶ Figure 3.3). All s subshells consist of a single orbital that is also designated by the letter s and further identified by the n value of the shell to which the subshell belongs. Thus, the $2s$ subshell mentioned earlier consists of a single $2s$ orbital. All p subshells consist of three p orbitals that also carry the n value of the shell. Thus, the $2p$ subshell of shell number 2 consists of three $2p$ orbitals. Since all the electrons in a subshell have the same energy, an electron in any one of the three $2p$ orbitals has the same energy, regardless of which orbital of the three it occupies. All d subshells contain five orbitals, and all f subshells contain seven orbitals. According to the quantum mechanical model, each orbital within a subshell can contain a maximum of two electrons.

shell A location and energy of electrons around a nucleus that is designated by a value for n, where $n = 1, 2, 3$, etc.

subshell A component of a shell that is designated by a letter from the group s, p, d, and f.

atomic orbital A volume of space around atomic nuclei in which electrons of the same energy move. Groups of orbitals with the same n value form subshells.

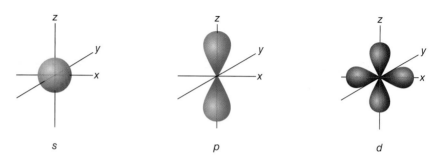

Figure 3.3 Shapes of typical s, p, and d orbitals.

s p d

The shapes of orbitals such as those given in Figure 3.3 must not be interpreted incorrectly. The fact that *s* orbitals are spherical in shape does not mean that the electrons move around on the spherical surface. According to the quantum mechanical model, electrons in *s* orbitals move around inside the spherical volume of the orbital in paths that cannot be determined. All that can be determined about their behavior within the orbital is the probability of finding them in a specific location, which is what Figure 3.3 depicts. Thus, if it were determined that an electron had a 2% probability of being at a specific location, it would simply mean that the electron could be found at that location within an orbital 2 times out of every 100 times we looked for it there. Similarly, electrons in *p* or *d* orbitals do not move on the surface of the dumbbell- or cloverleaf-shaped orbitals; they move *within* the three-dimensional dumbbell- or cloverleaf-shaped volumes.

The energy of electrons located in an orbital within a subshell and shell is determined by two factors. As mentioned earlier, the higher the value of *n*, the higher the energy. In addition, if *n* is the same, but the subshell is different, the energy of the contained electrons increases in the order *s, p, d, f*. Thus, a 3*p* electron (an electron in a 3*p* subshell) has a higher energy than a 3*s* electron (an electron in a 3*s* subshell). ▶ Figure 3.4 is a diagrammatic representation of the fourth shell of an atom completely filled with electrons.

Figure 3.4 The fourth shell of an atom, completely filled with electrons.

▶ Example 3.3

Determine the following for the third shell of an atom:

a. The number of subshells
b. The designation for each subshell
c. The number of orbitals in each subshell
d. The maximum number of electrons that can be contained in each subshell
e. The maximum number of electrons that can be contained in the shell

Solution

a. The number of subshells is the same as the number used to designate the shell. Therefore, the third shell contains three subshells.
b. Subshells increase in energy according to the order *s, p, d, f*. The subshells in the shell are therefore designated 3*s*, 3*p*, and 3*d*.
c. The number of orbitals in the subshells is 1, 3, and 5 because *s* subshells always contain a single orbital. *p* subshells always contain three orbitals, and *d* subshells always contain five orbitals.
d. Each atomic orbital can contain a maximum of two electrons, independent of the type of orbital under discussion. Therefore, the 3*s* subshell (one orbital) can hold a maximum of two electrons, the 3*p* subshell (three orbitals) a maximum of 6 electrons, and the 3*d* subshell (five orbitals) a maximum of 10 electrons.
e. The maximum number of electrons that can be contained in the shell is simply the sum of the maximum number for each subshell, $2 + 6 + 10 = 18$.

▶ LEARNING CHECK 3.3　In each of the following, what is the maximum number of electrons that can be found?

a. a 4*p* orbital　　**b.** a 5*d* subshell　　**c.** shell number 1

It might seem to you at this point that the modifications to the Bohr theory have created a number of hard-to-remember relationships between shells, subshells, orbitals, and electrons. However, some patterns help make the relationships easy to remember. ▶ Table 3.1 summarizes them for the first four shells of an atom.

TABLE 3.1 Relationships Between Shells, Subshells, Orbitals, and Electrons

Shell Number (n)	Number of Subshells in Shell	Subshell Designation	Number of Orbitals in Subshell	Orbital Designation	Maximum Number of Electrons in Subshell	Maximum Number of Electrons in Shell
1	1	1s	1	1s	2	2
2	2	2s	1	2s	2	
		2p	3	2p	6	8
3	3	3s	1	3s	2	
		3p	3	3p	6	
		3d	5	3d	10	18
4	4	4s	1	4s	2	
		4p	3	4p	6	
		4d	5	4d	10	
		4f	7	4f	14	32

3.3 The Shell Model and Chemical Properties

Learning Objective _____

3. Determine the number of valence shell electrons and the electronic structure for atoms, and relate this information to the location of elements in the periodic table.

The arrangement of electrons into orbitals, subshells, and shells provides an explanation for the similarities in chemical properties of various elements. ▶ Table 3.2 gives the number of electrons in each shell for the first 20 elements of the periodic table.

TABLE 3.2 Electron Occupancy of Shells

Element	Belongs to Group	Symbol	Atomic Number	Electrons in Shell Number			
				1	2	3	4
Hydrogen	IA(1)	H	1	1			
Helium	Noble gas(18)	He	2	2			
Lithium	IA(1)	Li	3	2	1		
Beryllium	IIA(2)	Be	4	2	2		
Boron	IIIA(13)	B	5	2	3		
Carbon	IVA(14)	C	6	2	4		
Nitrogen	VA(15)	N	7	2	5		
Oxygen	VIA(16)	O	8	2	6		
Fluorine	VIIA(17)	F	9	2	7		
Neon	Noble gas(18)	Ne	10	2	8		
Sodium	IA(1)	Na	11	2	8	1	
Magnesium	IIA(2)	Mg	12	2	8	2	
Aluminum	IIIA(13)	Al	13	2	8	3	
Silicon	IVA(14)	Si	14	2	8	4	
Phosphorus	VA(15)	P	15	2	8	5	
Sulfur	VIA(16)	S	16	2	8	6	
Chlorine	VIIA(17)	Cl	17	2	8	7	
Argon	Noble gas(18)	Ar	18	2	8	8	
Potassium	IA(1)	K	19	2	8	8	1
Calcium	IIA(2)	Ca	20	2	8	8	2

Figure 3.5 Left to right: Magnesium, calcium, and strontium, members of group IIA(2) of the periodic table, have similar chemical properties and appearances.

In Table 3.2, notice that the third shell stops filling when 8 electrons are present, even though the shell can hold a maximum of 18 electrons. The reasons for this are discussed in Section 3.4. Also, note that all the elements in a specific group of the periodic table have the same number of electrons in the outermost occupied shell. This outermost occupied shell (the one of highest energy) is called the **valence shell.** Similarities in elemental chemical properties result from identical numbers of electrons in the valence shells of the atoms (see ◗ Figure 3.5).

Magnesium(Mg)	Calcium(Ca)	Strontium(Sr)

© Spencer L. Seager

◗ Example 3.4

Referring to Table 3.2, indicate the number of electrons in the valence shell of elements in groups IA(1), IIA(2), IIIA(13), and IVA(14).

Solution

According to Table 3.2, the elements in group IA(1) are hydrogen, lithium, sodium, and potassium. Each element has one electron in the valence shell. Hydrogen belongs in group IA(1) on the basis of its electronic structure, but its properties differ significantly from other group members.

The group IIA(2) elements are beryllium, magnesium, and calcium. Each has two electrons in the valence shell. The group IIIA(13) elements are boron and aluminum. Both have three electrons in the valence shell. The group IVA(14) elements are carbon and silicon; each has four valence-shell electrons.

◗ LEARNING CHECK 3.4
Referring to Table 3.2, indicate the number of electrons in the valence shell of elements in groups VA(15), VIA(16), and VIIA(17), and the noble gases(18).

Example 3.4 and Learning Check 3.4 emphasize the fact that elements belonging to the same periodic table group have the same number of electrons in the valence shell (helium, the first element in the noble gases, is an exception). Notice also that the number of electrons in the valence shell is identical to the roman numeral that designates the group number. Elements of group IIIA(13), for example, have three electrons in the valence shell. It is also apparent that the n value for the valence shell increases by 1 with each heavier member of a group, that is, $n = 2$ for Li, $n = 3$ for Na, and $n = 4$ for K.

◗ Example 3.5

Using the periodic table and Example 3.4 and Learning Check 3.4, determine the n value for the valence shell and the number of electrons in the valence shell for the following elements:

a. Ba   **b.** Br   **c.** element number 50
d. The third element of group VIA(16)

Solution

a. Ba is the fifth element of group IIA(2), and because $n = 4$ for Ca (the third element of the group), $n = 6$ for the fifth element. The number of valence-shell electrons is two, the same as the group number.

b. In a similar way, Br, the third element of group VIIA(17), has $n = 4$ and seven valence-shell electrons.

c. Element number 50 is tin (Sn), which is the fourth element of group IVA(14). Therefore, $n = 5$ and the number of valence-shell electrons is four.

d. The third element of group VIA(16) is selenium (Se), for which $n = 4$ and the number of valence-shell electrons is six.

▶ **LEARNING CHECK 3.5** How many electrons will be found in the following?

a. The valence shell of Sr
b. The valence shell of the third element in group IVA(14)
c. The valence shell of the fifteenth element in period 4

3.4 Electronic Configurations

Learning Objective

4. Determine the following for elements: the electronic configuration of atoms, the number of unpaired electrons in atoms, and the identity of atoms based on provided electronic configurations.

According to Section 3.3, similarities in chemical properties between elements are related to the number of electrons that occupy the valence shells of their atoms. We now look at the electronic arrangements of atoms in more detail. These detailed arrangements, called **electronic configurations,** indicate the number of electrons that occupy each subshell and orbital of an atom.

Imagine that electrons are added one by one to the orbitals that belong to subshells and shells associated with a nucleus. The first electron will go to as low an energy state as possible, which is represented by the $1s$ orbital of the $1s$ subshell of the first shell. The second electron will join the first and completely fill the $1s$ orbital, the $1s$ subshell, and the first shell (remember, an orbital is filled when it contains two electrons). The third electron will have to occupy the lowest-energy subshell ($2s$) of the second shell. The fourth electron will also occupy (and fill) the $2s$ subshell. The fifth electron must seek out the next–highest-energy subshell, which is the $2p$. The $2p$ subshell contains three $2p$ orbitals, so the sixth electron can either join the fifth in one of the $2p$ orbitals or go into an empty $2p$ orbital. It will go into an empty orbital in compliance with **Hund's rule,** which states: Electrons will not join other electrons in an orbital if an empty orbital of the same energy is available.

It has been found that electrons behave as if they spin on an axis, and only electrons spinning in opposite directions (indicated by ↑ and ↓) can occupy the same orbital. This principle, known as the **Pauli exclusion principle,** explains why orbitals can contain a maximum of two electrons. Hund's rule and the Pauli exclusion principle can be combined: Electrons will pair with other electrons in an orbital only if there is no empty orbital of the same energy available and if there is one electron with opposite spin already in the orbital.

The seventh added electron will occupy the last empty $2p$ orbital, and the eighth, ninth, and tenth electrons will pair up with electrons already in the $2p$ orbitals. The tenth electron fills the $2p$ subshell, thus completing the second shell. This filling order is illustrated in ▶ Figure 3.6.

electronic configurations The detailed arrangement of electrons indicated by a specific notation, $1s^2 2s^2 2p^4$, etc.

Hund's rule A statement of the behavior of electrons when they occupy orbitals: Electrons will not join other electrons in an orbital if an empty orbital of the same energy is available for occupancy.

Pauli exclusion principle A statement of the behavior of electrons when they occupy orbitals: Only electrons spinning in opposite directions can simultaneously occupy the same orbital.

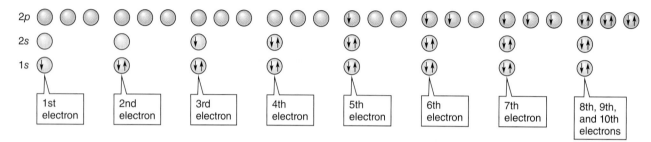

Figure 3.6 The filling order for the first 10 electrons.

The eleventh electron, with no empty orbital available that has the same energy as a 2p and no partially filled orbitals, will occupy the empty lowest-energy subshell of the third shell. The filling order for electrons beyond the tenth follows the pattern given in ▶ Figure 3.7, where shells are indicated by large rectangles, subshells by colored rectangles, and orbitals by circles. The filling order is obtained by following the arrows.

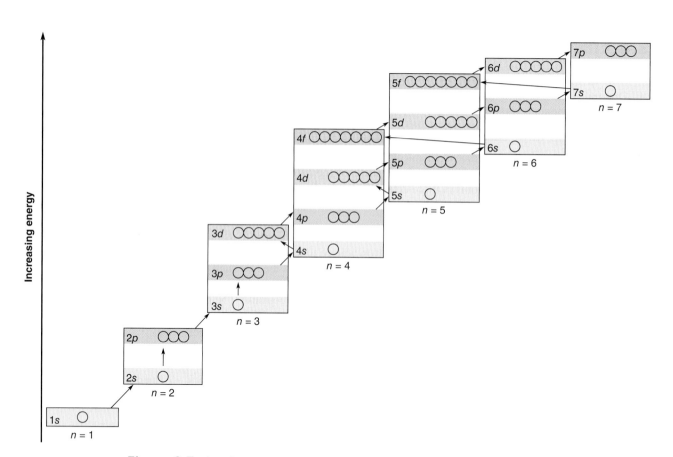

Figure 3.7 The relative energies and electron-filling order for shells and subshells.

As shown in Figure 3.7, some low-energy subshells of a specific shell have energies lower than the upper subshell of a preceding shell. For example, the 4s subshell has a lower energy than, and fills before, the 3d subshell. Figure 3.7 indicates that the third shell will not accept more than 8 electrons until the 4s subshell is complete. Thus, electrons 21 through 30 go into the 3d subshell and complete the filling of the third shell.

It is often convenient to represent the electronic configuration of an atom in a concise way. This is done by writing the subshells in the correct filling order and then indicating the number of electrons in each subshell by a superscript.

▶ Example 3.6

Write the electronic configurations for the following, and indicate the number of unpaired electrons in each case:

a. An atom that contains 7 electrons
b. An atom that contains 17 electrons
c. An atom of element number 22
d. An atom of arsenic (As)

Solution

The correct filling order of subshells from Figure 3.7 is

$$1s,\ 2s,\ 2p,\ 3s,\ 3p,\ 4s,\ 3d,\ 4p,\ 5s,\ 4d,\ 5p,\ 6s,\ 4f,\ 5d,\ 6p,\ 7s,\ 5f,\ 6d,\ 7p$$

a. Even though s subshells can hold 2 electrons, p subshells 6, d subshells 10, and f subshells 14, only enough subshells to hold 7 electrons will be used. Therefore, the configuration is written as shown below, starting on the left, with circles representing orbitals and arrows representing electrons. It is apparent that no subshells beyond $2p$ are needed because that subshell contains only 3 electrons. Note that both the $1s$ and $2s$ subshells are full; the $2p$ subshell is half-full, with one electron in each of the three orbitals (Hund's rule). These three electrons are unpaired.

b. Similarly, the configuration for 17 electrons is shown below. Here the $1s$, $2s$, $2p$, and $3s$ subshells are full, as are the first and second shells. The $3p$ subshell is not full and can accept 1 more electron. One electron in the $3p$ subshell is unpaired.

c. An atom of element number 22 contains 22 protons in the nucleus and must therefore contain 22 electrons. The electronic configuration is

Note here that only two of the 3d orbitals of the 3d subshell are occupied, and each of these orbitals contains a single electron. Thus, there are two unpaired electrons.

d. Arsenic (As) is element number 33 and therefore contains 33 electrons. The electronic configuration is

$$1s^2 2s^2 2p^6 3s^2 3p^6 4s^2 3d^{10} 4p^3$$

This time, the circles and arrows have not been used to indicate orbitals and electrons. You should satisfy yourself that the following facts are clear: The first, second, and third shells are full. The fourth shell is partially full, with the $4s$ subshell being full and the $4p$ subshell being half-full. The $4p$ subshell contains 3 unpaired electrons.

▶ **LEARNING CHECK 3.6** Write the electronic configurations for the following, and indicate the number of unpaired electrons in each case:

a. Element number 9
b. Mg
c. The element found in group VIA(16) and period 3
d. An atom that contains 23 protons

Nano World

The title given above might remind you of the title of a far out science-fiction movie. However, a quick review of Table 1.2 reminds us that *nano* is a prefix used to designate a quantity that is one-billionth (10^{-9}) the size of another quantity. Thus, a nanometer is a length that is one-billionth the length of a meter, or a length equal to the distance from one end to the other of a row of five to ten atoms laid side by side. We are living in a world that is becoming more and more influenced by the ability of scientists, engineers, and technologists to produce, manipulate, and commercially manufacture particles and other tiny objects whose sizes are measured in nanometers.

The design and creation of materials at this tiny scale is referred to as nanotechnology. Consumer products containing materials produced by nanotechnology began showing up in the mid-1990s. Two common applications at that time were the inclusion of nanometer-sized particles (called nanoparticles) in cosmetics and sunscreen products. Today, nanoparticles are found in a broad range of consumer products ranging from food packaging to sporting goods. The sport of tennis has benefited significantly from nanotechnology. One company injects nanoparticles of silicon dioxide into voids in the graphite frame of their tennis rackets. The result is a stronger frame that allows more power to be delivered to the ball with each stroke. The same company has developed a tennis ball that is coated on the inside with a very thin coating of a clay-like nanoparticle material that makes the balls retain air better, and extends their useful life. Nanoparticle-based textile treatments have revolutionized the textile industry by making possible products such as quick-drying, waterproof, wrinkle-free, and stain-resistant clothing. Other products resulting from nanotechnology include self-cleaning windows, glare-reducing and fog-resistant coatings for eyeglasses and automobile windshields, and lighter, stronger components for automobiles. Despite the fact that nanoparticles cannot be seen with the naked eye, they exert very visible influences on products we all use, and this influence will surely increase with the passage of time.

Both the clothing worn by these tennis players and the rackets they are using are improved by nanotechnology.

Although Figure 3.7 gives the details of subshell-filling order, a more concise diagram is available and easy to remember. It is shown in ❭ Figure 3.8, where the subshells are first arranged as shown on the left and then diagonal arrows are drawn as shown on the right. To get the correct subshell-filling order, follow the arrows from top to bottom, going from the head of one arrow to the tail of the next.

Figure 3.8 An aid to remembering subshell-filling order.

1s			
2s	2p		
3s	3p	3d	
4s	4p	4d	4f
5s	5p	5d	5f
6s	6p	6d	
7s			

The electronic configurations described to this point provide details of the shells, subshells, and orbitals involved but are somewhat cumbersome. In some applications, these details are not needed, and simplified representations are used that emphasize the electrons in the valence shell.

We see from Table 3.2 that the noble gases neon and argon both have electronic configurations that end with a completely filled p subshell. This is true for all noble gases except helium, which ends with a filled $1s$ subshell. In Chapter 4, we will see that these **noble gas configurations** are important in understanding the bonding that occurs between atoms.

Noble gas configurations can be used to write abbreviated electronic configurations. Instead of writing the configurations in their entirety, the symbols for the noble gases are used in brackets to represent the electrons found in their configurations. Electrons that are present in addition to those of the noble gases are written following the symbol. For example, the electronic configuration for sodium can be represented as $1s^2 2s^2 2p^6 3s^1$ or $[Ne]3s^1$.

noble gas configuration An electronic configuration in which the last eight electrons occupy and fill the s and p subshells of the highest-occupied shell.

▶ Example 3.7

Write abbreviated electronic configurations for the following:

a. An atom that contains 7 electrons
b. An atom that contains 17 electrons
c. An atom of element number 22
d. An atom of arsenic (As)

Solution

All these configurations were written in conventional form in Example 3.6.

a. From Example 3.6, the configuration is $1s^2 2s^2 2p^3$. We see that the two electrons in the $1s$ subshell represent the noble gas configuration of helium (He), so we can write the configuration as $[He]2s^2 2p^3$.
b. From Example 3.6, the configuration is $1s^2 2s^2 2p^6 3s^2 3p^5$. We see that the $1s^2 2s^2 2p^6$ portion is the configuration of neon (Ne), so we can write the configuration as $[Ne]3s^2 3p^5$.
c. From Example 3.6, the configuration is $1s^2 2s^2 2p^6 3s^2 3p^6 4s^2 3d^2$. We see that the first 18 electrons represented by $1s^2 2s^2 2p^6 3s^2 3p^6$ correspond to the electrons of argon (Ar). Thus, we can write the configuration as $[Ar]4s^2 3d^2$.
d. From Example 3.6, the configuration of arsenic is $1s^2 2s^2 2p^6 3s^2 3p^6 4s^2 3d^{10} 4p^3$. Once again, the first 18 electrons can be represented by the symbol for argon. The configuration is $[Ar]4s^2 3d^{10} 4p^3$.

▶ LEARNING CHECK 3.7 Write abbreviated electronic configurations for the following. These are the same elements used in Learning Check 3.6.

a. An atom of element number 9
b. An atom of magnesium (Mg)
c. An atom of the element found in group VIA(16) and period 3
d. An atom that contains 23 protons

3.5 Another Look at the Periodic Table

Learning Objective

5. Determine the shell and subshell locations of the distinguishing electrons in elements, and based on their location in the periodic table, classify elements into the categories given in Figures 3.10 and 3.12.

Now that you know more about electronic configurations, you can better understand the periodic law and table. First, consider the electronic configurations of the elements belonging to the same group of the periodic table. Group IA(1), for example, contains Li, Na, K, Rb, Cs, and Fr, with electronic configurations for the first four elements as shown below:

Element Symbol	Conventional Form	Abbreviated Form
Li	$1s^2 2s^1$	$[He]2s^1$
Na	$1s^2 2s^2 2p^6 3s^1$	$[Ne]3s^1$
K	$1s^2 2s^2 2p^6 3s^2 3p^6 4s^1$	$[Ar]4s^1$
Rb	$1s^2 2s^2 2p^6 3s^2 3p^6 4s^2 3d^{10} 4p^6 5s^1$	$[Kr]5s^1$

Notice that each of these elements has a single electron in the valence shell. Further, each of these valence-shell electrons is located in an *s* subshell. Elements belonging to other groups also have valence-shell electronic configurations that are similar for all members of the group, and all members have similar chemical properties. It has been determined that the similar chemical properties of elements in the same group result from similar valence-shell electronic configurations.

The periodic table becomes more useful when we interpret it in terms of the electronic configurations of the elements in various areas. One relationship is shown in ▶ Figure 3.9, where the periodic table is divided into four areas on the basis of the type of subshell occupied by the highest-energy electron in the atom. This last electron added to an atom is called the **distinguishing electron.** Note that the *s* area is 2 columns (elements) wide, the *p* area is 6 columns wide, the *d* area is 10 columns wide, and the *f* area is 14 columns wide—exactly the number of electrons required to fill the *s*, *p*, *d*, and *f* subshells, respectively.

distinguishing electron The last and highest-energy electron found in an element.

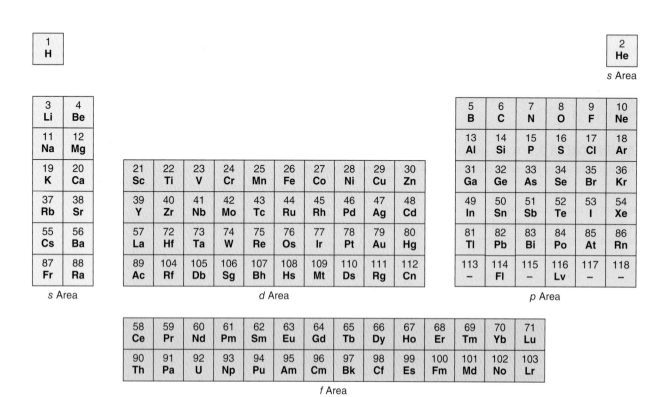

Figure 3.9 The periodic table divided by distinguishing electrons of the elements.

Electronic configurations are also used to classify elements of the periodic table, as shown in ❱ Figure 3.10.

	Representative elements		Transition elements										Representative elements					Noble gases

Figure 3.10 Elemental classification on the basis of electronic configurations.

The *noble gases* make up the group of elements found on the extreme right of the periodic table. They are all gases at room temperature and are unreactive with most other substances (hence, the group name). With the exception of helium, the first member of the group, noble gases are characterized by filled *s* and *p* subshells of the highest occupied shell.

Representative elements are those found in the *s* and *p* areas of the periodic table, not including the noble gases. The distinguishing electrons of representative elements partially or completely fill an *s* subshell—groups IA(1) and IIA(2)—or partially fill a *p* subshell—groups IIIA(13), IVA(14), VA(15), VIA(16), and VIIA(17). Most of the common elements are representative elements.

The *d* area of the periodic table contains the **transition elements** (Figure 3.10) in which the distinguishing electron is found in a *d* subshell. Some transition elements are used for everyday applications (❱ Figure 3.11). **Inner-transition elements** are those in the *f* area of the periodic table, and the distinguishing electron is found in an *f* subshell.

representative element An element in which the distinguishing electron is found in an *s* or a *p* subshell.

transition element An element in which the distinguishing electron is found in a *d* subshell.

inner-transition element An element in which the distinguishing electron is found in an *f* subshell.

Figure 3.11 Transition elements, such as gold, silver, copper, nickel, platinum, and zinc, are often used in coins and medals. List some chemical and physical properties that would be desirable in metals used for such purposes.

A new concept is often made easier to understand by relating it to something familiar. The concept of electronic configurations is very likely new to you, but you are probably familiar with hotels. The way electrons fill up orbitals, subshells, and shells around a nucleus can be compared to the way rooms, floors, and hotels located near a convention center will fill with convention delegates. To make our analogy work, imagine that the hotels are located on a street that runs uphill from the convention center, as shown. Further imagine that none of the hotels has elevators, so the only way to get to upper floors is to climb the stairs.

In this analogy, the convention center is equivalent to the nucleus of an atom, and each hotel represents a shell, each floor represents a subshell, and each room represents an orbital. If you were a delegate who wanted to describe to a friend where you were staying, you would indicate the hotel (shell), floor (subshell), and room (orbital) assigned to you. Electronic configurations such as $1s^2 2s^2 2p^1$ give similar information for each electron: The numbers preceding each letter indicate the shells, the letters indicate the subshells, and the superscripts coupled with Hund's rule indicate which orbitals are occupied.

Three more assumptions allow us to extend the analogy: (1) No more than two delegates can be assigned to a room, (2) delegates prefer not to have roommates if an empty room is available, and (3) all delegates want to use as little energy as possible when they walk from the convention center to their rooms (remember, there are no elevators).

With these assumptions in mind, it is obvious that the first delegate to check in will choose to stay in the single room of the s floor of Hotel One (a small but very exclusive hotel). The second delegate will also choose this same floor and room of Hotel One and will fill the hotel to capacity. The third delegate will have to go to Hotel Two and will choose the one room on the s floor. The fourth delegate will also choose the one room of the s floor, and fill that floor. The fifth delegate will choose a room on the p floor of Hotel Two because the s floor is full. The sixth delegate will also choose a room on the p floor of Hotel Two but will choose one that is not occupied. The seventh delegate will also choose an empty room on the p floor of Hotel Two. Delegates eight, nine, and ten can either pair up with the delegates already in the rooms of the p floor of Hotel Two or go uphill to Hotel Three. They choose to save energy by walking up the stairs and staying in rooms with roommates on the p floor of Hotel Two. Additional arriving delegates will occupy the floors of Hotels Three and Four in the order dictated by these same energy and pairing considerations. Thus, we see that the delegates in their rooms are analogous to the electrons in their orbitals.

▶ Example 3.8

Use the periodic table and Figures 3.9 and 3.10 to determine the following for Ca, Fe, S, and Kr:

a. The type of distinguishing electron
b. The classification based on Figure 3.10

Solution

a. On the basis of the location of each element in Figure 3.9, the distinguishing electrons are of the following types:

Ca: s Fe: d S: p Kr: p

b. The classifications based on Figure 3.10 are: Ca, representative element; Fe, transition element; S, representative element; Kr, noble gas.

▶ **LEARNING CHECK 3.8** Determine the following for element numbers 38, 47, 50, and 86:

a. The type of distinguishing electron
b. The classification according to Figure 3.10

The elements can also be classified into the categories of metals, nonmetals, and metalloids. This approach, used in ▶ Figure 3.12, shows that most elements are classified as metals. It is also apparent from Figure 3.10 that all nonmetals and all metalloids are representative elements.

	(1) I A													(13) III A	(14) IV A	(15) V A	(16) VI A	(17) VII A	(18) VIII A
1	1 H	(2) II A		Metals		Metalloids	Nonmetals													2 He
2	3 Li	4 Be												5 B	6 C	7 N	8 O	9 F	10 Ne
3	11 Na	12 Mg	(3) III B	(4) IV B	(5) V B	(6) VI B	(7) VII B	(8)	(9) VIII B	(10)	(11) I B	(12) II B		13 Al	14 Si	15 P	16 S	17 Cl	18 Ar
4	19 K	20 Ca	21 Sc	22 Ti	23 V	24 Cr	25 Mn	26 Fe	27 Co	28 Ni	29 Cu	30 Zn		31 Ga	32 Ge	33 As	34 Se	35 Br	36 Kr
5	37 Rb	38 Sr	39 Y	40 Zr	41 Nb	42 Mo	43 Tc	44 Ru	45 Rh	46 Pd	47 Ag	48 Cd		49 In	50 Sn	51 Sb	52 Te	53 I	54 Xe
6	55 Cs	56 Ba	57 La	72 Hf	73 Ta	74 W	75 Re	76 Os	77 Ir	78 Pt	79 Au	80 Hg		81 Tl	82 Pb	83 Bi	84 Po	85 At	86 Rn
7	87 Fr	88 Ra	89 Ac	104 Rf	105 Db	106 Sg	107 Bh	108 Hs	109 Mt	110 Ds	111 Rg	112 Cn		113 –	114 Fl	115 –	116 Lv	117 –	118 –

(Period is labeled along the left vertical axis)

58 Ce	59 Pr	60 Nd	61 Pm	62 Sm	63 Eu	64 Gd	65 Tb	66 Dy	67 Ho	68 Er	69 Tm	70 Yb	71 Lu
90 Th	91 Pa	92 U	93 Np	94 Pu	95 Am	96 Cm	97 Bk	98 Cf	99 Es	100 Fm	101 Md	102 No	103 Lr

Figure 3.12 Locations of metals, nonmetals, and metalloids in the periodic table of the elements.

Most **metals** have the following properties. (Are these properties physical or chemical?)

High thermal conductivity—they transmit heat readily.
High electrical conductivity—they transmit electricity readily.
Ductility—they can be drawn into wires.
Malleability—they can be hammered into thin sheets.
Metallic luster—they have a characteristic "metallic" appearance.

Nonmetals, the elements found in the right one-third of the periodic table, generally have chemical and physical properties opposite those of metals. Under normal conditions, they often occur as brittle, powdery solids or as gases.

Metalloids, such as boron (B) and silicon (Si), are the elements that form a diagonal separation zone between metals and nonmetals in the periodic table. Metalloids have properties somewhat between those of metals and nonmetals, and they often exhibit some of the characteristic properties of each type.

metals Elements found in the left two-thirds of the periodic table. Most have the following properties: high thermal and electrical conductivities, high malleability and ductility, and a metallic luster.

nonmetals Elements found in the right one-third of the periodic table. They often occur as brittle, powdery solids or gases and have properties generally opposite those of metals.

metalloids Elements that form a narrow diagonal band in the periodic table between metals and nonmetals. They have properties somewhat between those of metals and nonmetals.

▶ **LEARNING CHECK 3.9** Classify each of the following elements as metal, nonmetal, or metalloid:

a. Xe   **b.** As   **c.** Hg   **d.** Ba   **e.** Th

Iron: Too Much of a Good Thing?

Iron (II) ions (Fe^{2+}) play a vital role in a number of body processes. Perhaps the most well known of these functions is the role of iron as a component of hemoglobin, the oxygen-transporting protein of blood. When blood is iron-poor, body tissues do not receive enough oxygen, and *anemia*, a general weakening of the body, results.

The U.S. RDA (recommended dietary allowance) for iron is 10–15 mg per day, depending on age and sex. Some individuals, primarily pregnant women, need iron supplements to meet these recommendations. However, a well-balanced diet that includes red meat, whole grains, and dark green vegetables will generally provide enough iron to satisfy the RDA for most women. In an attempt to enhance the effectiveness of the diet in providing iron, many foods are enriched, or fortified, with iron—primarily breads, other flour products, and cereals. At present, 25% of all dietary iron consumed in the United States comes from such foods. Cooking utensils can also contribute to the amount of iron in the diet. It has been shown that food cooked in cast iron or black iron cookware contains significantly larger amounts of iron than identical food cooked in other types of cookware.

There is a downside to this abundance of iron in the general food supply. Well over a million Americans suffer from an inherited genetic disorder called hemochromatosis, or iron overload disease. Such individuals accumulate excess iron in their bodies. The excess iron causes damage to major organs such as the liver, pancreas, and heart. The disease most often affects men between the ages of 30 and 50. Women with the disorder are often protected from developing all the complications of the disease because of their natural loss of iron through menstruation. However, they do develop some of the complications.

The abundance of iron in the food supply obviously presents serious problems for individuals with this disorder. In addition to avoiding iron-rich foods or using them in moderation, the only treatment at the present time is phlebotomy, the regular removal of blood. When performed regularly, the potential for damage to vital organs is decreased.

It has been estimated that the treatment of patients with the condition results in their donation of about 3 million pints of blood each year. These blood donations have sparked a controversy because the American Red Cross, which collects about 45% of all blood donations in the United States, will not accept blood from hemochromatosis patients. The reasons behind this controversial practice are unclear, but they are not based on any health risks associated with using the blood. According to the Red Cross and the U.S. Centers for Disease Control and Prevention (CDC), the blood is as safe to use as the blood from regular donors. In fact, the CDC and many hematologists suggest that the iron-rich blood of hemochromatosis patients might actually be better than regular blood for patients who need regular transfusions. As a final irony, the CDC estimates that only about 10% of the people with hemochromatosis have been diagnosed. This means that many donors with the disease are giving blood anyway as regular blood donors, and their donations go right into the blood supply.

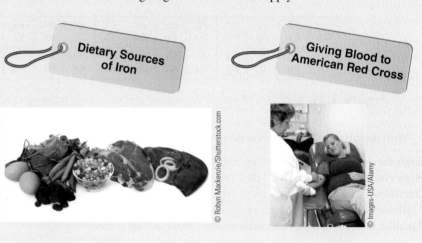

Dietary Sources of Iron

Giving Blood to American Red Cross

3.6 Property Trends within the Periodic Table

Learning Objective

6. Recognize property trends of elements within the periodic table, and use the trends to predict selected properties of the elements.

In Figure 3.12, the elements are classified into the categories of metal, metalloid, or nonmetal according to their positions in the table and properties such as thermal conductivity and metallic luster. It is generally true that within a period of the periodic table, the elements become less metallic as we move from left to right. Within a group, the metallic character increases from top to bottom.

Consider group VA(15) as an example. We see that the elements of the group consist of the two nonmetals nitrogen (N) and phosphorus (P), the two metalloids arsenic (As) and antimony (Sb), and the metal bismuth (Bi). We see the trend toward more metallic character from top to bottom that was mentioned above. According to what has been discussed concerning the periodic law and periodic table, these elements should have some similarities in chemical properties because they belong to the same group of the periodic table. Studies of their chemical properties show that they are similar but not identical. Instead, they follow trends according to the location of the elements within the group.

Certain physical properties also follow such trends, and some of these are easily observed, as shown in ▶ Figure 3.13. At room temperature and ordinary atmospheric pressure, nitrogen is a colorless gas, phosphorus is a white nonmetallic solid, arsenic is a brittle gray solid with a slight metallic luster, antimony is a brittle silver-white solid with a metallic luster, and bismuth is a lustrous silver-white metal.

Figure 3.13 The elements of group VA(15) of the periodic table. Phosphorus must be stored under water because it will ignite when exposed to the oxygen in air. It has a slight color because of reactions with air.

We see that these physical properties are certainly not identical, but we do see that they change in a somewhat regular way (they follow a trend) as we move from element to element down the group. Nitrogen is a gas, but the others are solids. Nitrogen is colorless, phosphorus is white, then the other three become more and more metallic-looking. Also, as mentioned previously, we note a change from nonmetal to metalloid to metal as we come down the group.

The trends in these properties occur in a regular way that would allow us to predict some of them for one element if they were known for the other elements in the group. For example, if the properties of bismuth were unknown, we would have predicted that it would have a silvery white color and a metallic luster based on the appearance of arsenic and antimony. The three elements from group VIIA(17) shown in ▶ Figure 3.14 also demonstrate some obvious predictable trends in physical properties.

Container of chlorine with molecular view of chlorine gas.

Container of Bromine and Bromine gas with molecular view of bromine (bottom), and its gas (top).

Container with iodine and iodine gas with molecular view of iodine(bottom), and its gas (top).

Figure 3.14 Chlorine, bromine, and iodine (left to right) all belong to group VIIA(17) of the periodic table. Their atoms all have the same number of electrons in the valence shell and therefore similar chemical properties. However, they do not have similar appearances. At room temperature and under normal atmospheric pressure, chlorine is a pale yellow gas, bromine is a dark red liquid that readily changes into a gas, and iodine is a gray-black solid that changes into a purple gas when heated slightly. Astatine is the next member of the group after iodine. Try to think like Mendeleev and predict its color and whether it will be a liquid, solid, or gas under normal conditions.

Trends in properties occur in elements that form periods across the periodic table as well as among those that form vertical groups. We will discuss two of these properties and their trends for representative elements. Our focus will be on the general trends, recognizing that some elements show deviations from these general behaviors. We will also propose explanations for the trends based on the electronic structure of atoms discussed in the chapter.

The first property we will consider is the size of the atoms of the representative elements. The size of an atom is considered to be the radius of a sphere extending from the center of the nucleus of the atom to the location of the outermost electrons around the nucleus. The behavior of this property across a period and down a group is shown in ❱ Figure 3.15. We see that the size increases from the top to the bottom of each group, and decreases from left to right across a period.

Figure 3.15 Scale drawings of the atoms of some representative elements enlarged about 60 million times. The numbers are the atomic radii in picometers (10^{-12} m).

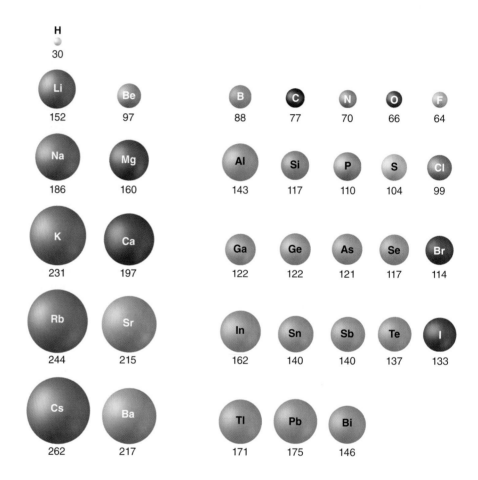

Remember that when we go from element to element down a group, a new electronic shell is being filled. Consider group IA(1). In lithium atoms (Li), the second shell ($n = 2$) is beginning to fill, in sodium atoms (Na), the third shell ($n = 3$) is beginning to fill, and so on for each element in the group until the sixth shell ($n = 6$) begins to fill in cesium atoms (Cs). As n increases, the distance of the electrons from the nucleus in the shell designated by n increases, and the atomic radius as defined above also increases.

The decrease in atomic radius across a period can also be understood in terms of the outermost electrons around the nucleus. Consider period 3, for example. The abbreviated electronic structure for sodium atoms (Na) is [Ne]$3s^1$, and the outermost electron is in the third shell ($n = 3$). In magnesium atoms (Mg), the electronic structure is [Ne]$3s^2$, and we see that the outermost electron is still in the third shell. In aluminum atoms (Al), the electronic structure is $[Ne]3s^23p^1$, and we see that, once again, the outermost electron is in the third shell. In fact, the outermost electrons in all the atoms across period 3 are in the third shell.

Because all these electrons are in the same third shell, they should all be the same distance from the nucleus, and the atoms should all be the same size. However, each time another electron is added to the third shell of these elements, another positively charged proton is added to the nucleus. Thus, in sodium atoms there are 11 positive nuclear charges attracting the electrons, but in aluminum atoms there are 13. The effect of the increasing nuclear charge attracting electrons in the same shell is to pull all the electrons of the shell closer to the nucleus and cause the atomic radii to decrease as the nuclear charge increases.

The chemical reactivity of elements is dependent on the behavior of the electrons of the atoms of the elements, especially the valence electrons. One property that is related to the behavior of the electrons of atoms is the ionization energy.

The *ionization energy* of an element is the energy required to remove an electron from an atom of the element in the gaseous state. The removal of one electron from an atom leaves the atom with a net $1+$ charge because the nucleus of the resulting atom contains one more proton than the number of remaining electrons. The resulting charged atom is called an *ion*. We will discuss ions and the ionization process in more detail in Sections 4.2 and 4.3. A reaction for the removal of one electron from an atom of sodium is

$$Na(g) \rightarrow Na^+(g) + e^-$$

Because this process represents the removal of the first electron from a neutral sodium atom, the energy necessary to accomplish the process is called the **first ionization energy.** If a second electron were removed, the energy required would be called the second ionization energy, and so forth. We will focus on only the first ionization energy for representative elements. ▶ Figure 3.16 contains values for the first ionization energy of a number of representative elements.

first ionization energy The energy required to remove the first electron from a neutral atom.

IA							VIIIA
H 1311	IIA	IIIA	IVA	VA	VIA	VIIA	**He** 2370
Li 521	**Be** 899	**B** 799	**C** 1087	**N** 1404	**O** 1314	**F** 1682	**Ne** 2080
Na 496	**Mg** 737	**Al** 576	**Si** 786	**P** 1052	**S** 1000	**Cl** 1245	**Ar** 1521
K 419	**Ca** 590	**Ga** 576	**Ge** 784	**As** 1013	**Se** 939	**Br** 1135	**Kr** 1351
Rb 402	**Sr** 549	**In** 559	**Sn** 704	**Sb** 834	**Te** 865	**I** 1007	**Xe** 1170
Cs 375	**Ba** 503	**Tl** 590	**Pb** 716	**Bi** 849	**Po** 791	**At** 926	**Rn** 1037

Figure 3.16 First ionization energies for selected representative elements. The values are given in kJ/mol. (Source: Data from Bard, A. J.; Parsons, R.; Jordan, J. *Standard Potentials in Aqueous Solution.* New York: Dekker, 1985, pp. 24–27.)

The general trend is seen to be a decrease from the top to the bottom of a group, and an increase from left to right across a period. The higher the value of the ionization energy, the more difficult it is to remove an electron from the atoms of an element. Thus, we see that in general, the electrons of metals are more easily removed than are the electrons of nonmetals. Also, the farther down a group a metal is located, the easier it is to remove an electron.

The metals of group IA(1) all react with ethyl alcohol, C_2H_5OH, to produce hydrogen gas, H_2, as follows, where M is a general representation of the metals of the group:

$$2M + 2C_2H_5OH \rightarrow 2C_2H_5OM + H_2$$

In this reaction, electrons are removed from the metal and transferred to the hydrogen. The reaction is shown in progress for the first three members of the group in ▶ Figure 3.17. The rate (speed) of the reaction is indicated by the amount of hydrogen gas being released.

Figure 3.17 The reaction of group IA(1) elements with ethyl alcohol. Left to right: Lithium (Li), sodium (Na), potassium (K). Each metal sample was wrapped in a wire screen to keep it from floating.

▶ **LEARNING CHECK 3.10** Refer to Figure 3.17, and do the following:

a. Arrange the three metals of the group vertically in order of the rate of the reaction with ethyl alcohol. Put the slowest reaction at the top and the fastest at the bottom.

b. Compare the trend in reaction rate coming down the group with the trend in ionization energy coming down the group.

c. Compare the trend in reaction rate coming down the group with the trend in ease of removing an electron from an atom of the metal coming down the group.

The trends in ionization energy and ease of removing an electron can be explained using arguments similar to those for explaining the sizes of atoms. As we come down a group, the valence electrons are located farther and farther away from the nucleus because they are located in higher-energy shells. The farther the electrons are away from the nucleus, the weaker is the attraction of the positively charged nucleus for the negatively charged electrons, and the easier it is to pull the electrons away.

Similarly, as we go from left to right in a period, the valence electrons are going into the same shell and therefore should be about the same distance from the nucleus. But as we saw before, the nuclear charge increases as well as the number of electrons. As a result, there is a greater nuclear charge attracting the electrons the farther we move to the right. Thus, a valence electron of an atom farther to the right is more difficult to remove than a valence electron of an atom farther to the left.

The general trends we have discussed are summarized in ❯ Figure 3.18. It is easiest to remember the trends by always going in the same direction in the table and noting how the properties change in that direction. We have chosen to summarize by always going from top to bottom for groups, and left to right for periods.

Figure 3.18 General trends for atomic size and ionization energy of representative elements.

Case Study Follow-up

The element iron (Fe) occupying the position of period 4, group VIIIB (8) on the periodic table and having the atomic number 26, is present in every cell in the human body. Iron is part of many enzymes and is required for numerous cell functions. (CDC, 1998)

A conservative definition of pica is "the ingestion of non-food substances frequently associated with a nutritional deficiency such as iron-deficiency anemia." Some health care practitioners include unusual food items in the definition. Pica is common among pregnant women and varies by culture. In the United States, pica is most often observed in African American women living in rural areas who also have a family history of the condition. Craving ice has been associated with anemia and is frequently classified as pica. In the same way, craving what is essentially plain cocoa by a toddler could be considered pica. Since it was an isolated instance in this case study, a meaningful correlation could not be made. The craving and the anemia might be a coincidence. However, the behavior served as a signal to the mother to assess her child. This case study highlights the importance of family members or closely associated caregivers having knowledge of a child's typical behavior and the caregiver's ability to detect unusual symptoms. It also calls attention to the role of education in public health. The fact that the older sister had some education in nutrition was an important factor in the diagnosis of the anemia.

Concept Summary

The Periodic Law and Table

The chemical properties of the elements tend to repeat in a regular (periodic) way when the elements are arranged in order of increasing atomic numbers. This periodic law is the basis for the arrangement of the elements called the periodic table. In this table, each element belongs to a vertical grouping, called a group or family, and a horizontal grouping, called a period. All elements in a group or family have similar chemical properties.

Objective 1, Exercise 3.4

Electronic Arrangements in Atoms

Niels Bohr proposed a theory for the electronic structure of hydrogen based on the idea that the electrons of atoms move around atomic nuclei in fixed circular orbits. Electrons change orbits only when they absorb or release energy. The Bohr model was modified as a result of continued research. It was found that precise Bohr orbits for electrons could not be determined. Instead, the energy and location of electrons could be specified in terms of shells, subshells, and orbitals, which are indicated by a notation system of numbers and letters.

Objective 2, Exercise 3.12

The Shell Model and Chemical Properties

The modified Bohr model, or shell model, of electronic structure provides an explanation for the periodic law. The rules governing electron occupancy in shells, subshells, and orbitals result in a repeating pattern of valence-shell electron arrangements. Elements with similar chemical properties turn out to be elements with identical numbers and types of electrons in their valence shells.

Objective 3, Exercises 3.18 and 3.22

Electronic Configurations

The arrangements of electrons in orbitals, subshells, and shells are called electronic configurations. Rules and patterns have been found that allow these configurations to be represented in a concise way. Written electronic configurations allow details of individual orbital, subshell, and shell electron occupancies to be seen readily. Also, the number of unpaired electrons in elements is easily determined. Electronic configurations can be represented in an abbreviated form by using noble gas symbols to represent some of the inner electrons.

Objective 4, Exercises 3.24 and 3.28

Another Look at the Periodic Table

Correlations between electronic configurations for the elements and the periodic table arrangement of elements make it possible to determine a number of details of electronic structure for an element simply on the basis of the location of the element in the periodic table. Special attention is paid to the last or distinguishing electron in an element. Elements are classified according to the type of subshell (s, p, d, f) occupied by this electron. The elements are also classified on the basis of other properties as metals, nonmetals, or metalloids.

Objective 5, Exercises 3.34 and 3.36

Property Trends within the Periodic Table

Chemical and physical properties of elements follow trends within the periodic table. These trends are described in terms of changes in properties of elements from the top to the bottom of groups, and from the left to the right of periods. The sizes of atoms and first ionization energies are two properties that show distinct trends.

Objective 6, Exercises 3.40 and 3.42

Key Terms and Concepts

Atomic orbital (3.2)
Distinguishing electron (3.5)
Electronic configuration (3.4)
First ionization energy (3.6)
Group or family of the periodic table (3.1)
Hund's rule (3.4)
Inner-transition element (3.5)

Metalloid (3.5)
Metal (3.5)
Noble gas configuration (3.4)
Nonmetal (3.5)
Pauli exclusion principle (3.4)
Periodic law (3.1)
Period of the periodic table (3.1)

Representative element (3.5)
Shell (3.2)
Subshell (3.2)
Transition element (3.5)
Valence shell (3.3)

Exercises

The Periodic Law and Table (Section 3.1)

3.1 Identify the group and period to which each of the following elements belongs:

   a. Si

   b. element number 21

   c. zinc

   d. element number 35

3.2 Identify the group and period to which each of the following elements belongs:

   a. Sb

   b. cadmium

   c. element number 31

   d. Zr

3.3 Write the symbol and name for the elements located in the periodic table as follows:

   a. Belongs to group VIA(16) and period 3

   b. The first element (reading down) in group VIB(6)

   c. The fourth element (reading left to right) in period 3

   d. Belongs to group IB(11) and period 5

3.4 Write the symbol and name for the elements located in the periodic table as follows:

   a. The noble gas belonging to period 4

   b. The fourth element (reading down) in group IVA(14)

   c. Belongs to group VIB(6) and period 5

   d. The sixth element (reading left to right) in period 6

3.5 **a.** How many elements are located in group VB(5) of the periodic table?

   b. How many elements are found in period 4 of the periodic table?

   c. How many total elements are in groups IIIA(13) and IVB(4) of the periodic table?

3.6 **a.** How many elements are located in group VIIB(7) of the periodic table?

   b. How many total elements are found in periods 1 and 2 of the periodic table?

   c. How many elements are found in period 5 of the periodic table?

3.7 The following statements either define or are closely related to the terms *periodic law, period,* and *group.* Match the terms to the appropriate statements.

   a. This is a vertical arrangement of elements in the periodic table

   b. The chemical properties of the elements repeat in a regular way as the atomic numbers increase

   c. The chemical properties of elements 11, 19, and 37 demonstrate this principle

   d. Elements 4 and 12 belong to this arrangement

3.8 The following statements either define or are closely related to the terms *periodic law, period,* and *group.* Match the terms to the appropriate statements.

   a. This is a horizontal arrangement of elements in the periodic table

   b. Element 11 begins this arrangement in the periodic table

   c The element nitrogen is the first member of this arrangement

   d. Elements 9, 17, 35, and 53 belong to this arrangement

Electronic Arrangements in Atoms (Section 3.2)

3.9 According to the Bohr theory, which of the following would have the higher energy?

   a. An electron in an orbit close to the nucleus

   b. An electron in an orbit located farther from the nucleus

3.10 What particles in the nucleus cause the nucleus to have a positive charge?

3.11 What is the maximum number of electrons that can be contained in each of the following?

   a. A 2s orbital

   b. A 2s subshell

   c. The first shell

3.12 What is the maximum number of electrons that can be contained in each of the following?

   a. A 2p orbital

   b. A 2p subshell

   c. The second shell

3.13 How many orbitals are found in the fourth shell? Write designations for the orbitals.

3.14 How many orbitals are found in the second shell? Write designations for the orbitals.

3.15 How many orbitals are found in a 3d subshell? What is the maximum number of electrons that can be located in this subshell?

3.16 How many orbitals are found in a 4f subshell? What is the maximum number of electrons that can be located in this subshell?

3.17 Identify the subshells found in the fourth shell; indicate the maximum number of electrons that can occupy each subshell and the total number of electrons that can occupy the shell.

The Shell Model and Chemical Properties (Section 3.3)

3.18 Look at the periodic table and tell how many electrons are in the valence shell of the following elements:

   a. element number 54

   b. The first element (reading down) in group VA(15)

c. Sn

d. The fourth element (reading left to right) in period 3

3.19 Look at the periodic table and tell how many electrons are in the valence shell of the following elements:

a. element number 51

b. the fifth element (reading left to right) in period 3

c. Al

d. the third element (reading down) in group VIA(16)

3.20 What period 6 element has chemical properties most like sodium? How many valence-shell electrons does this element have? How many valence-shell electrons does sodium have?

3.21 What period 2 element has chemical properties most like silicon? How many valence-shell electrons does this element have? How many valence-shell electrons does silicon have?

3.22 If you discovered an ore deposit containing copper, what other two elements might you also expect to find in the ore? Explain your reasoning completely.

3.23 Radioactive isotopes of strontium were produced by the explosion of nuclear weapons. They were considered serious health hazards because they were incorporated into the bones of animals that ingested them. Explain why strontium would be likely to be deposited in bones.

Electronic Configurations (Section 3.4)

3.24 Write an electronic configuration for each of the following elements, using the form $1s^2 2s^2 2p^6$, and so on. Indicate how many electrons are unpaired in each case.

a. element number 37

b. Si

c. titanium

d. Ar

3.25 Write an electronic configuration for each of the following elements, using the form $1s^2 2s^2 2p^6$, and so on. Indicate how many of the electrons are unpaired in each case.

a. Ca

b. nickel

c. element number 34

d. V

3.26 Write electronic configurations and answer the following:

a. How many total s electrons are found in an atom of potassium?

b. How many unpaired electrons are found in an atom of aluminum?

c. How many subshells are completely filled in an atom of magnesium?

3.27 Write electronic configurations and answer the following:

a. How many total electrons in Ge have a number designation (before the letters) of 4?

b. How many unpaired p electrons are found in sulfur? What is the number designation of these unpaired electrons?

c. How many $3d$ electrons are found in tin?

3.28 Write the symbol and name for each of the elements described. More than one element will fit some descriptions.

a. Contains only two $2p$ electrons

b. Contains an unpaired $3s$ electron

c. Contains two unpaired $3p$ electrons

d. Contains three $4d$ electrons

e. Contains three unpaired $3d$ electrons

3.29 Write the symbol and name for each of the elements described. More than one element will fit some descriptions.

a. Contains one unpaired $2p$ electron

b. Contains a half-filled $3s$ subshell

c. The last electron completes the $3d$ subshell

d. Contains one unpaired $5s$ electron

e. The last electron half-fills the $4p$ subshell

3.30 Write abbreviated electronic configurations for the following:

a. arsenic

b. An element that contains 25 electrons

c. silicon

d. element number 53

3.31 Write abbreviated electronic configurations for the following:

a. An element that contains 22 electrons

b. element number 23

c. bromine

d. iron

3.32 Refer to the periodic table and write abbreviated electronic configurations for all elements in which the noble gas symbol used will be [Ne].

3.33 Refer to the periodic table and determine how many elements have the symbol [Kr] in their abbreviated electronic configurations.

Another Look at the Periodic Table (Section 3.5)

3.34 Classify each of the following elements into the s, p, d, or f area of the periodic table on the basis of the distinguishing electron:

a. lead

b. element 27

c. Tb

d. Rb

3.35 Classify each of the following elements into the s, p, d, or f area of the periodic table on the basis of the distinguishing electron:

a. Cl

b. zinc

c. element 62

d. Sr

3.36 Classify the following elements as representative, transition, inner-transition, or noble gases:

a. iron

b. element 15

c. U

d. xenon

e. tin

3.37 Classify the following elements as representative, transition, inner-transition, or noble gases:

a. W

b. Cm

c. element 10

d. helium

e. barium

3.38 Classify the following as metals, nonmetals, or metalloids:

a. argon

b. element 3

c. Ge

d. boron

e. Pm

3.39 Classify the following as metals, nonmetals, or metalloids:

a. rubidium

b. arsenic

c. element 50

d. S

e. Br

Property Trends within the Periodic Table (Section 3.6)

3.40 Use trends within the periodic table to predict which member of each of the following pairs is more metallic:

a. K or Ti

b. As or Bi

c. Mg or Sr

d. Sn or Ge

3.41 Use trends within the periodic table to predict which member of each of the following pairs is more metallic:

a. C or Sn

b. Sb or In

c. Ca or As

d. Al or Mg

3.42 Use trends within the periodic table and indicate which member of each of the following pairs has the larger atomic radius:

a. Ga or Se

b. N or Sb

c. O or C

d. Te or S

3.43 Use trends within the periodic table and indicate which member of each of the following pairs has the larger atomic radius:

a. Mg or Sr

b. Rb or Ca

c. S or Te

d. I or Sn

3.44 Use trends within the periodic table and indicate which member of each of the following pairs gives up one electron more easily:

a. Si or Cl

b. Mg or Ba

c. F or Br

d. Ca or S

3.45 Use trends within the periodic table and indicate which member of each of the following pairs gives up one electron more easily:

a. Mg or Al

b. Ca or Be

c. S or Al

d. Te or O

Additional Exercises

3.46 How would you expect the chemical properties of isotopes of the same element to compare to each other? Explain your answer.

3.47 Bromine (Br) and mercury (Hg) are the only elements that are liquids at room temperature. All other elements in the periodic group that contains mercury are solids. Explain why mercury and bromine are not in the same group.

3.48 What would be the mass in mg of 3.0×10^{20} atoms that all have the same electronic configuration of $1s^2\, 2s^2\, 2p^4$?

3.49 Refer to Figure 3.15 and predict what would happen to the density of the metallic elements (purple color) as you go from left to right across a period of the periodic table. Explain your reasoning.

3.50 A 10.02-g sample of an element contains 0.250 mol of the element. Classify the element into the correct category of representative, transition, inner-transition, or noble gas. Will the element conduct electricity?

Allied Health Exam Connection

The following questions are from these sources:

1. *Nursing School Entrance Exam* © 2005, Learning Express, LLC.

2. *McGraw-Hill's Nursing School Entrance Exams* by Thomas A. Evangelist, Tamara B. Orr and Judy Unrein © 2009, The McGraw-Hill Companies, Inc.

3. *NSEE Nursing School Entrance Exams,* 3rd Edition © 2009, Kaplan Publishing.

4. *Cliffs Test Prep: Nursing School Entrance Exams* by Fred N. Grayson © 2004, Wiley Publishing, Inc.

5. *Peterson's Master the Nursing School and Allied Health Entrance Exams,* 18th Edition by Marion F. Gooding © 2008, Peterson's, a Nelnet Company.

3.51 The arrangement of the modern periodic table is based on atomic:

a. mass

b. number

c. radius

d. electronegativity

3.52 The horizontal rows of the periodic table are called:

    a. families

    b. groups

    c. representative elements

    d. periods

3.53 Which of the following is an example of a transition element?

    a. aluminum

    b. astatine

    c. nickel

    d. rubidium

3.54 Which two elements have chemical properties that are similar?

    a. H and He

    b. Fe and W

    c. Li and Be

    d. Mg and Ca

3.55 Which statement below is false?

    a. Hydrogen is a nonmetal

    b. Aluminum is a semimetal

    c. Calcium is a metal

    d. Argon is a gas at room temperature

3.56 Where on the periodic table are the nonmetals located?

    a. upper right

    b. upper left

    c. lower right

    d. lower left

3.57 What does the number 36 represent on the periodic table entry for krypton?

    a. atomic number

    b. relative atomic mass

    c. group number

    d. electron configuration

3.58 Which of the following is an alkali metal (group IA)?

    a. calcium

    b. sodium

    c. aluminum

    d. alkanium

3.59 Which of the following is an alkaline earth metal?

    a. Na

    b. Mg

    c. Sc

    d. Ti

3.60 What is the maximum number of electrons that each p orbital can hold?

    a. 8

    b. 2

    c. 6

    d. 4

3.61 From the periodic table, which of K and Br is larger?

    a. K is larger

    b. Br is larger

    c. They are the same size

    d. We cannot know which one is larger

3.62 The element with the smallest atomic radius of the following is:

    a. Sr

    b. Mg

    c. Ba

    d. Ra

3.63 Ionization energy is:

    a. The energy required to completely remove an electron from an atom or ion

    b. The energy created by an ion

    c. The same as kinetic energy

    d. The attraction between a proton and neutron

3.64 Which of the following has the largest first ionization energy?

    a. Cs

    b. Rb

    c. Ba

    d. Sr

3.65 Which elements conduct electricity?

    a. metals

    b. nonmetals

    c. metalloids

    d. ions

3.66 What term describes the electrons in the outermost principal energy level of an atom?

    a. vector

    b. core

    c. kernel

    d. valence

3.67 If the electron configuration of an element is written $1s^2\, 2s^2\, 2p_x^2\, 2p_y^2\, 2p_z^2\, 3s^1$, the element's atomic:

    a. number is 11

    b. number is 12

    c. weight is 11

    d. weight is 12

3.68 Identify the 2 atoms with the same number of electrons in their outermost energy level.

    a. Na/K

    b. K/Ca

    c. Na/Mg

    d. Ca/Na

3.69 The number of unpaired electrons in the outer subshell of a phosphorus atom (atomic number: 15) is:

   a. 2

   b. 0

   c. 3

   d. 1

3.70 How many valence electrons are needed to complete the outer valence shell of sulfur?

   a. 1

   b. 2

   c. 3

   d. 4

3.71 An atom that has five $3p$ electrons in its ground state is:

   a. Si

   b. P

   c. Cl

   d. O

Chemistry for Thought

3.72 Samples of three metals that belong to the same group of the periodic table are shown in Figure 3.5. When magnesium reacts with bromine, a compound with the formula $MgBr_2$ results. What would be the formulas of the compounds formed by reactions of bromine with each of the other metals shown? Explain your reasoning.

3.73 Answer the problem posed in Figure 3.14, then predict the same things for fluorine, the first member of the group. Explain the reasoning that led you to your answers.

3.74 Answer the problem posed in Figure 3.11. What property that makes gold suitable for coins and medals also makes it useful in electrical connectors for critical electronic parts such as computers in spacecraft?

3.75 Calcium metal reacts with cold water as follows:

$$Ca + 2H_2O \rightarrow Ca(OH)_2 + H_2$$

Magnesium metal does not react with cold water. What behavior toward cold water would you predict for strontium and barium? Write equations to represent any predicted reactions.

3.76 Refer to the hotels analogy in Study Skills 3.1 and determine the number of floors and the number of rooms on the top floor of Hotel Five.

3.77 A special sand is used by a company as a raw material. The company produces zirconium metal that is used to contain the fuel in nuclear reactors. What other metals are likely to be produced from the same raw material? Explain your answer.

Case Study

Isabelle was a perfectionist and a straight-A student. She kept her room clean, followed the rules, and never gave her parents any trouble. Isabelle was a pretty girl; but having gone through puberty early, by sixth grade she was larger than her peers. Isabelle was not emotionally prepared to accept the weight gain that accompanies puberty. By the beginning of seventh grade she began counting calories. By eighth grade she started to binge and purge. She was 5'6" tall and felt that it was unacceptable to be over 100 lbs. Her parents weren't concerned. They saw Isabelle sometimes consuming large amounts of food—but they just thought she had a high metabolic rate. Isabelle moved on to college, and her parents saw her infrequently. They were shocked to receive a call one day from a hospital informing them that Isabelle had collapsed during a choir rehearsal. Her heart rate was elevated, her blood pressure was low, and her eyes appeared sunken. Isabelle was dehydrated and had an electrolyte imbalance. She was given an IV and referred to the Behavioral Health Unit to receive treatment for her eating disorder.

How do anorexia and bulimia differ? Can a patient have both at the same time? How does family involvement influence patient outcomes?

Follow-up to this Case Study appears at the end the chapter before the Concept Summary.

Learning Objectives

When you have completed your study of this chapter, you should be able to:

1 Draw correct Lewis structures for atoms of representative elements. (Section 4.1)

2 Use electronic configurations to determine the number of electrons gained or lost by atoms as they achieve noble gas electronic configurations. (Section 4.2)

3 Use the octet rule to correctly predict the ions formed during the formation of ionic compounds, and write correct formulas for binary ionic compounds containing a representative metal and a representative nonmetal. (Section 4.3)

4 Correctly name binary ionic compounds. (Section 4.4)

5 Determine formula weights for ionic compounds. (Section 4.5)

6 Draw correct Lewis structures for covalent molecules. (Section 4.6)

7 Draw correct Lewis structures for polyatomic ions. (Section 4.7)

8 Use VSEPR theory to predict the shapes of molecules and polyatomic ions. (Section 4.8)

9 Use electronegativities to classify covalent bonds of molecules, and determine whether covalent molecules are polar or nonpolar. (Section 4.9)

10 Write correct formulas for ionic compounds containing representative metals and polyatomic ions, and correctly name binary covalent compounds and compounds containing polyatomic ions. (Section 4.10)

11 Relate melting and boiling points of pure substances to the strength and type of interparticle forces present in the substances. (Section 4.11)

I n the discussion to this point, we have emphasized that matter is composed of tiny particles. However, we have not yet discussed the forces that hold these particles together to form the matter familiar to us. It is these forces that produce many of the properties associated with various types of matter—properties such as the electrical conductivity of copper wire, the melting point of butter, and the boiling point of water.

4.1 Noble Gas Configurations

Learning Objective

1. Draw correct Lewis structures for atoms of representative elements.

The octet rule, proposed in 1916 by G. N. Lewis (an American) and Walter Kossel (a German), was the first widely accepted theory to describe bonding between atoms. The basis for the rule was the observed chemical stability of the noble gases. Because the noble gases did not react readily with other substances and because chemical reactivity depends on electronic structure, chemists concluded that the electronic structure of the noble gases represented a stable (low-energy) configuration. This noble gas configuration (see Section 3.4) is characterized by two electrons in the valence shell of helium and eight valence-shell electrons for the other members of the group (Ne, Ar, Kr, Xe, and Rn).

Use the techniques described in Section 3.4 to write the electronic configurations for the noble gases. Identify the valence-shell electrons, and verify the noble gas configurations just mentioned.

Solution

The subshell-filling order in Figure 3.7 gives the electronic configurations shown below. Valence-shell electrons are in darker type, and the noble gas configurations are verified.

		Number of Valence-shell Electrons
He:	$\mathbf{1s^2}$	2
Ne:	$1s^2\mathbf{2s^2 2p^6}$	8
Ar:	$1s^2 2s^2 2p^6 \mathbf{3s^2 3p^6}$	8
Kr:	$1s^2 2s^2 2p^6 3s^2 3p^6 \mathbf{4s^2} 3d^{10} \mathbf{4p^6}$	8
Xe:	$1s^2 2s^2 2p^6 3s^2 3p^6 4s^2 3d^{10} 4p^6 \mathbf{5s^2} 4d^{10} \mathbf{5p^6}$	8
Rn:	$1s^2 2s^2 2p^6 3s^2 3p^6 4s^2 3d^{10} 4p^6 5s^2 4d^{10} 5p^6 \mathbf{6s^2} 4f^{14} 5d^{10} \mathbf{6p^6}$	8

▶ **LEARNING CHECK 4.1** Write the electronic configurations for F and K. What would have to be done to change these configurations to noble gas configurations (how many electrons would have to be added or removed)?

Lewis structure A representation of an atom or ion in which the elemental symbol represents the atomic nucleus and all but the valence-shell electrons. The valence-shell electrons are represented by dots arranged around the elemental symbol.

A simplified way to represent the valence-shell electrons of atoms was invented by G. N. Lewis. In these representations, called electron-dot formulas or **Lewis structures,** the symbol for an element represents the nucleus and all electrons around the nucleus except those in the valence shell. Valence-shell electrons are shown as dots around the symbol. Thus, rubidium is represented as Rb·.

When a Lewis structure is to be written for an element, it is necessary to determine the number of valence-shell electrons in the element. This can be done by writing the electronic configuration using the methods described in Chapter 3 and identifying the valence-shell electrons as those with the highest n value. A simpler alternative for representative elements is to refer to the periodic table and note that the number of valence-shell electrons in the atoms of an element is the same as the number of the group in the periodic table to which the element belongs. The group number used must be the one that precedes the letter A, not the one in parentheses as given in the periodic table inside the front cover of this book. Using this method, we see that rubidium belongs to group IA(1) and therefore has one valence-shell electron.

▶ **Example 4.2**

Draw Lewis structures for atoms of the following:

a. element number 4
b. cesium (Cs)
c. aluminum (Al)
d. selenium (Se)

The accepted procedure is to write the element's symbol and put a dot for each valence electron in one of four equally spaced locations around the symbol. Imagine a square around the symbol. Each side of the square represents one of the four locations. An element with four valence electrons would have one dot in each of the four locations. A fifth electron would be represented by one additional dot in one of the locations. Each location can have a maximum of two dots.

Solution

a. Element number 4 is beryllium; it belongs to group IIA(2) and thus has two valence-shell electrons. The Lewis structure is Be·.

b. Cesium is in group IA(1), has one valence-shell electron, and has the Lewis structure Cs·.

c. Aluminum is in group IIIA(13). It has three valence-shell electrons and is represented by the following Lewis structure: Al·.

d. Selenium is in group VIA(16) and thus has six valence-shell electrons. The Lewis structure is :Se·.

▶ **LEARNING CHECK 4.2** Draw Lewis structures for atoms of the following:

a. element number 9
b. magnesium (Mg)
c. sulfur (S)
d. krypton (Kr)

It is important to clearly understand the relationship between Lewis structures and electronic configurations for atoms. The following example and learning check will help you review this relationship.

▶ Example 4.3

Represent the following using abbreviated electronic configurations and Lewis structures:

a. F   **b.** K   **c.** Mg   **d.** Si

Solution

a. Fluorine (F) contains 9 electrons, with 7 of them classified as valence electrons (those in the 2s and 2p subshells). The configuration and Lewis structure are

$$[He]2s^2 2p^5 \quad \text{and} \quad :\overset{..}{\underset{..}{F}}\cdot$$

b. Potassium (K) contains 19 electrons, with 1 of them classified as a valence electron (the 1 in the 4s subshell). The configuration and Lewis structure are

$$[Ar]4s^1 \quad \text{and} \quad K\cdot$$

c. Magnesium (Mg) contains 12 electrons, with 2 of them classified as valence electrons (the 2 in the 3s subshell). The configuration and Lewis structure are

$$[Ne]3s^2 \quad \text{and} \quad \overset{.}{M}g\cdot$$

d. Silicon (Si) contains 14 electrons, with 4 of them classified as valence electrons (those in the 3s and 3p subshells). The configuration and Lewis structure are

$$[Ne]3s^2 3p^2 \quad \text{and} \quad \cdot\overset{.}{Si}\cdot$$

▶ **LEARNING CHECK 4.3** Using the two methods illustrated in Example 4.3, write electronic configurations and Lewis structures for the following atoms:

a. Li   **b.** Br   **c.** Sr   **d.** S

4.2 Ionic Bonding

Learning Objective

2. Use electronic configurations to determine the number of electrons gained or lost by atoms as they achieve noble gas electronic configurations.

octet rule A rule for predicting electron behavior in reacting atoms. It says that atoms will gain or lose sufficient electrons to achieve an outer electron arrangement identical to that of a noble gas. This arrangement usually consists of eight electrons in the valence shell.

simple ion An atom that has acquired a net positive or negative charge by losing or gaining electrons.

ionic bond The attractive force that holds together ions of opposite charge.

According to the **octet rule** of Lewis and Kossel, atoms tend to interact through electronic rearrangements that produce a noble gas electronic configuration for each atom involved in the interaction. Except for the lowest-energy shell, this means that each atom ends up with eight electrons in the valence shell. There are exceptions to this octet rule, but it is still used because of the amount of information it provides. It is especially effective in describing reactions between the representative elements of the periodic table (see Figure 3.10).

During some chemical interactions, the octet rule is satisfied when electrons are transferred from one atom to another. As a result of the transfers, neutral atoms acquire net positive or negative electrical charges and become attracted to one another. These charged atoms are called **simple ions,** and the attractive force between oppositely charged atoms constitutes an **ionic bond.** A second type of interaction that also satisfies the octet rule is discussed in Section 4.6.

▶ Example 4.4

Show how the following atoms can achieve a noble gas configuration and become ions by gaining or losing electrons:

a. Na   **b.** Cl

Solution

a. The electronic structure of sodium (Na) is represented below using an abbreviated configuration and a Lewis structure:

$$[Ne]3s^1 \quad \text{and} \quad Na\cdot$$

The first representation makes it obvious that the Ne configuration with 8 electrons in the valence shell would result if the Na atom lost the single electron located in the $3s$ subshell. The loss is represented by the following equation:

$$Na \rightarrow Na^+ + 1e^-$$

Notice that the removal of a single negative electron from a neutral Na atom leaves the atom with 11 positive protons in the nucleus and 10 negative electrons. This gives the atom a net positive charge. The atom has become a positive ion.

b. The electronic structure for chlorine (Cl) is shown below using an abbreviated configuration and a Lewis structure:

$$[Ne]3s^23p^5 \quad \text{and} \quad :\overset{..}{\underset{..}{Cl}}\cdot$$

The Cl atom can achieve an Ne configuration by losing the 7 valence electrons. However, it is energetically much more favorable to achieve the configuration of argon (Ar) by adding 1 electron to the valence shell. This electron would complete an octet and change a Cl atom into a negative chloride ion as represented by the following equation, where Cl^- represents a Cl atom with 17 protons and 18 electrons, giving the atom a net negative charge:

$$Cl + 1e^- \rightarrow Cl^-$$

▶ LEARNING CHECK 4.4
In Learning Check 4.1, you described what had to be done to change the electronic configurations of F and K to noble gas configurations. Write equations to illustrate these changes.

As a general rule, metals lose electrons and nonmetals gain electrons during ionic bond formation. The number of electrons lost or gained by a single atom rarely exceeds three, and this number can be predicted accurately for representative elements by using the periodic table. The number of electrons easily lost by a representative metal atom is the same as the group number (represented by roman numerals preceding the A in the periodic table). The number of electrons that tend to be gained by a representative nonmetal atom is equal to eight minus the group number. However, the nonmetal hydrogen is an exception. In most reactions, it loses one electron, consistent with its placement in group IA(1). In other, less common, reactions it gains one electron just as group VIIA(17) elements do.

▶ Example 4.5

Use the periodic table to predict the number of electrons lost or gained by atoms of the following elements during ionic bond formation. Write an equation to represent the process in each case.

a. Li
b. Any group IIA(2) element, represented by the symbol M
c. element number 15
d. carbon

Solution

a. Lithium (Li), a metal, is in group IA(1); therefore it will lose 1 electron per atom:

$$\text{Li} \rightarrow \text{Li}^+ + 1e^-$$

b. Any group IIA(2) metal will lose 2 electrons; therefore,

$$\text{M} \rightarrow \text{M}^{2+} + 2e^-$$

c. Element number 15 is phosphorus (P), a nonmetal of group VA(15). It will gain $8 - 5 = 3$ electrons. The equation is

$$\text{P} + 3e^- \rightarrow \text{P}^{3-}$$

d. Carbon (C) is in group IVA(14) and is a nonmetal. It should, therefore, gain $8 - 4 = 4$ electrons. However, no more than 3 electrons generally become involved in ionic bond formation, so we conclude that carbon will not react readily to form ionic bonds.

▶ **LEARNING CHECK 4.5** Predict the number of electrons that would be gained or lost during ionic bond formation for each of the following elements. Write an equation to represent each predicted change.

a. element number 34   **b.** Rb   **c.** element number 18   **d.** In

While your attention is focused on noble gas configurations, it is appropriate to emphasize the following point: The attainment of a noble gas electronic configuration by an atom does not mean that the atom is converted into a noble gas. Instead, as seen above, the atoms are converted into simple ions (charged atoms). This point is emphasized by the comparison given in ▶ Table 4.1. Notice that the electronic configuration of Mg^{2+} is the same as that of Ne, but the number of protons (the atomic number) of Mg^{2+} is still the same as that of Mg. Atoms and simple ions that have identical electronic configurations are said to be **isoelectronic.**

isoelectronic A term that literally means "same electronic," used to describe atoms or ions that have identical electronic configurations.

TABLE 4.1 A Metal Atom, a Metal Ion, and a Noble Gas Atom Compared

Particle	Symbol	Number of Protons in Nucleus	Number of Electrons Around Nucleus	Net Charge on Particle
Magnesium atom	Mg	12	12	0
Magnesium ion	Mg^{2+}	12	10	+2
Neon atom	Ne	10	10	0

4.3 Ionic Compounds

Learning Objective

3. Use the octet rule to correctly predict the ions formed during the formation of ionic compounds, and write correct formulas for binary ionic compounds containing a representative metal and a representative nonmetal.

So far the discussion has focused on the electron-transfer process as it occurs for isolated atoms. In reality, the electrons lost by a metal are the same ones gained by the nonmetal with which it is reacting. Substances formed from such reactions are called *ionic compounds*. No atom can lose electrons unless another atom is available to accept them, as shown in the formulas used to represent ionic compounds. These formulas represent the combining ratio of the positive and negative ions found in the compounds. This ratio is determined by the charges on the ions, which are determined by the number of electrons transferred.

▶ Example 4.6

Represent the electron-transfer process that takes place when the following pairs of elements react ionically. Determine the formula for each resulting ionic compound.

a. Na and Cl   **b.** Mg and F

Solution

a. Sodium (Na) is in group IA(1), and we can write

$$Na \rightarrow Na^+ + 1e^-$$

Chlorine (Cl), in group VIIA(17), reacts as follows:

$$Cl + 1e^- \rightarrow Cl^-$$

The resulting ions, Na^+ and Cl^-, will combine in a 1:1 ratio because the total positive and total negative charges in the final formula must add up to zero. The formula then is NaCl. Note that the metal is written first in the formula, a generally followed practice. Also, the formula is written using the smallest number of each ion possible—in this case, one of each.

The actual electron-transfer process and the achievement of octets can more easily be visualized as follows:

$$[Ne]3s^1 + [Ne]3s^23p^5 \rightarrow [Ne] + [Ne]3s^23p^6$$
$$\quad Na \qquad\quad Cl \qquad\quad Na^+ \qquad Cl^-$$

In this representation, it must be remembered that the $[Ne]3s^23p^6$ electronic configuration of Cl^- is the same configuration as argon, the noble gas that follows chlorine in the periodic table.

Fight Hypertension with Potassium

In the United States, hypertension (high blood pressure) is the primary reason people visit doctor's offices, and more prescriptions are written for its treatment than any other health problem. In addition to the use of prescription drugs, hypertension is also usually treated by reducing or eliminating the dietary intake of sodium in the form of table salt (sodium chloride). Recently released research results indicate that combining an increase in dietary potassium intake with a reduction in sodium intake is probably the most important dietary decision (after excess weight loss) people can make to reduce cardiovascular diseases, including hypertension.

Studies reveal that in societies with diets rich in fruits and vegetables, only 1% of the population suffers from hypertension. By contrast, 33% of adults have hypertension in industrialized societies where the diet contains larger amounts of processed foods, which often contain added salt. The typical diet in the United States contains about twice the sodium and only half the 4700 milligrams per day of potassium currently recommended by the American Heart Association.

It might seem that taking a daily supplement is the only way to ensure a daily potassium intake as high as 4.7 grams, but that is not the case. Nature provides many potassium-rich foods such as squash, potatoes, tomatoes, carrots, spinach, beans, bananas, apricots, prunes, melons, peaches, halibut, tuna, trout, and low-fat dairy products. Specific examples of the potassium content of a few dietary potassium sources are given in the following table:

banana, 1 medium—422 mg
© iStockphoto.com/ ZoneCreative

sweet potato, 1 baked—694 mg
© iStockphoto.com/ EddWestmacott

white potato, 1 baked—610 mg
© iStockphoto.com/Viktor Kitaykin

white beans, canned, ½ cup—595 mg
© iStockphoto.com/Stephan Hoerold

yogurt, plain, nonfat, 8 oz—595 mg
© Foodcollection/PhotoLibrary

halibut, cooked, 3 oz—490 mg
© iStockphoto.com/Stacie Peterson

tuna, yellowfin, cooked, 3 oz—484 mg
© iStockphoto.com/ Eugene Bochkarev

lima beans, cooked, ½ cup—484 mg
© iStockphoto.com/ Joan Kimball

winter squash, cooked, ½ cup—448 mg
© iStockphoto.com/ velvetsnow

In addition to potassium, some studies have shown that the minerals magnesium and calcium may also have a positive influence in maintaining healthy blood pressure. The fruits and vegetables that provide potassium in the diet are also good sources of these two minerals. So, the parental directions traditionally given to children to "eat your fruits and vegetables and drink your milk" have been given scientific validity for all of us as a way to help maintain healthy levels of blood pressure.

---

b. Magnesium (Mg) of group IIA(2) will lose two electrons per atom, whereas fluorine (F) of group VIIA(17) will gain one electron per atom. Therefore, we can write

$$Mg \rightarrow Mg^{2+} + 2e^-$$
$$F + 1e^- \rightarrow F^-$$

It is apparent that two fluorine atoms will be required to accept the electrons from one magnesium atom. From another point of view, two F^- ions will be needed to balance the charge of a single Mg^{2+} ion. Both observations lead to the formula MgF_2 for the compound. Note the use of subscripts to indicate the number of ions involved in the formula. The subscript 1, on Mg, is understood and never written.

▶ **LEARNING CHECK 4.6** Write equations to represent the formation of ions for each of the following pairs of elements. Write a formula for the ionic compound that would form in each case.

a. Mg and O   **b.** K and S   **c.** Ca and Br

binary compound A compound made up of two different elements.

Figure 4.1 The reaction of sodium metal and chlorine gas. The vigorous reaction releases energy that heats the flask contents to a high temperature. Would you expect a similar reaction to occur between potassium metal and fluorine gas?

Ionic compounds of the types used in Example 4.6 and Learning Check 4.6 are called **binary compounds** because each contains only two different kinds of atoms, Na and Cl in one case (see ❯ Figure 4.1) and Mg and F in the other.

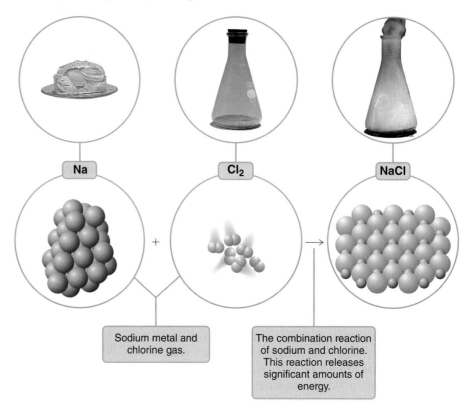

Na   Cl₂   NaCl

Sodium metal and chlorine gas.

The combination reaction of sodium and chlorine. This reaction releases significant amounts of energy.

4.4 Naming Binary Ionic Compounds

Learning Objective

4. Correctly name binary ionic compounds.

Names for binary ionic compounds are easily assigned when the names of the two elements involved are known. The name of the metallic element is given first, followed by the stem of the nonmetallic elemental name to which the suffix *-ide* has been added.

$$\text{name} = \text{metal} + \text{nonmetal stem} + \text{-ide}$$

The stem of the name of a nonmetal is the name of the nonmetal with the ending dropped. ❯ Table 4.2 gives the stem of the names of some common nonmetallic elements.

TABLE 4.2 Stem Names and Ion Formulas of Common Nonmetallic Elements

Element	Stem	Formula of Ion
Bromine	brom-	Br^-
Chlorine	chlor-	Cl^-
Fluorine	fluor-	F^-
Iodine	iod-	I^-
Nitrogen	nitr-	N^{3-}
Oxygen	ox-	O^{2-}
Phosphorus	phosph-	P^{3-}
Sulfur	sulf-	S^{2-}

▶ Example 4.7

Name the following binary ionic compounds:

a. KCl   **b.** SrO   **c.** Ca_3N_2

Solution

a. The metal is potassium (K), and the nonmetal is chlorine (Cl). Thus, the compound name is potassium <u>chlor</u>ide (the stem of the nonmetallic name is underlined).
b. Similarly, strontium (Sr) and oxygen (O) give the compound name strontium <u>ox</u>ide.
c. The elements are calcium (Ca) and nitrogen (N), hence the name calcium <u>nitr</u>ide.

▶ **LEARNING CHECK 4.7**   Assign names to the binary compounds whose formulas you wrote in Learning Check 4.6.

Names for the constituent ions of a binary compound are obtained in the same way as the compound name. Thus, K^+ is a potassium ion, whereas Cl^- is a chloride ion.

Some metal atoms, especially those of transition and inner-transition elements, form more than one type of charged ion. Copper, for example, forms both Cu^+ and Cu^{2+}, and iron forms Fe^{2+} and Fe^{3+}. The names of ionic compounds containing such elements must indicate which ion is present in the compound. A nomenclature system that does this well indicates the ionic charge of the metal ion by a roman numeral in parentheses following the name of the metal. Thus, CuCl is copper(I) chloride and $CuCl_2$ is copper(II) chloride. These names are expressed verbally as "copper one chloride" and "copper two chloride."

An older system is still in use but works only for naming compounds of metals that can form only two different charged ions. In this method, the endings *-ous* and *-ic* are attached to the stem of the metal name. For metals with elemental symbols derived from non-English names, the stem of the non-English name is used. The *-ous* ending is always used with the ion of lower charge and the *-ic* ending with the ion of higher charge. Thus, CuCl is cuprous chloride, and $CuCl_2$ is cupric chloride. In the case of iron, $FeCl_2$ is ferrous chloride, and $FeCl_3$ is ferric chloride (see ▶ Figure 4.2). Notice that the Cu^{2+} ion was called cupric, whereas the Fe^{2+} was called ferrous. The designations do not tell the actual ionic charge, but only which of the two is higher (Cu^{2+} and Fe^{3+}) or lower (Cu^+ and Fe^{2+}) for the metal in question.

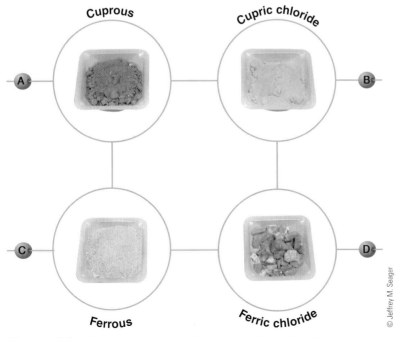

© Jeffrey M. Seager

Figure 4.2 Chloride compounds of copper and iron. Top: Cuprous and cupric chloride. Bottom: Ferrous and ferric chloride.

▶ Example 4.8

Write formulas for ionic compounds that would form between the following simple ions. Note that the metal forms two different simple ions, and name each compound two ways.

a. Cr^{2+} and S^{2-} **b.** Cr^{3+} and S^{2-}

Solution

In each case, the metal ion is from the metal chromium (Cr).

a. Because the ionic charges are equal in magnitude but opposite in sign, the ions will combine in a 1:1 ratio. The formula is CrS. Chromium forms simple ions with 2+ and 3+ charges. This one is 2+, the lower of the two; thus, the names are chromium(II) sulfide and chromous sulfide.

b. In this case, the charges on the combining ions are 3+ and 2−. The smallest combining ratio that balances the charges is two Cr^{3+} (a total of 6+ charge) and three S^{2-} (a total of 6− charge). The formula is Cr_2S_3. The names are chromium(III) sulfide and chromic sulfide.

▶ **LEARNING CHECK 4.8** Write formulas for the ionic compounds that would form between the following simple ions. The metal is cobalt, and it forms only the two simple ions shown. Name each compound, using two methods.

a. Co^{2+} and Br^- **b.** Co^{3+} and Br^-

4.5 The Smallest Unit of Ionic Compounds

Learning Objective

5. Determine formula weights for ionic compounds.

In Section 1.3, a molecule was defined as the smallest unit of a pure substance capable of a stable, independent existence. As you will see later, some compound formulas are used to represent single molecules. True molecular formulas represent the precise numbers of atoms of each element that are found in a molecule. However, as we have seen, formulas for ionic compounds represent only the simplest combining ratio of the ions in the compounds.

The stable form of an ionic compound is not a molecule, but a crystal in which many ions of opposite charge occupy **lattice sites** in a rigid three-dimensional arrangement called a **crystal lattice**. For example, the lattice of sodium chloride (ordinary table salt) represented in ▶ Figure 4.3 is the stable form of pure sodium chloride.

lattice site The individual location occupied by a particle in a crystal lattice.

crystal lattice A rigid three-dimensional arrangement of particles.

Figure 4.3 Crystal lattice for sodium chloride (table salt).

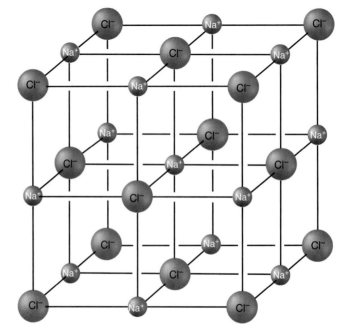

Even though the formulas of ionic compounds do not represent true molecular formulas, they are often used as if they did. This is especially true when equations representing chemical reactions are written, or when the mole concept is applied to chemical formulas (see Section 2.7). When the atomic weights of the atoms making up a true molecular formula are added together, the result is called the molecular weight of the compound (Section 2.4). A similar quantity obtained by adding up the atomic weights of the atoms shown in the formula of an ionic compound is called a **formula weight.** The mole concept is applied to formula weights in a manner similar to the way it is applied to molecular weights.

formula weight The sum of the atomic weights of the atoms shown in the formula of an ionic compound.

▶ Example 4.9

Carbon dioxide, CO_2, is a molecular compound, whereas magnesium chloride, $MgCl_2$, is an ionic compound.

a. Determine the molecular weight for CO_2 and the formula weight for $MgCl_2$ in atomic mass units.
b. Determine the mass in grams of 1.00 mol of each compound.
c. Determine the number of CO_2 molecules in 1.00 mol and the number of Mg^{2+} and Cl^- ions in 1.00 mol of $MgCl_2$.

Solution

a. For CO_2, the molecular weight is the sum of the atomic weights of the atoms in the formula:

$$MW = (1)(\text{at. wt. C}) + (2)(\text{at. wt. O}) = (1)(12.0 \text{ u}) + (2)(16.0 \text{ u})$$
$$MW = 44.0 \text{ u}$$

For $MgCl_2$, the formula weight is also equal to the sum of the atomic weights of the atoms in the formula:

$$FW = (1)(\text{at. wt. Mg}) + (2)(\text{at. wt. Cl}) = (1)(24.3 \text{ u}) + (2)(35.5 \text{ u})$$
$$FW = 95.3 \text{ u}$$

b. According to Section 2.6, 1.00 mol of a molecular compound has a mass in grams equal to the molecular weight of the compound. Thus, 1.00 mol CO_2 = 44.0 g CO_2.

For ionic compounds, 1.00 mol of compound has a mass in grams equal to the formula weight of the compound. Thus, 1.00 mol $MgCl_2$ = 95.3 g $MgCl_2$.

c. In Section 2.6 we learned that 1.00 mol of a molecular compound contains Avogadro's number of molecules. Thus, 1.00 mol CO_2 = 6.02×10^{23} molecules of CO_2.

In the case of ionic compounds, 1.00 mol of compound contains Avogadro's number of formula units. That is, 1.00 mol of magnesium chloride contains Avogadro's number of $MgCl_2$ units, where each unit represents one Mg^{2+} ion and two Cl^- ions. Thus,

$$1.00 \text{ mol } MgCl_2 = 6.02 \times 10^{23} \text{ } MgCl_2 \text{ units}$$

or

$$1.00 \text{ mol } MgCl_2 = 6.02 \times 10^{23} \text{ } Mg^{2+} \text{ ions} + 1.20 \times 10^{24} \text{ } Cl^- \text{ ions}$$

Note that the number of Cl^- ions is simply twice Avogadro's number.

▶ **LEARNING CHECK 4.9** Determine the same quantities that were determined in Example 4.9, but use the molecular compound hydrogen sulfide, H_2S, and the ionic compound calcium oxide, CaO.

4.6 Covalent Bonding

Learning Objective

6. Draw correct Lewis structures for covalent molecules.

In Example 4.5, we found that carbon, a nonmetal, would have to gain four electrons in order to form ionic bonds. We concluded that carbon does not generally form ionic bonds. Yet carbon is known to form many compounds with other elements. Also, electron transfers would not be expected to take place between two atoms of the same element because such an exchange would not change the electronic configuration of the atoms involved. Yet molecules of elements containing two atoms, such as chlorine (Cl_2), oxygen (O_2), and nitrogen (N_2), are known to exist and, in fact, represent the stable form in which these elements occur in nature. What is the nature of the bonding in these molecules?

Again, G. N. Lewis provided an answer by suggesting that the valence-shell electrons of the atoms in such molecules are shared in a way that satisfies the octet rule for each of the atoms. This process is symbolized below for fluorine, (F_2), using Lewis structures:

$$:\!\ddot{F}\!\cdot + \cdot\ddot{F}\!: \longrightarrow :\!\ddot{F}\!:\!\ddot{F}\!:$$

shared pair (counted in octet of each atom)

Sometimes the shared electron pairs are shown as straight lines, and sometimes the nonshared pairs are not shown, as illustrated in Example 4.10.

▶ Example 4.10

Represent the following reactions using Lewis structures:

a. $Cl + Cl \rightarrow Cl_2$ **b.** $H + H \rightarrow H_2$

Solution

a. Each chlorine (Cl) atom has seven valence-shell electrons. Because one electron from each atom is shared, the octet rule is satisfied for each atom

$$:\!\ddot{C}l\!\cdot + \cdot\ddot{C}l\!: \longrightarrow :\!\ddot{C}l\!:\!\ddot{C}l\!:$$

shared pair

The structure of Cl_2 can also be represented as

$$:\!\ddot{C}l\!-\!\ddot{C}l\!: \quad or \quad Cl\!-\!Cl$$

b. Each hydrogen (H) atom has one valence-shell electron. By sharing the two electrons, each H atom achieves the helium (He) structure:

$$H\cdot + \cdot H \longrightarrow H\!:\!H \quad or \quad H\!-\!H$$

▶ LEARNING CHECK 4.10 Represent the following reactions using Lewis structures:

a. $N + N \rightarrow N_2$ (more than two electrons will be shared)   **b.** $Br + Br \rightarrow Br_2$

To understand the origin of the attractive force between atoms that results from electron sharing, consider the H_2 molecule and the concept of atomic orbital overlap. Suppose two H atoms are moving toward each other. Each atom has a single electron in a spherical $1s$ orbital. While the atoms are separated, the orbitals are independent of each other; but as the atoms get closer together, the orbitals overlap and blend to create an orbital common to both atoms called a *molecular orbital*. The two shared electrons then move throughout the overlap region but have a high probability of being found somewhere between the two

Water: One of Earth's Special Compounds

An adequate supply of clean water is essential to our health and well-being. We can live without food for many days, but life would end in only a few days without water. Our circulatory system is an aqueous stream that distributes an amazing variety of substances throughout the body. Our cells are filled with water solutions in which the chemical reactions of life take place. Without water and its unique properties, life would not be possible on Earth.

Three-fourths of Earth's surface is covered with water, which gives the planet a blue color when it is viewed from space. The oceans represent the world's largest liquid solution. Water participates in most of the chemical reactions that occur in nature.

Most water used for human consumption comes from reservoirs, lakes, rivers, and wells. Much of this water is used and reused by numerous cities as it travels downstream. Such reused water may become seriously polluted with waste from previous users and with pathogenic microorganisms. As a safety precaution, most of the water we use undergoes chemical and physical treatment to purify it. This treatment process includes settling, whereby specific materials are added to bring down suspended solids. This is followed by filtration through sand and gravel to remove still more suspended matter. The filtered water may then be aerated by spraying it into the air. This part of the treatment process removes some odors and improves the taste of the water. Charcoal filtration, a treatment process that is gaining in use, also removes odors along with colored materials. In a

final step, chlorine or another disinfectant is added to kill any remaining bacteria.

The treatment of water before use is an important safety precaution.

nuclei. As a result, both of the positive nuclei are attracted toward the negative pair of electrons and hence toward each other. This process, represented in ▶ Figure 4.4, produces a net attractive force between the nuclei. This attractive force is the **covalent bond** that holds the atoms together. Because forces cannot be seen, we represent only the presence of the shared electron pair between the atoms (by a pair of dots or a line) and remember that an attractive force is also there.

covalent bond The attractive force that results between two atoms that are both attracted to a shared pair of electrons.

A Separated atoms   B Orbitals touch   C Orbitals overlap; a covalent bond is formed

Figure 4.4 Orbital overlap during covalent bond formation.

Covalent bonding between like atoms has been described, but it also occurs between unlike atoms. Examples of both types are given in ▶ Table 4.3. Lewis structures, like those in Table 4.3, are most easily drawn by using a systematic approach such as the one represented by the following steps:

1. Use the molecular formula to determine how many atoms of each kind are in the molecule.
2. Use the given connecting pattern of the atoms to draw an initial structure for the molecule, with the atoms arranged properly.
3. Determine the total number of valence-shell electrons contained in the atoms of the molecule.

TABLE 4.3 Examples of Covalent Bonding

Molecule	Atomic Lewis structure	Sharing pattern	Molecular Lewis structure
Nitrogen gas (N_2)	$\cdot\ddot{N}\cdot$ $(1s^2 2s^2 2p^3)$	:N· → ← ·N:	:N::N:
Carbon dioxide (CO_2) (each O is bonded to the C)	$\cdot\dot{C}\cdot$ $(1s^2 2s^2 2p^2)$ $:\ddot{O}\cdot$ $(1s^2 2s^2 2p^4)$	:O C :O:	:O::C::O:
Formaldehyde (H_2CO) (each H and the O are bonded to the C)	$\cdot\dot{C}\cdot$ $(1s^2 2s^2 2p^2)$ $:\ddot{O}\cdot$ $(1s^2 2s^2 2p^4)$ $H\cdot$ $(1s^1)$	H· / C O	H·C::O:
Methane (CH_4) (each H is bonded to the C)	$\cdot\dot{C}\cdot$ $(1s^2 2s^2 2p^2)$ $H\cdot$ $(1s^1)$	H C H H	H:C:H H
Ammonia (NH_3) (each H is bonded to the N)	$\cdot\ddot{N}\cdot$ $(1s^2 2s^2 2p^3)$ $H\cdot$ $(1s^1)$	H N H H	H:N:H H
Water (H_2O) (each H is bonded to the O)	$:\ddot{O}\cdot$ $(1s^2 2s^2 2p^4)$ $H\cdot$ $(1s^1)$	:O H H	:O:H H
Oxygen diflouride (OF_2) (each F is bonded to the O)	$:\ddot{O}\cdot$ $(1s^2 2s^2 2p^4)$ $:\ddot{F}:$ $(1s^2 2s^2 2p^5)$	:O F: :F:	:O:F: :F:

4. Put one pair of electrons between each bonded pair of atoms in the initial structure drawn in Step 2. Subtract the number of electrons used in this step from the total number determined in Step 3. Use the remaining electrons to complete the octets of all atoms in the structure, beginning with the atoms that are present in greatest number in the molecule. Remember, hydrogen atoms require only one pair to achieve the electronic configuration of helium.

5. If all octets cannot be satisfied with the available electrons, move nonbonding pairs (those that are not between bonded atoms) to positions between bonded atoms to complete octets. This will create **double** or **triple bonds** between some atoms.

double and triple bonds The bonds resulting from the sharing of two and three pairs of electrons, respectively.

▶ Example 4.11

Draw Lewis structures for the following molecules:

a. NH_3 **b.** SO_3 **c.** C_2H_2

Solution

a. Step 1. The formula indicates that the molecule contains 1 nitrogen (N) atom and 3 hydrogen (H) atoms.

Step 2. The connecting pattern will have to be given to you in some form. Table 4.3 shows that each H atom is connected to the N atom; thus, we draw

$$H\ N\ H$$
$$H$$

Step 3. Nitrogen is in group VA(15) of the periodic table and so has 5 valence electrons; hydrogen is in group IA(1) and has 1 valence electron. The total is 8 (5 from one N atom and 3 from the three H atoms).

Step 4. We put one pair of electrons between each H atom and the N atom in the initial structure drawn in Step 2:

$$H:\overset{..}{N}:H$$
$$\overset{..}{H}$$

This required 6 of the 8 available electrons. The remaining pair is used to complete the octet of nitrogen. Remember, hydrogen achieves the noble gas configuration of helium with only 2 electrons:

$$H:\overset{..}{N}:H \qquad \text{or} \qquad H-\overset{..}{N}-H$$
$$\overset{}{H} \qquad\qquad\qquad\qquad |$$
$$\qquad\qquad\qquad\qquad\qquad H$$

Step 5. All octets are satisfied by Step 4, so nothing more needs to be done.

b. Step 1. The formula indicates that 1 sulfur (S) and 3 oxygen (O) atoms are present in a molecule.

Step 2. Each O atom is bonded only to the S atom; thus, we draw

$$O\ S\ O$$
$$O$$

Step 3. Sulfur and oxygen are both in group VIA(16), and so each atom has 6 valence electrons. The total is 24 (6 from the one S atom and 18 from the three O atoms).

Step 4. We put one pair of electrons between each O atom and the S atom:

$$O:\overset{..}{S}:O$$
$$\overset{..}{O}$$

This required 6 of the 24 available electrons. The remaining 18 are used to complete the octets, beginning with the O atoms:

$$:\overset{..}{O}:\overset{..}{S}:\overset{..}{O}:$$
$$:\overset{..}{O}:$$

Step 5. We see that the octet of sulfur is not completed, even though all available electrons have been used. If one nonbonding pair of any one of the three O atoms is moved to a bonding position between the oxygen and sulfur, it will help satisfy the octet of both atoms. The final Lewis structure is given below. Note that it contains a double bond between one of the oxygens and the sulfur:

$$:\overset{..}{O}:\overset{..}{S}::\overset{..}{O}: \qquad \text{or} \qquad :\overset{..}{O}-\overset{}{S}=\overset{..}{O}:$$
$$:\overset{..}{O}: \qquad\qquad\qquad\qquad |$$
$$\qquad\qquad\qquad\qquad\qquad :\overset{..}{O}:$$

If a nonbonding pair of electrons from either of the other two oxygens is used, the following structures result. These are just as acceptable as the preceding structures:

$$:\overset{..}{O}=\overset{}{S}-\overset{..}{O}: \qquad\qquad :\overset{..}{O}-\overset{}{S}-\overset{..}{O}:$$
$$\quad\quad |\qquad\qquad\qquad\qquad\qquad ||$$
$$\quad :\overset{..}{O}:\qquad\qquad\qquad\qquad\overset{..}{O}:$$

c. Step 1. The formula indicates 2 carbon (C) atoms and 2 hydrogen (H) atoms per C_2H_2 molecule.

Step 2. The C atoms are bonded to each other, and 1 H atom is bonded to each C atom.

$$H \, C \, C \, H$$

Step 3. Carbon is in group IVA(14) and has 4 valence electrons; hydrogen is in group IA(1) and has 1 valence electron. The total electrons available is 10 (8 from the two C atoms and 2 from the two H atoms).

Step 4. We put one pair of electrons between the two C atoms and between each C atom and H atom.

$$H : C : C : H$$

This required 6 of the available 10 electrons. The remaining 4 electrons are used to complete the C octets (remember, hydrogen needs only 2 electrons):

$$H : C : \overset{..}{\underset{..}{C}} : H$$

Step 5. All electrons are used in Step 4, but the octet of 1 C atom is still incomplete. It needs two more pairs. If both nonbonding pairs on the second C atom were shared with the first C atom, both carbons would have complete octets. The result is

$$H : C ::: C : H \qquad \text{or} \qquad H - C \equiv C - H$$

This molecule contains a triple bond (three shared pairs) between the C atoms.

▶ **LEARNING CHECK 4.11** Draw Lewis structures for the following molecules:

a. CH_4 (each H atom is bonded to the C atom)
b. H_2CO (the O atom and two H atoms are each bonded to the C atom)
c. HNO_3 (the O atoms are each bonded to the N atom, and the H atom is bonded to one of the O atoms)

4.7 Polyatomic Ions

Learning Objective

7. Draw correct Lewis structures for polyatomic ions.

polyatomic ions Covalently bonded groups of atoms that carry a net electrical charge.

An interesting combination of ionic and covalent bonding is found in compounds that contain **polyatomic ions.** These ions are covalently bonded groups of atoms that carry a net electrical charge. With the exception of the ammonium ion, NH_4^+, the common polyatomic ions are negatively charged.

Lewis structures can be drawn for polyatomic ions with just a slight modification in the steps given earlier for covalently bonded molecules. In Step 3, the total number of electrons available is obtained by first determining the total number of valence-shell electrons contained in the atoms of the ion. To this number are added electrons representing the number required to give the negative charge to the ion.

For example, in the SO_4^{2-} ion, the total number of valence-shell electrons is 30, with 6 coming from the sulfur atom and 6 from each of the four oxygen atoms. To this number we add 2 because two additional electrons are required to give the group of neutral atoms the charge of 2− found on the ion. This gives 32 total available electrons.

The only exception to this procedure for the common polyatomic ions is for the positively charged ammonium ion, NH_4^+. In this case, the number of electrons available is the total number of valence electrons minus 1, because one electron would have to be removed from the neutral group of atoms to produce the + charge on the ion.

▶ Example 4.12

Draw Lewis structures for the following polyatomic ions. The connecting patterns of the atoms to each other are indicated.

a. SO_4^{2-} (each O atom is bonded to the S atom)
b. NO_3^- (each O atom is bonded to the N atom)
c. $H_2PO_4^-$ (each O atom is bonded to the P atom, and each H atom is bonded to an O atom)

Solution

a. The only difference between drawing Lewis structures for molecules and polyatomic ions comes in Step 3, where the total number of valence-shell electrons is determined.

Step 1. The formula indicates the ion contains one sulfur (S) atom and four oxygen (O) atoms.

Step 2. The given bonding relationships lead to the following initial structure:

$$\begin{array}{c} O \\ O\ S\ O \\ O \end{array}$$

Step 3. Both the S and O atoms are in group VIA(16), so each atom contributes 6 valence electrons. The total number of valence-shell electrons from the atoms is therefore $5 \times 6 = 30$. However, the ion has a 2− charge. This charge comes from the presence of 2 electrons in the ion in addition to the electrons from the constituent atoms. These two additional electrons must be included in the total valence-shell electrons. Thus, the total number of electrons is $30 + 2 = 32$.

Step 4. The following is obtained when the 32 electrons are distributed among the atoms of the initial structure of Step 2:

$$\begin{array}{c} :\overset{..}{O}: \\ :\overset{..}{O}:\overset{..}{\underset{..}{S}}:\overset{..}{O}: \\ :\overset{..}{O}: \end{array}$$

Step 5. All octets are satisfied by Step 4. However, it is conventional to enclose Lewis structures of polyatomic ions in brackets and to indicate the net charge on the ion. Thus, the Lewis structures are

$$\left[\begin{array}{c} :\overset{..}{O}: \\ :\overset{..}{O}:\overset{..}{\underset{..}{S}}:\overset{..}{O}: \\ :\overset{..}{O}: \end{array}\right]^{2-} \quad \text{or} \quad \left[\begin{array}{c} O \\ | \\ O-S-O \\ | \\ O \end{array}\right]^{2-}$$

where the second structure does not show nonbonding (unshared) electron pairs, and the bonding pairs are represented by lines.

b. Step 1. Three oxygen (O) atoms and one nitrogen (N) atom are found in the ion.
Step 2.

$$\begin{array}{c} O \\ O\ N\ O \end{array}$$

Step 3. Electrons from oxygen $= 3 \times 6 = 18$.
Electrons from nitrogen $= 1 \times 5 = 5$.
One electron comes from the ionic charge $= 1$.
Total electrons $= 18 + 5 + 1 = 24$.

Step 4.

$$\begin{array}{c} :\overset{..}{O}: \\ :\overset{..}{O}:\overset{..}{N}:\overset{..}{O}: \end{array}$$

Step 5. The octet of nitrogen is not satisfied by Step 4, so an unshared pair of electrons on an O atom must be shared with the N atom. This will form a double bond in the ion. The resulting Lewis structures are

$$\left[\begin{array}{c} :\ddot{O}: \\ :\ddot{O}:\ddot{N}::\ddot{O}: \end{array}\right]^{-} \quad \text{or} \quad \left[\begin{array}{c} O \\ | \\ O-N=O \end{array}\right]^{-}$$

c. Step 1. Two hydrogen (H) atoms, one phosphorus (P) atom, and four oxygen (O) atoms are found in the ion.

Step 2.

$$\begin{array}{c} O \\ O \; P \; O \; H \\ O \\ H \end{array}$$

Step 3. Electrons from hydrogen = $2 \times 1 = 2$.
Electrons from oxygen = $4 \times 6 = 24$.
Electrons from phosphorus = $1 \times 5 = 5$.
Electron from ionic charge = 1.
Total electrons = $2 + 24 + 5 + 1 = 32$.

Step 4.

Step 5. All octets are satisfied by Step 4 (remember, H atoms are satisfied by a single pair of electrons). The resulting Lewis structures are

$$\left[\begin{array}{c} :\ddot{O}: \\ :\ddot{O}:\ddot{P}:\ddot{O}:H \\ :\ddot{O}: \\ H \end{array}\right]^{-} \quad \text{or} \quad \left[\begin{array}{c} O \\ | \\ O-P-O-H \\ | \\ O \\ | \\ H \end{array}\right]^{-}$$

▶ **LEARNING CHECK 4.12** Draw Lewis structures for the following polyatomic ions:

a. PO_4^{3-} (each O atom is bonded to the P atom)
b. SO_3^{2-} (each O atom is bonded to the S atom)
c. NH_4^+ (each H atom is bonded to the N atom)

4.8 Shapes of Molecules and Polyatomic Ions

Learning Objective

8. Use VSEPR theory to predict the shapes of molecules and polyatomic ions.

Most molecules and polyatomic ions do not have flat, two-dimensional shapes like those implied by the molecular Lewis structures of Table 4.3. In fact, the atoms of most molecules and polyatomic ions form distinct three-dimensional shapes. Being able to predict the shape is important because the shape contributes to the properties of the molecule or ion. This can be done quite readily for molecules composed of representative elements.

To predict molecular or ionic shapes, first draw Lewis structures for the molecules or ions using the methods discussed in Sections 4.6 and 4.7. Once you have drawn the Lewis structure for a molecule or ion, you can predict its shape by applying a simple theory to the structure. The theory is called the *valence-shell electron-pair repulsion theory,* or **VSEPR theory** (sometimes pronounced "vesper" theory). According to VSEPR theory, electron pairs in the valence shell of an atom are repelled by other electron pairs and get as far away from one another as possible. Any atom in a molecule or ion that is bonded to two or more other atoms is called a *central atom.* When the VSEPR theory is applied to the valence-shell electrons of central atoms, the shape of the molecule or ion containing the atoms can be predicted. Two rules are followed:

VSEPR theory A theory based on the mutual repulsion of electron pairs. It is used to predict molecular shapes.

1. All valence-shell electron pairs around the central atom are counted equally, regardless of whether they are bonding or nonbonding pairs.
2. Double or triple bonds between atoms are treated like a single pair of electrons when predicting shapes.

The electron pairs around a central atom will become oriented in space to get as far away from one another as possible. Thus, two pairs will be oriented with one pair on each opposite side of the central atom. Three pairs will form a triangle around the central atom, and four pairs will be located at the corners of a regular tetrahedron with the central atom in the center (see ❱ Figure 4.5 and ❱ Figure 4.6). The VSEPR theory can be used to predict shapes of molecules and ions with five or more pairs on the central atom, but we will not go beyond four pairs in this book.

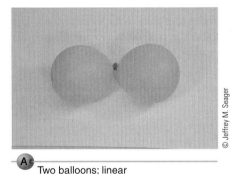

Two electron pairs   Three electron pairs   Four electron pairs

Figure 4.5 Arrangements of electron pairs around a central atom (E).

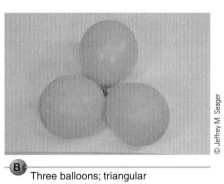

A⏺ Two balloons; linear   B⏺ Three balloons; triangular   C⏺ Four balloons; tetrahedral

Figure 4.6 When balloons of the same size and shape are tied together, they will assume positions like those taken by pairs of valence electrons around a central atom. Where would the central atom be located in these balloon models?

❱ Example 4.13

Draw Lewis structures for the following molecules, apply the VSEPR theory, and predict the shape of each molecule:

a. CO_2 (each O atom is bonded to the C atom)
b. C_2Cl_4 (the C atoms are bonded together, and two Cl atoms are bonded to each C atom)
c. NH_3 (each H atom is bonded to the N atom)
d. H_2O (each H atom is bonded to the O atom)

Solution

a. The Lewis structure is $:\ddot{O}=C=\ddot{O}:$. The central atom is carbon (C) and it has two pairs of electrons in the valence shell surrounding it (remember, each double bond is treated like a single pair of electrons in the VSEPR theory). The two pairs will be located on opposite sides of the C atom, so the molecule has the shape drawn above with the O, C, and O atoms in a line; it is a linear molecule.

b. The Lewis structure is

Both C atoms are central atoms in this molecule, and each one has three electron pairs in the valence shell (again, the double bond is treated as only one pair). The three pairs around each C atom will be arranged in the shape of a triangle:

Thus, the molecule has the shape of two flat triangles in the same plane connected together at one point:

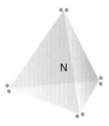

c. The Lewis structure is

$$H - \ddot{N} - H$$
$$|$$
$$H$$

The central atom is nitrogen (N), and it has four electron pairs in the valence shell. The four pairs will be located at the corners of a tetrahedron with the N atom in the middle:

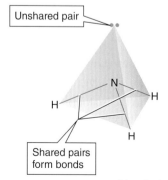

The shape of the molecule is determined only by the positions of the atoms, not by the position of the unshared pair of electrons. Thus, the NH_3 molecule has the shape of a pyramid with a triangular base. The N atom is at the peak of the pyramid, and an H atom is at each corner of the base:

Unshared pair

Shared pairs form bonds

d. The Lewis structure is H $\overset{..}{:}\overset{}{O}\overset{..}{:}$ H. The central atom is oxygen (O), and it is seen to have four pairs of electrons in its valence shell. The four pairs will be located at the corners of a tetrahedron with the O atom in the middle:

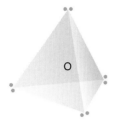

Once again, the shape of the molecule is determined by the locations of the atoms. Thus, H_2O is seen as having a bent or angular shape:

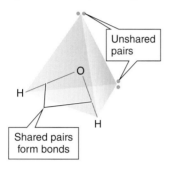

▶ **LEARNING CHECK 4.13** Predict the shapes of the following molecules by applying the VSEPR theory:

a. BF_3 (the Lewis structure is

$$
\begin{array}{c}
:\overset{..}{F}: \\
\overset{..}{B}:\overset{..}{F}: \\
:\overset{..}{F}:
\end{array}
$$

for this molecule, which does not obey the octet rule).

b. CCl_4 (each Cl atom is bonded to the C atom)

c. SO_2 (each O atom is bonded to the S atom)

d. CS_2 (each S atom is bonded to the C atom)

▶ Example 4.14

Use the VSEPR theory to predict the shapes of the following polyatomic ions. The Lewis structures were drawn in Example 4.12.

a. SO_4^{2-}

b. NO_3^{-}

Solution

a. The Lewis structure from Example 4.12 is

$$
\left[
\begin{array}{c}
:\overset{..}{O}: \\
:\overset{..}{O}:\overset{..}{S}:\overset{..}{O}: \\
:\overset{..}{O}:
\end{array}
\right]^{2-}
$$

We see that sulfur is the central atom, and it has four electron pairs around it. The four pairs will be located at the corners of a tetrahedron with the S in the middle.

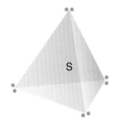

The shape of the ion is determined by the location of the oxygen atoms. Thus, the ion has a tetrahedral shape:

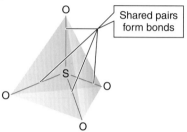

b. The Lewis structure from Example 4.12 is

$$\left[\begin{array}{c} :\ddot{O}: \\ :\ddot{O}:N::\ddot{O}: \end{array}\right]^{-}$$

We see that nitrogen is the central atom, and it has three electron pairs around it. Remember, the double bond formed by the two shared electron pairs between the nitrogen and one of the oxygens counts as only one pair in the VSEPR theory. The three pairs will form a triangle around the N.

An oxygen is located at each corner of the triangle, so the ion has a flat triangular shape:

▶ **LEARNING CHECK 4.14** Use the VSEPR theory to predict the shapes of the following polyatomic ions. The Lewis structures were drawn in Learning Check 4.12.

a. PO_4^{3-} **b.** SO_3^{2-} **c.** NH_4^{+}

4.9 The Polarity of Covalent Molecules

Learning Objective _____

9. Use electronegativities to classify covalent bonds of molecules, and determine whether covalent molecules are polar or nonpolar.

_____

Molecules of chlorine (Cl—Cl) and hydrogen chloride (H—Cl) have a number of similarities. For example, they contain two atoms each and are therefore diatomic

Versatile Zinc Oxide

While the jury is still out on the usefulness of zinc compounds in treating the symptoms of the common cold, the usefulness of zinc oxide as a topical astringent, antiseptic, and skin protectant is well established. Zinc oxide, with the simple formula ZnO, is a coarse white or grayish powder that is insoluble in water or in alcohol. Because of this insolubility, it is often blended with other ingredients to produce a suspension of the solid in such forms as lotions, creams, or salves. The solid compound in a finely divided form is sometimes added to products such as baby powders.

In these forms, zinc oxide is used topically on the skin as a sunscreen to block harmful ultraviolet rays, and as a shield from various chemical irritants. In addition, products containing the compound are used to relieve a number of common skin conditions, including diaper rash; poison ivy, oak, and sumac rashes; eczema; external hemorrhoids; and insect bites. In these applications its antiseptic property helps prevent infections, and it is thought to promote healing by attracting protein to the affected areas of the skin, thereby encouraging new tissue growth. In these topical applications, the insolubility and consistency of zinc oxide prevent it from being absorbed into the skin. As a result, side effects or allergic reactions to the compound are rare, even with long-term use.

In addition to its valuable medical uses, this versatile compound is also found in a wide variety of commercial applications. It functions as a pigment and reinforcing agent in some rubber products; as a pigment and mold-growth inhibitor in paints; as a pigment in ceramics, floor tile, and glass; as a thickener in cosmetics; as a feed additive for cattle; and as a dietary supplement for humans.

Zinc oxide in the form of an ointment is convenient and easy to use.

© Michael C. Slabaugh

molecules, and the atoms are held together by a single covalent bond. Some differences also exist, the most obvious being the homoatomic nature of the chlorine molecule (an element) and the heteroatomic nature of the hydrogen chloride molecule (a compound). This difference influences the distribution of the shared electrons in the two molecules.

An electron pair shared by two identical atoms is attracted equally to each of the atoms, and the probability of finding the electrons close to an atom is equal for the two atoms. Thus, on average, the electrons spend exactly the same amount of time associated with each atom. Covalent bonds of this type are called **nonpolar covalent bonds.**

Different atoms generally have different tendencies to attract the shared electrons of a covalent bond. A measurement of this tendency is called **electronegativity.** The electronegativity of the elements is a property that follows trends in the periodic table. It increases from left to right across a period and decreases from top to bottom of a group. These trends are shown in ▶ Table 4.4, which contains electronegativity values for the common representative elements. As a result of electronegativity differences, the bonding electrons shared by two different atoms are shared unequally. The electrons spend more of their time near the atom with the higher electronegativity. The resulting shift in average location of bonding electrons is called **bond polarization** and produces a **polar covalent bond.**

As a result of the unequal electron sharing, the more electronegative atom acquires a partial negative charge ($\delta-$), and the less electronegative atom has a partial positive charge ($\delta+$). The molecule as a whole has no net charge, just an uneven charge distribution.

nonpolar covalent bond A covalent bond in which the bonding pair of electrons is shared equally by the bonded atoms.

electronegativity The tendency of an atom to attract shared electrons of a covalent bond.

bond polarization A result of shared electrons being attracted to the more electronegative atom of a bonded pair of atoms.

polar covalent bond A covalent bond that shows bond polarization; that is, the bonding electrons are shared unequally.

TABLE 4.4 Electronegativities for the Common Representative Elements

Increasing electronegativity →

		H				
		2.1				
Li	Be		B	C	N	O	F
1.0	1.5		2.0	2.5	3.0	3.5	4.0
Na	Mg		Al	Si	P	S	Cl
0.9	1.2		1.5	1.8	2.1	2.5	3.0
K	Ca		Ga	Ge	As	Se	Br
0.8	1.0		1.6	1.8	2.0	2.4	2.8
Rb	Sr		In	Sn	Sb	Te	I
0.8	1.0		1.7	1.8	1.9	2.1	2.5
Cs	Ba					
0.7	0.9					

Decreasing electronegativity ↓

▶ Example 4.15

Use only the periodic table to determine the following for the diatomic covalent molecules listed below: (1) the more electronegative element, (2) the direction of bond polarization, and (3) the charge distribution resulting from the polarization.

a. I—Cl   **b.** Br—Br   **c.** C≡O

Solution

a. Both chlorine (Cl) and iodine (I) belong to group VIIA(17); chlorine is higher in the group and therefore is the more electronegative. The bond polarization (shift in average bonding-electron location) will be toward chlorine, as indicated by the arrow shown below the bond. The result is a partial negative charge on Cl and a partial positive charge on I:

$$\overset{\delta+}{I}\underset{\rightarrow}{-}\overset{\delta-}{Cl}$$

b. Since both bromine (Br) atoms have the same electronegativity, no bond polarization or unequal charge distribution results. The molecule is nonpolar covalent and is represented as Br—Br.

c. Oxygen (O) and carbon (C) both belong to the second period. Oxygen, being located farther to the right in the period, is the more electronegative. Thus, the molecule is represented as

$$\overset{\delta+}{C}\underset{\rightarrow}{\equiv}\overset{\delta-}{O}$$

▶ LEARNING CHECK 4.15 Use only the periodic table and show the direction of any bond polarization and resulting charge distribution in the following molecules:

a. N≡N   **b.** I—Br   **c.** H—Br

The extent of bond polarization depends on electronegativity differences between the bonded atoms and forms one basis for a classification of bonds. When the electronegativity difference (ΔEN) between the bonded atoms is 0.0, the bond is classified as nonpolar covalent (the electrons are shared equally). When ΔEN is 2.1 or greater, the bond is classified as ionic (the electrons are transferred). When ΔEN is between 0.0 and 2.1, the bond is classified as polar covalent (the electrons are unequally shared, and the bond is polarized). Note, however, that the change from purely covalent to completely ionic compounds is gradual and continuous as ΔEN values change. Thus, a bond between hydrogen (H) and boron (B) ($\Delta EN = 0.1$) is classified as polar, but the electron sharing is only slightly unequal and the bond is only slightly polar (see ❱ Figure 4.7). As the values in Table 4.4 indicate, compounds formed by a reaction of a metal with a nonmetal are generally ionic, while those between nonmetals are either nonpolar or polar covalent.

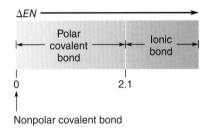

Figure 4.7 The darker the shading, the larger ΔEN is for the bonded atoms. Larger ΔEN values correspond to greater inequality in the sharing of the bonding electrons.

❱ Example 4.16

Using Table 4.4, classify the bonds in the following compounds as nonpolar covalent, ionic, or polar covalent:

a. ClF   **b.** MgO   **c.** PI_3

Solution

a. The electronegativities of chlorine (Cl) and fluorine (F) are 3.0 and 4.0, respectively. Obtain the ΔEN value by subtracting the smaller from the larger, regardless of the order of the elements in the formula. Thus,

$$\Delta EN = 4.0 - 3.0 = 1.0$$

and the bond is polar covalent.

b. Similarly,

$$\Delta EN = 3.5 - 1.2 = 2.3$$

and the bond is ionic.

c. The ΔEN is determined for each phosphorus–iodine bond:

$$\Delta EN = 2.5 - 2.1 = 0.4$$

Thus, each bond is classified as polar covalent.

❱ **LEARNING CHECK 4.16** Using Table 4.4, classify the bonds in the following compounds as nonpolar covalent, ionic, or polar covalent:

a. KF   **b.** NO   **c.** AlN   **d.** K_2O

You can now predict the polarity of bonds within molecules, but it is also important to be able to predict the polar nature of the molecules themselves. The terms *polar* and *nonpolar,* when used to describe molecules, indicate their electrical charge symmetry (see ❱ Figure 4.8). In **polar molecules** (often called dipoles), the charge distribution resulting from bond polarizations is nonsymmetric. In **nonpolar molecules,** any charges caused by bond polarization are symmetrically distributed through the molecule. ❱ Table 4.5 illustrates polar and nonpolar molecules. The molecular shapes were determined by using the VSEPR theory.

polar molecule A molecule that contains polarized bonds and in which the resulting charges are distributed nonsymmetrically throughout the molecule.

nonpolar molecule A molecule that contains no polarized bonds, or a molecule containing polarized bonds in which the resulting charges are distributed symmetrically throughout the molecule.

Figure 4.8 Polar and nonpolar molecules are affected differently by electrical charges. The ΔEN between C and Cl is 0.5. Propose a reason why CCl_4 molecules are nonpolar.

A stream of polar water molecules is attracted by a charged balloon.

B A stream of nonpolar carbon tetrachloride (CCl_4) molecules is not affected by a charged balloon.

TABLE 4.5 The Polarity of Molecules

Molecular formula	Geometric structure	Bond polarization and molecular charge distribution	Geometry of charge distribution	Classification of charge distribution	Classification of molecule
H_2	H — H	H — H	No charges	No charges	Nonpolar
HCl	H — Cl	$\overset{\delta+}{H} \Rightarrow \overset{\delta-}{Cl}$	+ −	Nonsymmetric	Polar
CO_2	O = C = O	$\overset{\delta-}{O} \rightleftharpoons \overset{\delta2+}{C} \equiv \overset{\delta-}{O}$	− 2+ −	Symmetric	Nonpolar
BF_3	F | B F F	$\overset{\delta-}{F}$ ↿ $\overset{\delta3+}{B}$ $\overset{\delta-}{F}$ $\overset{\delta-}{F}$	− 3+ − −	Symmetric	Nonpolar
N_2O	N ≡ N — O	$N \equiv \overset{\delta+}{N} \Rightarrow \overset{\delta-}{O}$	+ −	Nonsymmetric	Polar
H_2O	H — O \ H	$\overset{\delta+}{H} \Rightarrow \overset{\delta2-}{O} \nwarrow \overset{\delta+}{H}$	+ 2− +	Nonsymmetric	Polar

4.10 More about Naming Compounds

Learning Objective

10. Write correct formulas for ionic compounds containing representative metals and polyatomic ions, and correctly name binary covalent compounds and compounds containing polyatomic ions.

In Section 4.4, we discussed the rules for naming binary ionic compounds. In this section, we expand the rules to include the naming of binary covalent compounds and ionic compounds that contain polyatomic ions.

It is apparent from Table 4.4 that covalent bonds (including polar covalent) form most often between representative elements classified as nonmetals. It is not possible to predict formulas for these compounds in a simple way, as was done for ionic compounds in Section 4.4. However, simple rules do exist for naming binary covalent compounds on the basis of their formulas. The rules are similar to those used to name binary ionic compounds: (1) Give the name of the less electronegative element first (the element given first in the formula), (2) give the stem of the name of the more electronegative element next and add the suffix -ide, and (3) indicate the number of each type of atom in the molecule by means of the Greek prefixes listed in ▶ Table 4.6.

TABLE 4.6 Greek Numerical Prefixes

Number	Prefix
1	*mono-*
2	*di-*
3	*tri-*
4	*tetra-*
5	*penta-*
6	*hexa-*
7	*hepta-*
8	*octa-*
9	*nona-*
10	*deca-*

▶ Example 4.17

Name the following binary covalent compounds:

a. CO_2 **b.** CO **c.** NO_2 **d.** N_2O_5 **e.** CS_2

Solution

a. The elements are carbon (C) and oxygen (O), and carbon is the less electronegative. Because the stem of oxygen is *ox-*, the two portions of the name will be carbon and oxide. Only one C atom is found in the molecule, but the prefix *mono-* is not used when it appears at the beginning of a name. The two O atoms in the molecule are indicated by the prefix *di-*. The name therefore is carbon dioxide.
b. Similarly, we arrive at the name carbon monoxide for CO.
c. NO_2 is assigned the name nitrogen dioxide.
d. N_2O_5 is assigned the name dinitrogen pentoxide. Note: The a is dropped from the *penta-* prefix for oxygen. This is done to avoid the pronunciation problem created by having two vowels next to each other in a name (dinitrogen pentaoxide).
e. CS_2 is named carbon disulfide.

▶ LEARNING CHECK 4.17   Name the following binary covalent compounds:

a. SO_3 b. BF_3 c. S_2O_7 d. CCl_4

The formulas and names of some common polyatomic ions are given in ▶ Table 4.7. From this information, the formulas and names for compounds containing polyatomic ions can be written. The rules are essentially the same as those used earlier for binary ionic compounds. In the formulas, the metal (or ammonium ion) is written first, the positive and

TABLE 4.7   Some Common Polyatomic Ions

Very Common		Common	
NH_4^+	ammonium	CrO_4^{2-}	chromate
$C_2H_3O_2^-$	acetate	$Cr_2O_7^{2-}$	dichromate
CO_3^{2-}	carbonate	NO_2^-	nitrite
ClO_3^-	chlorate	MnO_4^-	permanganate
CN^-	cyanide	SO_3^{2-}	sulfite
HCO_3^-	hydrogen carbonate (bicarbonate)	ClO^-	hypochlorite
OH^-	hydroxide	HPO_4^{2-}	hydrogen phosphate
NO_3^-	nitrate	$H_2PO_4^-$	dihydrogen phosphate
PO_4^{3-}	phosphate	HSO_4^-	hydrogen sulfate (bisulfate)
SO_4^{2-}	sulfate	HSO_3^-	hydrogen sulfite (bisulfite)

When concepts are closely related, it is very easy to become confused. In this chapter, the terms *polar* and *nonpolar* are used to describe both bonds and molecules. What is the correct way to apply the terms?

It may be helpful to remember the numbers 3 and 2: three types of bonds, but only two types of molecules. The bonds are nonpolar covalent ($\Delta EN = 0$), polar covalent ($\Delta EN = 0.1$–2.0), and ionic ($\Delta EN = 2.1$ or greater). Remember that compounds bonded by ionic bonds do not form molecules, so there are only two types of molecules: nonpolar and polar.

The polarity of a molecule depends on two factors: (1) the polarity of the bonds between the atoms of the molecule, and (2) the geometric arrangement of the atoms in space. First, determine whether any of the bonds between atoms are polar. If only nonpolar bonds are present, the molecule will be nonpolar regardless of the geometric arrangement of the atoms. However, if one or more polar bonds are found, the arrangement of the atoms must be considered as a second step.

For diatomic (two-atom) molecules, a polar bond between the atoms results in a polar molecule:

A—A $\Delta EN = 0$ because atoms are identical; nonpolar
A—B If $\Delta EN = 0$; nonpolar
A—B If $\Delta EN = 0.1$–2.0; polar

Thus, we see that for diatomic molecules, the type of bond between atoms and the resulting type of molecule is the same.

For a molecule that contains polar bonds and three or more atoms, the molecular geometry must be known and taken into account before the polar nature of the molecule can be determined. Remember, the unequal sharing of electrons between two atoms bonded by a polar bond will cause one of the bonded atoms to have a partial positive charge and one a partial negative charge. If these atoms with their charges are symmetrically arranged in space to form the molecule, the molecule will be nonpolar. Thus, each of the following planar (flat) molecules will be nonpolar even if the bonds between the atoms are polar:

On the other hand, each of the following molecules will be polar if they contain polar bonds because the resulting charges on the atoms are not distributed symmetrically in space:

negative charges must add up to zero, and parentheses are used around the polyatomic ions if more than one is used. In names, the positive metal (or ammonium) ion is given first, followed by the name of the negative polyatomic ion (see ▶ Figure 4.9). No numerical

Figure 4.9 Examples of compounds that contain polyatomic ions. The samples are shown as solids and as solids dissolved in water. Referring to Table 4.7, write formulas for these compounds (clockwise from top): potassium carbonate, potassium chromate, potassium phosphate, and potassium permanganate.

© Jeffrey M. Seager

Unless otherwise noted all art on this page © Cengage Learning 2014

prefixes are used except where they are a part of the polyatomic ion name. Names and formulas of acids (compounds in which hydrogen is bound to polyatomic ions) will be given in Chapter 9.

▶ **Example 4.18**

Write formulas and names for compounds composed of ions of the following metals and the polyatomic ions indicated:

a. Na and NO_3^- **c.** K and HPO_4^{2-}
b. Ca and ClO_3^- **d.** NH_4^+ and NO_3^-

Solution

a. Sodium (Na) is a group IA(1) metal and forms Na^+ ions. Electrical neutrality requires a combining ratio of one Na^+ for one NO_3^-. The formula is $NaNO_3$. The name is given by the metal name plus the polyatomic ion name: sodium nitrate.
b. Calcium (Ca), a group IIA(2) metal, forms Ca^{2+} ions. Electrical neutrality therefore requires a combining ratio of one Ca^{2+} for two ClO_3^- ions. The formula is $Ca(ClO_3)_2$. (NOTE: The use of parentheses around the polyatomic ion prevents the confusion resulting from writing $CaClO_{32}$, which implies that there are 32 oxygen atoms in the formula. Parentheses are always used when multiples of a specific polyatomic ion are used in a formula and are indicated by a subscript.) The name is calcium chlorate.
c. Potassium (K), a group IA(1) metal, forms K^+ ions. The required combining ratio of 2:1 gives the formula K_2HPO_4. The name is potassium hydrogen phosphate.
d. The NH_4^+ is not a metallic ion, but it behaves like one in numerous compounds. The 1:1 combining ratio gives the formula NH_4NO_3. (NOTE: The polyatomic ions are written separately, and the nitrogen atoms are not grouped to give a formula such as $N_2H_4O_3$.) The name is ammonium nitrate.

▶ **LEARNING CHECK 4.18** Write formulas and names for compounds containing ions of the following metals and the polyatomic ions indicated:

a. Ca and HPO_4^{2-} **c.** K and MnO_4^-
b. Mg and PO_4^{3-} **d.** NH_4^+ and $Cr_2O_7^{2-}$

4.11 Other Interparticle Forces

Learning Objective

11. Relate melting and boiling points of pure substances to the strength and type of interparticle forces present in the substances.

Ionic and covalent bonding can account for certain properties of many substances. However, some experimental observations can be explained only by proposing the existence of other types of forces between particles.

Earlier in this chapter, a crystal lattice was described in connection with ionic bonding (see Figure 4.3). Most pure substances (elements or compounds) in the solid state also exist in the form of a crystal lattice. However, in some of these solids, neutral atoms or molecules occupy the lattice sites instead of ions. When solids are melted, the forces holding the lattice particles in place are overcome, and the particles move about more freely in what is called the liquid state. The addition of more heat overcomes the attractive interparticle forces to a still greater extent, and the liquid is converted into a gas or vapor; the liquid boils. Particles in the vapor state move about very freely and are influenced only slightly by interparticle attractions. Thus, the temperatures at which melting and boiling take place give an indication of the strength of the interparticle forces that are being overcome. (These states of matter—solid, liquid, and gas—are discussed in more detail in Chapter 6.)

Suppose an experiment is carried out with several pure substances. Some are familiar to you, but others you probably have never seen in the solid state. The substances are listed in ▶ Table 4.8, together with some pertinent information.

TABLE 4.8 Some Characteristics of Selected Pure Substances

Substance	Formula or Symbol	Classification	Particles Occupying Lattice Sites
Sodium chloride	NaCl	Compound	Na$^+$ and Cl$^-$ ions
Water	H$_2$O	Compound	H$_2$O molecules
Carbon monoxide	CO	Compound	CO molecules
Quartz (pure sand)	SiO$_2$	Compound	Si and O atoms
Copper metal	Cu	Element	Cu atoms
Oxygen	O$_2$	Element	O$_2$ molecules

The experiment is started at the very low temperature of –220°C to have all substances in the solid state. The temperature is then increased slowly and uniformly to 2600°C, at which point all the substances will be in the gaseous state. See ▶ Table 4.9, which gives only those temperatures corresponding to a specific change in one of the substances.

TABLE 4.9 The Behavior of Selected Pure Substances in Response to Heating

Temperature (°C)	Behavior or State of Substance					
	Oxygen (O$_2$)	Carbon Monoxide (CO)	Water (H$_2$O)	Salt (NaCl)	Copper (Cu)	Quartz (SiO$_2$)
−220	Solid	Solid	Solid	Solid	Solid	Solid
−218	**Melts**	Solid	Solid	Solid	Solid	Solid
−199	Liquid	**Melts**	Solid	Solid	Solid	Solid
−192	Liquid	**Boils**	Solid	Solid	Solid	Solid
−183	**Boils**	Gas	Solid	Solid	Solid	Solid
0	Gas	Gas	**Melts**	Solid	Solid	Solid
100	Gas	Gas	**Boils**	Solid	Solid	Solid
801	Gas	Gas	Gas	**Melts**	Solid	Solid
1083	Gas	Gas	Gas	Liquid	**Melts**	Solid
1413	Gas	Gas	Gas	**Boils**	Liquid	Solid
1610	Gas	Gas	Gas	Gas	Liquid	**Melts**
2230	Gas	Gas	Gas	Gas	Liquid	**Boils**
2595	Gas	Gas	Gas	Gas	**Boils**	Gas
2600	Gas	Gas	Gas	Gas	Gas	Gas

The melting points of the substances used in this experiment show that the weakest interparticle forces are found in solid oxygen. The forces then increase in the order CO, H$_2$O, NaCl, Cu, and SiO$_2$. With two exceptions, the boiling points follow the same order. How do these melting and boiling points relate to the lattice particles of these substances?

Rare Earths Are All Around You

The label "rare earth" is almost always used when referring to the metals yttrium, lanthanum, and the metallic elements found in the first row of the inner transition elements of the periodic table (elements number 58 through 71). The discovery of these metals began in the late eighteenth century. In nature, they are found in the form of metal oxides, or "earths." The word "rare" in the name would seem to indicate that these metals occur naturally in very small amounts. In fact, the rarest of these rare earths is nearly 200 times more abundant in nature than gold! The name actually refers to the rarity of natural deposits large enough and concentrated enough to be mined profitably.

Most people would probably find it difficult to name even one product or use of rare earth elements that significantly influence their lives. However, the list of such products is extensive. One significant example is permanent magnets. The inclusion of rare earth metals in their formulation makes it possible to manufacture tiny magnets that are much more powerful than conventional magnets of similar size. The availability of these small but powerful magnets makes possible the production of familiar electronic devices such as smart phones and MP3 players. Very large rare earth magnets also prove to be important in such applications as massive electricity-producing wind turbines. Some large turbines are made using magnets containing 500 pounds or more of the rare earth neodymium. The use of rare earth magnets has also made possible the availability of small electrically powered portable tools such as drills, grinders, and saws used in the construction trades.

Hybrid automobiles could not function well without the use of rare earth metals. Lanthanum is used in their very efficient energy-storing batteries, and powerful neodymium-containing magnets are used in their electric motors.

Military uses of rare earth metals are numerous. Night-vision goggles contain lanthanum, gadolinium, and yttrium. Heat-resistant magnets that contain samarium are found in the control systems of drone aircraft and some cruise missiles.

Rare earth metals are also found in numerous commonly used everyday items. Some sunglasses are tinted by the addition of praseodymium, erbium, and neodymium to the glass mixture. The presence of the rare earths in the glass provides color to reduce glare from visible light, and it also protects the

eyes from invisible but potentially damaging ultraviolet radiation. Fluorescent light bulbs contain small amounts of yttrium and terbium. The red and green screen colors of the liquid crystal displays (LCDs) used on many small visual devices such as smart phones and DVD players are produced by europium and terbium.

In recent years, the characteristics of the rare earth market place have become a concern. The United States currently has 13 percent of the world's rare earth reserves and was the largest supplier of rare earth metals from about 1965 until 1985, when China, which owns 48 percent of the world reserves, entered the world market. China now supplies 97 percent of the rare earth needs worldwide. In 2010, China shook up global rare earth markets when it cut off rare earth shipments to Japan for a month because of a diplomatic dispute. The power of one nation to essentially control rare earth availability to all other nations is seen by many as a serious issue.

In addition to the supply issue, the steady increase in the use of rare earth elements in the manufacturing of high-tech goods has sent prices soaring. Dysprosium, which sold for $6.77 a pound in 2002, sold for $212.00 a pound in 2011. During just two months in the summer of 2010, the price of cerium increased by 450%. Some experts predicted that world demand would probably exceed supply in the near future. With such serious supply, demand, and pricing issues before us, the rare earths might, indeed, live up to their name and become rare.

A young woman using an iPad.

The lattice particles in solid silicon dioxide are individual atoms of silicon and oxygen. They are held together in the lattice by covalent bonds. Solids of this type are called **network solids,** and when such solids are melted or vaporized, strong covalent bonds must be broken.

The individual atoms that are the lattice particles of copper metal are held together in the lattice by what is called a **metallic bond.** As described in Section 4.2, the atoms of metals lose valence-shell electrons readily. Imagine a large number of metal atoms occupying lattice sites. Now imagine that the valence-shell electrons of each of the atoms move readily throughout the lattice. The attraction of the positive kernels (the metal atom nuclei

network solid A solid in which the lattice sites are occupied by atoms that are covalently bonded to each other.

metallic bond An attractive force responsible for holding solid metals together. It originates from the attraction between positively charged atomic kernels that occupy the lattice sites and mobile electrons that move freely through the lattice.

plus low-level electrons) to the mobile electrons, and hence to one another, constitutes a metallic bond. The mobile electrons of the metallic bond are responsible for a number of the observed properties of metals, including high thermal conductivity, high electrical conductivity, and the characteristic metallic luster.

The bonds broken when sodium chloride (NaCl) melts or boils are ionic bonds resulting from attractions between positive (Na^+) and negative (Cl^-) ions, the lattice particles in Figure 4.3. As shown by Table 4.9, these bonds are generally quite strong.

The lattice particles of solid carbon monoxide (CO) are the nonsymmetric CO molecules. These molecules are polar because of the nonsymmetric distribution of charges between the carbon and oxygen atoms. They are held in the solid lattice by **dipolar forces** resulting from the attraction of the positive end of one polar molecule to the negative end of another polar molecule. These forces are usually weak, and melting and vaporization take place at very low temperatures. Such substances are usually thought of as gases because that is their normal state at room temperature.

Water molecules, the lattice particles of ice, are also held in place by dipolar forces. However, these forces are stronger than those of the dipoles for CO. In water molecules, the hydrogens carry a partial positive charge, and the oxygen has a partial negative charge (see Table 4.5). Thus, the hydrogens of one molecule are attracted to the oxygens of other molecules. This attraction, called **hydrogen bonding,** is stronger than most other dipolar attractions because of the small size of the hydrogen atom and the high electronegativity of oxygen (see ▶ Figure 4.10). Hydrogen bonding occurs in gases, liquids, and solids composed of polar molecules in which hydrogen atoms are covalently bonded to highly electronegative elements (generally O, N, or F).

dipolar force The attractive force that exists between the positive end of one polar molecule and the negative end of another.

hydrogen bonding The result of attractive dipolar forces between molecules in which hydrogen atoms are covalently bonded to very electronegative elements (O, N, or F).

A Polar water molecule

B Hydrogen bonding in liquid water

C Hydrogen bonding in solid water

Figure 4.10 Hydrogen bonding in water. Hydrogen bonds are dotted, and covalent bonds are solid.

Water is the most well-known substance in which hydrogen bonding dramatically influences the properties. These properties make water useful in many processes, including those characteristic of living organisms. Because of this widespread use, water is not often thought of as being a peculiar substance. However, it is often the peculiarities that make it so useful.

As we saw in Example 4.13d, water molecules are angular. Because of this shape and the large difference in electronegativity between hydrogen and oxygen, water molecules are polar, with partial positive charges on the hydrogens and a partial negative charge on the oxygen, as shown in Figure 4.10a. The attractions between these oppositely charged parts of the molecules result in hydrogen bonding. This is represented for the liquid and solid states of water in Figures 4.10b and c.

Let's look at some of the peculiar properties of water. As shown in Table 4.9, water has a normal boiling point of 100°C. This is much higher than would be predicted from the measured boiling points of compounds containing hydrogen and the other members of group VIA(16) of the periodic table. The boiling point of H_2Te is –2.2°C, of H_2Se is –41.3°C, and of H_2S is –60.3°C. Thus, it appears that the boiling point decreases with decreasing compound molecular weight. On the basis of this trend, water should boil at approximately –64°C. The 164-degree difference between predicted and measured boiling points is caused by strong hydrogen bonds between water molecules in the liquid. The other three compounds do not have strong hydrogen bonds between molecules because the electronegativites of the elements to which hydrogen is covalently bonded are too low. This deviation from prediction makes water a liquid at normal temperatures and allows it to be used in many of the ways familiar to us.

It is also common knowledge that solid water (ice) floats on liquid water (see ▶ Figure 4.11). Nothing peculiar here—or is there? In fact, there is. Water, like most other liquids, increases in density as it is cooled. This means that a specific mass of liquid decreases in volume as its temperature is lowered. Unlike most liquids, water behaves this way only until it reaches a temperature of about 4°C, as shown by the data in ▶ Table 4.10. Then, its density decreases as it is cooled further. When its temperature reaches 0°C, water freezes and, in the process, undergoes a dramatic decrease in density. Put another way, a quantity of liquid water at 0°C expands significantly when it freezes to solid at 0°C. Thus, a specific volume of ice has a lower mass than the same volume of liquid water, so the ice floats on the water. Hydrogen bonds between water molecules are, once again, responsible. The strong hydrogen bonds orient water molecules into a very open three-dimensional crystal lattice when it freezes (see Figure 4.10c). This open lattice of the solid occupies more space than is occupied by the molecules in the liquid state.

Even though containers or engines might be ruptured or cracked if water is allowed to freeze in them, the overall effect of this characteristic of water is beneficial. If ice did not float, it would form on natural waters in the winter and sink to the bottom. Gradually, ponds, lakes, and other bodies of water would fill with ice. In the spring, the ice would melt from the top down because the heavier solid ice would stay under the water. As a result, most natural waters in areas that experience freezing winters would contain significant amounts of ice year-round. It would be a strange (and possibly hostile) environment for us and for wildlife.

The amount of heat required to melt a quantity of solid and the amount of heat required to vaporize a quantity of liquid also depend on the attractive forces between molecules. As expected, these are high for water, compared with the hydrogen compounds of the three other elements in group VIA(16). The beneficial results of these high values for water will be discussed in Chapter 6.

The forces between O_2 molecules, the lattice particles of solid oxygen, do not fit into any of the classifications we have discussed to this point. The O_2 molecule is not polar and contains no ionic or metallic bonds, and solid oxygen melts at much too low a temperature to fit into the category of a network solid. The forces between O_2 molecules are called **dispersion forces** and result from momentary nonsymmetric electron distributions in the molecules.

The nonbonding electrons in an O_2 molecule can be visualized as being uniformly distributed. However, there is a small statistical chance that, during their normal movement, more of the electrons will momentarily be on one side of a molecule than on the other. This condition causes the molecule to become dipolar for an instant. The resulting negative side of the dipole will tend to repel electrons of adjoining molecules and induce them also to become dipolar (they become induced dipoles). The original (statistical) dipole and all induced dipoles are then attracted to one another. This happens many times per second throughout the solid or liquid oxygen. The net effect is a weak dispersion force of attraction between the molecules. The net force is weak because it represents the average result of many weak, short-lived attractions per second between molecules.

Figure 4.11 Icebergs move as a result of wind and ocean currents. They can be very dangerous to ships, so their positions are reported and their probable courses estimated by an International Ice Patrol. The patrol was established in 1914 following the sinking of the *Titanic*.

TABLE 4.10 Water Density as a Function of Temperature

Temperature (°C)	Water Density (g/mL)
100	0.9586
80	0.9719
60	0.9833
40	0.9923
20	0.9982
10	0.9997
5	0.9999
4 (actually 3.98)	1.0000
2	0.9999
0 (liquid H_2O)	0.9998
0 (solid H_2O)	0.9170

dispersion forces Very weak attractive forces acting between the particles of all matter. They result from momentary nonsymmetric electron distributions in molecules or atoms.

Dispersion forces exist in all matter, but because they are very weak, their contribution is negligible when other, stronger forces are also present. Thus, solid water is held together by dispersion forces and hydrogen bonds, but the properties are almost entirely the result of the hydrogen bonds. The ease with which a dipole can be induced increases with the size of the particle (atoms or molecules). The larger the particle, the stronger the resulting dipolar attraction (dispersion force), and the harder to separate the particles by melting or boiling. Thus, we expect to find the melting and boiling points of the elements increase as we move down a group of the periodic table.

Substance	Melting Point (°C)	Boiling Point (°C)
F_2	−223	−188
Cl_2	−103	−34.6
Br_2	−7.2	58.8
I_2	113.9	184.3

▶ Example 4.19

Illustrate the behavior of dispersion forces by using the information in ▶ Table 4.11.

Solution

The molecules increase in size in the order F_2, Cl_2, Br_2, I_2. The strength of dispersion forces increases in the same order, as shown in Table 4.11 by the increases in melting and boiling points.

▶ **LEARNING CHECK 4.19** Using only the periodic table, predict which member of each of the following pairs of elements would have the higher melting and boiling points:

a. O and Se   **b.** Sb and P   **c.** He and Ne

The melting and boiling points in Example 4.19 indicate the variation in magnitude of dispersion forces. Similar variations are found for the other forces described in this section; sometimes their strengths overlap. For example, some metallic bonds are weaker than ionic bonds, whereas others are stronger. This is illustrated in ▶ Figure 4.12, which summarizes the relative strengths of interparticle forces.

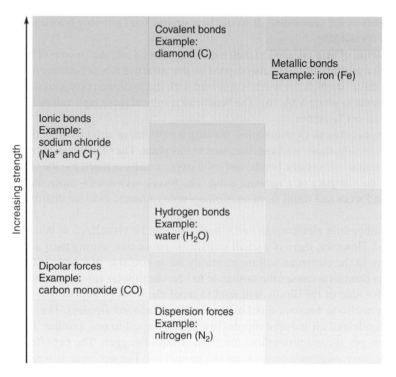

Figure 4.12 The relative strengths of interparticle forces.

Case Study Follow-up

Anorexia is characterized by failure to maintain a body weight of at least 85% of normal, whereas bulimia involves the intake of large quantities of food and then eliminating it by one or more of the following methods: forced vomiting, excessive use of laxatives, fasting, and exercise. Bulimic behaviors can result in electrolyte imbalances (Section 25.5). Many people with bulimia maintain a normal or slightly elevated weight. A patient who may "binge and purge" but also fasts and exercises to the point of failing to maintain 85% of recommended body weight would be considered anorexic, whereas a person who mainly binges and purges but maintains a normal weight would be considered bulimic. A 5' 6" tall woman who weighs below 102 lbs, with no other explanation for her low weight, would be considered anorexic. Both eating disorders occur more prevalently in females who come from highly controlling families where nurturance is lacking. When there is a family-environment component in the etiology of the eating disorder, it is helpful for the entire family to receive counseling in order to improve the functioning of family relationships and to create an environment where healthy thinking and behavior can thrive. If the person with the eating disorder receives treatment and subsequently returns to an unchanged family environment, chances of maintaining recovery are reduced.

Concept Summary

Noble Gas Configurations

The lack of reactivity for noble gases led to the proposal that the electronic configurations of the noble gases represented stable configurations. These configurations, usually consisting of eight electrons in the valence shell, can be represented in several useful ways.

Objective 1, Exercise 4.2

Ionic Bonding

Ionic compounds are formed when reacting atoms gain or lose electrons to achieve a noble gas configuration of eight electrons in the valence shell. This octet rule predicts that atoms will be changed into charged particles called simple ions. Ions of opposite charge are attracted to each other; the attractive force is called an ionic bond.

Objective 2, Exercise 4.12

Ionic Compounds

Oppositely charged ions group together to form compounds in ratios determined by the positive and negative charges of the ions. The formulas representing these ratios contain the lowest number of each ion possible in a proportion such that the total positive charges and total negative charges used are equal.

Objective 3, Exercises 4.20 and 4.22

Naming Binary Ionic Compounds

Binary ionic compounds contain a metal and a nonmetal. They are named by naming the metal, then adding the suffix -ide to the stem of the nonmetal.

Objective 4, Exercise 4.30

Concept Summary (continued)

The Smallest Unit of Ionic Compounds

Ionic compounds do not exist in the form of molecules but as three-dimensional arrangements of oppositely charged ions. The sum of the atomic weights of the elements in the formula of ionic compounds is called the formula weight and is used in calculations somewhat like the molecular weight of molecular compounds.

Objective 5, Exercise 4.38

Covalent Bonding

Elements with little or no tendency to gain or lose electrons often react and achieve noble gas electronic configurations by sharing electrons. Lewis structures are useful in representing electron sharing. Shared pairs of electrons exert an attractive force on both atoms that share them. The atoms are held together by this attraction to form a covalent bond.

Objective 6, Exercise 4.48

Polyatomic Ions

Polyatomic ions are groups of two or more covalently bonded atoms that carry a net electrical charge. They are conveniently represented using Lewis structures.

Objective 7, Exercise 4.50

Shapes of Molecules and Polyatomic Ions

The shapes of many molecules and polyatomic ions can be predicted by using the valence-shell electron-pair repulsion theory (VSEPR). According to the VSEPR theory, electron pairs in the valence shell of the central atom of a molecule or ion repel one another and become arranged so as to maximize their separation distances. The resulting arrangement determines the molecular or ionic shape when one or all of the electron pairs involved form bonds between the central atom and other atoms.

Objective 8, Exercises 4.52 and 4.54

The Polarity of Covalent Molecules

Shared electrons may be shared equally, or they may be attracted more strongly to one of the atoms they bond together. The tendency of a covalently bonded atom to attract shared electrons is called the electronegativity of the atom. Unequally shared bonding-electron pairs form polar covalent bonds. The extent of bond polarization can be estimated from the electronegativity differences between the bonded atoms. The higher the electronegativity difference, the more polar (or ionic) is the bond. Polar covalent bonds cause partial positive and negative charges to form within molecules. When these charges are symmetrically distributed in the molecule, it is said to be nonpolar. An unsymmetric distribution gives rise to a polar molecule.

Objective 9, Exercises 4.58 and 4.64

More about Naming Compounds

Binary covalent compounds are named using the name of the less electronegative element first, followed by the stem plus *-ide* of the more electronegative element. Greek prefixes are used to represent the number of each type of atom in molecules of the compounds. Ionic compounds that contain a metal ion (or ammonium ion) plus a polyatomic ion are named by first naming the metal (or ammonium ion) followed by the name of the polyatomic ion.

Objective 10, Assessment Exercises 4.66, 4.70, and 4.72

Other Interparticle Forces

Forces other than ionic and covalent bonds are also known to hold the particles of some pure substances together in the solid and liquid states. These forces include metallic bonds, dipolar attractions, hydrogen bonds, and dispersion forces. The strength of the predominant force acting in a substance is indicated by the melting and boiling points of the substance.

Objective 11, Exercises 4.78 and 4.80

Key Terms and Concepts

Binary compound (4.3)
Bond polarization (4.9)
Covalent bond (4.6)
Crystal lattice (4.5)
Dipolar force (4.11)
Dispersion force (4.11)
Double bond (4.6)
Electronegativity (4.9)
Formula weight (4.5)

Hydrogen bonding (4.11)
Ionic bond (4.2)
Isoelectronic (4.2)
Lattice site (4.5)
Lewis structure (4.1)
Metallic bond (4.11)
Network solid (4.11)
Nonpolar covalent bond (4.9)
Nonpolar molecule (4.9)

Octet rule (4.2)
Polar covalent bond (4.9)
Polar molecule (4.9)
Polyatomic ion (4.7)
Simple ion (4.2)
Triple bond (4.6)
VSEPR theory (4.8)

Exercises

Even-numbered exercises are answered in Appendix B.

Blue-numbered exercises are more challenging.

Noble Gas Configurations (Section 4.1)

4.1 Refer to the group numbers of the periodic table and draw Lewis structures for atoms of the following:

a. lithium

b. sodium

c. chlorine

d. boron

4.2 Refer to the group numbers of the periodic table and draw Lewis structures for atoms of the following:

a. iodine

b. strontium

c. tin

d. sulfur

4.3 Write abbreviated electronic configurations for the following:

a. iodine

b. element number 38

c. As

d. phosphorus

4.4 Write abbreviated electronic configurations for the following:

a. tin

b. Cs

c. element number 49

d. strontium

4.5 Draw Lewis structures for the elements given in Exercise 4.3.

4.6 Draw Lewis structures for the elements given in Exercise 4.4.

4.7 Use the symbol E to represent an element in a general way and draw Lewis structures for atoms of the following:

a. Any group IA(1)element

b. Any group IVA(14)element

4.8 Use the symbol E to represent an element in a general way and draw Lewis structures for atoms of the following:

a. Any group IIA(2) element

b. Any group VA(15) element

Ionic Bonding (Section 4.2)

4.9 Indicate both the minimum number of electrons that would have to be added and the minimum number that would have to be removed to change the electronic configuration of each element listed in Exercise 4.3 to a noble gas configuration.

4.10 Indicate both the minimum number of electrons that would have to be added and the minimum number that would have to be removed to change the electronic configuration of each element listed in Exercise 4.4 to a noble gas configuration.

4.11 Use the periodic table and predict the number of electrons that will be lost or gained by the following elements as they change into simple ions. Write an equation using elemental symbols, ionic symbols, and electrons to represent each change.

a. Ca

b. aluminum

c. fluorine

d. element number 34

4.12 Use the periodic table and predict the number of electrons that will be lost or gained by the following elements as they change into simple ions. Write an equation using elemental symbols, ionic symbols, and electrons to represent each change.

a. Cs

b. oxygen

c. element number 7

d. iodine

4.13 Write a symbol for each of the following ions:

a. A bromine atom that has gained one electron

b. A sodium atom that has lost one electron

c. A sulfur atom that has gained two electrons

4.14 Write a symbol for each of the following ions:

a. A phosphorus atom that has gained three electrons

b. A strontium atom that has lost two electrons

c. An iodine atom that has gained one electron

4.15 Identify the element in period 2 that would form each of the following ions. E is used as a general symbol for an element.

a. E^-

b. E^{2+}

c. E^{3-}

d. E^+

4.16 Identify the element in period 3 that would form each of the following ions. E is used as a general symbol for an element.

a. E^{2-}

b. E^{3+}

c. E^+

d. E^-

4.17 Identify the noble gas that is isoelectronic with each of the following ions:

a. Mg^{2+}

b. Te^{2-}

c. N^{3-}

d. Be^{2+}

4.18 Identify the noble gas that is isoelectronic with each of the following ions:

a. Ca^{2+}

b. S^{2-}

c. As^{3-}

d. Cs^+

Ionic Compounds (Section 4.3)

4.19 Write equations to represent positive and negative ion formation for the following pairs of elements. Then write a formula for the ionic compound that results when the ions combine.

   a. Mg and S

   b. strontium and nitrogen

   c. elements number 3 and 34

4.20 Write equations to represent positive and negative ion formation for the following pairs of elements. Then write a formula for the ionic compound that results when the ions combine.

   a. Ca and Cl

   b. lithium and bromine

   c. elements number 12 and 16

4.21 Write the formula for the ionic compound formed from Sr^{2+} and each of the following ions:

   a. S^{2-}

   b. Br^-

   c. N^{3-}

   d. Cl^-

4.22 Write the formula for the ionic compound formed from Ba^{2+} and each of the following ions:

   a. Se^{2-}

   b. P^{3-}

   c. I^-

   d. As^{3-}

4.23 Classify each of the following as a binary compound or not a binary compound:

   a. HF

   b. OF_2

   c. H_2SO_4

   d. H_2S

   e. $MgBr_2$

4.24 Classify each of the following as a binary compound or not a binary compound:

   a. CaS

   b. Na_3P

   c. $MgSO_4$

   d. $NaHCO_3$

   e. P_2O_5

Naming Binary Ionic Compounds (Section 4.4)

4.25 Name the following metal ions:

   a. Ca^{2+}

   b. K^+

   c. Al^{3+}

   d. Rb^+

4.26 Name the following metal ions:

   a. Li^+

   b. Mg^{2+}

   c. Ba^{2+}

   d. Cs^+

4.27 Name the following nonmetal ions:

   a. Cl^-

   b. N^{3-}

   c. S^{2-}

   d. Se^{2-}

4.28 Name the following nonmetal ions:

   a. Br^-

   b. O^{2-}

   c. P^{3-}

   d. Te^{2-}

4.29 Name the following binary ionic compounds:

   a. K_2O

   b. $SrCl_2$

   c. Al_2O_3

   d. LiBr

   e. CaS

4.30 Name the following binary ionic compounds:

   a. SrS

   b. CaF_2

   c. $BaCl_2$

   d. Li_2O

   e. MgO

4.31 Name the following binary ionic compounds, using a Roman numeral to indicate the charge on the metal ion:

   a. $CrCl_2$ and $CrCl_3$

   b. CoS and Co_2S_3

   c. FeO and Fe_2O_3

   d. $PbCl_2$ and $PbCl_4$

4.32 Name the following binary ionic compounds, using a Roman numeral to indicate the charge on the metal ion:

   a. PbO and PbO_2

   b. CuCl and $CuCl_2$

   c. Au_2S and Au_2S_3

   d. CoO and Co_2O_3

4.33 Name the binary compounds of Exercise 4.31 by adding the endings *-ous* and *-ic* to indicate the lower and higher ionic charges of the metal ion in each pair of compounds. The non-English root for lead (Pb) is *plumb-*.

4.34 Name the binary compounds of Exercise 4.32 by adding the endings *-ous* and *-ic* to indicate the lower and higher ionic charges of the metal ion in each pair of compounds. The non-English root for gold (Au) is *aur-*, and that of lead (Pb) is *plumb-*.

4.35 Write formulas for the following binary ionic compounds:

  a. manganese(II) chloride

  b. iron(III) sulfide

  c. chromium(II) oxide

  d. iron(II) bromide

  e. tin(II) chloride

4.36 Write formulas for the following binary ionic compounds:

  a. lead(IV) oxide

  b. cobalt(II) chloride

  c. copper(II) sulfide

  d. nickel(III) nitride

  e. platinum(II) phosphide

The Smallest Unit of Ionic Compounds (Section 4.5)

4.37 Determine the formula weight in atomic mass units for each of the following binary ionic compounds:

  a. Na_2O

  b. FeO

  c. PbS_2

  d. $AlCl_3$

4.38 Determine the formula weight in atomic mass units for each of the following binary ionic compounds:

  a. $NaBr$

  b. CaF_2

  c. Cu_2S

  d. Li_3N

4.39 Identify the ions that would occupy lattice sites in a solid sample of each compound given in Exercise 4.37.

4.40 Identify the ions that would occupy lattice sites in a solid sample of each compound given in Exercise 4.38.

4.41 Calculate the mass in grams of positive ions and negative ions contained in 1 mol of each compound given in Exercise 4.37.

4.42 Calculate the mass in grams of positive ions and negative ions contained in 1 mol of each compound given in Exercise 4.38.

4.43 Calculate the number of positive ions and negative ions contained in 1.00 mol of each compound given in Exercise 4.37.

4.44 Calculate the number of positive ions and negative ions contained in 1.00 mol of each compound given in Exercise 4.38.

Covalent Bonding (Section 4.6)

4.45 Represent the following reaction using Lewis structures:

$$I + I \rightarrow I_2$$

4.46 Represent the following reaction using Lewis structures:

$$8S \rightarrow S_8 \text{ (the atoms form a ring)}$$

4.47 Represent the following molecules by Lewis structures:

  a. HF

  b. IBr

c. PH_3 (each H atom is bonded to the P atom)

d. $HClO_2$ (the O atoms are each bonded to the Cl, and the H is bonded to one of the O atoms)

4.48 Represent the following molecules by Lewis structures:

  a. CH_4 (each H atom is bonded to the C atom)

  b. CO_2 (each O atom is bonded to the C atom)

  c. H_2Se (each H atom is bonded to the Se atom)

  d. NH_3 (each H atom is bonded to the N atom)

Polyatomic Ions (Section 4.7)

4.49 Draw Lewis structures for the following polyatomic ions:

  a. ClO_3^- (each O atom is bonded to the Cl atom)

  b. CN^-

  c. CO_3^{2-} (each O atom is bonded to the C atom)

4.50 Draw Lewis structures for the following polyatomic ions:

  a. NH_4^+ (each H atom is bonded to the N atom)

  b. PO_4^{3-} (each O atom is bonded to the P atom)

  c. SO_3^{2-} (each O atom is bonded to the S atom)

Shapes of Molecules and Polyatomic Ions (Section 4.8)

4.51 Draw Lewis structures for the following molecules:

  a. O_3 (the O atoms are bonded together, like beads on a string)

  b. CS_2 (each S atom is bonded to the C atom)

  c. SeO_2 (each O atom is bonded to the Se atom)

  d. H_2SO_3 (each O atom is bonded to the S atom, and one H atom is bonded to each of the two O atoms)

4.52 Predict the shape of each of the following molecules by first drawing a Lewis structure, then applying the VSEPR theory:

  a. H_2S (each H atom is bonded to the S atom)

  b. PCl_3 (each Cl atom is bonded to the P atom)

  c. OF_2 (each F atom is bonded to the O atom)

  d. SnF_4 (each F atom is bonded to the Sn atom)

4.53 Predict the shape of each of the following molecules by first drawing a Lewis structure, then applying the VSEPR theory:

  a. O_3 (see Exercise 4.51 for Lewis structure)

  b. SeO_2 (see Exercise 4.51 for Lewis structure)

  c. PH_3 (each H atom is bonded to the P atom)

  d. SO_3 (each O atom is bonded to the S atom)

4.54 Predict the shape of each of the following polyatomic ions by first drawing a Lewis structure, then applying the VSEPR theory:

  a. NO_2^- (each O is bonded to N)

  b. ClO_3^- (each O is bonded to Cl)

  c. CO_3^{2-} (each O is bonded to C)

  d. H_3O^+ (each H is bonded to O)

  Note the positive charge; compare with NH_4^+.

4.55 Predict the shape of each of the following polyatomic ions by first drawing a Lewis structure, then applying the VSEPR theory:

    a. NH_2^- (each H is bonded to N)

    b. PO_3^{3-} (each O is bonded to P)

    c. $BeCl_4^{2-}$ (each Cl is bonded to Be)

    d. ClO_4^- (each O is bonded to Cl)

The Polarity of Covalent Molecules (Section 4.9)

4.56 Use the periodic table and Table 4.4 to determine which of the following bonds will be polarized. Show the resulting charge distribution in those molecules that contain polarized bonds.

    a. H—I

    b.

    c.

4.57 Use the periodic table and Table 4.4 to determine which of the following bonds will be polarized. Show the resulting charge distribution in those molecules that contain polarized bonds.

    a. Cl—F

    b.

    c.
H, H, B—B, H, H

4.58 Use Table 4.4 and classify the bonds in the following compounds as nonpolar covalent, polar covalent, or ionic:

    a. LiBr

    b. HCl

    c. PH_3 (each H is bonded to P)

    d. SO_2 (each O is bonded to S)

    e. CsF

4.59 Use Table 4.4 and classify the bonds in the following compounds as nonpolar covalent, polar covalent, or ionic:

    a. MgI_2 (each I is bonded to Mg)

    b. NCl_3 (each Cl is bonded to N)

    c. H_2S (each H is bonded to S)

    d. RbF

    e. SrO

4.60 On the basis of the charge distributions you drew for the molecules of Exercise 4.56, classify each of the molecules as polar or nonpolar.

4.61 On the basis of the charge distributions you drew for the molecules of Exercise 4.57, classify each of the molecules as polar or nonpolar.

4.62 Use Table 4.4 and predict the type of bond you would expect to find in compounds formed from the following elements:

    a. magnesium and chlorine

    b. carbon and hydrogen

    c. phosphorus and hydrogen

4.63 Use Table 4.4 and predict the type of bond you would expect to find in compounds formed from the following elements:

    a. sulfur and oxygen

    b. aluminum and bromine

    c. C and Cl

4.64 Show the charge distribution in the following molecules, and predict which are polar molecules:

    a. C≡O

    b.

    c.

4.65 Show the charge distribution in the following molecules, and predict which are polar molecules:

    a. S=C=S

    b. H—C≡N

    c.
F—O, F

More about Naming Compounds (Section 4.10)

4.66 Name the following binary covalent compounds:

    a. PCl_3

    b. N_2O_5

    c. CCl_4

    d. BF_3

    e. CS_2

4.67 Name the following binary covalent compounds:

    a. SiO_2

    b. SiF_4

    c. P_2O_5

    d. $AlBr_3$

    e. CBr_4

4.68 Write formulas for the following binary covalent compounds:

    a. chlorine dioxide

    b. dinitrogen monoxide

    c. sulfur dioxide

    d. carbon tetrachloride

4.69 Write formulas for the following binary covalent compounds:

    a. disulfur monoxide

    b. sulfur hexafluoride

    c. silicon tetrachloride

    d. carbon diselenide

Unless otherwise noted all art on this page © Cengage Learning 2014

4.70 Write the formulas and names for compounds composed of ions of the following metals and the indicated polyatomic ions:

  a. calcium and the nitrite ion

  b. magnesium and the hypochlorite ion

  c. Cs and $Cr_2O_7^{2-}$

  d. K and SO_3^{2-}

4.71 Write the formulas and names for compounds composed of ions of the following metals and the indicated polyatomic ions:

  a. calcium and the phosphate ion

  b. sodium and the dichromate ion

  c. Li and CO_3^{2-}

  d. Na and PO_4^{3-}

4.72 Write formulas for the following compounds:

  a. barium hydroxide

  b. magnesium sulfite

  c. calcium carbonate

  d. ammonium sulfate

  e. lithium hydrogen carbonate

4.73 Write formulas for the following compounds:

  a. potassium permanganate

  b. calcium hydroxide

  c. calcium phosphate

  d. ammonium dihydrogen phosphate

  e. calcium hypochlorite

4.74 Write a formula for the following compounds, using M with appropriate charges to represent the metal ion:

  a. Any group IA(1) element and SO_3^{2-}

  b. Any group IA(1) element and $C_2H_3O_2^-$

  c. Any metal that forms M^{2+} ions and $Cr_2O_7^{2-}$

  d. Any metal that forms M^{3+} ions and PO_4^{3-}

  e. Any metal that forms M^{3+} ions and NO_3^-

4.75 Write a formula for the following compounds, using M with appropriate charges to represent the metal ion:

  a. Any group IIA(2) element and HSO_3^-

  b. Any group IIA(2) element and HPO_4^{2-}

  c. Any metal that forms M^+ ions and NO_2^-

  d. Any metal that forms M^{3+} ions and CO_3^{2-}

  e. Any metal that forms M^{2+} ions and HPO_4^{2-}

Other Interparticle Forces (Section 4.11)

4.76 The covalent compounds ethyl alcohol and dimethyl ether both have the formula C_2H_6O. However, the alcohol melts at $-117.3°C$ and boils at $78.5°C$, whereas the ether melts at $-138.5°C$ and boils at $-23.7°C$. How could differences in forces between molecules be used to explain these observations?

4.77 The following structural formulas represent molecules of ethyl alcohol and dimethyl ether. Assign the correct name to each formula and explain how your choice is consistent with your answer to Exercise 4.76.

4.78 Describe the predominant forces that exist between molecules of the noble gases. Arrange the noble gases in a predicted order of increasing boiling point (lowest first) and explain the reason for the order.

4.79 Use the concept of interparticle forces to propose an explanation for the fact that CO_2 is a soft, low-melting solid (dry ice), whereas SiO_2 is a hard solid (sand). Focus on the nature of the particles that occupy lattice sites in the solid.

4.80 Table sugar, sucrose, melts at about $185°C$. Which interparticle forces do you think are unlikely to be the predominant ones in the lattice of solid sucrose?

4.81 The formula for sucrose is $C_{12}H_{22}O_{11}$, where many of the hydrogens and oxygens are combined to form OH groups that are bonded to carbon atoms. What type of predominant interparticle bonding would you now propose for solid sucrose (see Exercise 4.80)?

Additional Exercises

4.82 Three atoms with an electronic configuration of $1s^1$ are covalently bonded to one atom with an electronic configuration of $1s^2\,2s^2\,2p^3$ to form a molecule. What is the formula of this molecule? Make a drawing to show how two of these molecules would hydrogen bond with each other.

4.83 Suppose an element from group IIA(2), and period 3 of the periodic table forms an ionic compound with the element that has an electronic configuration of $1s^2\,2s^2\,2p^5$. What would be the formula of the compound, and what would be the name of the compound?

4.84 What would be the mass in grams of 0.200 moles of the ionic compound formed when magnesium metal reacts with oxygen?

4.85 The ampere unit is used to describe the flow of electricity in an electrical circuit. One ampere is an amount of electricity corresponding to the flow of 6.2×10^{18} electrons past a point in a circuit in 1 second. In a hydrogen fuel cell, hydrogen atoms are dissociated into H^+ ions and electrons ($H \rightarrow H^+ + 1e^-$). How many grams of hydrogen must be dissociated each second in a fuel cell to produce 1 ampere of electricity?

4.86 If one atom of oxygen reacted with two atoms of nitrogen to form a molecule, what would be the formula of the molecule? Use Table 4.4 and Figure 4.7 to determine if the bond between the atoms is ionic or covalent. Also, name the compound that is formed.

Allied Health Exam Connection

The following questions are from these sources:

1. *Nursing School Entrance Exam* © 2005, Learning Express, LLC.

2. *McGraw-Hill's Nursing School Entrance Exams* by Thomas A. Evangelist, Tamara B. Orr and Judy Unrein © 2009, The McGraw-Hill Companies, Inc.

Even-numbered exercises answered in Appendix B.

Blue-numbered exercises are more challenging.

3. *NSEE Nursing School Entrance Exams*, 3rd Edition © 2009, Kaplan Publishing.

4. *Cliffs Test Prep: Nursing School Entrance Exams* by Fred N. Grayson © 2004, Wiley Publishing, Inc.

5. *Peterson's Master the Nursing School and Allied Health Entrance Exams*, 18th Edition by Marion F. Gooding © 2008, Peterson's, a Nelnet Company.

4.87 Noble gases:

    a. have low boiling points

    b. are all gases at room temperature

    c. are also called inert gases

    d. all of the above

4.88 The elements in group Zero of the periodic table are considered inert gases because each has how many electrons in its outermost energy level?

    a. 8

    b. 7

    c. 4

    d. 2

4.89 Name the type of bond that is formed when *electrons are shared* between two atoms.

    a. shared bond

    b. ionic bond

    c. covalent bond

    d. multiple bond

4.90 Which of the following is the correct name for Li_2SO_3?

    a. lithium sulfite

    b. lithium sulfide

    c. lithium sulfate

    d. lithium disulfate

4.91 An atom becomes an ion that possesses a negative charge. The atom must have:

    a. gained protons

    b. lost protons

    b. lost electrons

    d. gained electrons

4.92 When calcium reacts with chlorine to form calcium chloride, it:

    a. shares two electrons

    b. gains two electrons

    c. loses two electrons

    d. gains one electron

4.93 Which of the following is the probable charge for an ion formed from Ca?

    a. +1

    b. +2

    c. −1

    d. −2

4.94 A covalent bond is believed to be caused by a:

    a. transfer of electrons

    b. sharing of electrons

    c. release of energy

    d. none of the above

4.95 Which molecule below has a nonpolar bond in which the electrons are being shared equally?

    a. H_2O

    b. NH_3

    c. Cl_2

    d. CH_4

4.96 What is the formula for bismuth (III) hydroxide?

    a. Bi_3OH

    b. $Bi(OH)_3$

    c. $Bi(OH)_2$

    d. $BiOH$

4.97 Which of the following species will combine with a chloride ion to produce ammonium chloride?

    a. NH_3

    b. K^+

    c. NH_4^+

    d. Al^{+++}

4.98 What type of bond is created when bromine and magnesium are reacted to form $MgBr_2$?

    a. polar covalent

    b. metallic

    c. ionic

    d. nonpolar covalent

4.99 The parts of an atom directly involved in ionic bonding are the:

    a. protons in the nucleus

    b. neutrons in the nucleus

    c. electrons in the outer energy shell

    d. electrons in the innermost energy shell

4.100 In forming an ionic bond with an atom of chlorine, a sodium atom will:

    a. receive 1 electron from the chlorine atom

    b. receive 2 electrons from the chlorine atom

    c. give up 1 electron to the chlorine atom

    d. give up 2 electrons to the chlorine atom

4.101 In bonding, what would happen between the electrons of K and Br?

    a. they would be transferred

    b. they would be shared

    c. none of the above

    d. they would be both transferred and shared

4.102 Which compound contains a bond with **no** ionic character?

   a. CO

   b. CaO

   c. K_2O

   d. Na_2O

4.103 Which of the following molecules is nonpolar?

   a. NH_3

   b. H_2O, NO_2

   c. PCl_3

   d. N_2

4.104 Which molecule is nonpolar and contains a nonpolar covalent bond?

   a. F_2

   b. HI

   c. KCl

   d. NH_3

4.105 Which of the following is a nonpolar covalent bond?

   a. the bond between two carbons

   b. the bond between sodium and chloride

   c. the bond between two water molecules

   d. the bond between nitrogen and hydrogen

4.106 The type of bond that forms between two molecules of water is a:

   a. polar covalent bond

   b. hydrogen bond

   c. nonpolar covalent bond

   d. peptide bond

4.107 Which of the following is an example of hydrogen bonding?

   a. The bond between O and H in a single molecule of water

   b. The bond between the O of one water molecule and the H of a second water molecule

   c. The bond between the O of one water molecule and the O of a second water molecule

   d. The bond between the H of one water molecule and the H of a second water molecule

Chemistry for Thought

4.108 Refer to Figure 4.1, and answer the question in the caption. What other metals and nonmetals would you predict might react in a similar way?

4.109 The colors of some compounds, such as those shown in Figure 4.2, result from the presence of water in the compounds. Propose an experiment you could perform to see if this was true for the compounds shown in Figure 4.2.

4.110 Refer to Figure 4.6, and answer the question in the caption. Propose the shapes that would be assumed by a group of five balloons and by a group of six balloons.

4.111 Refer to Figure 4.9. Two of the compounds are highly colored (other than white). All the compounds consist of potassium and a polyatomic ion. If you have not yet done so, write formulas for the compounds and see if you can find a characteristic of the polyatomic ions of the colored compounds that is not found in the white compounds. Then refer to Table 4.7 and predict which of the other polyatomic anions would form colored compounds of potassium.

4.112 Recall how a metal atom changes to form a positively charged metal ion. How do you think the sizes of a metal atom and a positive ion of the same metal will compare?

4.113 Recall how a nonmetal atom such as chlorine changes to form a negatively charged ion. How do you think the size of a nonmetal atom and a negatively charged ion of the same nonmetal will compare?

4.114 Neon atoms do *not* combine to form Ne_2 molecules. Explain.

4.115 Refer to Figure 4.8, and answer the question in the caption. The balloon carries a negative charge. What would happen if a positively charged object was used in place of the balloon?

4.116 It has been found that NO is a vital biological molecule. Explain how NO forms when a fuel such as natural gas, CH_4, is burned in air at a high temperature.

© mangostock/Shutterstock.com

Case Study

Each new school year had the same drawback; the students brought every virus in town to the classroom! In spite of sanitizing wipes and hand gel, third-grade teacher Lateisha always managed to catch a cold. This year was worse than others, and the cold developed into an ear infection. Teaching was demanding enough without being sick! Lateisha went to an after-hours clinic and received a prescription for an antibiotic. The pharmacist asked if she had any questions; Lateisha said no and agreed to take the entire course of the medication. She saw a yellow sticker on the side of the bottle warning that antibiotics may interact with other medications but didn't take it seriously. However, after she finished the antibiotic, Lateisha felt worse—fatigue and nausea drove her to the clinic again. When the physician suggested a pregnancy test, Lateisha was surprised because she had been on birth-control pills for 10 years. Imagine her shock when the test was positive! Her entire life was going to change because she had failed to heed a warning label. Lateisha's drug interaction was not life threatening, but it was certainly life changing.

How proactive should pharmacists be in telling patients about interactions? What are the chemical processes involved in drug interactions?

Follow-up to this Case Study appears at the end the chapter before the Concept Summary.

Learning Objectives

When you have completed your study of this chapter, you should be able to:

1 Identify the reactants and products in written reaction equations, and balance the equations by inspection. **(Section 5.1)**

2 Assign oxidation numbers to elements in chemical formulas, and identify the oxidizing and reducing agents in redox reactions. **(Section 5.3)**

3 Classify reactions into the categories of redox or nonredox, then into the categories of decomposition,

combination, single replacement, or double replacement. **(Sections 5.4–5.6)**

4 Write molecular equations in total ionic and net ionic forms. **(Section 5.7)**

5 Classify reactions as exothermic or endothermic. **(Section 5.8)**

6 Use the mole concept to do calculations based on chemical reaction equations. **(Section 5.9)**

7 Use the mole concept to do calculations based on the limiting-reactant principle. **(Section 5.10)**

8 Use the mole concept to do percentage-yield calculations. **(Section 5.11)**

n previous chapters, we introduced the terms *molecule, element, compound,* and *chemical change.* In Chapter 1, you learned that chemical changes result in the transformation of one or more substances into one or more new substances. The processes involved in such changes are called *chemical reactions.* In this chapter, you will learn to write and read chemical equations that represent chemical reactions, to classify reactions, and to do calculations based on the application of the mole concept to chemical equations.

5.1 Chemical Equations

Learning Objective

1. Identify the reactants and products in written reaction equations, and balance the equations by inspection.

A simple chemical reaction between elemental hydrogen and oxygen has been used to power the engines of a number of spacecraft, including the space shuttle. The products are water and much heat. For the moment, we will focus only on the substances involved; we will deal with the heat later. The reaction can be represented by a word equation:

$$\text{hydrogen} + \text{oxygen} \rightarrow \text{water} \qquad (5.1)$$

The word equation gives useful information, including the reactants and products of the reaction. The **reactants** are the substances that undergo the chemical change; by convention,

reactants of a reaction The substances that undergo chemical change during the reaction. They are written on the left side of the equation representing the reaction.

products of a reaction The substances produced as a result of the reaction taking place. They are written on the right side of the equation representing the reaction.

law of conservation of matter Atoms are neither created nor destroyed in chemical reactions.

balanced equation An equation in which the number of atoms of each element in the reactants is the same as the number of atoms of that same element in the products.

these are written on the left side of the equation. The **products,** or substances produced by the chemical change, are written on the right side of the equation. The reactants and products are separated by an arrow that points to the products. A plus sign (+) is used to separate individual reactants and products.

Word equations convey useful information, but the chemical equations used by chemists convey much more:

$$2H_2(g) + O_2(g) \rightarrow 2H_2O(\ell) \tag{5.2}$$

In this equation, the reactants and products are represented by molecular formulas that tell much more than the names used in the word equation. For example, it is apparent that both hydrogen and oxygen molecules are diatomic. The equation is also consistent with a fundamental law of nature called the **law of conservation of matter.** According to this law, atoms are neither created nor destroyed in chemical reactions but are rearranged to form new substances. Thus, atoms are conserved in chemical reactions, but molecules are not. The numbers written as coefficients to the left of the molecular formulas make the equation consistent with this law by making the total number of each kind of atom equal in the reactants and products. Note that coefficients of 1 are never written but are understood. Equations written this way are said to be **balanced.** In addition, the symbol in parentheses to the right of each formula indicates the state or form in which the substance exists. Thus, in Equation 5.2, the reactants hydrogen and oxygen are both in the form of gases (g), and the product water is in the form of a liquid (ℓ). Other common symbols you will encounter are (s) to designate a solid and (aq) to designate a substance dissolved in water. The symbol (aq) comes from the first two letters of *aqua,* the Latin word for water.

▶ Example 5.1

Determine the number of atoms of each type on each side of Equation 5.2.

Solution

The coefficient 2 written to the left of H_2 means that two hydrogen molecules with the formula H_2 are reacted. Since each molecule contains two hydrogen atoms, a total of four hydrogen atoms is represented. The coefficient to the left of O_2 is 1, even though it is not written in the equation. The oxygen molecule contains two atoms, so a total of two oxygen atoms is represented. The coefficient 2 written to the left of H_2O means that two molecules of H_2O are produced. Each molecule contains two hydrogen atoms and one oxygen atom. Therefore, the total number of hydrogen atoms represented is four, and the number of oxygen atoms is two. These are the same as the number of hydrogen and oxygen atoms represented in the reactants.

▶ **LEARNING CHECK 5.1** Determine the number of atoms of each type on each side of the following balanced equation:

$$N_2(g) + 3H_2(g) \rightarrow 2NH_3(g)$$

When the identity and formulas of the reactants and products of a reaction are known, the reaction can be balanced by applying the law of conservation of matter.

▶ Example 5.2

Nitrogen dioxide (NO_2) is an air pollutant that is produced in part when nitric oxide (NO) reacts with oxygen gas (O_2). Write a balanced equation for the production of NO_2 by this reaction.

Solution

The reactants written to the left of the arrow will be NO and O_2. The product is NO_2:

$$NO(g) + O_2(g) \rightarrow NO_2(g) \tag{5.3}$$

A quick inspection shows that the reactants contain three oxygen atoms, whereas the product has just two. The number of nitrogen atoms is the same in both the reactants and the products. A practice sometimes resorted to by beginning chemistry students is to change the formula of oxygen gas to O by changing the subscript and to write

$$NO(g) + O(g) \rightarrow NO_2(g) \tag{5.4}$$

This is not allowed. The natural molecular formulas of any compounds or elements cannot be adjusted; they are fixed by the principles of chemical bond formation described in Chapter 4. All that can be done to balance a chemical equation is to change the coefficients of the reactants and products. In this case, inspection reveals that the following is a balanced form of the equation:

$$2NO(g) + O_2(g) \rightarrow 2NO_2(g) \tag{5.5}$$

Notice that the following forms of the equation are also balanced:

$$4NO(g) + 2O_2(g) \rightarrow 4NO_2(g) \tag{5.6}$$

and

$$NO(g) + \frac{1}{2}O_2(g) \rightarrow NO_2(g) \tag{5.7}$$

In Equation 5.6, the coefficients are double those used in Equation 5.5; in Equation 5.7, one coefficient is a fraction. The lowest possible whole-number coefficients are used in balanced equations. Thus, Equation 5.5 is the correct form.

▶ **LEARNING CHECK 5.2** Write and balance an equation that represents the reaction of sulfur dioxide (SO_2) with oxygen gas (O_2) to give sulfur trioxide (SO_3).

5.2 Types of Reactions

A large number of chemical reactions are known to occur; only a relatively small number of them will be studied in this book. This study is made easier by classifying the reactions according to certain characteristics. Such a classification scheme could be developed in a number of ways, but we have chosen to first classify reactions as being either oxidation–reduction (redox) reactions or nonredox reactions. As you will see, redox reactions are very important in numerous areas of study, including metabolism. Once reactions are classified as redox or nonredox, many can be further classified into one of several other categories, as shown in ▶ Figure 5.1. Notice that according to Figure 5.1, single-replacement, or substitution, reactions are redox reactions, whereas double-replacement, or metathesis, reactions are nonredox. Combination and decomposition reactions can be either redox or nonredox.

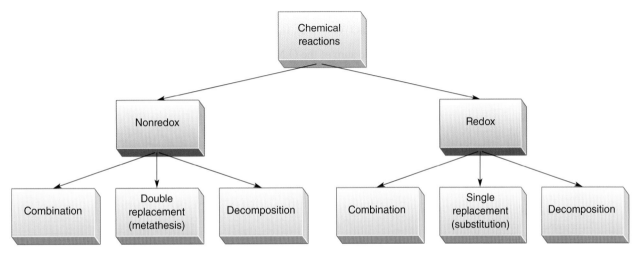

Figure 5.1 A classification of chemical reactions.

5.3 Redox Reactions

Learning Objective

2. Assign oxidation numbers to elements in chemical formulas, and identify the oxidizing and reducing agents in redox reactions.

Almost all elements react with oxygen to form oxides. The process is so common that the word *oxidation* was coined to describe it. Some examples are the rusting of iron,

$$4Fe(s) + 3O_2(g) \rightarrow 2Fe_2O_3(s) \tag{5.8}$$

and the burning of hydrogen,

$$2H_2(g) + O_2(g) \rightarrow 2H_2O(\ell) \tag{5.9}$$

The reverse process, reduction, originally referred to the technique of removing oxygen from metal oxide ores to produce the free metal. Some examples are

$$CuO(s) + H_2(g) \rightarrow Cu(s) + H_2O(\ell) \tag{5.10}$$

and

$$2Fe_2O_3(s) + 3C(s) \rightarrow 4Fe(s) + 3CO_2(g) \tag{5.11}$$

oxidation Originally, a process involving a reaction with oxygen. Today it means a number of things, including a process in which electrons are given up, hydrogen is lost, or an oxidation number increases.

reduction Originally, a process in which oxygen was lost. Today it means a number of things, including a process in which electrons are gained, hydrogen is accepted, or an oxidation number decreases.

oxidation numbers or oxidation states Positive or negative numbers assigned to the elements in chemical formulas according to a specific set of rules.

Today, the words **oxidation** and **reduction** are used in a rather broad sense. ▶ Table 5.1 contains most of the common meanings. To understand oxidation and reduction in terms of electron transfer and oxidation number (O.N.) change, you must become familiar with the concept of oxidation numbers.

Oxidation numbers, sometimes called **oxidation states,** are positive or negative numbers assigned to the elements in chemical formulas according to a set of rules. The following rules will be used; be sure to note that Rule 1 applies only to uncombined elements—that is, elements in their free state. Rules 2 through 7 apply to elements combined to form compounds or ions.

Rule 1. The oxidation number (O.N.) of any uncombined element is 0.
Examples: Al(0), $O_2(0)$, $Br_2(0)$, and Na(0)

Rule 2. The O.N. of a simple ion is equal to the charge on the ion.
Examples: $Na^+(+1)$, $Mg^{2+}(+2)$, $S^{2-}(-2)$, and $Br^-(-1)$

Rule 3. The O.N.s of combined group IA(1) and IIA(2) elements are +1 and +2, respectively.
Examples: $Na_2CO_3(Na = +1)$, $Sr(NO_3)_2$ (Sr = +2), and $CaCl_2$ (Ca = +2)

Rule 4. The O.N. of combined hydrogen is +1.
Example: HCl (H = +1) and H_3PO_4 (H = +1)

Rule 5. The O.N. of combined oxygen is −2 except in peroxides, where it is −1.
Examples: CaO (O = −2), H_2SO_4 (O = −2), H_2O (O = −2), and H_2O_2 (O = −1)

Rule 6. The algebraic sum of the O.N.s of all atoms in a complete compound formula equals zero.

Examples: K_2CO_3: 2(O.N. of K) + (O.N. of C) + 3(O.N. of O) = 0

$$
\begin{array}{cccc}
2(+1) & +4 & +3(-2) & = 0 \\
+2 & +4 & +(-6) & = 0
\end{array}
$$

HNO_2: (O.N. of H) + (O.N. of N) + 2(O.N. of O) = 0

$$
\begin{array}{cccc}
+1 & +3 & +2(-2) & = 0 \\
+1 & +3 & +(-4) & = 0
\end{array}
$$

TABLE 5.1 Common Uses of the Terms Oxidation and Reduction

Term	Meaning
Oxidation	To combine with oxygen
	To lose hydrogen
	To lose electrons
	To increase in oxidation number
Reduction	To lose oxygen
	To combine with hydrogen
	To gain electrons
	To decrease in oxidation number

Rule 7. The algebraic sum of the O.N.s of all atoms in a polyatomic ion is equal to the charge on the ion.

Examples: MnO_4^-: (O.N. of Mn) $+$ 4(O.N. of O) $= -1$

$$+7 \qquad\qquad +4(-2) \quad = -1$$
$$+7 \qquad\qquad +(-8) \quad = -1$$

HPO_4^{2-}: (O.N. of H) $+$ (O.N. of P) $+$ 4(O.N. of O) $= -2$

$$+1 \qquad\qquad +5 \qquad\qquad +4(-2) \quad = -2$$
$$+1 \qquad\qquad +5 \qquad\qquad +(-8) \quad = -2$$

▶ Example 5.3

a. Assign O.N.s to the blue element in each of the following:

$$CO_2, \qquad NO_2^-, \qquad H_2O, \qquad N_2, \qquad K^+$$

b. Assign O.N.s to each element in the following:

$$CO_2, \qquad Mg(NO_3)_2, \qquad NO_3^-, \qquad CH_2O$$

Solution

a. CO_2, O $= -2$ (Rule 5); NO_2^-, O $= -2$ (Rule 5); H_2O, H $= +1$ (Rule 4); N_2, N $= 0$ (Rule 1); K^+, K $= +1$ (Rule 2).

b. CO_2: The O.N. of O is -2 (Rule 5), and the O.N. of C can be calculated by using Rule 6 as follows:

$$\text{(O.N. of C)} + 2\text{(O.N. of O)} = 0$$
$$\text{(O.N. of C)} + 2(-2) = 0$$
$$\text{(O.N. of C)} + (-4) = 0$$

Therefore,

$$\text{O.N. of C} = +4$$

$Mg(NO_3)_2$: The O.N. of Mg is $+2$ (Rule 3), the O.N. of O is -2 (Rule 5), and the O.N. of N can be calculated by using Rule 6 as follows:

$$\text{(O.N. of Mg)} + 2\text{(O.N. of N)} + 6\text{(O.N. of O)} = 0$$
$$(+2) + 2\text{(O.N. of N)} + 6(-2) = 0$$
$$2\text{(O.N. of N)} + 2 + (-12) = 0$$
$$2\text{(O.N. of N)} - 10 = 0$$

Therefore,

$$\text{O.N. of N} = +5$$

NO_3^-: The O.N. of O is -2 (Rule 5), and the O.N. of N can be calculated by using Rule 7 as follows:

$$\text{(O.N. of N)} + 3\text{(O.N. of O)} = -1$$
$$\text{(O.N. of N)} + 3(-2) = -1$$
$$\text{(O.N. of N)} + (-6) = -1$$

Therefore,

$$\text{O.N. of N} = +5$$

Note that N has the same O.N. in NO_3^- and in $Mg(NO_3)_2$. This is expected because the polyatomic NO_3^- ion is present in $Mg(NO_3)_2$.

CH$_2$O: The O.N. of H is $+1$ (Rule 4), the O.N. of O is -2 (Rule 5), and the O.N. of C can be calculated by using Rule 6 as follows:

$$(\text{O.N. of C}) + 2(\text{O.N. of H}) + (\text{O.N. of O}) = 0$$
$$(\text{O.N. of C}) + 2(+1) + (-2) = 0$$
$$(\text{O.N. of C}) + 2 - 2 = 0$$

Therefore,

$$\text{O.N. of C} = 0$$

This example shows that an O.N. of 0 may be found for some combined elements as well as those in an uncombined state (Rule 1).

▶ **LEARNING CHECK 5.3** Assign O.N.s to each element in the following:

a. SO$_3$ **b.** Ca(ClO$_3$)$_2$ **c.** ClO$_4^-$

Figure 5.2 Combustion, the first reaction known to be carried out by humans, is a rapid redox reaction. Identify the oxidizing and reducing agents of the reaction.

reducing agent The substance that contains an element that is oxidized during a chemical reaction.

oxidizing agent The substance that contains an element that is reduced during a chemical reaction.

Now that you can assign O.N.s to the elements involved in reactions, you are ready to look at redox processes in terms of O.N.s and electron transfers. The reaction when sulfur is burned in oxygen is represented by the equation

$$S(s) + O_2(g) \rightarrow SO_2(g) \tag{5.12}$$

You can see that this reaction represents an oxidation because sulfur has combined with oxygen (remember Table 5.1). When oxidation numbers are assigned to sulfur, the reactant S has an oxidation number of 0, while the S in the SO$_2$ product has an oxidation number of $+4$. Thus, the oxidation that has taken place resulted in an increase in the oxidation number of sulfur. Oxidation always corresponds to an increase in oxidation number.

The same conclusion is arrived at by thinking of oxidation numbers as charges. Thus, a sulfur atom with 0 charge is oxidized as it acquires a $+4$ charge. This change in charge results when the sulfur atom releases four electrons that have a -1 charge each. This leads to another definition of oxidation: Oxidation takes place when electrons are lost.

The oxygen of Equation 5.12 undergoes an oxidation number change from 0 in O$_2$ to -2 in SO$_2$. This decrease in oxidation number corresponds to a reduction of the oxygen. In terms of electrons, this same change is accomplished when each uncharged oxygen atom in the O$_2$ molecule accepts two electrons. Thus, the four electrons lost by sulfur as it was oxidized is the same total number accepted by the oxygen as it was reduced.

An oxidation process must always be accompanied by a reduction process, and all the electrons released during oxidation must be accepted during reduction. Thus, oxidation and reduction processes always take place simultaneously, hence the term redox. In redox reactions, the substance that is oxidized (and releases electrons) is called the **reducing agent** because it is responsible for reducing another substance. Similarly, the substance that is reduced (accepts electrons) is called the **oxidizing agent** because it is responsible for oxidizing another material (see ▶ Figure 5.2). These characteristics are summarized in ▶ Table 5.2.

TABLE 5.2 Properties of Oxidizing and Reducing Agents

Oxidizing Agent	Reducing Agent
Gains electrons	Loses electrons
Oxidation number decreases	Oxidation number increases
Becomes reduced	Becomes oxidized

Antiseptics and Disinfectants

Antiseptics and disinfectants are both used for the same purpose: to kill bacteria. The difference between these two categories of bacteria killers is where they are used. Antiseptics are used to kill bacteria on living tissue, such as wounds. Disinfectants are used to kill bacteria on inanimate objects. Some antiseptics, such as iodine and hydrogen peroxide, operate by oxidizing and thus destroying compounds essential to the normal functioning of the bacteria. A solution containing 3% hydrogen peroxide dissolved in water is an antiseptic found in most pharmacies, and it is often used to treat minor cuts and abrasions. A 2% solution of iodine dissolved in alcohol, called tincture of iodine, is also generally available, and it is used in a way similar to hydrogen peroxide. One disadvantage of the iodine solution is that it stains the skin a yellow-brown color.

Oxidizing antiseptics are often regarded as being too harsh. They may damage skin and other normal tissue, as well as kill the bacteria. For this reason, they have been replaced in many products by antiseptics derived from phenol. Water solutions of phenol, called carbolic acid, were first introduced as hospital antiseptics in 1867 by the English surgeon Joseph Lister. Before that time, antiseptics had not been used, and very few patients survived even minor surgery because of postoperative infections. These phenolic derivatives can often be recognized on ingredient labels by the characteristic -*ol* ending of their names. Some examples are thymol, eucalyptol, and eugenol.

Because disinfectants are used on inanimate objects, there is much less concern about the damage they might do to living tissue, and many of them are oxidizing agents. Sodium hypochlorite is one of the most widely used disinfectant compounds. In 5% solutions, it is marketed as liquid laundry bleach. This solution is an effective disinfectant for sinks, toilets, and similar fixtures. A chemically similar compound called calcium hypochlorite is the active ingredient in bleaching powder, and it is also used in hospitals as a disinfectant for clothing and bedding. Chlorine gas and ozone gas are two widely used, strong, oxidizing disinfectants. Their most well-known use is water treatment; they are added in small quantities to municipal water supplies to kill any harmful bacteria that may be present.

Hydrogen peroxide is an antiseptic; bleach is a disinfectant.

A word of caution is appropriate here. Although electron transfers are a useful concept for understanding redox reactions involving covalent substances, it must be remembered that such transfers actually take place only during the formation of ionic compounds (see Section 4.3). In covalent substances, the electrons are actually shared (see Section 4.6). The oxidation number assignment rules that were given, as well as the electron-transfer idea, are based on the arbitrary practice of assigning shared electrons to the more electronegative element sharing them. However, it must be remembered that none of the atoms in covalent molecules actually acquire a net charge.

▶ Example 5.4

Determine oxidation numbers for each atom represented in the following equations and identify the oxidizing and reducing agents:

a. $4Al(s) + 3O_2(g) \rightarrow 2Al_2O_3(s)$

b. $CO(g) + 3H_2(g) \rightarrow H_2O(g) + CH_4(g)$

c. $S_2O_8^{2-}(aq) + 2I^-(aq) \rightarrow I_2(aq) + 2SO_4^{2-}(aq)$

Solution

The O.N. under each elemental symbol was calculated by using the methods demonstrated in Example 5.3.

a. $4Al + 3O_2 \rightarrow 2Al_2O_3$
$$\quad\ 0 \qquad 0 \qquad\quad +3\ -2$$

The O.N. of Al has changed from 0 to $+3$. Therefore, Al has been oxidized and is the reducing agent. The O.N. of O has decreased from 0 to -2. The oxygen has been reduced and is the oxidizing agent.

b. $C\ O + 3H_2 \rightarrow H_2\ O + C\ H_4$
$$+2\ -2 \quad\ 0 \qquad +1\ -2 \quad -4\ +1$$

The O.N. of H_2 increased from 0 to $+1$. H_2 has been oxidized and is the reducing agent. The O.N. of C has decreased from $+2$ to -4. Carbon has been reduced and could be called the oxidizing agent. However, when one element in a molecule or ion is the oxidizing or reducing agent, the convention is to refer to the entire molecule or ion by the appropriate term. Thus, carbon monoxide (CO) is the oxidizing agent.

c. $S_2\ O_8^{2-} + 2I^- \rightarrow I_2 + 2S\ O_4^{2-}$
$$+7\ -2 \qquad -1 \qquad\ 0 \qquad +6 \quad -2$$

The O.N. of I has increased from -1 to 0, and the I^- ion is the reducing agent. The O.N. of S in $S_2O_8^{2-}$ has decreased from $+7$ to $+6$. Thus, S has been reduced, and the $S_2O_8^{2-}$ is the oxidizing agent.

▶ **LEARNING CHECK 5.4** Assign oxidation numbers to each atom represented in the following equations and identify the oxidizing and reducing agents:

a. $Zn(s) + 2H^+(aq) \rightarrow Zn^{2+}(aq) + H_2(g)$
b. $2KI(aq) + Cl_2(aq) \rightarrow 2KCl(aq) + I_2(aq)$
c. $IO_3^-(aq) + 3HSO_3^-(aq) \rightarrow I^-(aq) + 3HSO_4^-(aq)$

5.4 Decomposition Reactions

Learning Objective

3. Classify reactions into the categories of redox or nonredox, then into the categories of decomposition, combination, single replacement, or double replacement.

decomposition reaction A chemical reaction in which a single substance reacts to form two or more simpler substances.

$2H_2O_2 \longrightarrow 2H_2O + O_2$

Figure 5.3 A decomposition reaction.

In **decomposition reactions,** a single substance is broken down to form two or more simpler substances, as shown in ▶ Figure 5.3. In this and other "box" representations of reactions, the number of molecules in the boxes will not match the coefficients of the reaction equation. However, they will be in the correct proportions. In Figure 5.3, for example, the box on the left contains the same number of H_2O_2 molecules as the number of H_2O molecules on the right. Also, the number of H_2O molecules on the right is twice the number of O_2 molecules. The general form of the equation for a decomposition reaction is

$$A \rightarrow B + C \qquad (5.13)$$

Some decomposition reactions are also redox reactions, whereas others are not. Examples of decomposition reactions are given by Equations 5.14 and 5.15:

$$2HgO(s) \rightarrow 2Hg(\ell) + O_2(g) \qquad (5.14)$$

$$CaCO_3(s) \rightarrow CaO(s) + CO_2(g) \qquad (5.15)$$

Equation 5.14 represents the redox reaction that takes place when mercury(II) oxide (HgO) is heated. Mercury metal (Hg) and oxygen gas (O_2) are the products. This reaction was used by Joseph Priestley in 1774 when he discovered oxygen. Equation 5.15 represents a nonredox reaction used commercially to produce lime (CaO) by heating limestone ($CaCO_3$) to a high temperature. The decomposition of H_2O_2 represented by Figure 5.3 is shown in ▶ Figure 5.4.

© Jeffrey M. Seager   © Jeffrey M. Seager

5.5 Combination Reactions

Learning Objective

3. Classify reactions into the categories of redox or nonredox, then into the categories of decomposition, combination, single replacement, or double replacement.

Combination reactions are sometimes called *addition* or *synthesis reactions*. Their characteristic is that two or more substances react to form a single substance (see ▶ Figure 5.5). The reactants can be any combination of elements or compounds, but the product is always a compound. The general form of the equation is

$$A + B \rightarrow C \qquad (5.16)$$

combination reaction A chemical reaction in which two or more substances react to form a single substance.

$$H_2 \quad + \quad F_2 \longrightarrow 2HF$$

At high temperatures, a number of metals will burn and give off very bright light. This burning is a redox combination reaction, represented for magnesium metal by Equation 5.17 and shown in ▶ Figure 5.6.

$$2Mg(s) + O_2(g) \rightarrow 2MgO(s) \qquad (5.17)$$

A nonredox combination reaction that takes place in the atmosphere contributes to the acid rain problem. An air pollutant called sulfur trioxide (SO_3) reacts with water vapor and forms sulfuric acid. The reaction is represented by Equation 5.18:

$$SO_3(g) + H_2O(\ell) \rightarrow H_2SO_4(aq) \qquad (5.18)$$

Figure 5.5 A combination reaction.

© Cengage Learning/Charles D. Winters

Figure 5.6 Magnesium metal burns in air to form magnesium oxide.

5.6 Replacement Reactions

Learning Objective

3. Classify reactions into the categories of redox or nonredox, then into the categories of decomposition, combination, single replacement, or double replacement.

Single-replacement reactions, also called *substitution reactions*, are always redox reactions and take place when one element reacts with a compound and displaces another element from the compound. A single-replacement reaction is represented in ▶ Figure 5.7

single-replacement reaction A chemical reaction in which an element reacts with a compound and displaces another element from the compound.

and demonstrated in ▶ Figure 5.8. The general equation for the reaction is shown in Equation 5.19.

$$A + BX \rightarrow B + AX \tag{5.19}$$

Figure 5.7 A single-replacement reaction.

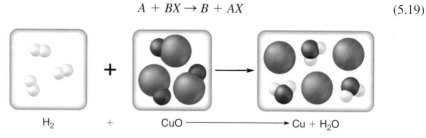

$$H_2 \quad + \quad CuO \longrightarrow Cu + H_2O$$

Figure 5.8 When a piece of copper wire (Cu) is placed in a solution of silver nitrate ($AgNO_3$) in water, crystals of silver metal (Ag) form on the wire, and the liquid solution that was originally colorless turns blue as copper nitrate [$Cu(NO_3)_2$] forms in solution. Write a balanced equation for this single-replacement reaction.

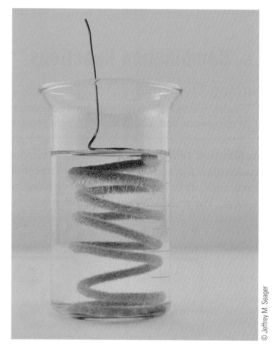

This type of reaction is useful in a number of processes used to obtain metals from their oxide ores. Iron, for example, can be obtained by reacting iron(III) oxide ore (Fe_2O_3) with carbon. The carbon displaces the iron from the oxide, and carbon dioxide is formed. The equation for the reaction is

$$3C(s) + 2Fe_2O_3(s) \rightarrow 4Fe(s) + 3CO_2(g) \tag{5.20}$$

double-replacement reaction
A chemical reaction in which two compounds react and exchange partners to form two new compounds.

Double-replacement reactions, also called *metathesis reactions,* are never redox reactions. These reactions, represented in ▶ Figure 5.9, often take place between substances dissolved in water. The general Equation 5.21 shows the partner-swapping characteristic of these reactions:

$$AX + BY \rightarrow BX + AY \tag{5.21}$$

Figure 5.9 A double-replacement, or metathesis, reaction.

$$NaOH \quad + \quad HF \longrightarrow NaF + H_2O$$

The reaction that takes place when a base is used to neutralize an acid is a good example of a double-replacement reaction:

$$HCl(aq) + NaOH(aq) \rightarrow NaCl(aq) + H_2O(\ell) \tag{5.22}$$

Classify each of the reactions represented by the following equations as redox or nonredox. Further classify them as decomposition, combination, single-replacement, or double-replacement reactions.

a. $SO_2(g) + H_2O(\ell) \rightarrow H_2SO_3(aq)$
b. $2K(s) + 2H_2O(\ell) \rightarrow 2KOH(aq) + H_2(g)$
c. $N_2(g) + 3H_2(g) \rightarrow 2NH_3(g)$
d. $BaCl_2(aq) + Na_2CO_3(aq) \rightarrow BaCO_3(s) + 2NaCl(aq)$

Solution
The O.N. under each elemental symbol was calculated by using the methods demonstrated in Example 5.3.

a. $\underset{+4}{S}\ \underset{-2}{O_2} + \underset{+1}{H_2}\ \underset{-2}{O} \rightarrow \underset{+1}{H_2}\ \underset{+4}{S}\ \underset{-2}{O_3}$

No O.N. changes take place; therefore, the reaction is nonredox. Because two substances combine to form a third, the reaction is a combination reaction.

b. $\underset{0}{2K} + \underset{+1\ -2}{2H_2\ O} \rightarrow \underset{+1\ -2\ +1}{2K\ O\ H} + \underset{0}{H_2}$

The O.N. of K increases from 0 to +1 and that of H decreases from +1 to 0. The reaction is a redox reaction. Because K displaces H, it is a single-replacement reaction.

c. $\underset{0}{N_2} + \underset{0}{3H_2} \rightarrow \underset{-3\ +1}{2N\ H_3}$

The O.N.s of both N and H change; therefore, the reaction is a redox reaction. Two substances combine to form a third, so it is a combination reaction.

d. $\underset{+2\ -1}{Ba\ Cl_2} + \underset{+1\ +4\ -2}{Na_2\ C\ O_3} \rightarrow \underset{+2\ +4\ -2}{Ba\ C\ O_3} + \underset{+1\ -1}{2Na\ Cl}$

No changes in O.N. occur. The reaction is nonredox and is an example of a double-replacement, or partner-swapping, reaction.

► **LEARNING CHECK 5.5** Classify each of the reactions represented by the following equations as redox or nonredox. Further classify them as decomposition, combination, single-replacement, or double-replacement reactions.

a. $2HI(g) \rightarrow H_2(g) + I_2(g)$
b. $2H_2O_2(aq) \rightarrow 2H_2O(\ell) + O_2(g)$ (NOTE: H_2O_2 is a peroxide)
c. $NaCl(aq) + AgNO_3(aq) \rightarrow AgCl(s) + NaNO_3(aq)$
d. $4P(s) + 5O_2(g) \rightarrow 2P_2O_5(s)$
e. $2NaI(aq) + Cl_2(aq) \rightarrow 2NaCl(aq) + I_2(aq)$

5.7 Ionic Equations

Learning Objective

4. Write molecular equations in total ionic and net ionic forms.

Many of the reactions of interest in the course you are taking occur between compounds or elements dissolved in water. Ionic compounds and some polar covalent compounds break apart (dissociate) into ions when they are dissolved in water. Thus, a water solution of sodium hydroxide (NaOH), an ionic compound, does not contain molecules of NaOH but, rather, contains equal numbers of sodium ions (Na^+) and hydroxide ions (OH^-). Covalently bonded hydrogen chloride, HCl, dissolves readily in water to form H^+ and Cl^- ions. Equations for reactions between substances that form ions in solution can be written in several ways. For example, Equation 5.22 contains three substances that form ions, HCl, NaOH, and NaCl. Equation 5.22 is written in the form of a **molecular equation** in which each compound is represented by its formula. This same reaction, when represented by a **total ionic equation,** becomes

$$H^+(aq) + Cl^-(aq) + Na^+(aq) + OH^-(aq) \rightarrow$$
$$Na^+(aq) + Cl^-(aq) + H_2O(\ell) \qquad (5.23)$$

molecular equation An equation written with each compound represented by its formula.

total ionic equation An equation written with all soluble ionic substances represented by the ions they form in solution.

The Importance of Color in Your Diet

Scientific evidence accumulated during the 1990s suggested that diets rich in fruits and vegetables had a protective effect against a number of different types of cancer. Studies showed that simply increasing the levels of vitamins and minerals in the diet did not provide the increased protection. This led to research into the nature of other substances found in fruits and vegetables that are important for good health. As a result of this research, a number of chemical compounds found in plants and called *phytonutrients* have been shown to be involved in the maintenance of healthy tissues and organs. The mechanism for their beneficial action in the body is not understood for all phytonutrients, but a significant number are known to work as antioxidants that stop harmful oxidation reactions from occurring.

The colors of fruits and vegetables help identify those containing beneficial compounds. The table below contains a list of some of the more well-known phytonutrients together with sources, colors, and beneficial actions. The amount of evidence supporting the existence of benefits from phytonutrients is not the same for all those listed in the table. In some cases, the experimental evidence is extensive (e.g., the cancer-blocking behavior of isothiocyanates), while in other cases the listed benefits are based on a limited amount of research and more studies are being done (e.g., the contribution to eye health by anthocyanins).

Fruit/Vegetable Color	Fruit/Vegetable Examples	Phytonutrients	Possible Benefits
Red	Tomatoes, watermelon, pink grapefruit	Lycopene (a carotenoid)	Protect against prostate, cervical, and pancreatic cancer and heart and lung disease
Red/purple	Red and blue grapes, blueberries, strawberries, beets, eggplant, red cabbage, red peppers, plums, red apples	Anthocyanins (flavonoids)	Antioxidants; block formation of blood clots and help maintain good eye health
Orange	Carrots, mangoes, sweet potatoes, cantaloupe, winter squash	Alpha- and beta-carotenes	Cancer fighters; protect skin against free radicals, promote repair of damaged DNA
Orange/yellow	Oranges, peaches, papaya, nectarines	Beta-cryptoxanthin	May help prevent heart disease
Yellow/green	Spinach, collards, corn, green peas, avocado, honeydew	Lutein and zeaxanthin (both are carotenoids)	Reduce risk of cataracts and age-related macular degeneration
Green	Broccoli, brussels sprouts, cabbage, kale, bok choy	Sulforaphane, isothiocyanates, and indoles	Cancer blocking
White/green	Onions, leeks, garlic, celery, asparagus, pears, green grapes, cauliflower, mushrooms	Allicin (in onions) and flavanoids	Antitumor agent; antioxidants

In this equation each ionic compound is shown dissociated into ions, the form it takes when it is dissolved in water. Some of the ions appear as both reactants and products. These so-called **spectator ions** do not actually undergo any changes in the reaction. Because of this, they are dropped from the equation when it is written as a **net ionic equation:**

$$H^+(aq) + OH^-(aq) \rightarrow H_2O(\ell) \qquad (5.24)$$

The net ionic equation makes the partner-swapping characteristics of this double-replacement reaction less obvious, but it does emphasize the actual chemical changes that take place. ▶ Figure 5.10 shows an experiment in which an ionic reaction occurs.

spectator ions The ions in a total ionic reaction that are not changed as the reaction proceeds. They appear in identical forms on the left and right sides of the equation.

net ionic equation An equation that contains only un-ionized or insoluble materials and ions that undergo changes as the reaction proceeds. All spectator ions are eliminated.

© Joel Gordon

Figure 5.10 The liquid in the large container is a solution of solid sodium chloride (NaCl) in water. The liquid being added is a solution of solid silver nitrate (AgNO₃) in water. When the two liquids are mixed, an insoluble white solid forms. The solid is silver chloride (AgCl). Write the molecular, total ionic, and net ionic equations for the reaction.

▶ Example 5.6

Write equations for the following double-replacement reaction in total ionic and net ionic forms. Note that barium sulfate ($BaSO_4$) does not dissolve in water and should not be written in dissociated form. All other compounds are ionic and soluble in water.

$$Na_2SO_4(aq) + BaCl_2(aq) \rightarrow BaSO_4(s) + 2NaCl(aq)$$

Solution
Total ionic:

$$2Na^+(aq) + SO_4^{2-}(aq) + Ba^{2+}(aq) + 2Cl^-(aq) \rightarrow$$
$$BaSO_4(s) + 2Na^+(aq) + 2Cl^-(aq)$$

Net ionic:

$$SO_4^{2-}(aq) + Ba^{2+}(aq) \rightarrow BaSO_4(s)$$

The Na^+ and Cl^- ions appear in equal numbers as reactants and products. Thus, they are spectator ions and are not shown in the net ionic equation.

▶ LEARNING CHECK 5.6
Write equations for the following reactions in total ionic and net ionic forms. Consider all ionic compounds to be soluble except $CaCO_3$ and $BaSO_4$, and remember that covalent molecules do not form ions when they dissolve.

a. $2NaI(aq) + Cl_2(aq) \rightarrow 2NaCl(aq) + I_2(aq)$
b. $CaCl_2(aq) + Na_2CO_3(aq) \rightarrow 2NaCl(aq) + CaCO_3(s)$
c. $Ba(OH)_2(aq) + H_2SO_4(aq) \rightarrow 2H_2O(\ell) + BaSO_4(s)$

5.8 Energy and Reactions

Learning Objective

5. Classify reactions as exothermic or endothermic.

Besides changes in composition, energy changes accompany all chemical reactions. The equation for the reaction between hydrogen and oxygen given at the beginning of this chapter (Equation 5.2) can also be written

$$2H_2(g) + O_2(g) \rightarrow 2H_2O(g) + \text{energy} \tag{5.25}$$

Most of the energy of this reaction appears as heat, but the energy released or absorbed during chemical changes can take many forms including sound, electricity, light, high-energy chemical bonds, and motion. Some of these forms are discussed in detail in later chapters, but here the focus will be on energy in general, and it will be expressed in calories or joules, as if it all took the form of heat.

On the basis of heat, Equation 5.25 can be written

$$2H_2(g) + O_2(g) \rightarrow 2H_2O(g) + 115.6 \text{ kcal } (483.7 \text{ kJ}) \tag{5.26}$$

This equation assumes that the hydrogen and oxygen are gases when they react and that the water produced is also a gas (vapor). According to Equation 5.26, whenever 2 mol of water vapor is formed from 2 mol of hydrogen gas and 1 mol of oxygen gas, 115.6 kcal, or 483.7 kJ, of energy is also released.

exothermic reaction A reaction that liberates heat.

endothermic reaction A reaction that absorbs heat.

The reaction represented by Equation 5.26 is an example of an **exothermic** (heat out) **reaction,** in which heat is released as the reaction takes place. In **endothermic** (heat in) **reactions,** heat is absorbed. A reaction of this type is utilized in emergency cold packs, which consist of a small plastic pouch of liquid sealed inside a larger pouch that also contains a solid substance. When the large pouch is squeezed firmly, the small pouch of liquid breaks, the liquid and solid react, heat is absorbed, and the liquid contents of the larger pouch become cold. This reaction is described in more detail in Chapter 7.

5.9 The Mole and Chemical Equations

Learning Objective

6. Use the mole concept to do calculations based on chemical reaction equations.

The mole concept introduced in Section 2.6 and applied to chemical formulas in Section 2.7 can also be used to calculate mass relationships in chemical reactions. The study of such mass relationships is called **stoichiometry,** a word derived from the Greek *stoicheion* (element) and *metron* (measure).

stoichiometry The study of mass relationships in chemical reactions.

Stoichiometry calculations require that balanced equations be used for the reactions being studied. Consider the following equation for the redox reaction that takes place when methane (CH_4), a major constituent of natural gas, is burned:

$$CH_4(g) + 2O_2(g) \rightarrow CO_2(g) + 2H_2O(\ell) \tag{5.27}$$

Remember, coefficients in balanced equations refer to the formula that follows the coefficient. With this in mind, note that the following statement is consistent with this balanced equation:

1. 1 CH_4 molecule + 2 O_2 molecules $\rightarrow$ 1 CO_2 molecule + 2 H_2O molecules

This statement indicates that one molecule of CH_4 gas reacts with two molecules of O_2 gas and produces one molecule of CO_2 gas plus two molecules of H_2O liquid. Put another way, according to this reaction equation, the number of O_2 molecules that react with CH_4 will be equal to twice the number of CH_4 molecules that react. We remember that one mole of any kind of molecule is equal to 6.02×10^{23} molecules. Thus, one mole of CH_4 is equal to 6.02×10^{23} CH_4 molecules. As shown earlier, the number of O_2 molecules that will react

Redox Reactions and Energy for Life

All living organisms require energy to carry out the processes necessary to sustain life. This energy is obtained through redox (reduction-oxidation) reactions of various types. The most familiar to most of us are the redox reactions involved in the complicated process called *metabolism*. An example of a metabolic redox process that provides energy is the oxidation of a carbohydrate such as glucose, $C_6H_{12}O_6$. In the body, the overall process consists of several reactions, but it can be represented by the following equation:

$$C_6H_{12}O_6 + 6O_2 \rightarrow 6CO_2 + 6H_2O + energy$$

Organisms that depend on such processes for energy must obtain the carbohydrate from a food source, and the primary source of carbohydrates for most animals is plants. Plants obtain the energy needed to form carbohydrates and carry out their life processes from the sun in the process called *photosynthesis*. Photosynthesis also consists of many individual reactions, but the overall process can be represented in a simplified way by the following equation:

$$sunlight + CO_2 + H_2O \rightarrow carbohydrates + O_2$$

Thus, we see that the energy used by animals is actually solar energy stored in the form of carbohydrates by plants. As seen in the above equation, during photosynthesis plants not only store energy in the form of carbohydrates but they also use up CO_2 and regenerate oxygen. The oxygen can be used by animals in the oxidation process of carbohydrates.

Some bacteria that live in soil as well as fresh and marine waters use other redox reactions as sources of energy. For example, soil-dwelling nitrifying bacteria convert both ammonia, NH_3, and nitrites, NO_2^-, to nitrates, NO_3^-. During this oxidation process energy is released that can be used by the bacteria, and soil fertility is increased by the nitrate that is also produced.

In 1977 scientists discovered a large number of previously unknown animals living so deep in the ocean that sunlight cannot reach them. Organisms living in more shallow water where sunlight can penetrate depend on photosynthesis for their food and energy supplies, but what provides energy to these bottom dwellers? It was discovered that these organisms lived in the vicinity of hydrothermal vents in the ocean floor. These vents are chimney-like structures that form over fissures or cracks in the ocean floor. Mineral-rich hot water flows upward through the fissures into the cold surrounding ocean water.

The hot water that pours from the vents forms when cold water seeps downward through cracks in the ocean floor and comes in contact with hot lava. The cold water contains sulfate, SO_4^{2-}, which is reduced to hydrogen sulfide, H_2S, under the conditions of high temperature and high pressure that are created as the water is heated. The heated water containing dissolved H_2S is then forced up through the vents.

It was discovered that the water surrounding the vents is inhabited by large numbers of bacteria that oxidize the H_2S back to sulfate, SO_4^{2-} and use energy derived from the oxidation process to convert dissolved carbon dioxide, CO_2, in the water into organic nutrients. These nutrients are used as food by animals living in the vicinity of the vents. Thus, the energy to sustain life in these dark regions comes from hot lava in the form of high-energy bonds in H_2S. This energy is released and used to form food when the H_2S is oxidized to SO_4^{2-}.

Ocean hydrothermal vents.

with CH_4 will be twice the number of CH_4 molecules that react. Thus, the number of O_2 molecules that will react with 6.02×10^{23} CH_4 molecules is $2(6.02 \times 10^{23})$. Remembering that 6.02×10^{23} molecules is 1 mole of molecules leads to the conclusion that 1 mole of CH_4 molecules will react with 2 moles of O_2 molecules. We note that the numbers 1 and 2 are identical to the coefficients of the CH_4 and O_2 in the balanced equation. This leads to the conclusion that the coefficients in reaction equations can be interpreted as the number of moles of reactants and products involved in the reaction, which leads to the following statement for the reaction of CH_4 with O_2:

2. $1 \text{ mol } CH_4(g) + 2 \text{ mol } O_2(g) \rightarrow 1 \text{ mol } CO_2(g) + 2 \text{ mol } H_2O(\ell)$

In general, any balanced reaction equation can be interpreted this way, where the coefficients of the equation become the number of moles of reactants and products involved in the reaction.

The fact used earlier that 1 mole of any substance is equal to 6.02×10^{23} particles of the substance leads to statement 3 given below for the reaction of CH_4 with O_2:

3. 6.02×10^{23} CH_4 molecules $+ 2(6.02 \times 10^{23})$ O_2 molecules $\rightarrow$
6.02×10^{23} CO_2 molecules $+ 2(6.02 \times 10^{23})$ H_2O molecules

Application of the fact that one mole of a compound has a mass in grams equal to the molecular weight of the compound leads to statement 4 given below for the reaction:

4. $16.0 \text{ g CH}_4 + 2(32.0) \text{ g O}_2 \rightarrow 44.0 \text{ g CO}_2 + 2(18.0) \text{ g H}_2\text{O}$

Statements 2, 3, and 4 are all based on the mole definition, and are very useful in solving numerical problems involving balanced reaction equations and the factor-unit method described earlier in Section 1.9. Any two quantities from statements 2, 3, and 4 can be used to form factors that can be used to solve problems. For example, the following factors are just four of the many that can be obtained from statements 2, 3, and 4 by combining various quantities from the statements:

$$\frac{2 \text{ mol O}_2}{1 \text{ mol CH}_4} \qquad \frac{16.0 \text{ g CH}_4}{44.0 \text{ g CO}_2} \qquad \frac{2(18.0) \text{ g H}_2\text{O}}{6.02 \times 10^{23} \text{ CH}_4 \text{ molecules}} \qquad \frac{1 \text{ mol CH}_4}{44.0 \text{ g CO}_2}$$

Suppose you were asked this question: How many moles of CO_2 could be formed by reacting together 2 mol of CH_4 and 4 mol of O_2? Statement 2 can be used to solve this problem quickly. We see that the amounts reacted correspond to twice the amounts represented by Statement 2. Thus, 2 mol of CO_2 would be formed, which is also twice the amount represented by Statement 2. However, many stoichiometric problems cannot be solved quite as readily, so it is helpful to learn a general approach that works well for many problems of this type. This approach is based on the factor-unit method described earlier in Section 1.9. The needed factors are obtained from Statements 2, 3, and 4.

▶ Example 5.7

The following questions are based on Equation 5.27.

a. How many moles of oxygen gas (O_2) will be required to react with 1.72 mol of methane (CH_4)?
b. How many grams of H_2O will be produced when 1.09 mol of CH_4 reacts with an excess of O_2?
c. How many grams of O_2 must react with excess CH_4 to produce 8.42 g of carbon dioxide (CO_2)?

Solution

a. The steps to follow in the factor-unit method are as follows (see Section 1.9):

1. Write down the known or given quantity. It will include both a number and a unit.
2. Leave working space, and set the known quantity equal to the unit of the unknown quantity.
3. Multiply the known quantity by a factor such that the unit of the known quantity is canceled and the unit of the unknown quantity is generated.
4. After the units on each side of the equation match, do the necessary arithmetic to get the final answer.

In this problem, the known quantity is 1.72 mol of CH_4, and the unknown quantity has the unit of mol O_2.

Step 1. 1.72 mol CH_4

Step 2. 1.72 mol CH_4 $\qquad\qquad = \text{mol O}_2$

Step 3. 1.72 $\cancel{\text{mol } CH_4} \times \dfrac{2 \text{ mol O}_2}{1 \cancel{\text{mol } CH_4}} = \text{mol O}_2$

The factor

$$\frac{2 \text{ mol O}_2}{1 \text{ mol CH}_4}$$

was used because it properly canceled the unit of the known and generated the unit of the unknown.

Step 4. $(1.72)\left(\dfrac{2 \text{ mol } O_2}{1}\right) = 3.44 \text{ mol } O_2$

The answer contains the proper number of significant figures because the 2 and 1 of the factor are exact numbers (see Section 1.8).

b.

Step 1. 1.09 mol CH_4

Step 2. 1.09 mol CH_4 = g H_2O

Step 3. 1.09 mol CH_4 × $\dfrac{2(18.0) \text{ g } H_2O}{1 \text{ mol } CH_4}$ = g H_2O

Step 4. $1.09 \times \dfrac{2(18.0) \text{ g } H_2O}{1} = 39.24 \text{ g } H_2O = 39.2 \text{ g } H_2O$

The answer was rounded to 3 significant figures to match the three in 1.09 and 18.0 g of H_2O. The 1 and 2 are exact counting numbers.

c.

Step 1. 8.42 g CO_2

Step 2. 8.42 g CO_2 = g O_2

Step 3. 8.42 g CO_2 × $\dfrac{2(32.0) \text{ g } O_2}{44.0 \text{ g } CO_2}$ = g O_2

Step 4. $8.42 \times \dfrac{2(32.0) \text{ g } O_2}{44.0} = 12.2472 \text{ g } O_2 = 12.2 \text{ g } O_2$

The answer was rounded to 3 significant figures to match the three in 8.42, 32.0 g of O_2 and the 44.0. The 2 is an exact counting number.

▶ **LEARNING CHECK 5.7** The following reaction equation is balanced:

$$N_2(g) + 3H_2(g) \rightarrow 2NH_3(g)$$

a. Write statements equivalent to statements 2, 3, and 4 based on this equation.
b. Calculate the number of mol of NH_3 that would be produced if 2.11 mol of H_2 was reacted with excess N_2.
c. Calculate the number of g of N_2 required to react with 9.47 g H_2.

5.10 The Limiting Reactant

Learning Objective

7. Use the mole concept to do calculations based on the limiting-reactant principle.

Nearly everyone is familiar with two characteristics of automobiles. They stop running when they run out of gasoline, and they stop running when the air intake to the engine becomes clogged. Most people also know that gasoline and air are mixed, and the mixture is burned inside the automobile engine. However, the combustion reaction occurs only as long as both reactants, gasoline and air, are present. The reaction (and engine) stops when gasoline is absent, despite the presence of ample amounts of air. On the other hand, the reaction stops when air is absent, even though plenty of gasoline is available.

The behavior of the auto engine illustrates the **limiting-reactant principle** that a chemical reaction will take place only as long as all necessary reactants are present. The reaction will stop when one of the reactants, the **limiting reactant,** is used up. According to this principle, the amount of product formed depends on the amount of limiting reactant present because the reaction will stop (and no more product will form) when the limiting reactant is used up. This is represented in ▶ Figure 5.11 for the reaction $F_2(g) + H_2(g) \rightarrow 2HF(g)$.

limiting-reactant principle The maximum amount of product possible from a reaction is determined by the amount of reactant present in the least amount, based on its reaction coefficient and molecular weight.

limiting reactant The reactant present in a reaction in the least amount, based on its reaction coefficients and molecular weight. It is the reactant that determines the maximum amount of product that can be formed.

H_2 + F_2 ⟶ Mixture of reactants ⟶ HF + unreacted H_2

Figure 5.11 The limiting reactant is used up in a reaction.

▶ Example 5.8

A mixture containing 20.0 g of CH_4 and 100 g of O_2 is ignited and burns according to Equation 5.27, which is repeated here: $CH_4(g) + 2O_2(g) \rightarrow CO_2(g) + 2H_2O(\ell)$. What substances will be found in the mixture after the reaction stops?

Solution

The task is to identify the limiting reactant. It will be completely used up, and the other reactant will be found mixed with the products (CO_2 and H_2O) after the reaction stops. A simple way to identify the limiting reactant is to find out which reactant will give the least amount of products. To do this, two problems must be solved, one in which the known quantity is 20.0 g of CH_4 and one in which the known quantity is 100.0 g of O_2. In each case, the theoretical amount of CO_2 produced is calculated, although the same information would be obtained by solving for the amount of water produced.

$$20.0 \text{ g } CH_4 \times \frac{44.0 \text{ g } CO_2}{16.0 \text{ g } CH_4} = 55.0 \text{ g } CO_2$$

$$100 \text{ g } O_2 \times \frac{44.0 \text{ g } CO_2}{2(32.0) \text{ g } O_2} = 68.75 \text{ g } CO_2 = 68.8 \text{ g } CO_2$$

Since the 20.0 g of CH_4 produces the smaller amount of CO_2, CH_4 is the limiting reactant. Therefore, the final mixture will contain the products CO_2 and H_2O and the leftover O_2.

▶ Example 5.9

The reaction shown in Figure 5.8 was carried out by a student who wanted to produce some silver metal. The equation for the reaction is

$$Cu(s) + 2AgNO_3(aq) \rightarrow 2Ag(s) + Cu(NO_3)_2(aq)$$

The student dissolved 17.0 g of $AgNO_3$ in distilled water, then added 8.50 g of copper metal, Cu, to the resulting solution. What mass of silver metal, Ag, was produced?

Solution

The limiting reactant must be identified because it will determine the amount of silver produced. To find the limiting reactant, we calculate the theoretical amount of silver produced by each reactant. The reactant that will produce the smallest mass of silver is the limiting reactant, and the mass of silver calculated for it is the amount produced by the reaction.

$$17.0 \text{ g } AgNO_3 \times \frac{2(108) \text{ g } Ag}{2(170) \text{ g } AgNO_3} = 10.8 \text{ g } Ag$$

$$8.50 \text{ g } Cu \times \frac{2(108) \text{ g } Ag}{63.6 \text{ g } Cu} = 28.8679 \text{ g } Ag = 28.9 \text{ g } Ag$$

We see that the $AgNO_3$ is the limiting reactant, and it will produce 10.8 g of Ag.

▶ LEARNING CHECK 5.8

Refer to the reaction used for Learning Check 5.7. Assume that 2.00 mol of H_2 and 15.5 g of N_2 are reacted. What is the maximum mass of ammonia (NH_3) that can be produced? Which reactant is the limiting reactant?

Air Bag Chemistry

Air bags installed in automobiles have proven to be effective, yet somewhat controversial, safety devices. According to the National Transportation Safety Administration, air bags have saved an estimated 1100 lives and prevented many more serious injuries from automobile accidents. This effectiveness is tempered by the fact that air bags have caused the deaths of a number of small children and infants. Short adults have also died as a result of being hit in the face rather than the chest by an expanding air bag. For this reason, it is now generally recommended that infants, small children, and short adults not occupy the front passenger seat. Also, most newer model automobiles are equipped with a switch that allows the air bag to be deactivated.

An air bag is essentially a nylon fabric bag that fills very rapidly with a gas when a collision occurs. Inflated air bags cushion and protect the driver and front-seat passenger. The gas that inflates an air bag is nitrogen, N_2, that is produced in a gas generator by a redox reaction of a toxic, explosive material called sodium azide, NaN_3. When a collision occurs, an electrical impulse triggers the rapid decomposition of the sodium azide according to the following equation:

$$2NaN_3(s) \rightarrow 2Na(g) + 3N_2(g)$$

The sodium vapor produced by this reaction would be very hazardous to the occupants of the automobile, but potassium nitrate, KNO_3, also included in the gas-generating mixture, reacts with the sodium vapor to produce potassium oxide, K_2O, sodium oxide, Na_2O, and some additional nitrogen gas. The equation for the reaction is

$$10Na(g) + 2KNO_3(s) \rightarrow K_2O(s) + 5Na_2O(s) + N_2(g)$$

The potassium oxide and sodium oxide are also hazardous materials; they are both basic oxides that react with water to form very caustic potassium hydroxide and sodium hydroxide, respectively. However, silicon dioxide, SiO_2, a third substance included in the gas-generating mixture of chemicals, is an acidic oxide that reacts with the basic potassium and sodium oxides, neutralizes their caustic characteristics, and converts them into a safe silicate-glass powder.

Thus, we see that from a chemical standpoint, the products of an air bag inflation are rendered safe by a series of carefully designed reactions. However, most air bags are never inflated, which means that old cars sent to scrap yards still contain a small amount of very toxic and explosive sodium azide. This is a recycling or disposal problem that is yet to be resolved.

Air bags: Protection provided by chemical reactions.

© Don Johnston/Getty Images

5.11 Reaction Yields

Learning Objective

8. Use the mole concept to do percentage-yield calculations.

In Sections 5.9 and 5.10, calculations were done to determine how much product could be obtained from the reaction of specified amounts of reactants. In many instances, the amount determined by such calculations would be greater than the amount produced by reacting the specified reactants in a laboratory. Does this mean matter is destroyed when reactions are actually done in the laboratory? No, this would violate the law of conservation of matter, a fundamental law of nature. Less product might be obtained because some of the reactants form compounds other than the desired product. Such reactions are called **side reactions.** Carbon dioxide gas, CO_2, is produced when carbon-containing fuels burn in an ample supply of oxygen. However, a side reaction produces toxic carbon monoxide gas, CO, when the oxygen supply is limited.

Another cause of less product than expected might be poor laboratory technique resulting in such things as losses when materials are transferred from one container to another.

side reactions Reactions that do not give the desired product of a reaction.

Assigning oxidation numbers is a task that seems to be straightforward, but mistakes caused by carelessness often show up when it is done on exams. You can minimize such mistakes by using a systematic approach to the task. First, have the seven rules for assigning oxidation numbers well in mind. Then, use an organized method such as the boxlike approach (shown below) to assign an oxidation number to the Mn in Mn_2O_3. In this approach, we begin with an oxidation number that is given by one of the rules (O is –2 according to Rule 5):

Each Mn must be +3 because two Mn give +6 total.

Mn_2 O_3
+3

Rule 5: O.N. of O = –2.

–2

Rule 6: Sum of total O.N. of Mn and O = 0.

+6 ← –6

Total O.N. for 3 O's = –6

Follow the arrows in a clockwise direction to solve for the O.N. of each atom of Mn (+3 is correct). Remember that oxidation numbers are reported for individual atoms. For example, the total oxidation number due to Mn atoms is +6, but because there are two atoms of Mn in the formula, the oxidation number for each atom is just +3.

percentage yield The percentage of the theoretical amount of a product actually produced by a reaction.

Whatever the cause, the mass of product obtained in an experiment is called the *actual yield,* and the mass calculated according to the methods of Sections 5.9 and 5.10 is called the *theoretical yield.* The **percentage yield** is the actual yield divided by the theoretical yield multiplied by 100 (see Section 1.10):

$$\% \text{ yield} = \frac{\text{actual yield}}{\text{theoretical yield}} \times 100 \qquad (5.28)$$

▶ Example 5.10

A chemist wants to produce urea (N_2CH_4O) by reacting ammonia (NH_3) and carbon dioxide (CO_2). The balanced equation for the reaction is

$$2NH_3(g) + CO_2(g) \rightarrow N_2CH_4O(s) + H_2O(\ell)$$

The chemist reacts 5.11 g of NH_3 with excess CO_2 and isolates 3.12 g of solid N_2CH_4O. Calculate the percentage yield of the experiment.

Solution

First, the theoretical yield must be calculated. Because this involves only masses of reactants and products, statements 2 and 4 based on the balanced equation will be used:

$$2NH_3(g) + CO_2(g) \rightarrow N_2CH_4O(s) + H_2O(\ell)$$

2. $\quad 2 \text{ mol } NH_3(g) + 1 \text{ mol } CO_2(g) \rightarrow 1 \text{ mol } N_2CH_4O(s) + 1 \text{ mol } H_2O(\ell)$

4. $\quad 2(17.0) \text{ g } NH_3 + 44.0 \text{ g } CO_2 \rightarrow 60.0 \text{ g } N_2CH_4O + 18.0 \text{ g } H_2O$

The known quantity is 5.11 g of NH_3, and the unit of the unknown quantity is g N_2CH_4O. The needed factor will come from statement 4.

Step 1. 5.11 g NH_3

Step 2. 5.11 g NH_3 $\qquad\qquad\qquad\qquad\qquad$ = g N_2CH_4O

Step 3. 5.11 g NH_3 $\times \dfrac{60.0 \text{ g } N_2CH_4O}{2(17.0) \text{ g } NH_3}$ $\qquad\qquad$ = g N_2CH_4O

Step 4. $5.11 \times \dfrac{60.0 \text{ g } N_2CH_4O}{2(17.0)}$ = 9.0176 g N_2CH_4O = 9.02 g N_2CH_4O

The actual yield was 3.12 g, so the percentage yield is

$$\% \text{ yield} = \frac{\text{actual yield}}{\text{theoretical yield}} \times 100 = \frac{3.12 \text{ g}}{9.02 \text{ g}} \times 100 = 34.5898\% = 34.6\%$$

▶ **LEARNING CHECK 5.9**

a. A chemist isolates 17.43 g of product from a reaction that has a calculated theoretical yield of 21.34 g. What is the percentage yield?

b. Lime (CaO) is produced by heating calcium carbonate. The equation for the reaction is $CaCO_3(s) \rightarrow CaO(s) + CO_2(g)$. Suppose 510 g of $CaCO_3$ is heated, and 235 g of CaO is isolated after the reaction mixture cools. What is the percentage yield for the reaction?

Case Study Follow-up

Most drug interactions result in a change in the absorption, metabolism, or elimination of a drug when it is taken at the same time as another drug. Interactions may also occur when two drugs have either additive or canceling effects and are administered together. Drug interactions are complex and are sometimes unpredictable. Known interactions might not occur in every individual. The frequency of interactions increases with the number of medications, so individuals should work with their healthcare providers to eliminate unnecessary medications (Ogbrue, 1997).

Pharmacists have the responsibility to carefully and safely dispense medications. The responsibility for safety is shared to some degree with the patient and his or her physician, who should inform the pharmacist which other drugs the patient is taking, confirm that the dosage is correct, and so on. According to the American Association of the Colleges of Pharmacy, pharmacists should ensure that drugs do not interact in a harmful way with other medications the patient might be taking. Patients expect pharmacists to look out for their interests and to warn them about possible problems with their medications. Pharmacists are highly educated drug experts who serve as important sources of information and guidance (AACP, 2011).

Concept Summary

Chemical Equations

Chemical reactions are conveniently represented by equations in which reacting substances, called reactants, and the substances produced, called products, are written in terms of formulas. Coefficients are placed before reactant and product formulas to balance the equation. A balanced equation satisfies the law of conservation of matter.

Objective 1, Exercises 5.2 and 5.6

Types of Reactions

To facilitate their study, reactions are classified into the general category redox or nonredox, and further classified as decomposition, combination, single-replacement, or double-replacement reactions.

Redox Reactions

Reactions in which reactants undergo oxidation or reduction are called redox reactions. Oxidation and reduction are indicated conveniently by the oxidation number changes that occur. Oxidation numbers are assigned according to a specific set of rules. A substance is oxidized when the oxidation number of a constituent element increases, and it is reduced when the oxidation number of a constituent element decreases.

Objective 2, Exercises 5.10 and 5.15

Decomposition Reactions

Decomposition reactions are characterized by one substance reacting to give two or more products. Decomposition reactions can be redox or nonredox.

Objective 3, Exercise 5.20

Combination Reactions

Combination reactions, also called addition or synthesis reactions, are characterized by two or more reactants that form a single compound as a product. Combination reactions can be redox or nonredox.

Objective 3, Exercise 5.20

Replacement Reactions

Single-replacement reactions, sometimes called substitution reactions, are always redox reactions. One element reacts with a compound and displaces another element from the compound. Double-replacement reactions, also called metathesis reactions, are always nonredox reactions. They can be recognized by their partner-swapping characteristics.

Objective 3, Exercise 5.20

Ionic Equations

Many water-soluble compounds separate (dissociate) into ions when dissolved in water. Reactions of such materials can be represented by molecular equations in which no ions are shown, total ionic equations in which all ions are shown, or net ionic equations in which only ions actually undergoing a change are shown.

Objective 4, Exercise 5.30a, b, c

Energy and Reactions

Energy changes accompany all chemical reactions. The energy can appear in a variety of forms, but heat is a common one. Exothermic reactions liberate heat, and endothermic reactions absorb it.

Objective 5, Exercise 5.34

The Mole and Chemical Equations

The mole concept, when applied to chemical equations, yields relationships that can be used to obtain factors for doing factor-unit calculations.

Objective 6, Exercise 5.42

The Limiting Reactant

The limiting reactant is the reactant present in a reaction in an amount that determines the maximum amount of product that can be made. Factor-unit calculations can be used to determine which reactant is limiting.

Objective 7, Exercise 5.52

Reaction Yields

The mass of product isolated after a reaction is often less than the mass theoretically possible. The ratio of the actual isolated mass (actual yield) to the calculated theoretical yield multiplied by 100 is the percentage yield.

Objective 8, Exercise 5.56

Key Terms and Concepts

Balanced equation (5.1)
Combination reaction (5.5)
Decompostion reaction (5.4)
Double-replacement reaction (5.6)
Endothermic reaction (5.8)
Exothermic reaction (5.8)
Law of conservation of matter (5.1)
Limiting reactant (5.10)

Limiting-reactant principle (5.10)
Molecular equation (5.7)
Net ionic equation (5.7)
Oxidation (5.3)
Oxidation number (oxidation state) (5.3)
Oxidizing agent (5.3)
Percentage yield (5.11)
Product of a reaction (5.1)

Reactant of a reaction (5.1)
Reducing agent (5.3)
Reduction (5.3)
Side reaction (5.11)
Single-replacement reaction (5.6)
Spectator ion (5.7)
Stoichiometry (5.9)
Total ionic equation (5.7)

Key Equations

1.	Decomposition reaction—one substance changes to two or more new substances (Section 5.4):	$A \rightarrow B + C$	Equation 5.13
2.	Combination reaction—two or more substances react to produce one new substance (Section 5.5):	$A + B \rightarrow C$	Equation 5.16
3.	Single-replacement reaction—one element reacts with a compound to produce a new compound and new element (Section 5.6):	$A + BX \rightarrow B + AX$	Equation 5.19
4.	Double-replacement reaction—partner-swapping reaction (Section 5.6):	$AX + BY \rightarrow BX + AY$	Equation 5.21
5.	Percentage yield (Section 5.11):	$\% \text{ yield} = \dfrac{\text{actual yield}}{\text{theoretical yield}} \times 100$	Equation 5.28

Exercises

Even-numbered exercises are answered in Appendix B.

Blue-numbered exercises are more challenging.

Chemical Equations (Section 5.1)

5.1 Identify the reactants and products in each of the following reaction equations:

a. $BaO_2(s) + H_2SO_4(\ell) \rightarrow BaSO_4(s) + H_2O_2(\ell)$

b. $2H_2O_2(aq) \rightarrow 2H_2O(\ell) + O_2(g)$

c. methane + water → carbon monoxide + hydrogen

d. copper(II) oxide + hydrogen → copper + water

5.2 Identify the reactants and products in each of the following reaction equations:

a. $H_2(g) + Cl_2(g) \rightarrow 2HCl(g)$

b. $2KClO_3(s) \rightarrow 2KCl(s) + 3O_2(g)$

c. magnesium oxide + carbon →
magnesium + carbon monoxide

d. ethane + oxygen → carbon dioxide + water

5.3 Identify which of the following are consistent with the law of conservation of matter. For those that are not, explain why they are not.

a. $4Al(s) + 3O_2(g) \rightarrow 2Al_2O_3(s)$

b. $P_4(s) + O_2(g) \rightarrow P_4O_{10}(s)$

c. 3.20 g oxygen + 3.21 g sulfur → 6.41 g sulfur dioxide

d. $CH_4(g) + 2O_2(g) \rightarrow CO_2(g) + 2H_2O(g)$

5.4 Identify which of the following are consistent with the law of conservation of matter. For those that are not, explain why they are not.

a. $ZnS(s) + O_2(g) \rightarrow ZnO(s) + SO_2(g)$

b. $Cl_2(aq) + 2I^-(aq) \rightarrow I_2(aq) + 2Cl^-(aq)$

c. 1.50 g oxygen + 1.50 g carbon → 2.80 g carbon monoxide

d. $2C_2H_6(g) + 7O_2(g) \rightarrow 4CO_2(g) + 6H_2O(g)$

5.5 Determine the number of atoms of each element on each side of the following equations and decide which equations are balanced:

a. $H_2S(aq) + I_2(aq) \rightarrow 2HI(aq) + S(s)$

b. $KClO_3(s) \rightarrow KCl(s) + O_2(g)$

c. $SO_2(g) + H_2O(\ell) \rightarrow H_2SO_3(aq)$

d. $Ba(ClO_3)_2(aq) + H_2SO_4(aq) \rightarrow 2HClO_3(aq) + BaSO_4(s)$

5.6 Determine the number of atoms of each element on each side of the following equations and decide which equations are balanced:

  a. $Ag(s) + Cu(NO_3)_2(aq) \rightarrow Cu(s) + AgNO_3(aq)$

  b. $2N_2O(g) + 3O_2(g) \rightarrow 4NO_2(g)$

  c. $Mg(s) + O_2(g) \rightarrow 2MgO(s)$

  d. $H_2SO_4(aq) + Ca(OH)_2(aq) \rightarrow CaSO_4(s) + 2H_2O(\ell)$

5.7 Balance the following equations:

  a. $Cl_2(aq) + NaBr(aq) \rightarrow NaCl(aq) + Br_2(aq)$

  b. $CaF_2(s) + H_2SO_4(aq) \rightarrow CaSO_4(s) + HF(g)$

  c. $Cl_2(g) + NaOH(aq) \rightarrow NaOCl(aq) + NaCl(aq) + H_2O(\ell)$

  d. $KClO_3(s) \rightarrow KClO_4(s) + KCl(s)$

  e. dinitrogen monoxide $\rightarrow$ nitrogen + oxygen

  f. dinitrogen pentoxide $\rightarrow$ nitrogen dioxide + oxygen

  g. $P_4O_{10}(s) + H_2O(\ell) \rightarrow H_4P_2O_7(aq)$

  h. $CaCO_3(s) + HCl(aq) \rightarrow CaCl_2(aq) + H_2O(\ell) + CO_2(g)$

5.8 Balance the following equations:

  a. $KClO_3(s) \rightarrow KCl(s) + O_2(g)$

  b. $C_2H_6(g) + O_2(g) \rightarrow CO_2(g) + H_2O(\ell)$

  c. nitrogen + oxygen $\rightarrow$ dinitrogen pentoxide

  d. $MgCl_2(s) + H_2O(g) \rightarrow MgO(s) + HCl(g)$

  e. $CaH_2(s) + H_2O(\ell) \rightarrow Ca(OH)_2(s) + H_2(g)$

  f. $Al(s) + Fe_2O_3(s) \rightarrow Al_2O_3(s) + Fe(s)$

  g. aluminum + bromine $\rightarrow$ aluminum bromide

  h. $Hg_2(NO_3)_2(aq) + NaCl(aq) \rightarrow HgCl(s) + NaNO_3(aq)$

Redox Reactions (Section 5.3)

5.9 Assign oxidation numbers to the blue element in each of the following formulas:

  a. Cl_2O_5

  b. $KClO_4$

  c. Ba^{2+}

  d. F_2

  e. $H_4P_2O_7$

  f. H_2S

5.10 Assign oxidation numbers to the blue element in each of the following formulas:

  a. $HClO$

  b. $NaNO_2$

  c. N_2

  d. $Ca(ClO)_2$

  e. Al^{3+}

  f. N_2O_5

5.11 Find the element with the highest oxidation number in each of the following formulas:

  a. N_2O_5

  b. $KHCO_3$

  c. $NaOCl$

  d. $NaNO_3$

  e. $HClO_4$

  f. $Ca(NO_3)_2$

5.12 Find the element with the highest oxidation number in each of the following formulas:

  a. $Na_2Cr_2O_7$

  b. $K_2S_2O_3$

  c. HNO_3

  d. P_2O_5

  e. $Mg(ClO_4)_2$

  f. $HClO_2$

5.13 For each of the following equations, indicate whether the blue element has been oxidized, reduced, or neither oxidized nor reduced.

  a. $2Mg(s) + O_2(g) \rightarrow 2MgO(s)$

  b. $CuO(s) + H_2(g) \rightarrow Cu(s) + H_2O(g)$

  c. $Ag^+(aq) + Cl^-(aq) \rightarrow AgCl(s)$

  d. $BaCl_2(aq) + H_2SO_4(aq) \rightarrow BaSO_4(s) + 2HCl(aq)$

  e. $Zn(s) + 2H^+(aq) \rightarrow Zn^{2+}(aq) + H_2(g)$

5.14 For each of the following equations, indicate whether the blue element has been oxidized, reduced or neither oxidized nor reduced.

  a. $4Al(s) + 3O_2(g) \rightarrow 2Al_2O_3(s)$

  b. $SO_2(g) + H_2O(\ell) \rightarrow H_2SO_3(aq)$

  c. $2KClO_3(s) \rightarrow 2KCl(s) + 3O_2(g)$

  d. $2CO(g) + O_2(g) \rightarrow 2CO_2(g)$

  e. $2Na(s) + 2H_2O(\ell) \rightarrow 2NaOH(aq) + H_2(g)$

5.15 Assign oxidation numbers to each element in the following equations and identify the oxidizing and reducing agents:

  a. $H_2(g) + Cl_2(g) \rightarrow 2HCl(g)$

  b. $H_2O(g) + CH_4(g) \rightarrow CO(g) + 3H_2(g)$

  c. $CuO(s) + H_2(g) \rightarrow Cu(s) + H_2O(g)$

  d. $B_2O_3(s) + 3Mg(s) \rightarrow 2B(s) + 3MgO(s)$

  e. $Fe_2O_3(s) + CO(g) \rightarrow 2FeO(s) + CO_2(g)$

  f. $Cr_2O_7^{2-}(aq) + 2H^+(aq) + 3Mn^{2+}(aq) \rightarrow$
     $\quad 2Cr^{3+}(aq) + 3MnO_2(s) + H_2O(\ell)$

5.16 Assign oxidation numbers to each element in the following equations and identify the oxidizing and reducing agents:

  a. $2Cu(s) + O_2(g) \rightarrow 2CuO(s)$

  b. $Cl_2(aq) + 2KI(aq) \rightarrow 2KCl(aq) + I_2(aq)$

  c. $3MnO_2(s) + 4Al(s) \rightarrow 2Al_2O_3(s) + 3Mn(s)$

  d. $2H^+(aq) + 3SO_3^{2-}(aq) + 2NO_3^-(aq) \rightarrow$
     $\quad 2NO(g) + H_2O(\ell) + 3SO_4^{2-}(aq)$

  e. $Mg(s) + 2HCl(aq) \rightarrow MgCl_2(aq) + H_2(g)$

  f. $4NO_2(g) + O_2(g) \rightarrow 2N_2O_5(g)$

5.17 The tarnish of silver objects is a coating of silver sulfide (Ag_2S), which can be removed by putting the silver in contact with aluminum metal in a dilute solution of baking soda or salt. The equation for the cleaning reaction is

$$3Ag_2S(s) + 2Al(s) \rightarrow 6Ag(s) + Al_2S_3(s)$$

The sulfur in these compounds has a -2 oxidation number. What are the oxidizing and reducing agents in the cleaning reaction?

5.18 Aluminum metal reacts rapidly with highly basic solutions to liberate hydrogen gas and a large amount of heat. This reaction is utilized in a popular solid drain cleaner that is composed primarily of lye (sodium hydroxide) and aluminum granules. When wet, the mixture reacts as follows:

$$6NaOH(aq) + 2Al(s) \rightarrow 3H_2(g) + 2Na_3AlO_3(aq) + heat$$

The liberated H_2 provides agitation that, together with the heat, breaks the drain stoppage loose. What are the oxidizing and reducing agents in the reaction?

5.19 Identify the oxidizing and reducing agents in the Haber process for producing ammonia from elemental nitrogen and hydrogen. The equation for the reaction is

$$N_2(g) + 3H_2(g) \rightarrow 2NH_3(g)$$

Decomposition, Combination, and Replacement Reactions (Sections 5.4–5.6)

5.20 Classify each of the reactions represented by the following equations, first as a redox or nonredox reaction. Then further classify each redox reaction as a decomposition, single-replacement, or combination reaction, and each nonredox reaction as a decomposition, double-replacement, or combination reaction.

a. $K_2CO_3(s) \rightarrow K_2O(s) + CO_2(g)$

b. $Ca(s) + 2H_2O(\ell) \rightarrow Ca(OH)_2(s) + H_2(g)$

c. $BaCl_2(aq) + H_2SO_4(aq) \rightarrow BaSO_4(s) + 2HCl(aq)$

d. $SO_2(g) + H_2O(\ell) \rightarrow H_2SO_3(aq)$

e. $2NO(g) + O_2(g) \rightarrow 2NO_2(g)$

f. $2Zn(s) + O_2(g) \rightarrow 2ZnO(s)$

5.21 Classify each of the reactions represented by the following equations, first as a redox or nonredox reaction. Then further classify each redox reaction as a decomposition, single-replacement, or combination reaction, and each nonredox reaction as a decomposition, double-replacement, or combination reaction:

a. $N_2O_5(g) + H_2O(\ell) \rightarrow 2HNO_3(aq)$

b. $Cr_2O_3(s) + 2Al(s) \rightarrow 2Cr(s) + Al_2O_3(s)$

c. $CaO(s) + SiO_2(s) \rightarrow CaSiO_3(s)$

d. $H_2CO_3(aq) \rightarrow CO_2(g) + H_2O(\ell)$

e. $PbCO_3(s) \rightarrow PbO(s) + CO_2(g)$

f. $Zn(s) + Cl_2(g) \rightarrow ZnCl_2(s)$

5.22 Baking soda ($NaHCO_3$) can serve as an emergency fire extinguisher for grease fires in the kitchen. When heated, it liberates CO_2, which smothers the fire. The equation for the reaction is

$$2NaHCO_3(s) \xrightarrow{\text{Heat}} Na_2CO_3(s) + H_2O(g) + CO_2(g)$$

Classify the reaction into the categories used in Exercises 5.20 and 5.21.

5.23 Baking soda may serve as a source of CO_2 in bread dough. It causes the dough to rise. The CO_2 is released when $NaHCO_3$ reacts with an acidic substance:

$$NaHCO_3(aq) + H^+(aq) \rightarrow Na^+(aq) + H_2O(\ell) + CO_2(g)$$

Classify the reaction as redox or nonredox.

5.24 Many homes are heated by the energy released when natural gas (represented by CH_4) reacts with oxygen. The equation for the reaction is

$$CH_4(g) + 2O_2(g) \rightarrow CO_2(g) + 2H_2O(g)$$

Classify the reaction as redox or nonredox.

5.25 Hydrogen peroxide will react and liberate oxygen gas. In commercial solutions, the reaction is prevented to a large degree by the addition of an inhibitor. The equation for the oxygen-liberating reaction is

$$2H_2O_2(aq) \rightarrow 2H_2O(\ell) + O_2(g)$$

Classify the reaction as redox or nonredox.

5.26 Chlorine, used to treat drinking water, undergoes the reaction in water represented by the following equation:

$$Cl_2(aq) + H_2O(\ell) \rightarrow HOCl(aq) + HCl(aq)$$

Classify the reaction as redox or nonredox.

5.27 Triple superphosphate, an ingredient of some fertilizers, is prepared by reacting rock phosphate (calcium phosphate) and phosphoric acid. The equation for the reaction is

$$Ca_3(PO_4)_2(s) + 4H_3PO_4(aq) \rightarrow 3Ca(H_2PO_4)_2(s)$$

Classify the reaction into the categories used in Exercises 5.20 and 5.21.

Ionic Equations (Section 5.7)

5.28 Consider all of the following ionic compounds to be water soluble, and write the formulas of the ions that would be formed if the compounds were dissolved in water. Table 4.7 will be helpful.

a. $LiNO_3$

b. Na_2HPO_4

c. $Ca(ClO_3)_2$

d. KOH

e. $MgBr_2$

f. $(NH_4)_2SO_4$

5.29 Consider all the following ionic compounds to be water soluble, and write the formulas of the ions that would be formed if the compounds were dissolved in water. Table 4.7 will be helpful.

a. $K_2Cr_2O_7$

b. H_2SO_4

c. NaH_2PO_4

d. Na_3PO_4

e. NH_4Cl

f. $KMnO_4$

5.30 Reactions represented by the following equations take place in water solutions. Write each molecular equation in total ionic form, then identify spectator ions and write the equations in net ionic form. Solids that do not dissolve are designated by (s), gases that do not dissolve are designated by (g), and substances that dissolve but do not dissociate appear in blue.

a. $SO_2(aq) + H_2O(\ell) \rightarrow H_2SO_3(aq)$

b. $CuSO_4(aq) + Zn(s) \rightarrow Cu(s) + ZnSO_4(aq)$

c. $2KBr(aq) + 2H_2SO_4(aq) \rightarrow$
$$Br_2(aq) + SO_2(aq) + K_2SO_4(aq) + 2H_2O(\ell)$$

d. $AgNO_3(aq) + NaOH(aq) \rightarrow AgOH(s) + NaNO_3(aq)$

e. $BaCO_3(s) + 2HNO_3(aq) \rightarrow$
$$Ba(NO_3)_2(aq) + CO_2(g) + H_2O(\ell)$$

f. $N_2O_5(aq) + H_2O(\ell) \rightarrow 2HNO_3(aq)$

5.31 Reactions represented by the following equations take place in water solutions. Write each molecular equation in total ionic form, then identify spectator ions and write the equations in net ionic form. Solids that do not dissolve are designated by (s), gases that do not dissolve are designated by (g), and substances that dissolve but do not dissociate appear in blue.

a. $H_2O(\ell) + Na_2SO_3(aq) + SO_2(aq) \rightarrow 2NaHSO_3(aq)$

b. $3Cu(s) + 8HNO_3(aq) \rightarrow$
$$3Cu(NO_3)_2(aq) + 2NO(g) + 4H_2O(\ell)$$

c. $2HCl(aq) + CaO(s) \rightarrow CaCl_2(aq) + H_2O(\ell)$

d. $CaCO_3(s) + 2HCl(aq) \rightarrow CaCl_2(aq) + CO_2(aq) + H_2O(\ell)$

e. $MnO_2(s) + 4HCl(aq) \rightarrow MnCl_2(aq) + Cl_2(aq) + 2H_2O(\ell)$

f. $2AgNO_3(aq) + Cu(s) \rightarrow Cu(NO_3)_2(aq) + 2Ag(s)$

5.32 The following molecular equations all represent neutralization reactions of acids and bases. These reactions will be discussed further in Chapter 9. Write each equation in total ionic form, identify the spectator ions, then write the net ionic equation. Water is the only substance that does not dissociate. What do you notice about all the net ionic equations?

a. $HNO_3(aq) + KOH(aq) \rightarrow KNO_3(aq) + H_2O(\ell)$

b. $H_3PO_4(aq) + 3NH_4OH(aq) \rightarrow (NH_4)_3PO_4(aq) + 3H_2O(\ell)$

c. $HI(aq) + NaOH(aq) \rightarrow NaI(aq) + H_2O(\ell)$

5.33 The following molecular equations all represent neutralization reactions of acids and bases. These reactions will be discussed further in Chapter 9. Write each equation in total ionic form, identify the spectator ions, then write the net ionic equation. Water is the only substance that does not dissociate. What do you notice about all the net ionic equations?

a. $HBr(aq) + RbOH(aq) \rightarrow RbBr(aq) + H_2O(\ell)$

b. $H_2SO_4(aq) + 2LiOH(aq) \rightarrow Li_2SO_4(aq) + 2H_2O(\ell)$

c. $HCl(aq) + CsOH(aq) \rightarrow CsCl(aq) + H_2O(\ell)$

Energy and Reactions (Section 5.8)

5.34 In addition to emergency cold packs, emergency hot packs are available that heat up when water is mixed with a solid. Is the process that takes place in such packs exothermic or endothermic? Explain.

5.35 In refrigeration systems, the area to be cooled has pipes running through it. Inside the pipes, a liquid evaporates and becomes a gas. Is the evaporation process exothermic or endothermic? Explain.

5.36 An individual wants to keep some food cold in a portable picnic cooler. A piece of ice is put into the cooler, but it is wrapped in a thick insulating blanket to slow its melting. Comment on the effectiveness of the cooler in terms of the direction of heat movement inside the cooler.

5.37 The human body cools itself by the evaporation of perspiration. Is the evaporation process endothermic or exothermic? Explain.

The Mole and Chemical Equations (Section 5.9)

5.38 For the reactions represented by the following equations, write statements equivalent to Statements 2, 3, and 4 given in Section 5.9.

a. $S(s) + O_2(g) \rightarrow SO_2(g)$

b. $Sr(s) + 2H_2O(\ell) \rightarrow Sr(OH)_2(s) + H_2(g)$

c. $2H_2S(g) + 3O_2(g) \rightarrow 2H_2O(g) + 2SO_2(g)$

d. $4NH_3(g) + 5O_2(g) \rightarrow 4NO(g) + 6H_2O(g)$

e. $CaO(s) + 3C(s) \rightarrow CaC_2(s) + CO(g)$

5.39 For the reactions represented by the following equations, write statements equivalent to Statements 2, 3, and 4 given in Section 5.9.

a. $N_2(g) + 3H_2(g) \rightarrow 2NH_3(g)$

b. $2Na(s) + Cl_2(g) \rightarrow 2NaCl(s)$

c. $BaCl_2(aq) + H_2SO_4(aq) \rightarrow BaSO_4(s) + 2HCl(aq)$

d. $2H_2O_2(aq) \rightarrow 2H_2O(\ell) + O_2(g)$

e. $2C_3H_6(g) + 9O_2(g) \rightarrow 6CO_2(g) + 6H_2O(g)$

5.40 For the following equation, write statements equivalent to Statements 2, 3, and 4 given in Section 5.9. Then write at least six factors (including numbers and units) that could be used to solve problems by the factor-unit method.
$$2SO_2(g) + O_2(g) \rightarrow 2SO_3(g)$$

5.41 Calculate the number of grams of SO_2 that must react to produce 350 g of SO_3. Use the statements written in Exercise 5.40 and express your answer using the correct number of significant figures.

5.42 Calculate the mass of limestone ($CaCO_3$) that must be decomposed to produce 500 g of lime (CaO). The equation for the reaction is
$$CaCO_3(s) \rightarrow CaO(s) + CO_2(g)$$

5.43 Calculate the number of moles of CO_2 generated by the reaction of Exercise 5.42 when 500 g of CaO is produced.

5.44 Calculate the number of grams of bromine (Br_2) needed to react exactly with 50.1 g of aluminum (Al). The equation for the reaction is
$$2Al(s) + 3Br_2(\ell) \rightarrow 2AlBr_3(s)$$

5.45 Calculate the number of moles of $AlBr_3$ produced by the process of Exercise 5.44.

5.46 How many grams of $AlBr_3$ are produced by the process in Exercise 5.44?

5.47 In Exercise 5.17 you were given the following equation for the reaction used to clean tarnish from silver:
$$3Ag_2S(s) + 2Al(s) \rightarrow 6Ag(s) + Al_2S_3(s)$$

a. How many grams of aluminum would need to react to remove 0.250 g of Ag_2S tarnish?

b. How many moles of Al_2S_3 would be produced by the reaction described in (a)?

5.48 Pure titanium metal is produced by reacting titanium(IV) chloride with magnesium metal. The equation for the reaction is
$$TiCl_4(s) + 2Mg(s) \rightarrow Ti(s) + 2MgCl_2(s)$$

How many grams of Mg would be needed to produce 1.00 kg of pure titanium?

5.49 An important metabolic process of the body is the oxidation of glucose to water and carbon dioxide. The equation for the reaction is
$$C_6H_{12}O_6(aq) + 6O_2(aq) \rightarrow 6CO_2(aq) + 6H_2O(\ell)$$

Even-numbered exercises answered in Appendix B.

Unless otherwise noted all art on this page © Cengage Learning 2014

a. What mass of water in grams is produced when the body oxidizes 1.00 mol of glucose?

b. How many grams of oxygen are needed to oxidize 1.00 mol of glucose?

5.50 Caproic acid is oxidized in the body as follows:

$$C_6H_{12}O_2(aq) + 8O_2(aq) \rightarrow 6CO_2(aq) + 6H_2O(\ell)$$

How many grams of oxygen are needed to oxidize 1.00 mol of caproic acid?

The Limiting Reactant (Section 5.10)

5.51 A sample of 4.00 g of methane (CH_4) is mixed with 15.0 g of chlorine (Cl_2).

a. Determine which is the limiting reactant according to the following equation:

$$CH_4(g) + 4Cl_2(g) \rightarrow CCl_4(\ell) + 4HCl(g)$$

b. What is the maximum mass of CCl_4 that can be formed?

5.52 Nitrogen and oxygen react as follows:

$$N_2(g) + 2O_2(g) \rightarrow 2NO_2(g)$$

Suppose 1.25 mol of N_2 and 50.0 g of O_2 are mixed together.

a. Which one is the limiting reactant?

b. What is the maximum mass in grams of NO_2 that can be produced from the mixture?

5.53 Suppose you want to use acetylene (C_2H_2) as a fuel. You have a cylinder that contains 500 g of C_2H_2 and a cylinder that contains 2000 g of oxygen (O_2). Do you have enough oxygen to burn all the acetylene? The equation for the reaction is

$$2C_2H_2(g) + 5O_2(g) \rightarrow 4CO_2(g) + 2H_2O(g)$$

5.54 Ammonia, carbon dioxide, and water vapor react to form ammonium bicarbonate as follows:

$$NH_3(aq) + CO_2(aq) + H_2O(\ell) \rightarrow NH_4HCO_3(aq)$$

Suppose 50.0 g of NH_3, 80.0 g of CO_2, and 2.00 mol of H_2O are reacted. What is the maximum number of grams of NH_4HCO_3 that can be produced?

5.55 Chromium metal (Cr) can be prepared by reacting the oxide with aluminum. The equation for the reaction is

$$Cr_2O_3(s) + 2Al(s) \rightarrow 2Cr(s) + Al_2O_3(s)$$

What three substances will be found in the final mixture if 150 g of Cr_2O_3 and 150 g of Al are reacted?

Reaction Yields (Section 5.11)

5.56 The actual yield of a reaction was 11.74 g of product, while the calculated theoretical yield was 16.09 g. What was the percentage yield?

5.57 A product weighing 14.37 g was isolated from a reaction. The amount of product possible according to a calculation was 17.55 g. What was the percentage yield?

5.58 For a combination reaction, it was calculated that 7.59 g of *A* would exactly react with 4.88 g of *B*. These amounts were reacted, and 9.04 g of product was isolated. What was the percentage yield of the reaction?

5.59 A sample of calcium metal with a mass of 2.00 g was reacted with excess oxygen. The following equation represents the reaction that took place:

$$2Ca(s) + O_2(g) \rightarrow 2CaO(s)$$

The isolated product (CaO) weighed 2.26 g. What was the percentage yield of the reaction?

5.60 Upon heating, mercury(II) oxide undergoes a decomposition reaction:

$$2HgO(s) \rightarrow 2Hg(\ell) + O_2(g)$$

A sample of HgO weighing 7.22 g was heated. The collected mercury weighed 5.95 g. What was the percentage yield of the reaction?

Additional Exercises

5.61 Would you expect argon, Ar, to be involved in a redox reaction? Explain your reasoning.

5.62 Rewrite the following word equation using chemical formulas. Balance the equation and write the balanced equation in net ionic form:

barium chloride (aq) + sodium sulfate (aq) $\rightarrow$
sodium chloride (aq) + barium sulfate (s)

5.63 The element with an electron configuration of $1s^2 2s^2 2p^6 3s^1$ undergoes a combination reaction with the element that has 15 protons in its nucleus. Assume the product of the reaction consists of simple ions, and write a balanced chemical equation for the reaction.

5.64 Assuming a 100% reaction yield, it was calculated that 6.983 g of naturally occurring elemental iron would be needed to react with another element to form 18.149 g of product. If iron composed only of the iron-60 isotope was used to form the product, how many grams of iron-60 would be required?

5.65 The decomposition of a sample of a compound produced 1.20×10^{24} atoms of nitrogen and 80.0 grams of oxygen atoms. What was the formula of the sample that was decomposed? What is the correct name of the decomposed sample?

Allied Health Exam Connection

The following questions are from these sources:

1. *Nursing School Entrance Exam* © 2005, Learning Express, LLC.

2. *McGraw-Hill's Nursing School Entrance Exams* by Thomas A. Evangelist, Tamara B. Orr and Judy Unrein © 2009, The McGraw-Hill Companies, Inc.

3. *NSEE Nursing School Entrance Exams,* 3rd Edition © 2009, Kaplan Publishing.

4. *Cliffs Test Prep: Nursing School Entrance Exams* by Fred N. Grayson © 2004, Wiley Publishing, Inc.

5. *Peterson's Master the Nursing School and Allied Health Entrance Exams,* 18th Edition by Marion F. Gooding © 2008, Peterson's, a Nelnet Company.

5.66 Balance the following redox reaction:

$$Mg(s) + H_2O(g) \rightarrow Mg(OH)_2(s) + H_2(g)$$

a. $Mg(s) + H_2O(g) \rightarrow Mg(OH)_2(s) + H_2(g)$

b. $Mg(s) + 4H_2O(g) \rightarrow Mg(OH)_2(s) + H_2(g)$

c. $Mg(s) + 2H_2O(g) \rightarrow Mg(OH)_2(s) + H_2(g)$

d. $Mg(s) + H_2O(g) \rightarrow Mg(OH)_2(s) + \frac{1}{2}H_2(g)$

5.67 Which one of the following equations is balanced?

a. $H_2O \rightarrow H_2 + O_2$

b. $Al + H_2SO_4 \rightarrow Al_2(SO_4)_3 + H_2$

c. $S + O_2 \rightarrow SO_3$

d. $2HgO \rightarrow 2Hg + O_2$

5.68 Which of the following equations is balanced?

a. $Mg + N_2 \rightarrow Mg_3N_2$

b. $Fe + O_2 \rightarrow Fe_2O_3$

c. $C_{12}H_{22}O_{11} \rightarrow 12C + 11H_2O$

d. $Ca + H_2O \rightarrow Ca(OH)_2 + H_2$

5.69 The coefficient of O_2 after the following equation is balanced is:
$\_CH_4 + \_O_2 \rightarrow \_CO_2 + \_H_2O$

a. 1

b. 2

c. 3

d. 4

5.70 Ammonia can be produced from nitrogen and hydrogen, according to the unbalanced equation

$$N_2(g) + H_2(g) = NH_3(g)$$

After balancing the equation, the coefficient before ammonia should be:

a. 1

b. 2

c. 3

d. 4

5.71 What is the oxidation number of sodium in the following reaction? $Pb(NO_3)_2(aq) + 2NaI(aq) \rightarrow PbI_2(s) + 2NaNO_3(aq)$

a. $+1$

b. $+2$

c. -1

d. -2

5.72 What is the oxidation number for nitrogen in HNO_3?

a. -2

b. $+5$

c. -1

d. -5

5.73 The oxidation number of sulfur in the ion SO_4^{-2} is:

a. -2

b. $+2$

c. $+6$

d. $+10$

5.74 Which of the following is the oxidation number of sulfur in the compound sodium thiosulfate, $Na_2S_2O_3$?

a. $+1$

b. -1

c. $+2$

d. -2

5.75 Which best describes the following redox reaction:
$Br^-(aq) + MnO_4^-(aq) \rightarrow Br_2(\ell) + Mn^{+2}(aq)$?

a. Br and Mn are both reduced

b. Br is oxidized and Mn is reduced

c. Br is oxidized and O is reduced

d. Br is reduced and Mn is oxidized

5.76 Which reactant is oxidized and which is reduced in the following reaction?

$$C_2H_4(g) + 3O_2(g) \rightarrow 2CO_2(g) + 2H_2O(g)$$

a. oxidized: $C_2H_4(g)$, reduced: $3O_2(g)$

b. oxidized: $C_2H_4(g)$, reduced: $2H_2O(g)$

c. oxidized: $C_2H_4(g)$, reduced: $2CO_2(g)$

d. oxidized: $2CO_2(g)$, reduced: $C_2H_4(g)$

5.77 Which of the following species is being oxidized in this redox reaction?
$Zn(s) + Cu^{2+}(aq) \rightarrow Zn^{2+}(aq) + Cu(s)$

a. $Zn(s)$

b. $Cu^{2+}(aq)$

c. $Zn^{2+}(aq)$

d. $Cu(s)$

5.78 Identify the oxidizing agent and the reducing agent in the following reaction:
$8H^+(aq) + 6Cl^-(aq) + Sn(s) + 4NO_3^-(aq) \rightarrow$
$SnCl_6^{2-}(aq) + 4NO_2(g) + 4H_2O(\ell)$

a. oxidizing agent: $8H^+(aq)$, reducing agent: $Sn(s)$

b. oxidizing agent: $4NO_3^-(aq)$, reducing agent: $Sn(s)$

c. oxidizing agent: $4NO_3^-(aq)$, reducing agent: $4NO_2(g)$

d. oxidizing agent: $4NO_3^-(aq)$, reducing agent: $8H^+(aq)$

5.79 The chemical reaction: $2Zn + 2HCl \rightarrow 2ZnCl + H_2$ is an example of:

a. double displacement

b. synthesis

c. analysis

d. single displacement

5.80 Identify the following as an oxidation, a reduction, a decomposition, or a dismutation reaction:
$Cl_2 + 2e^- \rightarrow 2Cl^-$

a. a reduction

b. an oxidation

c. a decomposition

d. a dismutation

5.81 The reaction $2NaI + Cl_2 \rightarrow 2NaCl + I_2$ demonstrates a:

a. decomposition reaction

b. synthesis reaction

c. single replacement reaction

d. double replacement reaction

Even-numbered exercises answered in Appendix B.

5.82 Classify the reaction as to the reaction type:
$$Mg(OH)_2 + 2HCl \rightarrow MgCl_2 + 2H_2O$$

   a. synthesis

   b. decomposition

   c. single replacement

   d. double replacement

5.83 Identify the statement which is NOT characteristic of exergonic reactions:

   a. They are downhill reactions

   b. They have a negative energy change ($-H$)

   c. They are uphill reactions

   d. The products have less energy than the reactants

5.84 What is the net ionic equation of the following transformations?
$$Ca(NO_3)_2(aq) + 2KCl(aq) \rightarrow CaCl_2(s) + 2KNO_3(aq)$$

   a. $2NO_3^-(aq) + 2K^+(aq) \rightarrow 2KNO_3(aq)$

   b. $Ca^{2+}(aq) + 2Cl^-(aq) \rightarrow CaCl_2(s)$

   c. $Ca^{2+}(aq) + 2NO_3^-(aq) + 2K^+(aq) + 2Cl^-(aq) \rightarrow$
   $$CaCl_2(s) + 2K^+(aq) + 2NO_3^-(aq)$$

   d. $Ca^{2+}(aq) + 2NO_3^-(aq) + 2K^+(aq) + 2Cl^-(aq) \rightarrow$
   $$Ca^{2+}(aq) + 2Cl^-(aq) + 2K^+(aq) + 2NO_3^-(aq)$$

5.85 In the Haber process, ammonia is produced according to the following equation:
$$N_2(g) + 3H_2(g) = 2NH_3(g)$$

How many moles of hydrogen gas are needed to react with one of nitrogen?

   a. 1

   b. 3

   c. 6

   d. 22.4

5.86 What mass of product would you expect given that you started with 17 g of NH_3 and 36.5 g of HCl?
$$NH_3 + HCl \rightarrow NH_4Cl$$

   a. 17 g

   b. 36.5 g

   c. 53.5 g

   d. 19.5 g

5.87 The number of grams of hydrogen formed by the action of 6 grams of magnesium (atomic weight = 24) on an appropriate quantity of acid is:

   a. 0.5

   b. 8

   c. 22.4

   d. 72

5.88 In the reaction $CaCl_2 + Na_2CO_3 \rightarrow CaCO_3 + 2NaCl$, if 0.5 mole of NaCl is to be formed, then:

   a. 1 mole of Na_2CO_3 is needed

   b. 0.5 mole of $CaCO_3$ is also formed

   c. 0.5 mole of Na_2CO_3 is needed

   d. 0.25 mole of $CaCl_2$ is needed

5.89 In the reaction $4Al + 3O_2 \rightarrow 2Al_2O_3$, how many grams of O_2 are needed to completely react with 1.5 moles of Al?

   a. 24 g

   b. 36 g

   c. 48 g

   d. 60 g

Chemistry for Thought

5.90 In experiments where students prepare compounds by precipitation from water solutions, they often report yields of dry product greater than 100%. Propose an explanation for this.

5.91 Refer to Figure 5.2 and follow the instructions. Then explain how the concept of limiting reactant is used to extinguish a fire.

5.92 Refer to Figure 5.4 and answer the question. Suggest another material that might provide the enzyme catalyst.

5.93 What do the observations of Figure 5.10 indicate to you about the abilities of the following solids to dissolve in water: NaCl, $AgNO_3$, and AgCl? What would you expect to observe if you put a few grams of solid AgCl into a test tube containing 3 mL of water and shook the mixture? What would you expect to observe if you repeated this experiment but used a few grams of $NaNO_3$ instead of AgCl?

5.94 The concentration of alcohol (CH_3CH_2OH) in the breath of an individual who has been drinking is measured by an instrument called a Breathalyzer. The breath sample is passed through a solution that contains the orange-colored $Cr_2O_7^{2-}$ ion. The equation for the net ionic redox reaction that occurs is:

$$2Cr_2O_7^{2-}(aq) + 3CH_3CH_2OH(aq) + 16H^+(aq) \rightarrow$$
$$4Cr^{3+}(aq) + 3CH_3COOH(aq) + 11H_2O(\ell)$$

The Cr^{3+} ion has a pale violet color in solution. Explain how color changes in the solution could be used to indicate the amount of alcohol in the breath sample.

5.95 Certain vegetables and fruits, such as potatoes and apples, darken quickly when sliced. Submerging the slices in water slows this process. Explain.

5.96 In an ordinary flashlight battery, an oxidation reaction and a reduction reaction take place at different locations to produce an electrical current that consists of electrons. In one of the reactions, the zinc container of the battery slowly dissolves as it is converted into zinc ions. Is this the oxidation or the reduction reaction? Is this reaction the source of electrons, or are electrons used to carry out the reaction? Explain your answers with a reaction equation.

Case Study

The combined sound of a seal barking and a child crying was coming from three-year-old Adrian's room at 2:00 AM. His father, Kyle, turned on the light to investigate. Adrian was blue around the lips, and when he inhaled his stomach looked as if a vacuum was pulling it in. When Kyle took his son to the car, he seemed to improve in the cold winter air. At the ER, Adrian was diagnosed with a classic case of croup. The doctor called for a respiratory therapist to administer a treatment of an aerosol containing racemic epinephrine and a shot of corticosteroids. After two hours in the ER, the respiratory therapist recommended oxygen and humidity treatment administered through a tent. Adrian was admitted to Pediatrics for treatment. Kyle asked if he could stay with his son, and, to his relief, the hospital encouraged parental accompaniment.

How important is it to allow parents to stay with hospitalized youngsters? What is an acceptable O_2 saturation level for a three year old? What are the signs that a cough is a medical emergency?

© Juriah Mosin/Shutterstock.com

Follow-up to this Case Study appears at the end the chapter before the Concept Summary.

Learning Objectives

When you have completed your study of this chapter, you should be able to:

1 Do calculations based on the property of density. **(Section 6.1)**

2 Demonstrate an understanding of the kinetic molecular theory of matter. **(Sections 6.2–6.4)**

3 Use the kinetic molecular theory to explain and compare the properties of matter in different states. **(Section 6.5)**

4 Do calculations to convert pressure and temperature values into various units. **(Section 6.6)**

5 Do calculations based on Boyle's law, Charles's law, and the combined gas law. **(Section 6.7)**

6 Do calculations based on the ideal gas law. **(Section 6.8)**

7 Do calculations based on Dalton's law. **(Section 6.9)**

8 Do calculations based on Graham's law. **(Section 6.10)**

9 Classify changes of state as exothermic or endothermic. **(Section 6.11)**

10 Demonstrate an understanding of the concepts of vapor pressure and evaporation. **(Section 6.12)**

11 Demonstrate an understanding of the process of boiling and the concept of boiling point. **(Section 6.13)**

12 Demonstrate an understanding of the processes of sublimation and melting. **(Section 6.14)**

13 Do calculations based on energy changes that accompany heating, cooling, or changing the state of a substance. **(Section 6.15)**

If you live in an area that has cold winters, you have probably seen water in the three different forms used to categorize the states in which matter occurs. On a cold day, you can usually find solid water (ice) floating in a pool of cold liquid water, and at the same time you can see a small cloud of tiny water droplets that forms when gaseous water condenses as you exhale into the cold air.

Most matter is not as easily observed in all three states as the water in the preceding example. In fact, most matter is classified as a solid, a liquid, or a gas on the basis of the form in which it is commonly observed. However, according to Section 4.11, the state of a substance depends on temperature. You will see in this chapter that the state also depends on pressure. Therefore, when a substance is classified as a solid, a liquid, or a gas, we are usually simply stating its form under normal atmospheric pressure and at a temperature near 25°C. ❯ Figure 6.1 gives the states of the elements under those conditions.

In this chapter, we will study the characteristics of the common states of matter, along with a theory that relates these states to molecular behavior. We will also investigate the energy relationships that accompany changes of state.

| Gases | Liquids | Solids |

1																	2
3	4											5	6	7	8	9	10
11	12											13	14	15	16	17	18
19	20	21	22	23	24	25	26	27	28	29	30	31	32	33	34	35	36
37	38	39	40	41	42	43	44	45	46	47	48	49	50	51	52	53	54
55	56	57	72	73	74	75	76	77	78	79	80	81	82	83	84	85	86
87	88	89	104	105	106	107	108	109	110	111	112	113	114	115	116	117	118

58	59	60	61	62	63	64	65	66	67	68	69	70	71
90	91	92	93	94	95	96	97	98	99	100	101	102	103

6.1 Observed Properties of Matter

Learning Objective

1. Do calculations based on the property of density.

Solids, liquids, and gases can easily be recognized and distinguished by differences in physical properties. Four properties that can be used are density, shape, compressibility, and thermal expansion.

Density was defined in Section 1.11 as the mass of a sample of matter divided by the volume of the same sample. A failure to remember this is the basis for the wrong answers given to the children's riddle, "Which is heavier, a pound of lead or a pound of feathers?" Of course, they weigh the same, but many people say lead is heavier because they think in terms of samples having the same volume. Of course, if you have lead and feather samples of the same volume, the lead sample will be much heavier because its density is much greater.

shape Shape depends on the physical state of matter.

The **shape** of matter is sometimes independent of a container (for solids), or it may be related to the shape of the container (for liquids and gases). In the case of liquids, the shape depends on the extent to which the container is filled. Gases always fill the container completely (see ❱ Figure 6.2).

1 Solids have a shape and volume that does not depend on the container.

© Spencer L. Seager

2 Liquids take the shape of the part of the container they fill. Each sample above has the same volume.

© Spencer L. Seager

© Spencer L. Seager

3 Gases completely fill and take the shape of their container. When the valve separating the two parts of the container is opened, the gas fills the entire container volume (bottom photo).

Figure 6.2 Characteristics of the states of matter.

Compressibility, the change in volume resulting from a pressure change, is quite high for gases. Compressibility allows a lot of gas to be squeezed into a small volume if the gas is put under sufficient pressure—think of automobile tires, or a tank of compressed helium used to fill toy balloons.

Thermal expansion, the change in volume resulting from temperature changes, is a property used in liquid-in glass thermometers. As the temperature of the liquid increases, the liquid expands and fills more of the fine capillary tube on which the temperature scale is engraved. You see the liquid move up the tube and know the temperature is increasing.

These four properties are compared for the three states of matter in ▶ Table 6.1.

compressibility The change in volume of a sample resulting from a pressure change acting on the sample.

thermal expansion The change in volume of a sample resulting from a change in temperature of the sample.

TABLE 6.1 Physical Properties of Solids, Liquids, and Gases

Property	State		
	Solid	Liquid	Gas
Density	High	High—usually lower than that of corresponding solid	Low
Shape	Definite	Indefinite—takes shape of container to the extent it is filled	Indefinite—takes shape of container it fills
Compressibility	Small	Small—usually greater than that of corresponding solid	Large
Thermal expansion	Very small	Small	Moderate

▶ Example 6.1

Samples of plumber's solder, rubbing alcohol, and air are collected. The volume and mass of each sample are determined at 20°C as follows: solder—volume = 28.6 mL, mass = 268.8 g; alcohol—volume = 100.0 mL, mass = 78.5 g; air—volume = 500.0 mL, mass = 0.602 g. Calculate the density of each substance in grams per milliliter.

Solution

The density (d) is obtained by dividing the sample mass (in grams) by the sample volume (in milliliters):

solder: $\quad d = \dfrac{268.8}{28.6\ \text{mL}} = 9.40\ \text{g/mL}$

alcohol: $\quad d = \dfrac{78.5\ \text{g}}{100.0\ \text{mL}} = 0.785\ \text{g/mL}$

air: $\quad d = \dfrac{0.602\ \text{g}}{500.0\ \text{mL}} = 0.00120\ \text{g/mL}$

The calculated densities follow the pattern given in Table 6.1.

▶ **LEARNING CHECK 6.1** Samples of copper, glycerin, and helium are collected and weighed at 20°C. The results are as follows: copper—volume = 12.8 mL, mass = 114.2 g; glycerin—volume = 50.0 mL, mass = 63.0 g; helium—volume = 1500 mL, mass = 0.286 g.

a. Calculate the density of each sample at 20°C.
b. Describe qualitatively (increase, decrease, little change, no change, etc.) what happens to the following properties for each sample when the pressure on the sample is doubled: sample volume, sample mass, sample density (see Table 6.1).

The density calculations done in Example 6.1 used Equation 1.9, which is repeated here:

$$d = \frac{m}{V}$$

Remember, in this equation d is density, m is mass, and V is volume. This useful equation can be rearranged so that any one quantity in it can be calculated if the other two are known.

▶ Example 6.2

The sample of rubbing alcohol used in Example 6.1 is heated to 50°C. The density of the sample is found to decrease to a value of 0.762 g/mL. What is the volume of the sample at 50°C?

Solution

It is assumed that no alcohol is lost by evaporation, so the sample mass has the same value of 78.5 g it had in Example 6.1. Equation 1.9 is rearranged and solved for volume:

$$V = \frac{m}{d} \tag{6.1}$$

The new volume is

$$V = \frac{m}{d} = \frac{78.5 \; \cancel{g}}{0.762 \; \cancel{g}/\text{mL}} = 103 \; \text{mL}$$

Thus, the heating causes the liquid to expand from a volume of 100 mL to 103 mL.

▶ LEARNING CHECK 6.2
Calculate the mass in grams of a 1200-mL air sample at a temperature where the air has a density of 1.18×10^{-3} g/mL. NOTE: Equation 6.1 can be rearranged to $m = d \times V$.

6.2 The Kinetic Molecular Theory of Matter

Learning Objective

2. Demonstrate an understanding of the kinetic molecular theory of matter.

The long title of this section is the name scientists have given to a model or theory used to explain the behavior of matter in its various states. Some theories, including this one, are made up of a group of generalizations or postulates. This useful practice makes it possible to study and understand each postulate individually, instead of the entire theory.

The postulates of the kinetic molecular theory are

1. Matter is composed of tiny particles called molecules.
2. The particles are in constant motion and therefore possess kinetic energy.
3. The particles possess potential energy as a result of attracting or repelling each other.
4. The average particle speed increases as the temperature increases.
5. The particles transfer energy from one to another during collisions in which no net energy is lost from the system.

kinetic energy The energy a particle has as a result of its motion. Mathematically, it is KE $= \frac{1}{2}mv^2$.

These postulates contain two new terms, kinetic energy and potential energy. **Kinetic energy** is the energy a particle has as a result of its motion. Mathematically, kinetic energy is calculated as

$$\text{KE} = \frac{1}{2} mv^2 \tag{6.2}$$

where m is the particle mass and v is its velocity. Thus, if two particles of different mass are moving at the same velocity, the heavier particle will possess more kinetic energy than the other particle. Similarly, the faster moving of two particles with equal masses will have more kinetic energy.

▶ Example 6.3

Calculate the kinetic energy of two particles with masses of 2.00 g and 3.00 g if they are both moving with a velocity of 15.0 cm/s.

Solution

The kinetic energy of the 2.00-g particle is

$$ \text{KE} = \frac{1}{2}(2.00\ \text{g})\left(15.0\ \frac{\text{cm}}{\text{s}}\right)^2 = 225\ \frac{\text{g cm}^2}{\text{s}^2} $$

The kinetic energy of the 3.00-g particle is

$$ \text{KE} = \frac{1}{2}(3.00\ \text{g})\left(15.0\ \frac{\text{cm}}{\text{s}}\right)^2 = 337.5\ \frac{\text{g cm}^2}{\text{s}^2} $$

Rounding gives 338 g cm²/s²; thus, the more massive particle has more kinetic energy.

▶ **LEARNING CHECK 6.3** Calculate the kinetic energy of two 3.00-g particles if one has a velocity of 10.0 cm/s and the other has a velocity of 20.0 cm/s.

Potential energy results from attractions or repulsions of particles. A number of these interactions are familiar, such as the gravitational attraction of Earth and the behavior of the poles of two magnets brought near each other. In each of these examples, the size of the force and the potential energy depend on the separation distance. The same behavior is found for the potential energy of atomic-sized particles. The potential energy of attraction increases as separation increases, whereas the potential energy of repulsion decreases with increasing separation.

The kinetic molecular theory provides reasonable explanations for many of the observed properties of matter. An important factor in these explanations is the relative influence of cohesive forces and disruptive forces. **Cohesive forces** are the attractive forces associated with potential energy, and **disruptive forces** result from particle motion (kinetic energy). Disruptive forces tend to scatter particles and make them independent of each other; cohesive forces have the opposite effect. Thus, the state of a substance depends on the relative strengths of the cohesive forces that hold the particles together and the disruptive forces tending to separate them. Cohesive forces are essentially temperature-independent because they involve interparticle attractions of the type described in Chapter 4. Disruptive forces increase with temperature because they arise from particle motion, which increases with temperature (Postulate 4). This explains why temperature plays such an important role in determining the state in which matter is found.

potential energy The energy a particle has as a result of attractive or repulsive forces acting on it.

cohesive force The attractive force between particles; it is associated with potential energy.

disruptive force The force resulting from particle motion; it is associated with kinetic energy.

6.3 The Solid State

Learning Objective

3. Use the kinetic molecular theory to explain and compare the properties of matter in different states.

In the solid state, the cohesive forces are stronger than the disruptive forces (see ▶ Figure 6.3A). Each particle of a crystalline solid occupies a fixed position in the crystal lattice (see Section 4.11). Disruptive kinetic energy causes the particles to vibrate about

A Solid state: The particles are close together and held in fixed positions; they do not need a container.

B Liquid state: The particles are close together but not held in fixed positions; they take the shape of the container.

C Gaseous state: The particles are far apart and completely fill the container.

Figure 6.3 A kinetic molecular view of solids, liquids, and gases.

their fixed positions, but the strong cohesive forces prevent the lattice from breaking down. The properties of solids in Table 6.1 are explained by the kinetic theory as follows:

High density. The particles of solids are located as closely together as possible. Therefore, large numbers of particles are contained in a small volume, resulting in a high density.

Definite shape. The strong cohesive forces hold the particles of solids in essentially fixed positions, resulting in a definite shape.

Small compressibility. Because there is very little space between particles of solids, increased pressure cannot push them closer together, and it will have little effect on the volume.

Very small thermal expansion. Increased temperature increases the vibrational motion of the particles and the disruptive forces acting on them. Each particle vibrates with an increased amplitude and "occupies" a slightly larger volume. However, there is only a slight expansion of the solid because the strong cohesive forces prevent this effect from becoming very large.

6.4 The Liquid State

Learning Objective

3. Use the kinetic molecular theory to explain and compare the properties of matter in different states.

Particles in the liquid state are randomly packed and relatively close to each other (see Figure 6.3B). They are in constant, random motion, sliding freely over one another, but without sufficient kinetic energy to separate completely from each other. The liquid state is a situation in which cohesive forces dominate slightly. The characteristic properties of liquids are also explained by the kinetic theory:

High density. The particles of liquids are not widely separated; they essentially touch each other. There will, therefore, be a large number of particles per unit volume and a high density.

Indefinite shape. Although not completely independent of each other, the particles in a liquid are free to move over and around each other in a random manner, limited only by the container walls and the extent to which the container is filled.

Small compressibility. Because the particles in a liquid essentially touch each other, there is very little space between them. Therefore, increased pressure cannot squeeze the particles much more closely together.

Small thermal expansion. Most of the particle movement in a liquid is vibrational because the particles can move only a short distance before colliding with a neighbor. Therefore, the increased particle velocity that accompanies a temperature increase results only in increased vibration. The net effect is that the particles push away from each other a little more, thereby causing a slight volume increase in the liquid.

6.5 The Gaseous State

Learning Objective

3. Use the kinetic molecular theory to explain and compare the properties of matter in different states.

Disruptive forces completely overcome cohesive forces between particles in the gaseous or vapor state. As a result, the particles of a gas move essentially independently of one another in a totally random way (see Figure 6.3C). Under ordinary pressure, the particles are relatively far apart except when they collide with each other. Between collisions with each other or with the container walls, gas particles travel in straight lines. The particle velocities and resultant collision frequencies are quite high for gases, as shown in ❱ Table 6.2.

TABLE 6.2 Some Numerical Data Related to the Gaseous State

Gas	Average Speed at 0°C	Average Distance Traveled Between Collisions at 1 atm of Pressure	Number of Collisions of 1 Molecule in 1 s at 1 atm of Pressure
Hydrogen (H_2)	169,000 cm/s (3700 mi/h)	1.12×10^{-5} cm	1.6×10^{10}
Nitrogen (N_2)	45,400 cm/s (1015 mi/h)	0.60×10^{-5} cm	0.80×10^{10}
Carbon dioxide (CO_2)	36,300 cm/s (811 mi/h)	0.40×10^{-5} cm	0.95×10^{10}

The kinetic theory explanation of gaseous-state properties follows the same pattern seen earlier for solids and liquids:

Low density. The particles of a gas are widely separated. There are relatively few of them in a given volume, which means there is little mass per unit volume.

Indefinite shape. The forces of attraction between particles have been overcome by kinetic energy, and the particles are free to travel in all directions. The particles, therefore, completely fill the container and assume its inner shape.

Large compressibility. The gas particles are widely separated, so that a gas sample is mostly empty space. When pressure is applied, the particles are easily pushed closer together, decreasing the amount of empty space and the gas volume, as shown by ❱ Figure 6.4.

Moderate thermal expansion. Gas particles move in straight lines except when they collide with each other or with container walls. An increase in temperature causes the particles to collide with more energy. Thus, they push each other away more strongly, and at constant pressure this causes the gas itself to occupy a significantly larger volume.

It must be understood that the size of the particles is not changed during expansion or compression of gases, liquids, or solids. The particles are merely moving farther apart or closer together, and the space between them is changed.

Gas at low pressure | Gas at higher pressure

Figure 6.4 The compression of a gas.

6.6 The Gas Laws

Learning Objective

4. Do calculations to convert pressure and temperature values into various units.

The states of matter have not yet been discussed in quantitative terms. We have pointed out that solids, liquids, and gases expand when heated, but the amounts by which they expand have not been calculated. Such calculations for liquids and solids are beyond the intended scope of this text, but gases obey relatively simple quantitative relationships that were discovered during the 17th, 18th, and 19th centuries. These relationships, called **gas laws,** describe in mathematical terms the behavior of gases as they are mixed, subjected to pressure or temperature changes, or allowed to diffuse.

To use the gas laws, you must clearly understand the units used to express pressure, and you must utilize the Kelvin temperature scale introduced in Section 1.6. **Pressure** is defined as force per unit area. However, most of the units commonly used to express pressure reflect a relationship to barometric measurements of atmospheric pressure. The mercury barometer was invented by Italian physicist Evangelista Torricelli (1608–1647). Its essential components are shown in ❱ Figure 6.5. A glass tube, sealed at one end, is filled with mercury, stoppered, and inverted so the stoppered end is under the surface of a pool of mercury. When the stopper is removed, the mercury in the tube falls until its weight is just balanced by the weight of air pressing on the mercury pool. The pressure of the atmosphere is then expressed in terms of the height of the supported mercury column.

Figure 6.5 Setting up a simple mercury barometer. Standard atmospheric pressure is 760 mm of mercury.

Mercury

760 mm

Mercury cup

Despite attempts to standardize measurement units, you will probably encounter a number of different units of pressure in your future studies and employment. Some of the more common are standard atmosphere (atm), torr, millimeters of mercury (mmHg), inches of mercury (in. Hg), pounds per square inch (psi), bar, and kilopascals (kPa). One **standard atmosphere** is the pressure needed to support a 760-mm column of mercury in a barometer tube, and 1 **torr** (named in honor of Torricelli) is the pressure needed to support a 1-mm column of mercury in a barometer tube. The relationships of the various units to the standard atmosphere are given in ❱ Table 6.3. Note that the values of 1 atm and 760 torr (or 760 mmHg) are exact numbers based on definitions and do not limit the number of significant figures in calculated numbers.

TABLE 6.3 Units of Pressure

Unit	Relationship to One Standard Atmosphere	Typical Application
Atmosphere	—	Gas laws
Torr	760 torr = 1 atm	Gas laws
Millimeters of mercury	760 mmHg = 1 atm	Gas laws
Pounds per square inch	14.7 psi = 1 atm	Compressed gases
Bar	1.01 bar = 29.9 in. Hg = 1 atm	Meteorology
Kilopascal	101 kPa = 1 atm	Gas laws

▶ Example 6.4

The gauge on a cylinder of compressed oxygen gas reads 1500 psi. Express this pressure in terms of (a) atm, (b) torr, and (c) mmHg.

Solution

We will use the factor-unit method of calculation from Section 1.9, and the necessary factors from Table 6.3.

a. It is seen from Table 6.3 that 14.7 psi = 1 atm. Therefore, the known quantity, 1500 psi, is multiplied by the factor atm/psi in order to generate units of atm. The result is

$$1500 \text{ psi} \left(\frac{1 \text{ atm}}{14.7 \text{ psi}} \right) = 102 \text{ atm}$$

b. Table 6.3 does not give a direct relationship between psi and torr. However, both psi and torr are related to atm. Therefore, the atm is used somewhat like a "bridge" between psi and torr:

$$(1500 \text{ psi}) \left(\frac{1 \text{ atm}}{14.7 \text{ psi}} \right) \left(\frac{760 \text{ torr}}{1 \text{ atm}} \right) = 77{,}551 \text{ torr} = 7.76 \times 10^4 \text{ torr}$$

Note how the "bridge" term canceled out.

c. Because the torr and mmHg are identical, the problem is worked as it was in part b:

$$(1500 \text{ psi}) \left(\frac{1 \text{ atm}}{14.7 \text{ psi}} \right) \left(\frac{760 \text{ mmHg}}{1 \text{ atm}} \right) = 7.76 \times 10^4 \text{ mmHg}$$

▶ **LEARNING CHECK 6.4** A barometer has a pressure reading of 670 torr. Convert this reading into (a) atm and (b) psi.

According to the kinetic molecular theory, a gas expands when it is heated at constant pressure because the gaseous particles move faster at the higher temperature. It makes no difference to the particles what temperature scale is used to describe the heating. However, gas law calculations are based on the Kelvin temperature scale (Section 1.6) rather than the Celsius or Fahrenheit scales. The only apparent difference between the Kelvin and Celsius scales is the location of the zero reading (see Figure 1.11). It should be noted that a 0 reading on the Kelvin scale has a great deal of significance. It is the lowest possible temperature and is called **absolute zero.** It represents the temperature at which particles have no kinetic energy because all motion stops. The Kelvin and Celsius temperature scales have the same size degree, but the 0 of the Kelvin scale is 273 degrees below the freezing point of water, which is the 0 of the Celsius scale. Thus, a Celsius reading can be converted to a Kelvin reading simply by adding 273.

absolute zero The temperature at which all motion stops; a value of 0 on the Kelvin scale.

Huffing: A Potential Introduction of Children to Drug Abuse

"Huffing" is a general term used to describe various types of inhalant abuse. The inhalants involved are usually liquid chemicals with relatively high vapor pressures. The chemicals are often found in ordinary household products. For example, the propellants or solvents found in such products as hair spray, deodorant spray, spray paint, or cooking oil spray are often used for huffing. Some relatively pure liquids such as paint thinner, cleaning fluid, or fingernail polish remover are also commonly used for huffing.

Parents should be on the lookout for the use of certain words by their children which could indicate that inhalant abuse is taking place. While "huffing" is a general word for inhalant abuse, it also describes a specific technique in which a rag is soaked in a liquid inhalant and the rag is then pressed against the mouth and nose area of the face while the fumes are inhaled. "Sniffing" refers to the practice of sniffing, or snorting fumes sprayed directly from an aerosol container. In some cases the contents of an aerosol container might by sprayed directly into the nostrils or mouth. The term "bagging" is used to describe the practice of inhaling fumes from a product that has been sprayed or poured into a plastic or paper bag.

The first effect of huffing, sniffing, or bagging is generally a sense of euphoria. However, if an inhalant is abused for a longer time, the initial euphoria is replaced by dizziness, slurred speech, and loss of coordination, inhibitions and control. Some individuals become irritable or agitated, and some have delusions or hallucinations. Other potential risks associated with huffing include a rapid, irregular heartbeat that might trigger lethal heart failure. Chronic huffing can result in weakness and fatigue and serious liver and kidney damage. Permanent brain damage and hearing loss are also possible.

Some of the chemicals frequently used in huffing along with their common sources given in parentheses are as follows: acetone (solvent—paint or hardware store), butane (disposable cigar and cigarette lighters), propane (outdoor grill fuel, spray paint propellant), toluene (solvent—paint or hardware store), methylene chloride (paint stripper—paint or hardware store), Freon (propellent—compressed gas duster), and xylene (solvent—paint or hardware store).

Propane is often used as a fuel for outdoor grills and space heaters.

Numerous common products contain solvents used for huffing.

▶ Example 6.5

Refer to Figure 1.11 for the necessary conversion factors, and make the following temperature conversions (remember, Kelvin temperatures are expressed in kelvins, K):

a. 37°C (body temperature) to K   **b.** −50°C to K   **c.** 400 K to °C

Solution

a. Celsius is converted to Kelvin by adding 273, K = °C + 273. Therefore,

$$K = 37°C + 273 = 310 \text{ K (body temperature)}$$

b. Again, 273 is added to the Celsius reading. However, in this case the Celsius reading is negative, so the addition must be done with the signs in mind. Therefore,

$$K = −50°C + 273 = 223 \text{ K}$$

c. Because K = °C + 273, an algebraic rearrangement shows that °C = K −273. Therefore,

$$°C = 400\ K - 273 = 127°C$$

▶ **LEARNING CHECK 6.5** Convert the following Celsius temperatures into kelvins, and the Kelvin (K) temperatures into degrees Celsius:

a. 27°C   **b.** 0°C   **c.** 0 K   **d.** 100 K

6.7 Pressure, Temperature, and Volume Relationships

Learning Objective

5. Do calculations based on Boyle's law, Charles's law, and the combined gas law.

Experimental investigations into the behavior of gases as they were subjected to changes in temperature and pressure led to several gas laws that could be expressed by simple mathematical equations. In 1662 Robert Boyle, an Irish chemist, reported his discovery of a relationship between the pressure and volume of a gas sample kept at constant temperature. This relationship, known as **Boyle's law,** is

$$P = \frac{k}{V} \tag{6.3}$$

or

$$PV = k \tag{6.4}$$

where k is an experimentally determined constant, and (remember) the measurements are made without changing the temperature of the gas. Mathematically, the pressure and volume are said to be related by an inverse relationship because the change in one is in a direction opposite (inverse) to the change in the other. That is, as the pressure on a gas sample is increased, the volume of the gas sample decreases.

Another gas law was discovered in 1787 by Jacques Charles, a French scientist. He studied the volume behavior of gas samples kept at constant pressure as they were heated. He found that at constant pressure, the volume of a gas sample was directly proportional to its temperature expressed in kelvins. In other words, if the temperature was doubled, the sample volume doubled as long as the pressure was kept constant (see ▶ Figure 6.6). This behavior, known as **Charles's law,** is represented mathematically as

$$V = k'T \tag{6.5}$$

Boyle's law A gas law that describes the pressure and volume behavior of a gas sample kept at constant temperature. Mathematically, it is $PV = k$.

Charles's law A gas law that describes the temperature and volume behavior of a gas sample kept at constant pressure. Mathematically, it is $V/T = k'$.

Figure 6.6 A balloon collapses when the gas it contains is cooled in liquid nitrogen (temperature = −196°C, or 77 K). Assume the inflated yellow balloon had a volume of 4.0 L at room temperature (25°C) and the prevailing pressure. Assume the pressure remained constant and calculate the volume of the gas in the balloon at the temperature of liquid nitrogen.

or

$$\frac{V}{T} = k' \tag{6.6}$$

where k' is another experimentally determined constant.

Boyle's law and Charles's law can be combined to give a single law called the combined gas law, which provides a relationship between the pressure, volume, and temperature of gases. The **combined gas law** is written as

$$\frac{PV}{T} = k'' \tag{6.7}$$

where k'' is another experimentally determined constant.

Equation 6.7 can be put into a useful form by the following line of reasoning. Suppose a gas sample is initially at a pressure and temperature of P_i and T_i and has a volume of V_i. Now suppose that the pressure and temperature are changed to some new (final) values represented by P_f and T_f and that the volume changes to a new value of V_f. According to Equation 6.7,

$$\frac{P_i V_i}{T_i} = k''$$

and

$$\frac{P_f V_f}{T_f} = k''$$

Since both quotients on the left side are equal to the same constant, they can be set equal to each other, and we get

$$\frac{P_i V_i}{T_i} = \frac{P_f V_f}{T_f} \tag{6.8}$$

combined gas law A gas law that describes the pressure, volume, and temperature behavior of a gas sample. Mathematically, it is $PV/T = k''$.

▶ Example 6.6

a. A sample of helium gas has a volume of 5.00 L at 25°C and a pressure of 0.951 atm. What volume will the sample have if the temperature and pressure are changed to 50°C and 1.41 atm?

b. A sample of gas has a volume of 3.75 L at a temperature of 25°C and a pressure of 1.15 atm. What will the volume be at a temperature of 35°C and a pressure of 620 torr?

Solution

a. Equation 6.8 will be used to solve this problem. An important step is to identify the quantities that are related and will thus have the same subscripts in Equation 6.8. In this problem, the 5.00-L volume, 25°C temperature, and 0.951 atm pressure are all related and will be used as the initial conditions. The final conditions are the 50°C temperature, the 1.41 atm pressure, and the final volume V_f, which is to be calculated. Substitution of these quantities into Equation 6.8 gives

$$\frac{(0.951 \text{ atm})(5.00 \text{ L})}{(298 \text{ K})} = \frac{(1.41 \text{ atm})V_f}{(323 \text{ K})}$$

Note that the Celsius temperatures have been changed to kelvins. The desired quantity, V_f, can be isolated by multiplying both sides of the equation by 323 K and dividing both sides by 1.41 atm:

$$\frac{(323 \text{ K})(0.951 \text{ atm})(5.00 \text{ L})}{(298 \text{ K})(1.41 \text{ atm})} = \frac{(323 \text{ K})(1.41 \text{ atm})V_f}{(323 \text{ K})(1.41 \text{ atm})}$$

or

$$V_f = \frac{(323 \text{ K})(0.951 \text{ atm})(5.00 \text{ L})}{(298 \text{ K})(1.41 \text{ atm})} = 3.66 \text{ L}$$

b. This is the same as the problem in part a, except that the initial and final pressures are given in different units. As we saw in part a, the pressure units must cancel in the final calculation, so the initial and final units must be the same. The units of pressure used make no difference as long as they are the same, so we will convert the initial pressure into torr. The necessary conversion factor was obtained from Table 6.3:

$$P_i = (1.15 \text{ atm})\left(\frac{760 \text{ torr}}{1 \text{ atm}}\right) = 874 \text{ torr}$$

The quantities are substituted into Equation 6.8 to give

$$\frac{(874 \text{ torr})(3.75 \text{ L})}{(298 \text{ K})} = \frac{(620 \text{ torr})(V_f)}{(308 \text{ K})}$$

or

$$V_f = \frac{(308 \text{ K})(874 \text{ torr})(3.75 \text{ L})}{(298 \text{ K})(620 \text{ torr})} = 5.46 \text{ L}$$

▶ Example 6.7

a. A steel cylinder has a bursting point of 10,000 psi. It is filled with gas at a pressure of 2500 psi when the temperature is 20°C. During transport, the cylinder is allowed to sit in the hot sun, and its temperature reaches 100°C. Will the cylinder burst?
b. A sample of gas has a volume of 3.00 ft³ at a temperature of 30°C. If the pressure on the sample remains constant, at what Celsius temperature will the volume be half the volume at 30°C?

Solution

a. The volume of the gas is constant (unless the cylinder bursts), so $V_i = V_f$. With this in mind, Equation 6.8 is used with the following values:
$P_i = 2500$ psi, $V_i = V_f$, $T_i = 20°C + 273$ K, $P_f =$ unknown pressure, and $T_f = 100°C + 273$ K. Substitution into Equation 6.8 gives

$$\frac{(2500 \text{ psi})(V_i)}{293 \text{ K}} = \frac{(P_f)(V_f)}{373 \text{ K}}$$

or

$$P_f = \frac{(2500 \text{ psi})(V_f)(373 \text{ K})}{(V_f)(293 \text{ K})} = 3.18 \times 10^3 \text{ psi}$$

The cylinder will not burst.
b. In this case, $P_i = P_f$ because the pressure remains constant. Also, $V_i = 3.00 \text{ ft}^3$; $V_f = \frac{1}{2}V_i$, or 1.50 ft³; $T_i = 30°C + 273 = 303$ K; and T_f is to be determined. Substitution into Equation 6.8 gives

$$\frac{(P_i)(3.00 \text{ ft}^3)}{(303 \text{ K})} = \frac{(P_f)(1.50 \text{ ft}^3)}{T_f}$$

Cutting Drug Costs with Generics

Most individuals in their youthful years think very little (if at all) about the cost of medications. If a young person searches the couch cushions for change, it is generally for gas money or a soft drink. However, as we age we might find ourselves searching under couch cushions once again, this time in hopes of finding folding money to help pay for one of our numerous prescriptions.

Why are prescriptions so costly, and is there any way to get them at a lower cost? The development of new drugs usually requires years of research and substantial funding. As a result, any newly developed drugs are under patent protection for approximately 20 years in the U.S., and their production and sale are limited to those who own the patent. Once the patent expires, others have the right to produce and market the drug in what is called a generic version. Regulations set by the Food and Drug Administration (FDA) ensure that generic drugs must contain the same active ingredients as the original patented drug. However, the generic versions can be different in other ways such as the color, shape, and fillers. Generic versions are generally less expensive than the original, with savings to consumers of 30 to 90 percent of the purchase price of the original drug. The amount of savings often depends on the number of generic brands on the market. To benefit from the savings, ask your doctor or pharmacist if the medications you require are available in a less expensive generic form.

Generic drugs are sometimes different colors than the patented version.

The quantity to be calculated, T_f, is in the denominator of the right side. We put it into the numerator by inverting both sides of the equation:

$$\frac{(303\text{ K})}{(P_i)(3.00\text{ ft}^3)} = \frac{T_f}{(P_f)(1.50\text{ ft}^3)}$$

The desired quantity, T_f, is now isolated by multiplying both sides of the equation by $(P_f)(1.50\text{ ft}^3)$:

$$\frac{(P_f)(1.50\text{ ft}^3)(303\text{ K})}{(P_i)(3.00\text{ ft}^3)} = \frac{(P_f)(1.50\text{ ft}^3)(T_f)}{(P_f)(1.50\text{ ft}^3)}$$

or

$$T_f = \frac{(P_f)(1.50\text{ ft}^3)(303\text{ K})}{(P_i)(3.00\text{ ft}^3)} = 152\text{ K}$$

Remember, P_f and P_i canceled because they were equal (the pressure was kept constant). The answer is given in kelvins, but we want it in degrees Celsius:

$$K = °C + 273$$

or

$$°C = K - 273 = 152 - 273 = -121°C$$

a. A sample of argon gas is confined in a 10.0-L container at a pressure of 1.90 atm and a temperature of 30°C. What volume would the sample have at 1.00 atm and −10.2°C?

b. A sample of gas has a volume of 500 mL at a temperature and pressure of 300 K and 800 torr. It is desired to compress the sample to a volume of 250 mL at a pressure of 900 torr. What temperature in both kelvins and Celsius degrees will be required?

6.8 The Ideal Gas Law

Learning Objective

6. Do calculations based on the ideal gas law.

The combined gas law (Equation 6.8) works for samples only in which the mass of gas remains constant during changes in temperature, pressure, and volume. However, it is often useful to work with situations in which the amount of gas varies. The foundation for this kind of work was proposed in 1811 by Amadeo Avogadro, an Italian scientist. According to his proposal, which is now known as **Avogadro's law,** equal volumes of different gases measured at the same temperature and pressure contain equal numbers of gas molecules. According to this idea, two identical compressed gas cylinders of helium and oxygen at the same pressure and temperature would contain identical numbers of molecules of the respective gases. However, the mass of gas in the cylinders would *not* be the same because the molecules of the two gases have different molecular weights.

The actual temperature and pressure used do not influence the validity of Avogadro's law, but it is convenient to specify a standard set of values. Chemists have chosen 0°C (273 K) and 1.00 atm to represent what are called **standard conditions** for gas measurements. These conditions are often abbreviated STP (standard temperature and pressure).

As we have seen, the mole (defined in Section 2.6) is a convenient quantity of matter to work with. What volume does 1 mol of a gas occupy according to Avogadro's law? Experiments show that 1 mol of any gas molecules has a volume of 22.4 L at STP (see ▶ Figure 6.7).

A combination of Boyle's, Charles's, and Avogadro's laws leads to another gas law that includes the quantity of gas in a sample as well as the temperature, pressure, and volume of the sample. This law, known as the **ideal gas law,** is written as

$$PV = nRT \tag{6.9}$$

In this equation, P, V, and T are defined as they were in the gas laws given earlier. The symbol n stands for the number of moles of gas in the sample being used, and R is a constant known as the **universal gas constant.** The measured value for the volume of 1 mol of gas at STP allows R to be evaluated by substituting the values into Equation 6.9 after rearrangement to isolate R:

$$R = \frac{PV}{nT} = \frac{(1.00 \text{ atm})(22.4 \text{ L})}{(1 \text{ mol})(273 \text{ K})} = 0.0821 \frac{\text{L atm}}{\text{mol K}}$$

This value of R is the same for all gases under any conditions of temperature, pressure, and volume.

Avogadro's law Equal volumes of gases measured at the same temperature and pressure contain equal numbers of molecules.

standard conditions (STP) A set of specific temperature and pressure values used for gas measurements.

ideal gas law A gas law that relates the pressure, volume, temperature, and number of moles in a gas sample. Mathematically, it is $PV = nRT$.

Figure 6.7 The ball has a volume of 22.4 L, the volume of 1 mole of any gas at STP. The bottle has a volume of 1.00 quart (0.946 L). How many moles of gas would the bottle contain at STP?

universal gas constant The constant that relates pressure, volume, temperature, and number of moles of gas in the ideal gas law.

Use the ideal gas law to calculate the volume of 0.413 mol of hydrogen gas at a temperature of 20°C and a pressure of 1200 torr.

Solution

To use the ideal gas law (Equation 6.9), it is necessary that all units match those of R. Thus, the temperature will have to be expressed in kelvins and the pressure in atmospheres:

$$K = °C + 273 = 20°C + 273 = 293 \text{ K}$$

$$P = (1200 \text{ torr})\left(\frac{1 \text{ atm}}{760 \text{ torr}}\right) = 1.58 \text{ atm}$$

Because volume is the desired quantity, we isolate it by dividing both sides of the ideal gas law by P:

$$\frac{PV}{P} = \frac{nRT}{P} \qquad \text{or} \qquad V = \frac{nRT}{P}$$

Substitution of quantities gives

$$V = \frac{(0.413 \text{ mol})\left(0.0821\dfrac{\text{L atm}}{\text{mol K}}\right)(293 \text{ K})}{(1.58 \text{ atm})} = 6.29 \text{ L}$$

▶ **LEARNING CHECK 6.7** A 2.15-mol sample of sulfur dioxide gas (SO_2) occupies a volume of 12.6 L at 30°C. What is the pressure of the gas in atm?

Because R is a constant for all gases, it follows that if any three of the quantities P, V, T, or n are known for a gas sample, the fourth quantity can be calculated by using Equation 6.9. An interesting application of this concept makes it possible to determine the molecular weights of gaseous substances. If the mass of a sample is known, the number of moles in the sample is the mass in grams divided by the molecular weight. This fact is represented by Equation 6.10, where n is the number of moles in a sample that has a mass in grams of m and a molecular weight of MW:

$$n = \frac{m}{\text{MW}} \tag{6.10}$$

When $\dfrac{m}{\text{MW}}$ is substituted for n in Equation 6.9, the result is

$$PV = \frac{mRT}{\text{MW}} \tag{6.11}$$

A gas sample has a volume of 2.74 L, a mass of 16.12 g and is stored at a pressure of 4.80 atm at a temperature of 25°C. The gas might be CH_4, C_2H_6, or C_3H_{10}. Which gas is it?

Solution

The molecular weight of the gas is determined using Equation 6.11 and compared with the molecular weights of the three possibilities calculated from their formulas. Rearrangement of Equation 6.11 to isolate molecular weight gives

$$\text{MW} = \frac{mRT}{PV}$$

Substitution of the known quantities after conversion to proper units gives

$$MW = \frac{mRT}{PV} = \frac{(16.12 \text{ g})\left(0.0821 \frac{L \text{ atm}}{\text{mol K}}\right)(298 \text{ K})}{(4.80 \text{ atm})(2.74 \text{ L})} = 30.0 \text{ g/mol}$$

Thus, the molecular weight is 30.0 u. This matches the molecular weight of C_2H_6 calculated from atomic weights and the molecular formula.

▶ **LEARNING CHECK 6.8** A sample of unknown gas has a mass of 3.35 g and occupies 2.00 L at 1.21 atm and 27°C. What is the molecular weight of the gas? Is the gas H_2S, HBr, or NH_3?

The gas laws discussed in this chapter apply only to gases that are ideal, but interestingly, no ideal gases actually exist. If they did exist, ideal gases would behave exactly as predicted by the gas laws at all temperatures and pressures. Real gases deviate from the behavior predicted by the gas laws, but under normally encountered temperatures and pressures, the deviations are small for many real gases. This fact allows the gas laws to be used for real gases. Interparticle attractions tend to make gases behave less ideally. Thus, the gas laws work best for gases in which such forces are weak, that is, those made up of single atoms (the noble gases) or nonpolar molecules (O_2, N_2, etc.). Highly polar molecules such as water vapor, hydrogen chloride, and ammonia deviate significantly from ideal behavior.

6.9 Dalton's Law

Learning Objective

7. Do calculations based on Dalton's law.

John Dalton (1766–1844), an English schoolteacher, made a number of important contributions to chemistry. Some of his experiments led to the law of partial pressures, also called **Dalton's law.** According to this law, the total pressure exerted by a mixture of different gases kept at a constant volume and temperature is equal to the sum of the partial pressures of the gases in the mixture. The **partial pressure** of each gas in such mixtures is the pressure each gas would exert if it were confined alone under the same temperature and volume conditions as the mixture.

Imagine you have four identical gas containers, as shown in ▶ Figure 6.8. Place samples of three different gases (represented by △, ○, and □) into three of the containers, one to a

Dalton's law of partial pressures The total pressure exerted by a mixture of gases is equal to the sum of the partial pressures of the gases in the mixture.

partial pressure The pressure an individual gas of a mixture would exert if it were in the container alone at the same temperature as the mixture.

Figure 6.8 Dalton's law of partial pressures.

container, and measure the pressure exerted by each sample. Then place all three samples into the fourth container and measure the total pressure (P_t) exerted. The result is a statement of Dalton's law:

$$P_t = P_\triangle + P_\bigcirc + P_\square \tag{6.12}$$

where $P_\triangle$, $P_\bigcirc$, and $P_\square$ are the partial pressures of gases $\triangle$, $\bigcirc$, and $\square$, respectively.

▶ Example 6.10

A sample of air is collected when the atmospheric pressure is 742 torr. The partial pressures of nitrogen and oxygen in the sample are found to be 581 torr and 141 torr, respectively. Assume water vapor to be the only other gas present in the air sample, and calculate its partial pressure.

Solution

Dalton's law says $P_t = P_{O_2} + P_{N_2} + P_{H_2O}$. The total pressure of the sample is the atmospheric pressure of 742 torr. Therefore,

$$742 \text{ torr} = 141 \text{ torr} + 581 \text{ torr} + P_{H_2O}$$

and

$$P_{H_2O} = 742 - 141 - 581 = 20 \text{ torr}$$

▶ **LEARNING CHECK 6.9** A mixture is made of helium, nitrogen, and oxygen. Their partial pressures are, respectively, 310 torr, 0.200 atm, and 7.35 psi. What is the total pressure of the mixture in torr?

1 A mylar balloon (left) and rubber balloon freshly filled with helium.

2 The same balloons the next day.

Figure 6.9 Graham's law in action. Does helium diffuse more rapidly through rubber or plastic?

effusion A process in which a gas escapes from a container through a small hole.

diffusion A process that causes gases to spontaneously intermingle when they are brought together.

Graham's law A mathematical expression that relates rates of effusion or diffusion of two gases to the masses of the molecules of the two gases.

6.10 Graham's Law

Learning Objective

8. Do calculations based on Graham's law.

Effusion and diffusion are processes by which gases move and mix. **Effusion** is the escape of a gas through a small hole in its container. **Diffusion** is the process in which two or more gases spontaneously intermingle when brought together. Even though the processes appear to be different, they are related, and both follow a law proposed in 1828 by Thomas Graham to describe effusion (see ▶ Figure 6.9). For two gases, represented by A and B, **Graham's law** is

$$\frac{\text{effusion rate } A}{\text{effusion rate } B} = \sqrt{\frac{\text{molecular mass of } B}{\text{molecular mass of } A}} \tag{6.13}$$

▶ Example 6.11

Oxygen molecules weigh 16 times as much as hydrogen molecules. Which molecule will diffuse faster and how much faster?

Solution

Applying Equation 6.13 but recognizing that diffusion follows the equation as well as effusion, we get

$$\frac{\text{rate } H_2}{\text{rate } O_2} = \sqrt{\frac{\text{mass } O_2}{\text{mass } H_2}} = \sqrt{\frac{32}{2}} = \sqrt{16} = 4$$

Therefore, we conclude that hydrogen will diffuse four times faster than oxygen.

▶ **LEARNING CHECK 6.10** Which will diffuse faster, He molecules or Ne molecules? How much faster?

6.11 Changes in State

Learning Objective

9. Classify changes in state as exothermic or endothermic.

Matter can be changed from one state into another by processes such as heating, cooling, or changing the pressure. Heating and cooling are the processes most often used, and a change in state that requires an input of heat is called endothermic, whereas one in which heat is given up (or removed) is called exothermic (see Section 5.8). (These terms come from the Greek *endo* = in and *exo* = out). Endothermic changes are those in which particles are moved farther apart as cohesive forces are being overcome, such as the change of a solid to a liquid or gas. Exothermic changes are those in the opposite direction. See ▶ Figure 6.10.

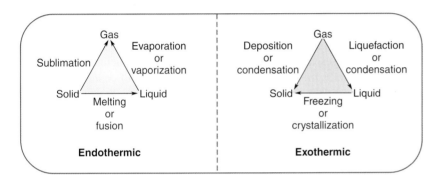

Figure 6.10 Endothermic and exothermic changes of state.

▶ **LEARNING CHECK 6.11** Classify the following processes as endothermic or exothermic:

a. Alcohol evaporates.

b. Water freezes.

c. Iron melts in a furnace.

6.12 Evaporation and Vapor Pressure

Learning Objective

10. Demonstrate an understanding of the concepts of vapor pressure and evaporation.

The evaporation of liquids is a familiar process. Water in a container will soon disappear (evaporate) if the container is left uncovered. **Evaporation, or vaporization,** is an endothermic process that takes place as a result of molecules leaving the surface of a liquid. The rate of evaporation depends on the temperature of the liquid and the surface area from which the molecules can escape. Temperature is an important factor because it is related directly to the speed and kinetic energy of the molecules and hence to their ability to break away from the attractive forces present at the liquid surface.

Evaporating molecules carry significant amounts of kinetic energy away from the liquid, and as a result, the temperature of the remaining liquid will drop unless heat flows in from the surroundings. This principle is involved in all evaporative cooling processes, including evaporative coolers for homes, the cooling of the human body by perspiration, and the cooling of a panting dog by the evaporation of saliva from the mucous membranes of its mouth.

evaporation or vaporization An endothermic process in which a liquid is changed to a gas.

Compare evaporation in a closed container (◗ Figure 6.11) with that in an open container. Evaporation occurs in both containers, as indicated by a drop in liquid level. But, unlike that in an open container, the liquid level in the closed container eventually stops dropping and becomes constant.

Figure 6.11 Liquid evaporation in a closed container. The drop in liquid level is greatly exaggerated for emphasis.

Initially   After some time   At equilibrium

Constant liquid level

What would explain this behavior? In a closed container the molecules of liquid that go into the vapor (gaseous) state are unable to move completely away from the liquid surface as they do in an open container. Instead, the vapor molecules are confined to a space immediately above the liquid, where they have many random collisions with the container walls, other vapor molecules, and the liquid surface. Occasionally, their collisions with the liquid result in condensation, and they are recaptured by the liquid. **Condensation** is an exothermic process in which a gas or vapor is converted to either a liquid or a solid. Thus, two processes—evaporation (escape) and condensation (recapture)—actually take place in the closed container. Initially, the rate of evaporation exceeds that of condensation, and the liquid level drops. However, the rates of the two processes eventually become equal, and the liquid level stops dropping because the number of molecules that escape in a given time is the same as the number recaptured.

condensation An exothermic process in which a gas or vapor is changed to a liquid or solid.

A system in which two opposite processes take place at equal rates is said to be in equilibrium (see Chapter 8). Under the equilibrium conditions just described, the number of molecules in the vapor state remains constant. This constant number of molecules will exert a constant pressure on the liquid surface and the container walls. This pressure exerted by a vapor in equilibrium with a liquid is called the **vapor pressure** of the liquid. The magnitude of a vapor pressure depends on the nature of the liquid (molecular polarity, mass, etc.) and the temperature of the liquid. These dependencies are illustrated in ◗ Tables 6.4 and 6.5.

vapor pressure The pressure exerted by vapor that is in equilibrium with its liquid.

TABLE 6.4 The Vapor Pressure of Various Liquids at 20°C

Liquid	Molecular Weight (u)	Polarity	Vapor Pressure (torr)
pentane (C_5H_{12})	72	Nonpolar	414.5
hexane (C_6H_{14})	86	Nonpolar	113.9
heptane (C_7H_{16})	100	Nonpolar	37.2
ethanol (C_2H_5—OH)	46	Polar (hydrogen bonds)	43.9
1-propanol (C_3H_7—OH)	60	Polar (hydrogen bonds)	17.3
1-butanol (C_4H_9—OH)	74	Polar (hydrogen bonds)	7.1

TABLE 6.5 Vapor Pressure of Water at Various Temperatures

Temperature (°C)	Vapor Pressure (torr)
0	4.6
20	17.5
40	55.3
60	149.2
80	355.5
100	760.0

The effect of molecular mass on vapor pressure is seen in both the pentane–heptane and ethanol–1-butanol series of compounds in Table 6.4. Within each series, the molecular polarities are the same for each member, but vapor pressure decreases as molecular weight increases. This shows the effect of stronger dispersion forces between heavier molecules. A comparison of pentane and 1-butanol shows the effects of polar attractions between

Blood Pressure: A Barometer of Health

Blood moves through the blood vessels of the body under pressure created by the pumping heart. The pressure is necessary to efficiently move the blood to all parts of the body. If blood pressure is too low, symptoms such as dizziness develop from a shortage of oxygen-carrying blood to the brain. On the other hand, blood pressure that is too high can cause damage to delicate organs such as the kidneys. Both high and low blood pressure can be treated and controlled. Because of this, a blood pressure measurement is routinely included as a part of medical examinations.

Blood pressure is reported as a ratio; for example, a typical healthy blood pressure is 120/80. The higher number is the pressure in millimeters of mercury (Hg) when the heart contracts, and the lower number is the pressure in millimeters of mercury when the heart relaxes. Both numbers are easily obtained by using a sphygmomanometer, which is essentially a barometer tube connected to an inflatable cuff by a hollow tube (see diagram). The cuff is wrapped around the upper arm, and the air pressure inside the cuff, connecting tube, and barometer tube is pumped up to a value of about 200 mm Hg. The inflated cuff temporarily cuts off the flow of blood through the vessels of the arm. A valve is then opened, and the air pressure in the cuff, connecting tube, and barometer is allowed to slowly decrease while a stethoscope is used to detect the flow of blood through the brachial artery of the arm. When a faint tapping sound is first heard, the level of mercury

in the barometer tube is noted. This is the higher, or systolic, blood pressure. The tapping sound becomes louder and then fades away as the cuff pressure is reduced. When the last sound is heard, the mercury level in the barometer tube is again read. This is the lower, or diastolic, blood pressure.

In young adults, the normal systolic range is 100–120 mm Hg and the normal diastolic range is 60–80 mm Hg. In older people, the corresponding normal ranges are 115–135 and 75–85 mm Hg.

A sphygmomanometer reports blood pressure as a ratio such as 105/80 (the higher or systolic blood pressure is shown here).

molecules. Molecules of these liquids have similar molecular weights, but the hydrogen-bonded 1-butanol has a much lower vapor pressure.

The effect of increasing the kinetic energy of molecules by heating is clearly evident by the vapor pressure behavior of water shown in Table 6.5.

> ▶ **LEARNING CHECK 6.12** Decide which member of each of the following pairs of compounds would have the higher vapor pressure. Explain your choice in each case.
>
> **a.** Methyl alcohol (CH_3OH) and propyl alcohol (C_3H_7OH)
> **b.** Liquid helium (He) and liquid nitrogen (N_2)
> **c.** Liquid HF and liquid neon (Ne)

6.13 Boiling and the Boiling Point

Learning Objective

11. Demonstrate an understanding of the process of boiling and the concept of boiling point.

As a liquid is heated, its vapor pressure increases, as shown for water in Table 6.5. If the liquid's temperature is increased enough, its vapor pressure will reach a value equal to that of the prevailing atmospheric pressure. Up to that temperature, all vaporization appears to take place at the liquid surface. However, when the vapor pressure becomes equal to atmospheric pressure, vaporization begins to occur beneath the surface as well.

boiling point The temperature at which the vapor pressure of a liquid is equal to the prevailing atmospheric pressure.

normal or standard boiling point The temperature at which the vapor pressure of a liquid is equal to 1 standard atmosphere (760 torr).

When bubbles of vapor form and rise rapidly to the surface, where the vapor escapes, the liquid is boiling. The **boiling point** of a liquid is the temperature at which the vapor pressure of the liquid is equal to the atmospheric pressure above the liquid. The **normal** or **standard boiling point** of a liquid is the temperature at which the vapor pressure is equal to 1 standard atmosphere (760 torr). Normal values were used in all examples of boiling points given earlier in this book.

Liquids boil at temperatures higher than their normal boiling points when external pressures are greater than 760 torr. When external pressures are less than 760 torr, liquids boil at temperatures below their normal boiling points. Thus, the boiling point of water fluctuates with changes in atmospheric pressure. Such fluctuations seldom exceed 2°C at a specific location, but there can be striking variations at different elevations, as shown by ▶ Table 6.6. (This is why cooking directions are sometimes different for different altitudes.)

TABLE 6.6 Variations in the Boiling Point of Water with Elevation

Location	Elevation (feet above sea level)	Boiling Point of Water (°C)
San Francisco, CA	Sea level	100.0
Salt Lake City, UT	4,390	95.6
Denver, CO	5,280	95.0
La Paz, Bolivia	12,795	91.4
Mount Everest	20,028	76.5

TABLE 6.7 The Boiling Point of Water in a Pressure Cooker

Pressure above Atmospheric		Boiling Point of Water (°C)
psi	torr	
5	259	108
10	517	116
15	776	121

The increase in boiling point caused by an increase in pressure is the principle used in the ordinary pressure cooker. Increasing the pressure inside the cooker causes the boiling point of the water in it to rise, as shown in ▶ Table 6.7. It then becomes possible to increase the temperature of the food-plus-water in the cooker above 100°C. Such an increase of just 10°C will make the food cook approximately twice as fast (see ▶ Figure 6.12).

Figure 6.12 A pressure cooker shortens the time required to cook food. A carrot cooks completely in boiling water (100°C) in 8 minutes. About how long would it take to cook in a pressure cooker that raised the boiling point of water to 110°C?

© Jgroup/Dreamstime.com

6.14 Sublimation and Melting

Learning Objective

12. Demonstrate an understanding of the processes of sublimation and melting.

Solids, like liquids, have vapor pressures. Although the motion of particles is much more restricted in solids, particles at the surface can escape into the vapor state if they acquire sufficient energy. However, the strong cohesive forces characteristic of the solid state usually cause the vapor pressures of solids to be quite low.

As expected, vapor pressures of solids increase with temperature. When the vapor pressure of a solid is high enough to allow escaping molecules to go directly into the vapor state without passing through the liquid state, the process is called **sublimation**

sublimation The endothermic process in which a solid is changed directly to a gas without first becoming a liquid.

(see ▶ Figure 6.13). Sublimation is characteristic of materials such as solid carbon dioxide (dry ice) and naphthalene (moth crystals). Frozen water also sublimes under appropriate conditions. Wet laundry hung out in freezing weather eventually dries as the frozen water sublimes. Freeze drying, a technique based on this process, is used to remove water from materials that would be damaged by heating (e.g., freeze-dried foods).

Even though solids have vapor pressures, most pure substances in the solid state melt before appreciable sublimation takes place. Melting involves the breakdown of a rigid, orderly, solid structure into a mobile, disorderly liquid state. This collapse of the solid structure occurs at a characteristic temperature called the **melting point.** At the melting point, the kinetic energy of solid particles is large enough to partially overcome the strong cohesive forces holding the particles together, and the solid and liquid states have the same vapor pressure.

In some instances, solids cannot be changed into liquids, or liquids into gases, by heating. The atoms making up the molecules of some solids acquire enough kinetic energy on heating to cause bonds within the molecules to break before the solid (or liquid) can change into another state. This breaking of bonds within molecules changes the composition of the original substance. When this **decomposition** occurs, the original substance is said to have decomposed. This is why cotton and paper, when heated, char rather than melt.

Figure 6.13 Solid CO_2 (left) and solid H_2O (right) both sublime. Contrast the endothermic processes involved when each is used in the most common way to keep things cold. Are the same processes involved?

melting point The temperature at which a solid changes to a liquid; the solid and liquid have the same vapor pressure.

decomposition A change in chemical composition that can result from heating.

6.15 Energy and the States of Matter

Learning Objective

13. Do calculations based on energy changes that accompany heating, cooling, or changing the state of a substance.

A pure substance in the gaseous state contains more energy than in the liquid state, which in turn contains more energy than in the solid state. Before we look at this, note the following relationships. Kinetic energy, the energy of particle motion, is related to heat. In fact, temperature is a measurement of the average kinetic energy of the particles in a system. Potential energy, in contrast, is related to particle separation distances rather than motion. Thus, we conclude that an increase in temperature on adding heat corresponds to an increase in kinetic energy of the particles, whereas no increase in temperature on adding heat corresponds to an increase in the potential energy of the particles.

Now let's look at a system composed of 1 g of ice at an initial temperature of $-20°C$. Heat is added at a constant rate until the ice is converted into 1 g of steam at $120°C$. The atmospheric pressure is assumed to be 760 torr throughout the experiment. The changes in the system take place in several steps, as shown in ▶ Figure 6.14.

Figure 6.14 The temperature behavior of a system during changes in state.

Therapeutic Uses of Oxygen Gas

A steady supply of oxygen is essential for the human body to function properly. The most common source of this gas is inhaled air, in which the partial pressure of oxygen is about 160 torr. In a healthy individual, this partial pressure is high enough to allow sufficient oxygen needed for body processes to be transported into the blood and distributed throughout the body. Patients suffering from a lung disease, such as pneumonia or emphysema, often cannot transport sufficient oxygen to the blood from the air they breathe unless the partial pressure of oxygen in the air is increased. This is done by mixing an appropriate amount of oxygen with the air by using an oxygen mask or nasal cannula.

Oxygen at very high partial pressures is used in other clinical applications based on a technique called *hyperbaric oxygenation*. In one application, patients infected by anaerobic bacteria, such as those that cause tetanus and gangrene, are placed in a hyperbaric chamber in which the partial pressure of oxygen is 3 to 4 standard atmospheres (2.2×10^3 to 3.0×10^3 torr). Because of the high partial pressure, body tissues pick up large amounts of oxygen, and the bacteria are killed.

Hyperbaric oxygenation may also be used to treat other abnormal conditions or injuries, such as certain heart disorders, carbon monoxide poisoning, crush injuries, certain hard-to-treat bone infections, smoke inhalation, near-drowning, asphyxia, and burns.

Hyperbaric chambers must be strongly built to resist significant internal gas pressure.

The solid is first heated from $-20°C$ to the melting point of $0°C$ (line *AB*). The temperature increase indicates that most of the added heat causes an increase in the kinetic energy of the molecules. Along line *BC*, the temperature remains constant at $0°C$ while the solid melts. The constant temperature during melting reveals that the added heat has increased the potential energy of the molecules without increasing their kinetic energy; the molecules are moved farther apart, but their motion is not increased. The addition of more heat warms the liquid water from $0°C$ to the normal boiling point of $100°C$ (line *CD*). At $100°C$, another change of state occurs as heat is added, and the liquid is converted into vapor (steam) at $100°C$ (line *DE*). The constant temperature of this process again indicates an increase in potential energy. Line *EF* represents heating the steam from $100°C$ to $120°C$ by adding more heat.

The addition of heat to this system resulted in two obviously different results: The temperature of water in a specific state was increased, or the water was changed from one state to another at constant temperature. The amount of heat required to change the temperature of a specified amount of a substance by $1°C$ is called the **specific heat** of the substance. In scientific work, this is often given in units of calories or joules (J) per gram degree. Thus, the specific heat of a substance is the number of calories or joules required to raise the temperature of 1 g of the substance by $1°C$. (NOTE: 1 cal = 4.184 J.) The specific heat is related to the amount of heat required to increase the temperature of a sample of substance by Equation 6.14.

specific heat The amount of heat energy required to raise the temperature of exactly 1 g of a substance by exactly 1°C.

$$\text{Heat} = (\text{sample mass})(\text{specific heat})(\text{temp. change}) \qquad (6.14)$$

Specific heats for a number of substances in various states are listed in ❱ Table 6.8. A substance with a high specific heat is capable of absorbing more heat with a small temperature change than substances with lower specific heats. Thus, substances to be used as heat transporters (e.g., in cooling and heating systems) should have high specific heats—note the value for liquid water.

TABLE 6.8 Specific Heats for Selected Substances

Substance and State	Specific Heat cal/g Degree	Specific Heat J/g Degree
Aluminum (solid)	0.24	1.0
Copper (solid)	0.093	0.39
Ethylene glycol (liquid)	0.57	2.4
Helium (gas)	1.25	5.23
Hydrogen (gas)	3.39	14.2
Lead (solid)	0.031	0.13
Mercury (liquid)	0.033	0.14
Nitrogen (gas)	0.25	1.1
Oxygen (gas)	0.22	0.92
Sodium (solid)	0.29	1.2
Sodium (liquid)	0.32	1.3
Water (solid)	0.51	2.1
Water (liquid)	1.00	4.18
Water (gas)	0.48	2.0

▶ Example 6.12

You notice that your car is running hot on a summer day. Someone tells you to drain out the ethylene glycol antifreeze and replace it with water. Will this allow the engine to run cooler?

Solution

The engine will run cooler because a given mass of water has a greater ability to carry heat from the engine to the radiator than does an equal mass of ethylene glycol. The coolant in your car is a mixture of ethylene glycol and water, but to simplify the following comparison, let us imagine it is pure ethylene glycol. We calculate the amount of heat absorbed by 1000 g each of pure ethylene glycol and pure water as the temperature changes from 20°C to 80°C.

$$\text{Heat absorbed} = (\text{mass})(\text{specific heat})(\text{temp. change})$$

ethylene glycol:
$$\text{Heat absorbed} = (1000 \text{ g})\left(\frac{0.57 \text{ cal}}{\text{g }°C}\right)(60° C)$$
$$= 34{,}200 \text{ cal} = 34.2 \text{ kcal}$$

water:
$$\text{Heat absorbed} = (1000 \text{ g})\left(\frac{1.00 \text{ cal}}{\text{g }°C}\right)(60° C)$$
$$= 60{,}000 \text{ cal} = 60.0 \text{ kcal}$$

Thus, the same amount of water will absorb (and transport) nearly twice as much heat for the same temperature increase.

▶ **LEARNING CHECK 6.13** Some nuclear reactors are cooled by gases. Calculate the number of calories that 1.00 kg of helium gas will absorb when it is heated from 25°C to 700°C. See Table 6.8 for the specific heat of He.

The amount of heat required to change the state of 1 g of a substance at constant temperature is called the **heat of fusion** (for melting) and the **heat of vaporization** (for boiling). The units used for these quantities are just calories or joules per gram because no temperature changes are involved. These heats represent the amount of energy required

heat of fusion The amount of heat energy required to melt exactly 1 g of a solid substance at constant temperature.

heat of vaporization The amount of heat energy required to vaporize exactly 1 g of a liquid substance at constant temperature.

to change 1 g of a substance to the liquid or vapor state at the characteristic melting or boiling point. The heats of fusion and vaporization for water are 80 and 540 cal/g, respectively. This explains why a burn caused by steam at 100°C is more severe than one caused by water at 100°C. Vaporization has added 540 cal to each gram of steam, and each gram will release these 540 cal when it condenses on the skin. Liquid water at 100°C would not have this extra heat with which to burn the skin (see ▶ Figure 6.15).

Figure 6.15 Water can be boiled in a paper cup. Explain why heat from the burner does not increase the temperature of the water-containing cup to the ignition temperature.

1 A paper cup filled with water does not reach the ignition temperature when heated by a burner, but the water boils.

2 An empty paper cup burns when heated by a burner because the heat from the burner increases the temperature of the cup to the ignition point.

Study Skills 6.1

Which Gas Law to Use

For many students, the biggest challenge of this chapter is solving problems using the gas laws. Once you recognize that you are faced with a gas law problem, you must decide which of the six gas laws will work to solve the problem. One aid to selecting the appropriate law is to look for key words, phrases, or ideas that are often associated with specific laws. Some of these are given here:

Gas Law	Equation	Key
Boyle's law	$PV = k$	T is constant
Charles's law	$\dfrac{V}{T} = k'$	P is constant
Combined gas law	$\dfrac{P_i V_i}{T_i} = \dfrac{P_f V_f}{T_f}$	P, V, and T all change
Dalton's law	$P_t = P_\triangle + P_\bigcirc + P_\square$	Two or more different gases
Graham's law	$\dfrac{\text{effusion rate } A}{\text{effusion rate } B} = \sqrt{\dfrac{\text{molecular mass of } B}{\text{molecular mass of } A}}$	Effusion and diffusion
Ideal gas law	$PV = nRT$	Moles of gas

For example, if you see the key word *diffusion*, it is likely that you are dealing with a Graham's law problem.

The first three gas laws (Boyle's, Charles's, and the combined) are very similar in that they all use the symbols P, V, and T. If you have narrowed down a gas law problem to one of these three, a simple approach is to use the combined gas law. It works in all three cases. If T is constant,

$$\frac{P_i V_i}{T_i} = \frac{P_f V_f}{T_f}$$

simplifies to $P_i V_i = P_f V_f$, a form of Boyle's law. If P is constant, it simplifies to

$$\frac{V_i}{T_i} = \frac{V_f}{T_f}$$

a form of Charles's law. If V is constant, the combined law becomes

$$\frac{P_i}{T_i} = \frac{P_f}{T_f}$$

an equation that is not named but is useful for certain problems.

A final point to remember is that temperatures must be expressed in kelvins to get correct answers using any of the gas laws that involve temperature.

▶ Example 6.13

Calculate the heat released when 5.0 kg of steam at 120°C condenses to water at 100°C in a radiator of a steam heating system.

Solution

Consider the process as taking place in two steps. The steam must cool from 120°C to 100°C and then condense to liquid water at 100°C. When the steam cools to 100°C,

$$\text{Heat released} = (\text{mass})(\text{specific heat})(\text{temp. change})$$

$$= (5.0 \times 10^3 \text{ g})\left(\frac{0.48 \text{ cal}}{\text{g °C}}\right)(20° \text{ C})$$

$$= 4.8 \times 10^4 \text{ cal} = 48 \text{ kcal}$$

When the steam condenses to liquid water at 100°C,

$$\text{Heat released} = (\text{mass})(\text{heat of vaporization})$$

$$= (5.0 \times 10^3 \text{ g})\left(\frac{540 \text{ cal}}{\text{g}}\right)$$

$$= 2.7 \times 10^6 \text{ cal} = 2.7 \times 10^3 \text{ kcal}$$

Note that the heat of vaporization was used even though the water was changing from the vapor to the liquid state. The only difference between vaporization and condensation is the direction of heat flow; the amount of heat involved remains the same for a specific quantity of material. (Accordingly, 1 g of liquid water will release 80 cal when it freezes.)

These results make it clear that most of the transported heat (98%) was carried in the form of potential energy, which was released when the steam condensed.

▶ **LEARNING CHECK 6.14** Suppose the radiator of your car overheats and begins to boil. Calculate the number of calories absorbed by 5.00 kg of water (about 1 gallon) as it boils and changes to steam.

Case Study Follow-up

Coughing is considered an emergency if it is accompanied by cyanosis (appearance of bluish skin), apnea (cessation of breathing), intercostal retractions (inward movement of muscles between the ribs), seizures or convulsions, high fever, or dehydration (NLM, 2009). During an emergency visit for a serious cough, a pulse oximeter (a device that clips onto the finger) may be used to measure actual oxygen saturation levels of the blood. Levels below 90% are considered indicative of hypoxemia (low blood-oxygen) (Mayo Clinic, 2010).

Parental presence in hospitals has been standard for many years. It alleviates the child's anxiety and reduces psychological suffering of the parents as well. Since 2000, parental presence has been supported even during resuscitation of pediatric patients (Meeks 2009). However, controversy exists over parental presence in some situations. If a parent is anxious, his or her anxiety may serve to increase the child's fears, and concerns about a distressed parent needing treatment and adding to the workload of pediatric nurses have also been raised (Buffett-jerrott et al., 2007). In the absence of complicating factors, a calm parent's presence can serve to alleviate the patient's fear and decrease the workload of pediatric nurses.

Concept Summary

Observed Properties of Matter

Matter in the solid, liquid, or gaseous state shows differences in physical properties such as density, shape, compressibility, and thermal expansion.

Objective 1, Exercise 6.2

The Kinetic Molecular Theory of Matter

Much of the behavior of matter in different states can be explained by the kinetic molecular theory, according to which all matter is composed of tiny molecules that are in constant motion and are attracted or repelled by each other.

Objective 2, Exercise 6.8

The Solid State

In the solid state, cohesive forces between particles of matter are stronger than disruptive forces. As a result, the particles of solids are held in rigid three-dimensional lattices in which the particle's kinetic energy takes the form of vibrations about each lattice site.

Objective 3, Exercises 6.12 and 6.16

The Liquid State

In the liquid state, cohesive forces between particles slightly dominate disruptive forces. As a result, particles of liquids are randomly arranged but relatively close to each other and are in constant random motion, sliding freely over each other but without enough kinetic energy to become separated.

Objective 3, Exercises 6.12 and 6.16

The Gaseous State

In the gaseous state, disruptive forces dominate and particles move randomly, essentially independent of each other. Under ordinary pressure, the particles are separated from each other by relatively large distances except when they collide.

Objective 3, Exercises 6.12 and 6.16

The Gas Laws

Mathematical relationships, called gas laws, describe the observed behavior of gases when they are mixed, subjected to pressure or temperature changes, or allowed to diffuse. When these relationships are used, it is necessary to express the volume and pressure in consistent units and the temperature in kelvins.

Objective 4, Exercises 6.20 and 6.22

Pressure, Temperature, and Volume Relationships

Gas laws discovered by Robert Boyle and Jacques Charles led to the development of the combined gas law. This law allows calculations to be made that relate temperature, pressure, and volume changes for gases.

Objective 5, Exercises 6.24, 6.32, and 6.34

The Ideal Gas Law

The application of a gas law discovered by Amadeo Avogadro to the combined gas law led to the ideal gas law, which allows calculations to be made that account for the number of moles of gas in a sample.

Objective 6, Exercise 6.46

Dalton's Law

John Dalton discovered experimentally that the total pressure exerted by a mixture of gases is equal to the sum of the partial pressures of the individual gases in the mixture.

Objective 7, Exercise 6.58

Graham's Law

Thomas Graham proposed a relationship that mathematically relates the rate of effusion or diffusion of a gas to the mass of the gas molecules:

$$\frac{\text{effusion rate } A}{\text{effusion rate } B} = \sqrt{\frac{\text{molecular mass of } B}{\text{molecular mass of } A}}$$

Objective 8, Exercise 6.60

Changes in State

Most matter can be changed from one state to another by heating, cooling, or changing pressure. State changes that give up heat are called exothermic, and those that absorb heat are called endothermic.

Objective 9, Exercise 6.64

Evaporation and Vapor Pressure

The evaporation of a liquid is an endothermic process and as a result is involved in many cooling processes. In a closed container, evaporation takes place only until the rate of escape of molecules from the liquid is equal to the rate at which they return to the liquid. The pressure exerted by the vapor, which is in equilibrium with the liquid, is called the vapor pressure of the liquid. Liquid vapor pressures increase as the liquid temperature increases.

Objective 10, Exercise 6.68

Boiling and the Boiling Point

At the boiling point of a liquid, its vapor pressure equals the prevailing atmospheric pressure, and bubbles of vapor form within the liquid and rise to the surface as the liquid boils. The boiling point of a liquid changes as the prevailing atmospheric pressure changes.

Objective 11, Exercise 6.70

Sublimation and Melting

Solids, like liquids, have vapor pressures that increase with temperature. Some solids have high enough vapor pressures to allow them to change to vapor without first becoming a liquid, a process called sublimation. Most solids change to the liquid state before they change to the vapor state. The temperature at which solids change to liquids is called the melting point.

Objective 12, Exercise 6.74

Energy and the States of Matter

Energy is absorbed or released when matter is changed in temperature or changed from one state to another. The amount of heat energy required to produce temperature changes is called the specific heat of the matter involved. For phase changes, the amount of heat required is called the heat of fusion, or vaporization.

Objective 13, Exercises 6.76 and 6.78

Key Terms and Concepts

Absolute zero (6.6)
Avogadro's law (6.8)
Boiling point (6.13)
Boyle's law (6.7)
Charles's law (6.7)
Cohesive force (6.2)
Combined gas law (6.7)
Compressibility (6.1)
Condensation (6.12)
Dalton's law of partial pressures (6.9)
Decomposition (6.14)
Diffusion (6.10)

Disruptive force (6.2)
Effusion (6.10)
Evaporation or vaporization (6.12)
Gas law (6.6)
Graham's law (6.10)
Heat of fusion (6.15)
Heat of vaporization (6.15)
Ideal gas law (6.8)
Kinetic energy (6.2)
Melting point (6.14)
Normal or standard boiling point (6.13)
Partial pressure (6.9)

Potential energy (6.2)
Pressure (6.6)
Shape (6.1)
Specific heat (6.15)
Standard atmosphere (6.6)
Standard conditions (STP) (6.8)
Sublimation (6.14)
Thermal expansion (6.1)
Torr (6.6)
Universal gas constant (6.8)
Vapor pressure (6.12)

Key Equations

1. Calculation of volume from mass and density (Section 6.1):	$V = \dfrac{m}{d}$	Equation 6.1
2. Calculation of kinetic energy of particles in motion (Section 6.2):	$KE = \dfrac{1}{2}mv^2$	Equation 6.2
3. Boyle's law (Section 6.7):	$P = \dfrac{k}{V}$	Equation 6.3
	$PV = k$	Equation 6.4
4. Charles's law (Section 6.7):	$V = k'T$	Equation 6.5
	$\dfrac{V}{T} = k'$	Equation 6.6
5. Combined gas law (Section 6.7):	$\dfrac{PV}{T} = k''$	Equation 6.7
	$\dfrac{P_iV_i}{T_i} = \dfrac{P_fV_f}{T_f}$	Equation 6.8
6. Ideal gas law (Section 6.8):	$PV = nRT$	Equation 6.9
7. Molecular weight determination (Section 6.8):	$PV = \dfrac{mRT}{MW}$	Equation 6.11
8. Dalton's law (Section 6.9):	$P_t = P_\triangle + P_\bigcirc + P_\square$	Equation 6.12
9. Graham's law of effusion (Section 6.10):	$\dfrac{\text{effusion rate } A}{\text{effusion rate } B} = \sqrt{\dfrac{\text{molecular mass of } B}{\text{molecular mass of } A}}$	Equation 6.13
10. Heat calculation (Section 6.15):	Heat = (sample mass)(specific heat)(temp. change)	Equation 6.14

Exercises

Even-numbered exercises are answered in Appendix B.

Blue-numbered exercises are more challenging.

Observed Properties of Matter (Section 6.1)

6.1 Calculate the volume of 125 g of the following liquids:

   a. Acetone $(d = 0.792$ g/mL)

   b. Olive oil $(d = 0.918$ g/mL)

   c. Chloroform $(d = 1.49$ g/mL)

6.2 Calculate the volume of 125 g of the following liquids:

   a. Sea water $(d = 1.03$ g/mL)

   b. Methyl alcohol $(d = 0.792$ g/mL)

   c. Concentrated sulfuric acid $(d = 1.84$ g/mL)

6.3 Copper metal has a density of 8.92 g/cm³ at 20.0°C and 8.83 g/cm³ at 100.0°C. Calculate the change in volume that occurs when a 10.0-cm³ piece of copper is heated from 20.0°C to 100.0°C.

6.4 Liquid water has a density of 1.00 g/mL at 10.0°C and 0.996 g/mL at 30.0°C. Calculate the change in volume that occurs when 500 mL of water is heated from 10.0°C to 30.0°C.

6.5 Gallium metal melts at 29.8°C. At the melting point, the density of the solid is 5.90 g/mL, and that of the liquid is 6.10 g/mL.

   a. Does solid gallium expand or contract when it is melted? Explain.

   b. What is the change in volume when 5.00 mL (cm³) of solid gallium is melted?

6.6 A 1.50-L rubber balloon is filled with carbon dioxide gas at a temperature of 0.00°C and a pressure of 1.00 atm. The density of the carbon dioxide gas under these conditions is 1.98 g/L.

   a. Will the density of the carbon dioxide gas increase or decrease when the balloon is heated?

   b. At 50.0°C, the balloon has a volume of 1.78 L. Calculate the carbon dioxide density at this temperature.

The Kinetic Molecular Theory of Matter (Section 6.2)

6.7 Describe the changes in form of energy (kinetic changes to potential, etc.) that occur for the energy of a rock dropped to the ground from a cliff. What form or forms do you suppose the energy takes when the rock hits the ground?

6.8 Suppose a toy ball is thrown into the air such that it goes straight up, then falls and is caught by the person who threw it. Describe the changes in the form of energy that occur for the ball from the time it is thrown until it is caught.

6.9 Suppose a 180-lb (81.8-kg) halfback running at a speed of 8.0 m/s collides head-on with a 260-lb (118.2-kg) tackle running at 3.0 m/s. Which one will be pushed back? That is, which one has more kinetic energy? If you're not familiar with football, check with someone who is for definition of terms.

6.10 At 25.0°C, He molecules (He) have an average velocity of 1.26×10^5 cm/s, and methane molecules (CH_4) have an average velocity of 6.30×10^4 cm/s. Calculate the kinetic energy of each type of molecule at 25.0°C and determine which is greater. Express molecular masses in u for this calculation.

6.11 Which have the greater kinetic energy, hydrogen molecules traveling with a velocity of $2v$, or helium molecules traveling with a velocity of v? Express molecular masses in u.

The Solid, Liquid, and Gaseous States (Sections 6.3–6.5)

6.12 Explain each of the following observations using the kinetic molecular theory of matter:

   a. A liquid takes the shape, but not necessarily the volume, of its container.

   b. Solids and liquids are practically incompressible.

   c. A gas always exerts uniform pressure on all walls of its container.

6.13 Explain each of the following observations using the kinetic molecular theory of matter:

   a. Gases have low densities.

   b. The densities of a substance in the solid and liquid states are nearly identical.

   c. Solids, liquids, and gases all expand when heated.

6.14 Discuss differences in kinetic and potential energy of the constituent particles for a substance in the solid, liquid, and gaseous states.

6.15 The following statements are best associated with the solid, liquid, or gaseous states of matter. Match the statements to the appropriate state of matter.

   a. This state is characterized by the lowest density of the three.

   b. This state is characterized by an indefinite shape and a high density.

   c. In this state, disruptive forces prevail over cohesive forces.

   d. In this state, cohesive forces are most dominant.

6.16 The following statements are best associated with the solid, liquid, or gaseous states of matter. Match the statements to the appropriate state of matter.

   a. Temperature changes influence the volume of this state substantially.

   b. In this state, constituent particles are less free to move about than in other states.

   c. Pressure changes influence the volume of this state more than that of the other two states.

   d. This state is characterized by an indefinite shape and a low density.

The Gas Laws (Section 6.6)

6.17 What is a gas law?

6.18 A weather reporter on TV reports the barometer pressure as 28.6 inches of mercury. Calculate this pressure in the following units:

   a. atm

   b. torr

   c. psi

   d. bars

6.19 The pressure of a gas sample is recorded as 615 torr. Calculate this pressure in the following units:

   a. atm

   b. in. Hg

   c. psi

   d. bars

6.20 An engineer reads the pressure gauge of a boiler as 210 psi. Calculate this pressure in the following units:

   a. atm

   b. bars

   c. mmHg

   d. in. Hg

6.21 A chemist reads a pressure from a manometer attached to an experiment as 17.6 cmHg. Calculate this pressure in the following units:

   a. atm

   b. mmHg

   c. torr

   d. psi

6.22 Convert each of the following temperatures from the unit given to the unit indicated:

   a. The melting point of potassium metal, 63.7°C, to kelvins.

   b. The freezing point of liquid hydrogen, 14.1 K, to degrees Celsius.

   c. The boiling point of liquid helium, −268.9°C, to kelvins.

6.23 Convert each of the following temperatures from the unit given to the unit indicated:

a. The melting point of gold, 1337.4 K, to degrees Celsius.

b. The melting point of tungsten, 3410°C, to kelvins.

c. The melting point of tin, 505 K, to degrees Celsius.

Pressure, Temperature, and Volume Relationships (Section 6.7)

6.24 Use the combined gas law (Equation 6.8) to calculate the unknown quantity for each gas sample described in the following table.

| | Sample | | |
	A	B	C
P_i	1.50 atm	2.35 atm	9.86 atm
V_i	2.00 L	1.97 L	11.7 L
T_i	300 K	293 K	500 K
P_f	?	1.09 atm	5.14 atm
V_f	3.00 L	?	9.90 L
T_f	450 K	310 K	?

6.25 A 200-mL sample of oxygen gas is collected at 26.0°C and a pressure of 690 torr. What volume will the gas occupy at STP (0°C and 760 torr)?

6.26 A 200-mL sample of nitrogen gas is collected at 45.0°C and a pressure of 650 torr. What volume in mL will the gas sample occupy at STP (0.0°C and 760 torr)?

6.27 A 3.00-L sample of helium at 0.00°C and 1.00 atm is compressed into a 0.50-L cylinder. What pressure will the gas exert in the cylinder at 50°C?

6.28 A 2.50-L sample of neon gas at 0.00°C and 1.00 atm is compressed into a 1.00-L cylinder. What pressure will the gas exert in the cylinder at 25.0°C?

6.29 What volume (in liters) of air measured at 1.00 atm would have to be put into a bicycle tire with a 1.00-L volume if the pressure in the bike tire is to be 65.0 psi? Assume the temperature of the gas remains constant.

6.30 What volume (in liters) of air measured at 1.00 atm would have to be put into a car tire with a volume of 12.9 L if the pressure in the car tire is to be 32.0 psi? Assume the temperature of the gas remains constant.

6.31 A sample of gas has a volume of 750 mL at a pressure of 700 torr. What volume will the gas occupy at the same temperature but at the standard atmospheric pressure, 760 torr?

6.32 A sample of gas has a volume of 800 mL at a pressure of 650 torr. What volume in mL will the gas occupy at the same temperature, but at standard atmospheric pressure, 760 torr?

6.33 A 3.0-L sample of gas at 1.0 atm and 0.0°C is heated to 85°C. Calculate the gas volume at the higher temperature if the pressure remains at 1.0 atm.

6.34 A 3.8-L sample of gas at 1.0 atm and 20°C is heated to 85°C. Calculate the gas volume in liters at the higher temperature if the pressure remains constant at 1.0 atm.

6.35 A sample of gas has a volume of 350 mL at 27°C. The gas is heated at a constant pressure until the volume is 500 mL. What is the new temperature of the gas in degrees Celsius?

6.36 What volume of gas in liters at 120°C must be cooled to 40°C if the gas volume at constant pressure and 40°C is to be 2.0 L?

6.37 A 5.00-L gas sample is collected at a temperature and pressure of 27.0°C and 1.20 atm. The gas is to be transferred to a 3.00-L container at a pressure of 1.00 atm. What must the Celsius temperature of the gas in the 3.00-L container be?

6.38 A 2500-L sample of oxygen gas is produced at 1.00 atm pressure. It is to be compressed and stored in a 25.0-L steel cylinder. Assume it is produced and stored at the same temperature and calculate the pressure in atm of the stored oxygen in the cylinder.

6.39 A steel tank with a volume of 6.25 L is full of gas at a pressure of 3.25 atm. What volume in liters would the gas occupy at a pressure of 0.250 atm if its temperature did not change?

6.40 A helium balloon was partially filled with 8000 ft³ of helium when the atmospheric pressure was 0.98 atm and the temperature was 23°C. The balloon rose to an altitude where the atmospheric pressure was 400 torr and the temperature was 5.3°C. What volume did the helium occupy at this altitude?

6.41 You have a 1.50-L balloon full of air at 30°C. To what Celsius temperature would you have to heat the balloon to double its volume if the pressure remained unchanged?

6.42 A gas has a volume of 200 mL at a pressure of 2.50 atm. What volume in mL would it occupy at the same temperature and a pressure of 60.0 kPa?

6.43 What minimum pressure in psi would a 250-mL aerosol can have to withstand if it were to contain 2.50 L of gas measured at 700 torr? Assume the temperature remains constant.

6.44 A 2.00-L sample of nitrogen gas at 760 torr and 0.0°C weighs 2.50 g. The pressure on the gas is increased to 4.00 atm at 0.0°C. Calculate the gas density at the new pressure in g/L.

The Ideal Gas Law (Section 6.8)

6.45 Use the ideal gas law and calculate the following:

a. The pressure exerted by 2.00 mol of oxygen confined to a volume of 500 mL at 20.0°C.

b. The volume of hydrogen gas in a steel cylinder if 0.525 mol of the gas exerts a pressure of 3.00 atm at a temperature of 10.0°C.

c. The temperature (in degrees Celsius) of a nitrogen gas sample that has a volume of 2.50 L and a pressure of 300 torr and contains 0.100 mol.

6.46 Use the ideal gas law and calculate the following:

a. The number of moles of argon in a gas sample that occupies a volume of 400 mL at a temperature of 90.0°C and has a pressure of 735 torr.

b. The pressure in atm exerted by 0.750 mol of hydrogen gas confined to a volume of 2.60 L at 50.0°C.

c. The volume in liters of a tank of nitrogen if 1.50 mol of the gas exerts a pressure of 4.32 atm at 20.0°C.

6.47 Suppose 0.246 mol of SO_2 gas was compressed into a 1.25-L steel cylinder at a temperature of 38.0°C. What pressure in atm would the gas exert on the walls of the cylinder?

6.48 Suppose 10.0 g of SO_2 gas was compressed into a 1.25-L steel cylinder at a temperature of 38.0°C. What pressure in atm would the gas exert on the walls of the cylinder?

6.49 Calculate the volume occupied by 8.75 g of oxygen gas (O_2) at a pressure of 0.890 atm and a temperature of 35.0°C.

6.50 The pressure gauge of a steel cylinder of methane gas (CH_4) reads 400 psi. The cylinder has a volume of 1.50 L and is at a temperature of 30.0°C. How many grams of methane does the cylinder contain?

6.51 Suppose 10.0 g of dry ice (solid CO_2) was placed in an empty 400-mL steel cylinder. What pressure would develop if all the solid sublimed at a temperature of 35.0°C?

6.52 An experimental chamber has a volume of 60 L. How many moles of oxygen gas will be required to fill the chamber at STP?

6.53 How many molecules of nitrogen gas (N_2) are present in a sample that fills a 10.0-L tank at STP?

6.54 A sample of gaseous methyl ether has a mass of 8.12 g and occupies a volume of 3.96 L at STP. What is the molecular weight of methyl ether?

6.55 A sample of a gaseous nitrogen oxide is found to weigh 0.525 g. The sample has a volume of 300 mL at a pressure of 708 torr and a temperature of 25.7°C. Is the gas NO or NO_2?

6.56 A sample of gas weighs 0.176 g and has a volume of 114.0 mL at a pressure and temperature of 640 torr and 20°C. Determine the molecular weight of the gas, and identify it as CO, CO_2, or O_2.

6.57 A 2.00-g sample of gas has a volume of 1.12 L at STP. Calculate its molecular weight and identify it as He, Ne, or Ar.

Dalton's Law (Section 6.9)

6.58 A steel cylinder contains a mixture of nitrogen, oxygen, and carbon dioxide gases. The total pressure in the tank is 2100 torr. The pressure exerted by the nitrogen and oxygen is 810 and 920 torr, respectively. What is the partial pressure in torr of the carbon dioxide in the mixture?

6.59 A 250-mL sample of oxygen gas is collected by water displacement. As a result, the oxygen is saturated with water vapor. The partial pressure of water vapor at the prevailing temperature is 22 torr. Calculate the partial pressure of the oxygen if the total pressure of the sample is 720 torr.

Graham's Law (Section 6.10)

6.60 Hydrogen gas (H_2) is found to diffuse approximately four times as fast as oxygen gas (O_2). Using this information, determine how the masses of hydrogen molecules and oxygen molecules compare. How do they compare based on information in the periodict table?

6.61 The mass of a bromine molecule is 160 u, and the mass of an argon molecule is 40 u. Compare the rates at which these gases will diffuse.

6.62 Two identical rubber balloons were filled with gas—one with helium and the other with nitrogen. After a time, it was noted that one of the balloons appeared to be going "flat." Which one do you think it was? Explain.

6.63 Assume the balloon in Exercise 6.62 that went flat first showed signs of "flatness" 12 hours after it was filled. How long would it take for the other balloon to begin to show signs of going flat?

Changes in State (Section 6.11)

6.64 Classify each of the following processes as endothermic or exothermic:

    a. condensation

    b. liquefaction

    c. boiling

6.65 Classify each of the following processes as endothermic or exothermic:

    a. freezing

    b. sublimation

    c. vaporization

6.66 Discuss what is meant by a change in state.

Evaporation and Vapor Pressure (Section 6.12)

6.67 The following are all nonpolar liquid hydrocarbon compounds derived from petroleum: butane (C_4H_{10}), pentane (C_5H_{12}), hexane (C_6H_{14}), and heptane (C_7H_{16}). Arrange these compounds in order of increasing vapor pressure (lowest first, highest last) and explain how you arrived at your answer.

6.68 Methylene chloride (CH_2Cl_2) was used at one time as a local anesthetic by dentists. It was sprayed onto the area to be anesthetized. Propose an explanation for how it worked.

6.69 Suppose a drop of methyl ether (C_2H_6O) was put on the back of one of your hands and a drop of ethyl ether ($C_4H_{10}O$) was put on your other hand. Propose a way you could tell which compound was which without smelling them.

Boiling and The Boiling Point (Section 6.13)

6.70 Each of two glass containers contains a clear, colorless, odorless liquid that has been heated until it is boiling. One liquid is water (H_2O) and the other is ethylene glycol ($C_2H_6O_2$). Explain how you could make one measurement of each boiling liquid, using the same device, and tell which liquid was which.

6.71 Suppose a liquid in an open container was heated to a temperature just 1 or 2 degrees below its boiling point, then insulated so it stayed at that temperature. Describe how the liquid would behave (what you would see happen) if the hot sample was suspended beneath a helium balloon and taken rapidly to higher altitudes.

6.72 Suppose you were on top of Mount Everest and wanted to cook a potato as quickly as possible. You left your microwave oven at home, so you could either boil the potato in water or throw it into a campfire. Explain which method you would use and why.

Sublimation and Melting (Section 6.14)

6.73 List three common substances that will sublime.

6.74 Solid iodine readily sublimes without melting when moderately heated. The hot vapor will condense back to the solid state when it cools. Describe a method that could be used to obtain pure solid iodine from a mixture of solid iodine and sand. Explain your reasoning.

6.75 A mixture was made of pure water and ice. The mixture was allowed to come to a constant temperature of 0.0°C, the melting point of solid water. The vapor pressure of the water was measured and found to be 4.58 torr. What is the vapor pressure of the ice in torr? Explain your answer.

Energy and The States of Matter (Section 6.15)

6.76 Using the specific heat data of Table 6.8, calculate the amount of heat (in calories) needed to increase the temperature of the following:

a. 50 g of aluminum from 25°C to 55°C

b. 2500 g of ethylene glycol from 80°C to 85°C

c. 500 g of steam from 110°C to 120°C

6.77 Using the specific heat data of Table 6.8, calculate the amount of heat (in calories) needed to increase the temperature of the following:

a. 210 g of copper from 40°C to 95°C

b. 150 g of mercury from 120°C to 300°C

c. 2500 g of helium gas from 250°C to 900°C

6.78 For solar energy to be effective, collected heat must be stored for use during periods of decreased sunshine. One proposal suggests that heat can be stored by melting solids that, upon solidification, would release the heat. Calculate the heat that could be stored by melting 1000 kg of each of the following solids. (Note: The water in each formula is included in the molecular weight.)

a. Calcium chloride ($CaCl_2 \cdot 6H_2O$): melting point = 30.2°C, heat of fusion = 40.7 cal/g

b. Lithium nitrate ($LiNO_3 \cdot 3H_2O$): melting point = 29.9°C, heat of fusion = 70.7 cal/g

c. Sodium sulfate ($Na_2SO_4 \cdot 10H_2O$): melting point = 32.4°C, heat of fusion = 57.1 cal/g

6.79 Why wouldn't a solid such as K_2SO_4 (melting point = 1069°C, heat of fusion = 50.3 cal/g) be suitable for use in a solar heat storage system? (See Exercise 6.78.)

6.80 Liquid Freon (CCl_2F_2) is used as a refrigerant. It is circulated inside the cooling coils of older refrigerators or freezers. As it vaporizes, it absorbs heat. How much heat can be removed by 2.00 kg of Freon as it vaporizes inside the coils of a refrigerator? The heat of vaporization of Freon is 38.6 cal/g.

6.81 Calculate the total amount of heat needed to change 500 g of ice at −10°C into 500 g of steam at 120°C. Do this by calculating the heat required for each of the following steps and adding to get the total:

Step 1. Ice (−10°C) → ice (0°C)

Step 2. Ice (0°C) → water (0°C)

Step 3. Water (0°C) → water (100°C)

Step 4. Water (100°C) → steam (100°C)

Step 5. Steam (100°C) → steam (120°C)

Additional Exercises

6.82 What is the density of argon gas in g/mL at STP?

6.83 Explosives react very rapidly and produce large quantities of heat and gaseous products. When nitroglycerine explodes, several gases are produced:

$$4C_3H_5O_9N_3(\ell) \rightarrow 12CO_2(g) + O_2(g) + 6N_2(g) + 10H_2O(g)$$

Suppose 10 g of nitroglycerine was sealed inside a 1.0-L soda bottle and detonated. Assume the bottle would not break, and the temperature immediately after detonation was 750 K. Calculate the pressure of the gases inside the bottle in atmospheres.

6.84 How many liters of oxygen gas, O_2, will it take to completely react with 2.31 L of hydrogen gas, H_2, to produce water? Assume both gases are at the same pressure and temperature. What type of reaction is this?

6.85 Review Table 6.4; then propose an explanation for the fact that hydrogen bonding decreases the vapor pressure of liquids made up of molecules with similar molecular weights.

6.86 Some people say they have a hard time breathing on top of a tall mountain because there is less oxygen in the air they breathe. Is this a true statement? Explain your answer.

Allied Health Exam Connection

The following questions are from these sources:

1. *Nursing School Entrance Exam* © 2005, Learning Express, LLC.

2. *McGraw-Hill's Nursing School Entrance Exams* by Thomas A. Evangelist, Tamara B. Orr and Judy Unrein © 2009, The McGraw-Hill Companies, Inc.

3. *NSEE Nursing School Entrance Exams,* 3rd Edition © 2009, Kaplan Publishing.

4. *Cliffs Test Prep: Nursing School Entrance Exams* by Fred N. Grayson © 2004, Wiley Publishing, Inc.

5. *Peterson's Master the Nursing School and Allied Health Entrance Exams,* 18th Edition by Marion F. Gooding © 2008, Peterson's, a Nelnet Company.

6.87 Definite shape and definite volume best describes a sample of:

a. $I_2(s)$

b. $Br_2(\ell)$

c. $Cl(g)$

d. $F_2(g)$

6.88 Which of the following is NOT characteristic of gases?

a. They have a definite volume and shape

b. They are low in density

c. They are highly compressible

d. They mix rapidly

Even-numbered exercises answered in Appendix B.
210 Blue-numbered exercises are more challenging.

Unless otherwise noted all art on this page © Cengage Learning 2014

6.89 Which of the following states has the highest average translational kinetic energy?

    a. solid

    b. liquid

    c. gas

    d. none of the above

6.90 In which of the following states of matter are molecules most likely to move freely?

    a. solid

    b. liquid

    c. gas

    d. all have similar freedom of movement

6.91 Which of the following indicates the relative randomness of molecules in the three states of matter?

    a. solid > liquid < gas

    b. liquid < solid < gas

    c. liquid > gas > solid

    d. gas > liquid > solid

6.92 In the kinetic molecular theory of gases, which of the following statements concerning average speeds is true?

    a. Most of the molecules are moving at the average speed.

    b. Any given molecule moves at the average speed most of the time.

    c. When the temperature increases, more of the molecules will move at the new average speed.

    d. When the temperature increases, fewer molecules will move at the new average speed.

6.93 All of the following statements underlie the kinetic molecular theory of gases EXCEPT:

    a. Gas molecules have no intermolecular forces.

    b. Gas particles are in random motion.

    c. Gas particles have no volume.

    d. The average kinetic energy is proportional to the temperature ($°C$) of the gas.

6.94 What are the differentiating factors between potential and kinetic energy?

    a. Properties—physical or chemical

    b. State—solid or liquid

    c. Temperature—high or low

    d. Activity—in motion or in storage

6.95 The transformation of a solid directly into a gas is called:

    a. vaporization

    b. ionization

    c. sublimation

    d. polarization

6.96 Evaporation can best be described as:

    a. the process in which molecules may have enough energy to leave the liquid phase and escape into the gaseous phase

    b. a heating process

    c. fusion

    d. the process in which molecules in the solid phase absorb enough energy to begin the liquid phase

6.97 When a vapor condenses into a liquid:

    a. it absorbs heat

    b. it generates heat

    c. its temperature rises

    d. its temperature drops

6.98 When solid iodine becomes gaseous iodine with no apparent liquid phase in between, the process is called:

    a. evaporation

    b. condensation

    c. sublimation

    d. precipitation

6.99 What does the term *pressure* mean when applied to a gas?

    a. weight

    b. how heavy the gas is

    c. mass divided by volume

    d. force exerted per unit area

6.100 When one liquid evaporates much faster than another liquid, the first liquid is said to be more:

    a. volatile

    b. transient

    c. viscous

    d. evaporative

6.101 Which of the following laws is related to this expression: $P_T = P_1 + P_2 + P_3$?

    a. Boyle's law

    b. Charles's law

    c. Gay-Lussac's law

    d. Dalton's law

6.102 Which law predicts that if the temperature (in Kelvin) doubles, the pressure will also double?

    a. Boyle's law

    b. Charles's law

    c. Gay-Lussac's law

    d. Dalton's law

6.103 The inhaling and exhaling of air by the human lungs is mainly an application of:

    a. Boyle's law—the inverse relationship between the pressure and the volume of a gas.

    b. The volume of a gas at standard temperature and pressure (STP).

    c. Charles's law—the direct relationship between the temperature and the volume of a gas.

    d. The number of O_2 and CO_2 particles per mole.

6.104 How much heat is required to raise the temperature of 100 grams of water from 25°C (near room temperature) to 100°C (its boiling point)? The specific heat of water is approximately 4.2 J per g-K.

a. 3.2×10^4 J

b. 32 J

c. 4.2×10^4 J

d. 76 J

6.105 How many calories are required to change the temperature of 2,000 grams of H_2O from 20°C to 38°C?

a. 36 calories

b. 24 calories

c. 18 calories

d. 12 calories

6.106 If temperature and pressure are held constant for a sample of gas, and the number of moles is doubled, in what manner will the volume change?

a. it will double

b. it will quadruple

c. it will be halved

d. there will be no change

6.107 A gas has a volume of 0.25 liter at a pressure of 1 atmosphere. If the volume increases to 0.50 liter and the temperature remains constant, the new pressure will be:

a. 1 atmosphere

b. 0.5 atmosphere

c. 0.25 atmosphere

d. 2 atmospheres

6.108 Given 3.0 moles of krypton gas, Kr(g), how many liters will this sample occupy at STP?

a. 11.2

b. 22.4

c. 44.8

d. 67.2

6.109 A sample of helium at 25°C occupies a volume of 725 mL at 730 mm Hg. What volume will it occupy at 25°C and 760 mm Hg?

a. 755 mL

b. 760 mL

c. 696 mL

d. 730 mL

6.110 One liter of a certain gas, under standard conditions, weighs 1.16 grams. A possible formula for the gas is:

a. C_2H_2

b. CO

c. NH_3

d. O_2

6.111 A mixture consisting of 8.0 g of oxygen and 14 g of nitrogen is prepared in a container such that the total pressure is 750 mm Hg. The partial pressure of oxygen in the mixture is:

a. 125 mm Hg

b. 500 mm Hg

c. 135 mm Hg

d. 250 mm Hg

6.112 A sample of nitrogen at 20°C in a volume of 875 mL has a pressure of 730 mm Hg. What will be its pressure at 20°C if the volume is changed to 955 mL?

a. 750 mm Hg

b. 658 mm Hg

c. 797 mm Hg

d. 669 mm Hg

6.113 The weight in grams of 22.4 liters of nitrogen (atomic weight = 14) is:

a. 3

b. 7

c. 14

d. 28

6.114 Water at sea level boils at what temperature?

a. 100°F

b. 180°F

c. 212°C

d. 373 K

6.115 Which gas law states that the volume of a gas is inversely proportional to the pressure?

a. Ideal gas law

b. Boyle's law

c. Combined gas law

d. Charles's law

6.116 When a liquid is at its boiling point, the vapor pressure of the liquid:

a. is less than the external pressure on the liquid

b. is equal to the external pressure on the liquid

c. is greater than the external pressure on the liquid

d. can be either less or greater than the external pressure on the liquid

Chemistry for Thought

6.117 As solids are heated, they melt. Explain this in terms of the effect of temperature on cohesive and disruptive forces.

6.118 Explain how a hot-air balloon works in terms of gas densities.

6.119 Which of the following gases would you expect to behave most ideally: He, Ar, or HCl? Explain.

6.120 Refer to Figure 6.6 and do the calculation. When a gas is heated, it expands. Explain how the gas in a hot-air balloon remains at constant pressure as the gas is heated. Hot-air balloons do *not* stretch.

6.121 Refer to Figure 6.7 and answer the question. Would a quart bottle full of gas actually contain the number of moles you calculated? Explain.

6.122 Refer to Figure 6.9 and answer the question. Propose an explanation for your answer in terms of the distances between molecules in rubber and plastic.

6.123 Refer to Figure 6.12 and answer the question. Water is sometimes made safe to drink by boiling. Explain why this might not work if you attempted to do it in an open pan on the summit of Mount Everest.

6.124 Refer to Figure 6.15 and answer the question. Suppose a sample of water was heated to the boiling point in a glass beaker, using a single burner. What would happen to the temperature of the boiling water if a second burner was added to help with the heating? Explain.

6.125 Suppose you put four 250-mL bottles of water into an ice chest filled with crushed ice. If the bottles were initially at a temperature of 23°C, calculate the number of grams of ice that would have to melt in order to cool the water in the bottles to 5°C. Assume the density of water is 1.00g/mL.

Case Study

Elizabeth was a "granola mom"—it was only natural and organic foods and therapies for her family. She belonged to a food co-op and ordered special foods and home remedies through the group. Elizabeth was well read and conscientious, and she had a remedy for everything. Little white pills that dissolved under the tongue treated everything from teething to bruising. When her 8-year-old daughter Samantha had a sore throat, Elizabeth had her gargle with colloidal silver—a remedy used before all the new and "dangerous" antibiotics. Samantha did get over her sore throat, but three weeks later, she broke out in a rash, complained that her joints hurt, and had swollen glands. This worried Elizabeth, and so she consulted a conventional medical doctor. Samantha was diagnosed with rheumatic fever, and the doctor prescribed antibiotic treatment. Elizabeth had read that antibiotics could damage the immune system, and was reluctant to give them to her daughter.

How can the physician encourage compliance with long-term antibiotic treatment without alienating Elizabeth? To what extent does a parent have the right to refuse medical advice for a child? When are alternative therapies helpful?

Follow-up to this Case Study appears at the end the chapter before the Concept Summary.

© Steve Wisbauer/Getty Images

Learning Objectives

When you have completed your study of this chapter, you should be able to:

1 Classify mixtures as solutions or nonsolutions based on their appearance. **(Section 7.1)**

2 Demonstrate your understanding of terms related to the solubility of solutes in solution. **(Section 7.2)**

3 Predict in a general way the solubilities of solutes in solvents on the basis of molecular polarity. **(Section 7.3)**

4 Calculate solution concentrations in units of molarity, weight/weight percent, weight/volume percent, and volume/volume percent. **(Section 7.4)**

5 Describe how to prepare solutions of specific concentration using pure solutes and solvent, or solutions of greater concentration than the one desired. **(Section 7.5)**

6 Do stoichiometric calculations based on solution concentrations. **(Section 7.6)**

7 Do calculations based on the colligative solution properties of boiling point, freezing point, and osmotic pressure. **(Section 7.7)**

8 Describe the characteristics of colloids. **(Section 7.8)**

9 Describe the process of dialysis, and compare it to the process of osmosis. **(Section 7.9)**

Earlier, homogeneous matter was classified into two categories—pure substances and mixtures (see Figure 1.5). Since then, the discussion has been limited to pure substances. We now look at homogeneous mixtures called solutions and their distant relatives, colloidal suspensions.

Solutions and colloidal suspensions are very important in our world. They bring nutrients to the cells of our bodies and carry away waste products. The ocean is a solution of water, sodium chloride, and many other substances (even gold). Many chemical reactions take place in solution—including most of those discussed in this book.

7.1 Physical States of Solutions

Learning Objective

1. Classify mixtures as solutions or nonsolutions based on their appearance.

Solutions are homogeneous mixtures of two or more substances in which the components are present as atoms, molecules, or ions. These uniformly distributed particles are too small to reflect light, and as a result solutions are transparent (clear); light passes through them. In addition, some solutions are colored. The component particles are in constant motion (remember the kinetic theory, Section 6.2) and do not settle under the influence of gravity.

solution A homogeneous mixture of two or more substances in which the components are present as atoms, molecules, or ions.

solvent The substance present in a solution in the largest amount.

solute One or more substances present in a solution in amounts less than that of the solvent.

dissolving A term used to describe the process of solution formation when one or more solutes are dispersed in a solvent to form a homogeneous mixture.

In most solutions, a larger amount of one substance is present compared to the other components. This most abundant substance in a solution is called the **solvent,** and any other components are called **solutes.** Most people normally think of solutions as liquids, but solutions in solid and gaseous forms are known as well. The state of a solution is often the same as the state of the solvent. This is illustrated by ▶ Table 7.1, which lists examples of solutions in various states. The original states of the solvents and solutes are given in parentheses. Solution formation takes place when one or more solutes **dissolve** in a solvent.

TABLE 7.1 Solutions in Various States

Solution	Solution State	Solvent	Solute
Salt water	Liquid	Water (liquid)	Sodium chloride (solid)
Alcoholic beverage	Liquid	Water (liquid)	Alcohol (liquid)
Carbonated water	Liquid	Water (liquid)	Carbon dioxide (gas)
Gold alloy (jewelry)	Solid	Gold (solid)	Copper (solid)
Gold amalgam	Solid	Gold (solid)	Mercury (liquid)
Hydrogen in palladium	Solid	Palladium (solid)	Hydrogen (gas)
Air	Gaseous	Nitrogen (gas)	Oxygen (gas)
Humid oxygen	Gaseous	Oxygen (gas)	Water (liquid)
Camphor in nitrogen	Gaseous	Nitrogen (gas)	Camphor (solid)

▶ Example 7.1

Identify the solvent and solute(s) in each of the following solutions:

a. A sample of natural gas contains 97% methane (CH_4), 1.5% ethane (C_2H_6), 1% carbon dioxide (CO_2), and 0.5% nitrogen (N_2).
b. The label on a bottle of Scotch whiskey says, among other things, 86 proof. This means the contents contain 43% ethyl alcohol (C_2H_5OH). Assume water to be the only other component.
c. The physiological saline solution used in hospitals contains 0.9 g NaCl for each 100 g water.

Solution

a. Methane is present in the largest amount; it is the solvent. All other components are solutes.
b. The whiskey is 43% alcohol and 57% water. Thus, water is the solvent and alcohol the solute. The whiskey actually contains small amounts of many other components, which impart flavor and color. These components are also solutes.
c. Water is the component present in the larger amount and is therefore the solvent.

▶ LEARNING CHECK 7.1 Identify the solvent and solute(s) in the following solutions:

a. White gold is a solid solution containing 58.3% gold, 17.0% copper, 7.7% zinc, and 17.0% nickel.
b. Clean, dry air contains about 78.1% nitrogen, 21.0% oxygen, 0.9% argon, and very small amounts of other gases.

7.2 Solubility

Learning Objective

2. Demonstrate your understanding of terms related to the solubility of solutes in solution.

A few experiments with water, sugar, cooking oil, and rubbing alcohol (isopropyl alcohol) illustrate some important concepts associated with solution formation. Imagine you have three drinking glasses, each containing 100 mL (100 g) of pure water. You begin by adding small amounts of the other three substances to the glasses. The mixtures are stirred well, with results shown in ▶ Figure 7.1A. The sugar and alcohol form homogeneous mixtures (solutions) with the water, while the oil forms a two-layer (heterogeneous) mixture. **Soluble substances** such as the sugar and alcohol dissolve completely in the solvent and form solutions. **Insoluble substances** do not dissolve in the solvent. The term **immiscible** is used to describe a liquid solute that does not dissolve in a liquid solvent.

soluble substance A substance that dissolves to a significant extent in a solvent.

insoluble substance A substance that does not dissolve to a significant extent in a solvent.

immiscible A term used to describe liquids that are insoluble in each other.

1 g of sugar → Homogeneous

10 drops of oil → Oil layer → Nonhomogeneous

10 drops of alcohol → Homogeneous

A Small amounts of substance added

300 g of sugar → Nonhomogeneous

Many drops of oil → Oil layer → Nonhomogeneous

Many drops of alcohol → Homogeneous

B Larger amounts of substance added

Figure 7.1 The homogeneity of solutions and mixtures of sugar (limited solubility), cooking oil (immiscible), and alcohol (completely soluble).

The experiments are continued by adding more sugar, alcohol, and oil to the water samples. This ultimately leads to the situation shown in Figure 7.1B. Regardless of the amount of alcohol added, a solution forms. In fact, other experiments could be done that show that water and isopropyl alcohol are completely soluble in each other and will mix in any proportion. Sugar behaves differently. About 204 g can be dissolved in the 100 mL of water (at 20°C), but any additional sugar simply sinks to the bottom of the glass and remains undissolved. Thus, the sugar has a solubility in water of 204 g/100 g H_2O. The term **solubility** refers to the maximum amount of solute that can dissolve in a specific amount of solvent at a specific temperature. With the oil–water mixture, any additional oil simply floats on the surface of the water along with the oil added initially.

The solubilities of a number of solutes in water are given in ❱ Table 7.2. The use of specific units, such as grams of solute per 100 g of water, makes it possible to compare solubilities precisely. However, such precision is often unnecessary, and when this is the case, the approximate terms defined in ❱ Table 7.3 will be used.

solubility The maximum amount of solute that can be dissolved in a specific amount of solvent under specific conditions of temperature and pressure.

TABLE 7.2 Examples of Solute Solubilities in Water (0°C)

Solute Name	Formula	Solubility (g solute/100 g H_2O)
Ammonium chloride	NH_4Cl	29.7
Ammonium nitrate	NH_4NO_3	118.3
Ammonium orthophosphate	$NH_4H_2PO_4$	22.7
Ammonium sulfate	$(NH_4)_2SO_4$	70.6
Calcium carbonate	$CaCO_3$	0.0012
Calcium chloride	$CaCl_2$	53.3
Calcium sulfate	$CaSO_4$	0.23
Potassium carbonate	K_2CO_3	101
Potassium chloride	KCl	29.2
Sodium bicarbonate	$NaHCO_3$	6.9
Sodium bromide	$NaBr$	111
Sodium carbonate	Na_2CO_3	7.1
Sodium chloride	$NaCl$	35.7
Sodium iodide	NaI	144.6
Ascorbic acid (vitamin C)	$C_6H_8O_6$	33
Ethyl alcohol	C_2H_5OH	∞[a]
Ethylene glycol (antifreeze)	$C_2H_4(OH)_2$	∞
Glycerin	$C_3H_5(OH)_3$	∞
Sucrose (table sugar)	$C_{12}H_{22}O_{11}$	179.2

[a]Soluble in all proportions.

TABLE 7.3 Approximate Solubility Terms

Solute Solubility (g solute/100 g H_2O)	Solubility Term
Less than 0.1	Insoluble
0.1–1	Slightly soluble
1–10	Soluble
Greater than 10	Very soluble

A solution in which the maximum amount of solute has been dissolved in a quantity of solvent is called a **saturated solution.** The final sugar solution described in the earlier experiments (204 g in 100 g H_2O) was a saturated solution. Solutions in which the amount of solute dissolved is greater than the solute solubility are called **supersaturated solutions.** Supersaturated solutions are usually prepared by forming a nearly saturated solution at a high temperature and then cooling the solution to a lower temperature at which the solubility is lower. Such solutions are not stable. The addition of a small amount of solid solute (or even a dust particle) will usually cause the excess solute to crystallize out of solution until the solution becomes saturated (see ❱ Figure 7.2). The temperature dependence of solute solubility is illustrated in ❱ Figure 7.3.

Whereas the solubility of most liquids and solids in water increases with temperature, the solubility of most gases in water decreases as the temperature increases (see the SO_2 curve in Figure 7.3). This is easily demonstrated for gaseous CO_2 by opening both

saturated solution A solution that contains the maximum amount possible of dissolved solute in a stable situation under the prevailing conditions of temperature and pressure.

supersaturated solution An unstable solution that contains an amount of solute greater than the solute solubility under the prevailing conditions of temperature and pressure.

a cold and a warm carbonated beverage. The solubility of gaseous solutes is also influenced significantly by pressure; the effect on liquid or solid solutes is minimal. It has been found that the solubility of many gases is directly proportional to the pressure of the gas above the solution at constant temperature. Thus, if the gas pressure is doubled, the solubility doubles.

① A supersaturated solution just after adding a small seed crystal.

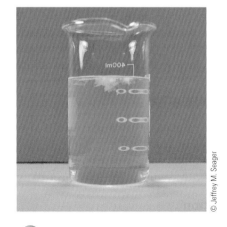
② Rapid crystallization begin on the seed crystal.

③ The excess solute has all crystallized, leaving behind a saturated solution.

Figure 7.2 Crystallization converts a supersaturated solution to a saturated solution.

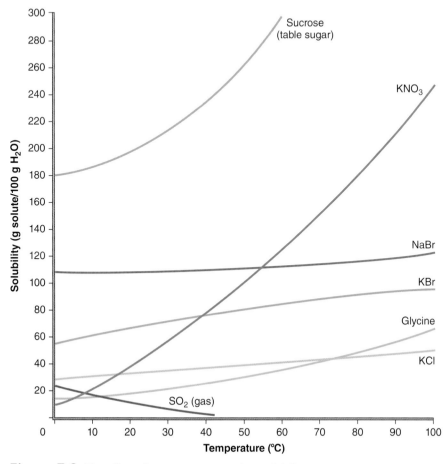

Figure 7.3 The effect of temperature on solute solubility.

The pressure dependence of gas solubility provides the "sparkle" for carbonated beverages. The cold beverage is saturated with CO_2 and capped under pressure. When the bottle is opened, the pressure is relieved, and the gas, now less soluble, comes out of solution as fine bubbles. A similar effect sometimes takes place in the bloodstream of deep-sea divers. While submerged, they inhale air under pressure that causes nitrogen to be more soluble in the blood than it is under normal atmospheric pressure. If the diver is brought to the lower pressure on the surface too quickly, the excess dissolved nitrogen comes out of solution and forms bubbles in the blood and joints. The result, called decompression sickness or the bends, is painful and dangerous. The chances of getting the bends are decreased by breathing a mixture of oxygen and helium rather than air (oxygen and nitrogen) because helium is less soluble in the body fluids than nitrogen.

▶ Example 7.2

A 260-g sample of table sugar is added to 100 g of water at 70°C. The sugar dissolves completely. The resulting solution is allowed to cool slowly. At 30°C, crystals of sugar begin to form and increase in size as the solution cools to 20°C. Refer to Figure 7.3 and describe the nature of the solution at 70°, 60°, 50°, 40°, 30°, and 20°C.

Solution
The solubility of sugar is greater than 260 g/100 g H_2O at all temperatures above 50°C. Therefore, the solution is unsaturated at 70°C and 60°C. At 50°C, the solubility is equal to the amount dissolved, so the solution is saturated. At 40°C, the solution contains more dissolved sugar than it should on the basis of solubility, and the solution is supersaturated. At 30°C, the excess sugar crystallizes from solution, and the resulting solution becomes saturated. From that point, excess sugar continues to crystallize from the solution, and the solution remains saturated to 20°C.

▶ **LEARNING CHECK 7.2** Refer to Figure 7.3. Saturated solutions of potassium nitrate(KNO_3) and sodium bromide (NaBr) are made at 80°C. Which solution contains more solute per 100 g of H_2O? The solutions are cooled to 50°C, and excess solute crystallizes out of each solution. Which solution at 50°C contains more solute per 100 g of H_2O? Are the solutions saturated at 50°C?

7.3 The Solution Process

Learning Objective

3. Predict in a general way the solubilities of solutes in solvents on the basis of molecular polarity.

The how and why of solution formation are the topics of this section. How are solute particles removed from the bulk solute and uniformly distributed throughout the solvent? Why are some solutes very soluble whereas others are not?

Consider the formation of a saltwater solution. Earlier we pointed out (Section 4.2) that solid ionic compounds are collections of ions held together by attractions between the opposite ionic charges. When an ionic compound dissolves, the orderly ionic arrangement is destroyed as the interionic attractions are overcome. Thus, the attractive forces between water molecules and ions must be stronger than the interionic attractions within the crystal.

The solution-forming process for an ionic solute is represented in ▶ Figure 7.4. When the solid ionic crystal is placed in water, the polar water molecules become oriented so that the negative oxygen portion points toward positive sodium ions, and the positive hydrogen portion points toward negative chloride ions. As the polar water molecules begin to surround ions on the crystal surface, they tend to create a shielding effect that reduces the attraction between the ion and the remainder of the crystal. As a result, the ion breaks away from the crystal surface and is surrounded by water molecules. Ions surrounded by water molecules in solution are called **hydrated ions.** As each ion leaves the surface, others are exposed to the water, and the crystal is picked apart ion by ion. Once in solution, the hydrated ions are uniformly distributed by stirring or by random collisions with other molecules or ions.

hydrated ion An ion in solution that is surrounded by water molecules.

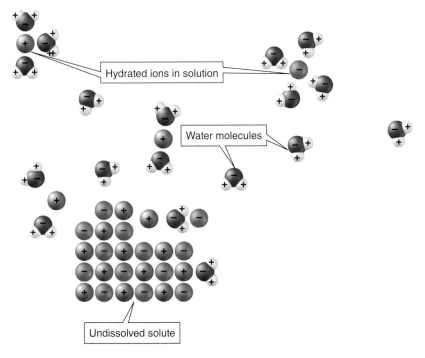

Figure 7.4 The dissolving of an ionic substance in water.

The random motion of solute ions in solution causes them to collide with one another, with solvent molecules, and occasionally with the surface of any undissolved solute. Ions undergoing such collisions occasionally stick to the solid surface and thus leave the solution. When the number of ions in solution is low, the chances for collision with the undissolved solute are low. However, as the number of ions in solution increases, so do the chances for such collisions, and more ions leave the solution and become attached once again to the solid. Eventually, the number of ions in solution reaches a level at which ions return to the undissolved solute at the same rate as other ions leave. At this point the solution is saturated, an equilibrium condition. Even though the processes of leaving and returning continue, no net changes in the number of ions in solution or the amount of undissolved solute can be detected as time passes, and an experimenter would observe that no more solid solute dissolves.

Supersaturated solutions form when there are no sites with which the excess solute particles can collide. The addition of such sites in the form of a seed crystal of solute causes the excess solute to crystallize from solution very quickly.

Polar but nonionic solutes such as sugar dissolve in water in much the same way as ionic solutes. The only difference is the attraction of polar water molecules to both poles of the solute molecules. The process is represented in ❯ Figure 7.5.

This solution-forming process also explains the low solubility of some solutes. A solute will not dissolve in a solvent if (1) the forces between solute particles are too strong to be overcome by interactions with solvent particles or (2) the solvent particles are more strongly attracted to each other than to solute particles.

The cooking oil of the earlier experiment did not dissolve in water because the polar water molecules were attracted to each other more strongly than they were to the nonpolar oil molecules. The cooking oil will dissolve in a nonpolar solvent such as gasoline or carbon tetrachloride (CCl_4). In these solvents, the weak forces between oil molecules and solvent molecules are no stronger than the weak forces between nonpolar solvent molecules. A good rule of thumb is "like dissolves like." Thus, polar solvents will dissolve polar or ionic solutes, and nonpolar solvents will dissolve nonpolar or nonionic solutes.

Figure 7.5 The dissolving of a polar solute in water.

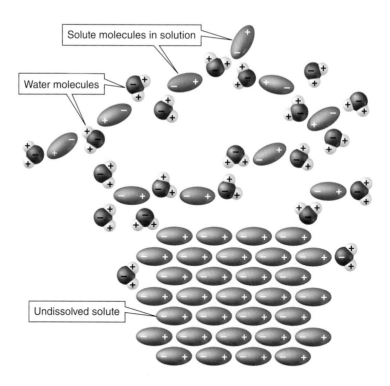

Solute molecules in solution

Water molecules

Undissolved solute

These generalizations apply best to nonionic compounds. Some ionic compounds (such as $CaCO_3$ and $CaSO_4$) have very low solubilities in water (see Table 7.2). Both the attractive forces between ions and the attraction of polar solvent molecules for ions are electrical and depend on such characteristics as ionic charge and size. Changes in ionic compounds that increase the attractive forces between ions also increase the forces between ions and polar solvent molecules, but not always by the same amount. Thus, simple rules for quantitatively predicting the water solubility of ionic compounds are not available. (We can't easily predict how many grams will dissolve in a specific quantity of water.) However, ionic compounds are generally insoluble in nonpolar solvents; they usually follow the solubility guidelines given in ▶ Table 7.4 when water is the solvent.

TABLE 7.4 General Solubilities of Ionic Compounds in Water

Compounds	Solubility	Exceptions
Group IA (Na^+, K^-, etc.) and NH_4^+	Soluble	
Nitrates (NO_3^-)	Soluble	
Acetates ($C_2H_3O_2^-$)	Soluble	
Chlorides (Cl^-)	Soluble	Chlorides of Ag^+, Pb^{2+}, Hg^+ (Hg_2^{2+})
Sulfates (SO_4^{2-})	Soluble	Sulfates of Ba^{2+}, Sr^{2+}, Pb^{2+}, Hg^+ (Hg_2^{2+})
Carbonates (CO_3^{2-})	Insoluble[a]	Carbonates of group IA and NH_4^+
Phosphates (PO_4^{3-})	Insoluble[a]	Phosphates of group IA and NH_4^+

[a]Many hydrogen carbonates (HCO_3^-) and phosphates (HPO_4^{2-}, $H_2PO_4^-$) are soluble.

▶ Example 7.3

Predict the solubility of the following solutes in the solvent indicated:

a. ammonia gas (NH_3) in water
b. oxygen gas (O_2) in water
c. $Ca(NO_3)_2$ in water
d. $Mg_3(PO_4)_2$ in water
e. paraffin wax (nonpolar) in CCl_4

Solution

a. Soluble: NH_3 is polar—like dissolves like. The actual solubility at 20°C is 51.8 g/100 g H_2O.
b. Insoluble: O_2 is nonpolar. The actual solubility at 20°C is 4.3×10^{-3} g/100 g H_2O.
c. Soluble: nitrates are soluble in water (Table 7.4).
d. Insoluble: phosphates are insoluble except those of group IA(1) and NH_4^+ (Table 7.4).
e. Soluble: CCl_4 is nonpolar—like dissolves like.

▶ LEARNING CHECK 7.3

a. Refer to Table 7.4 and decide which of the following salts you would use as a solute if you wanted to prepare a solution that contained as many barium ions (Ba^{2+}) as possible: $BaSO_4$, $Ba(NO_3)_2$, or $BaCO_3$. Explain your answer.
b. A common sight on the evening TV news in the late 1960s and 1970s was sea birds soaked with crude oil being cleaned by concerned people. Which of the following do you think was used to clean the oil from the birds? Explain your answer. Light mineral oil, purified water, seawater, or gasoline.

Be careful not to confuse solute solubility with the rate at which a solute dissolves. Under some conditions, even a very soluble solute will dissolve slowly—for example, a lump of rock candy (sugar) dissolves much more slowly than an equal weight of granulated sugar. The dissolving rate can be increased in a number of ways (see ▶ Figure 7.6):

1. **Crushing or grinding the solute**—small particles provide more surface area for solvent attack and dissolve more rapidly than larger particles.
2. **Heating the solvent**—solvent molecules move faster and have more frequent collisions with solute at higher temperatures.
3. **Stirring or agitating the solution**—stirring removes locally saturated solution from the vicinity of the solute and allows unsaturated solvent to take its place.

Heat is usually absorbed or released when a solute dissolves in a solvent. When heat is absorbed, the process is endothermic, and the solution becomes cooler. The fact that heat absorption leads to cooling might sound strange, and it needs to be explained. The heat is absorbed by the interacting solvent and solute molecules, and it is removed from the solvent molecules that are not involved in the actual attack on the solute. Since most of the solvent is in the latter category, the entire solution becomes cooled.

When heat is released, the process is exothermic, and the solution temperature increases, this time because the heat released by the interacting molecules is absorbed by the uninvolved solvent. This behavior is the more common one. These processes can be represented by equations:

$$\text{Endothermic: Solute + solvent + heat} \rightarrow \text{solution} \qquad (7.1)$$

$$\text{Exothermic: Solute + solvent} \rightarrow \text{solution + heat} \qquad (7.2)$$

An endothermic solution process is the basis for commercially available instant cold packs. Water is sealed in a thin plastic bag and placed inside a larger, stronger bag together with a quantity of solid solute (NH_4Cl or NH_4NO_3). When the inner bag is broken

Figure 7.6 Heat and agitation increase the rate at which solutes dissolve. Will sugar dissolve faster in hot tea or iced tea?

by squeezing, the solid dissolves in the water, heat is absorbed, and the mixture becomes quite cold (see ▶ Figure 7.7).

Figure 7.7 Exothermic (top) and endothermic (bottom) solution processes.

 Solid NaOH not yet added to water.

 Solution becomes hot when NaOH dissolves in water.

 Solid NH$_4$NO$_3$ not yet added to water.

 Solution becomes cool when NH$_4$NO$_3$ dissolves in water.

7.4 Solution Concentrations

Learning Objective

4. Calculate solution concentrations in units of molarity, weight/weight percent, weight/volume percent, and volume/volume percent.

In Chapter 5, quantitative calculations were done using equations for reactions such as the following:

$$CH_4(g) + 2O_2(g) \rightarrow CO_2(g) + 2H_2O(g) \tag{7.3}$$

The reaction represented by this equation, and most of the reactions we have discussed to this point in the book, involve pure substances as reactants and products. However, many of the reactions done in laboratories, and most of those that go on in our bodies, take place between substances dissolved in a solvent to form solutions. In our bodies, the solvent is almost always water. A double-replacement reaction of this type done in laboratories is represented by the following equation:

$$2AgNO_3(aq) + Na_2CO_3(aq) \rightarrow Ag_2CO_3(s) + 2NaNO_3(aq) \tag{7.4}$$

We remember from Chapter 5 that the coefficients in equations such as Equation 7.3 allow the relative number of moles of pure reactants and products involved in the reaction to be determined. These relationships coupled with the mole definition in terms of masses then yield factors that can be used to solve stoichiometric problems involving the reactants and products. Similar calculations can be done for reactions that take place between the solutes of solutions if the amount of solute contained in a specific quantity of the reacting solutions is known. Such relationships are known as solution **concentrations.** Solution concentrations may be expressed in a variety of units, but only two, molarity and percentage, will be discussed at this time.

concentration The relationship between the amount of solute and the specific amount of solution in which it is contained.

Oral Rehydration Therapy

Dehydration, the excessive loss of water from the body, can result from a number of causes, including severe episodes of diarrhea or vomiting and excessive sweating without proper fluid intake. The condition can be life threatening when it is severe enough and goes untreated. Dehydration is serious because the body loses electrolytes such as sodium, potassium, and chloride ions along with water. When the electrolyte balance in the body is upset, many organs, including the heart, cannot function properly. Small children are especially susceptible to dehydration caused by diarrhea because their small bodies do not have much of a fluid reserve, and it doesn't take much fluid loss to get their electrolytes out of balance.

Oral rehydration therapy (ORT) is a simple and effective way to treat or prevent dehydration and the accompanying electrolyte loss, especially if the dehydration is caused by diarrhea. Oral rehydration therapy was developed in the 1950s for use in developing countries where diarrhea-producing diseases like cholera, combined with unsanitary water and food, cause the death of an estimated 4 million children annually. The threat to children in developed countries is not nearly as great; an estimated 500 children die annually from diarrhea in the United States.

The materials used for ORT are simple mixtures of water, salts, and carbohydrates. These materials are regulated by the FDA as a medical food and are available in most grocery and drug stores. A list of ingredients from the label of the liquid form of a popular ORT product includes water; the two carbohydrate sugars dextrose and fructose; citric acid; and the salts potassium citrate, sodium citrate, and sodium chloride. This material is available in an unflavored form and in several flavors, such as berry and bubblegum, to appeal to children. The product also comes in the form of a powder that has to be dissolved in water before use. A third form consists of liquid sealed in plastic sleeves that can be frozen and then eaten like popular frozen treats.

The appropriate dosage for ORT products depends on the weight of the child; the directions included on the product labels should be read and followed. Many pediatricians suggest that parents of young children should include at least one bottle of oral rehydration fluid in the family medicine chest. The low cost of the commercially available materials makes it possible for most families to follow this prudent advice.

Oral rehydration therapy products are available in numerous forms and flavors.

The **molarity (M)** of a solution expresses the number of moles of solute contained in exactly 1 L of the solution:

$$M = \frac{\text{moles of solute}}{\text{liters of solution}} \qquad (7.5)$$

molarity (M) A solution concentration expressed in terms of the number of moles of solute contained in a liter of solution.

It is important to note that even though a concentration in molarity expresses the number of moles contained in 1 L of solution, the molarity of solutions that have total volumes different from 1 L can be calculated using Equation 7.5. We simply determine the number of moles of solute contained in a specified volume of solution, then express that volume in liters before substituting the values into Equation 7.5.

▶ Example 7.4

Express the concentration of each of the following solutions in terms of molarity:

a. 2.00 L of solution contains 1.50 mol of solute.

b. 150 mL of solution contains 0.210 mol of solute.

c. 315 mL of solution contains 10.3 g of isopropyl alcohol, C_3H_7OH.

Solution

a. Because the data are given in terms of moles of solute and liters of solution, the data may be substituted directly into Equation 7.5:

$$M = \frac{1.50 \text{ mol solute}}{2.00 \text{ L solution}} = 0.750 \frac{\text{mol solute}}{\text{L solution}}$$

The solution is 0.750 molar, or 0.750 M.

b. In this problem, the number of moles of solute is given, but the volume of solution is given in milliliters rather than liters. The volume must first be converted into liters, then the data may be substituted into Equation 7.5:

$$(150 \text{ mL solution})\left(\frac{1 \text{ L}}{1000 \text{ mL}}\right) = 0.150 \text{ L solution}$$

$$M = \frac{0.210 \text{ mol solute}}{0.150 \text{ L solution}} = 1.40 \frac{\text{mol solute}}{\text{L solution}}$$

The solution is 1.40 molar, or 1.40 M.

c. Before Equation 7.5 can be used, the number of moles of solute (isopropyl alcohol) must be determined. This is done as follows, where the factor

$$\frac{1 \text{ mol alcohol}}{60.1 \text{ g alcohol}}$$

comes from the calculated molecular weight of 60.1 u for isopropyl alcohol:

$$(10.3 \text{ g alcohol})\left(\frac{1 \text{ mol alcohol}}{60.1 \text{ g alcohol}}\right) = 0.171 \text{ mol alcohol}$$

Next, this number of moles of solute and the solution volume expressed in liters are substituted into Equation 7.5:

$$M = \frac{0.171 \text{ mol alcohol}}{0.315 \text{ L solution}} = 0.543 \frac{\text{mol alcohol}}{\text{L solution}}$$

The solution is 0.543 molar, or 0.543 M.

▶ **LEARNING CHECK 7.4** Express the concentrations of each of the following solutions in terms of molarity:

a. 2.50 L of solution contains 1.25 mol of solute.
b. 225 mL of solution contains 0.486 mol of solute.
c. 100 mL of solution contains 2.60 g of NaCl solute.

Sometimes a detailed knowledge of the actual stoichiometry of a process involving solutions is not needed, but some information about the solution concentrations would be useful. When this is true, solution concentrations are often expressed as percentages. In general, a concentration in **percent** gives the number of parts of solute contained in 100 parts of solution. You saw in Section 1.10 that it is convenient to use a formula for percentage calculations because we seldom work with exactly 100 units of anything. The general formula used is:

$$\% = \frac{\text{part}}{\text{total}} \times 100$$

percent A solution concentration that expresses the amount of solute in 100 parts of solution.

Three different percent concentrations for solutions are used. A **weight/weight percent** (abbreviated w/w) is the mass of solute contained in 100 mass units of solution. Thus, a 12.0% (w/w) sugar solution contains 12.0 grams of sugar in each 100 g of solution. In terms of this concentration, the general formula for percent calculations becomes

weight/weight percent A concentration that expresses the mass of solute contained in 100 mass units of solution.

$$\%(w/w) = \frac{\text{solute mass}}{\text{solution mass}} \times 100 \qquad (7.6)$$

The Risk of Dehydration During Vigorous Youth Activities

Generally, the participation of youngsters in organized sports or other physical activities is considered to be desirable from a health point of view. Such behavior helps prevent the couch potato syndrome and obesity in children that have become health concerns in recent years. However, vigorous exercise in hot, humid weather can cause dehydration, which can lead to mild or severe heat-related illnesses such as heat cramps, heat exhaustion or heat stroke.

Any child who exercises in hot weather may be at risk for dehydration. However, the concern is greatest for those participating in activities or organized sports such as football, soccer, or cross-country running that take place during the hot days of late summer. Exercising children are particularly vulnerable to heat-related illnesses during hot weather if they

1. Are just beginning to exercise after being relatively inactive for extended periods of time.
2. Are overweight or obese.
3. Have had a recent illness that caused vomiting or diarrhea.
4. Have had a previous heat-related illness.

Most heat-related illness occurs within the first few days of organized vigorous activity such as participation on a sports team. To avoid this, the activity should be relatively light during the first few days and should be increased gradually. For hot and humid conditions, the American Academy of Pediatrics gives the following recommendations:

1. Sessions of intense activity should not last longer than 15 minutes with cooling-off breaks between sessions.
2. Participants should drink plenty of fluids before the activities and during regularly scheduled breaks in the activities—even if they do not feel thirsty.
3. Where possible, the clothing of participants should consist of single layers of light-colored lightweight material.

Youthful participants as well as their parents, coaches, or other supervisors should learn to identify the following early signs and symptoms of dehydration: dry or sticky mouth, thirst, lethargy, irritability, headache, dizziness, cramps, and excessive fatigue. Young participants should be taught the importance of recognizing these signs and symptoms in themselves and the importance of reporting them to their parents, coach, or supervisor. Embarrassment should not be allowed to prevent the early detection and treatment of dehydration.

Regular fluid intake is a must for participants in vigorous physical activities.

Any mass units may be used, but the mass of solute and solution must be expressed in the same units.

A more commonly used percent concentration is **weight/volume percent** (abbreviated w/v), which is the grams of solute contained in 100 mL of solution. In these units, a 12.0% (w/v) sugar solution would contain 12.0 g of sugar in each 100 mL of solution. This percent concentration is normally used when the solute is a solid and the solvent and resulting solutions are liquids. The general formula for the calculation of percent concentrations in these units is

$$\%(\text{w/v}) = \frac{\text{grams of solute}}{\text{milliliters of solution}} \times 100 \qquad (7.7)$$

In weight/volume percent calculations, the solute amount is always given in grams, and the solution volume is always given in milliliters.

A percent concentration that is useful when the solute and solvent are either both liquids or both gases is **volume/volume percent** (abbreviated v/v). Concentrations given in

weight/volume percent
A concentration that expresses the grams of solute contained in 100 mL of solution.

volume/volume percent
A concentration that expresses the volume of solute contained in 100 volumes of solution.

these units express the number of volumes of solute found in 100 volumes of solution. For these units, the general percentage equation becomes

$$\%(v/v) = \frac{\text{solute volume}}{\text{solution volume}} \times 100 \qquad (7.8)$$

Any volume units may be used, but they must be the same for both the solute and the solution.

▶ Example 7.5

a. A solution contains 100 g of water and 1.20 g of solute. What is the %(w/w) concentration?
b. A solution is made by mixing 90.0 mL of alcohol with enough water to give 250 mL of solution. What is the %(v/v) concentration of alcohol in the solution?
c. A 150-mL sample of saltwater is evaporated to dryness. A residue of salt weighing 27.9 g is left behind. Calculate the %(w/v) of the original saltwater.

Solution

a. Equation 7.6 is used:

$$\%(w/w) = \frac{\text{solute mass}}{\text{solution mass}} \times 100 = \frac{1.20 \text{ g}}{101.2 \text{ g}} \times 100 = 1.19\% \ (w/w)$$

Note that the solution mass is the sum of the solvent (water) and solute masses.

b. Equation 7.8 is used:

$$\%(v/v) = \frac{\text{solute volume}}{\text{solution volume}} \times 100 = \frac{90 \text{ mL}}{250 \text{ mL}} \times 100 = 36.0\% \ (v/v)$$

c. Equation 7.7 is used after checking to make certain the amount of solute is given in grams and the solution volume is in mL:

$$\%(w/v) = \frac{\text{grams of solute}}{\text{milliliters of solution}} \times 100$$

$$= \frac{27.9 \text{ g}}{150 \text{ mL}} \times 100$$

$$= 18.6 \ \% \ (w/v)$$

▶ LEARNING CHECK 7.5

a. A solution is made by dissolving 0.900 g of salt in 100.0 mL of water. Assume that each milliliter of water weighs 1.00 g and that the final solution volume is 100.0 mL. Calculate the %(w/w) and %(w/v) for the solution using the assumptions as necessary.
b. An alcoholic beverage is labeled 90 proof, which means the alcohol concentration is 45% (v/v). How many milliliters of pure alcohol would be present in 1 oz (30 mL) of the beverage?

7.5 Solution Preparation

Learning Objective

5. Describe how to prepare solutions of specific concentration using pure solutes and solvent, or solutions of greater concentration than the one desired.

Solutions are usually prepared by mixing together proper amounts of solute and solvent or by diluting a concentrated solution with solvent to produce a solution of lower concentration. In the first method, the solute is measured out and placed in a container, and the correct amount of solvent is added. When the concentration is based on solution volume [%(v/v), %(w/v), and M], a volumetric flask or other container is used that, when filled to a specific mark, holds an accurately known volume (see ▶ Figure 7.8). When the concentration is based on solvent mass [%(w/w)], the correct mass of solvent is added. This mass is usually converted to a volume by using the density, so that a volume of solvent can be measured rather than an amount weighed on a balance.

Step 1. Put 0.125 mol of solute into a 250-mL flask.

Step 2. Add some water and dissolve the solute.

Step 3. Fill the flask to the mark with water. Mix thoroughly.

Figure 7.8 Preparation of a 0.500 M solution. Use the data given in the figure and show by a calculation that the resulting solution is 0.500 M.

▶ Example 7.6

Describe how you would prepare the following solutions from pure solute and water:

a. 1.00 L of 1.50 M $CoCl_2$ solution
b. 250 mL of 0.900% (w/v) NaCl solution
c. 500 mL of 8.00% (v/v) methyl alcohol solution

Solution

a. According to Equation 7.5, molarity is the moles of solute per (divided by) the liters of solution containing the solute:

$$M = \frac{\text{moles of solute}}{\text{liters of solution}}$$

This equation can be rearranged to calculate the moles of solute required:

$$(M)(\text{liters of solution}) = \text{moles of solute}$$

By substituting the molarity and volume of the desired solution, the required number of moles of solute is determined:

$$\left(1.50\,\frac{\text{mol}}{L}\right)(1.00\,L) = 1.50\,\text{mol}$$

The formula weight of $CoCl_2$ is 129.9 u. The mass of solute is calculated:

$$(1.50\,\text{mol}\,CoCl_2)\left(\frac{129.9\,\text{g}\,CoCl_2}{1\,\text{mol}\,CoCl_2}\right) = 195\,\text{g}\,CoCl_2$$

The solution is prepared by weighing out 195 g $CoCl_2$, putting it into a 1-L volumetric flask, and adding enough water to fill the flask to the mark (see Figure 7.8). It is a good practice to let the solute dissolve completely in part of the solvent before filling the flask to the mark.

b. Similarly, rearrange Equation 7.7, and use the result to calculate the amount of solute needed:

$$\%(w/v) = \frac{\text{grams of solute}}{\text{milliliters of solution}} \times 100$$

Therefore,

$$\frac{(\%)(\text{milliliters of solution})}{100} = \text{grams of solute}$$

Substitute percentage and solution volume:

$$\frac{(0.900\%)(250 \text{ mL})}{100} = 2.25 \text{ g NaCl}$$

Thus, the solution is prepared by putting 2.25 g NaCl into a 250-mL volumetric flask and adding water to the mark.

c. Rearrange Equation 7.8, and use the result to calculate the volume of solute needed:

$$\%(v/v) = \frac{\text{solute volume}}{\text{solution volume}} \times 100$$

Therefore,

$$\frac{(\%)(\text{solution volume})}{100} = \text{solute volume}$$

Substitute percentage and solution volume:

$$\frac{(8.00\%)(500 \text{ mL})}{100} = 40.0 \text{ mL}$$

Thus, 40.0 mL methyl alcohol is put into a 500-mL volumetric flask, and water is added up to the mark.

▶ **LEARNING CHECK 7.6** Describe how you would prepare the following, using pure solute and water:

a. 500 mL of 1.00 M $MgCl_2$ solution
b. 100 mL of 12.0% (w/v) $MgCl_2$ solution
c. 1.00 L of 20.0% (v/v) ethylene glycol solution

Solutions are often prepared by diluting a more concentrated solution with solvent (usually water) to produce a solution of lower concentration. Suppose, for example, that you want to prepare 250 mL of 0.100 M NaCl solution using a 2.00 M NaCl solution as the source of NaCl. First, calculate the number of moles of NaCl that would be contained in 250 mL of 0.100 M solution. Remember Equation 7.5,

$$M = \frac{\text{moles of solute}}{\text{liters of solution}}$$

which can be rearranged to

$$(M)(\text{liters of solution}) = \text{moles of solute}$$

So, the number of moles of NaCl contained in the desired solution is

$$\left(0.100 \frac{\text{mol}}{\text{L}}\right)(0.250 \text{ L}) = 0.0250 \text{ mol}$$

These moles of NaCl must be obtained from the 2.00 M NaCl solution. The volume of this solution that contains the desired number of moles can be obtained by rearranging Equation 7.5 a different way:

$$\text{liters of solution} = \frac{\text{moles of solute}}{\text{M}}$$

or

$$\text{liters of solution} = \frac{0.0250 \text{ mol}}{2.00 \text{ mol/L}} = 0.0125 \text{ L} = 12.5 \text{ mL}$$

Thus, the solution is prepared by putting 12.5 mL of 2.00 M NaCl solution into a 250-mL volumetric flask and adding water up to the mark. ▶ Figure 7.9 will help you understand this process.

Step 1. 20.0 mL of 0.200 M $K_2Cr_2O_7$ solution is withdrawn from a beaker using a pipette.

Step 2. The 20.0 mL of the 0.200 M $K_2Cr_2O_7$ solution is put into a 100-mL flask.

Step 3. Water is added (while the solution is swirled) to fill the flask to the mark.

Figure 7.9 Preparation of a 0.0400 M $K_2Cr_2O_7$ solution by dilution of a 0.200 M $K_2Cr_2O_7$ solution. Use the data and show by a calculation that the new solution is 0.0400 M.

The same result can be obtained by using a simplified calculation. Note that the product of concentration in molarity (M) and solution volume in liters will give the number of moles of solute in a sample of solution. In a dilution such as the one done above, the number of moles of solute taken from the concentrated solution and diluted with water is the same as the number of moles of solute in the resulting more dilute solution (see Figure 7.9). Thus, the following can be written:

$$M_c V_c = \begin{array}{c}\text{solute moles in} \\ \text{concentrated solution}\end{array} = \begin{array}{c}\text{solute moles in} \\ \text{dilute solution}\end{array} = M_d V_d$$

or

$$(C_c)(V_c) = (C_d)(V_d) \tag{7.9}$$

where the subscripts c and d refer to the more concentrated and dilute solutions, respectively, and C is used to represent any appropriate concentration. An advantage of using this equation is that any volume units may be used, as long as the same one is used for both

V_c and V_d. Also, Equation 7.9 is true for any solution concentration based on volume. This means that M, %(v/v), or %(w/v) concentrations can be used just as long as C_c and C_d are expressed in the same units.

▶ Example 7.7

Use Equation 7.9 and describe how to prepare 250 mL of 0.100 M NaCl solution using a 2.00 M NaCl solution as the source of NaCl.

Solution

Equation 7.9 can be used to determine the volume of 2.00 M NaCl needed. The relationships between the terms in Equation 7.9 and the quantities of the problem are C_c = 2.00 M, V_c = volume of 2.00 M NaCl needed, C_d = 0.100 M, and V_d = 250 mL. Substitution gives

$$(2.00 \text{ M})(V_c) = (0.100 \text{ M})(250 \text{ mL})$$

or

$$V_c = \frac{(0.100 \text{ M})(250 \text{ mL})}{(2.00 \text{ M})} = 12.5 \text{ mL}$$

Thus, we quickly find the volume of 2.00 M NaCl needs to be 12.5 mL, the same result found earlier. This volume of solution is put into a 250-mL flask, and water is added to the mark as before.

▶ **LEARNING CHECK 7.7** Use Equation 7.9 and describe how to prepare 500 mL of a 0.250 M NaOH solution from a 6.00 M NaOH solution.

7.6 Solution Stoichiometry

Learning Objective

6. Do stoichiometric calculations based on solution concentrations.

The stoichiometry calculations of solution reactions can be done using the factor-unit method. The sources of the needed factors will be the mole interpretation of the reactions introduced as statement 2 in Section 5.9, and the molarities of the solutions involved in the reactions. Each known solution molarity will provide two factors. For example, the following two factors can be obtained based on a 0.400 M HCl solution:

$$\frac{0.400 \text{ mol HCl}}{1 \text{ L HCl soln.}} \quad \text{and} \quad \frac{1 \text{ L HCl soln.}}{0.400 \text{ mol HCl}}$$

▶ Example 7.8

Solutions of NaOH and HCl react according to the following equation:

$$\text{NaOH(aq)} + \text{HCl(aq)} \rightarrow \text{NaCl(aq)} + \text{H}_2\text{O}(\ell)$$

a. What volume of a 0.250 M NaOH solution contains 0.110 moles of NaOH?
b. What volume of a 0.200 M NaOH solution is needed to exactly react with 0.150 moles of HCl?
c. What volume of a 0.185 M NaOH solution is needed to exactly react with 25.0 mL of 0.225 M HCl solution?

Solution

a. The known quantity is 0.110 mol of NaOH, and the unit of the unknown quantity is a volume which we will express as L NaOH soln.

Step 1. 0.110 mol NaOH

Step 2. 0.110 mol NaOH $= $ L NaOH soln.

Step 3. 0.110 ~~mol NaOH~~ $\times \dfrac{1 \text{ L NaOH soln.}}{0.250 \text{ ~~mol NaOH~~}} = $ L NaOH soln.

The factor $\dfrac{1 \text{ L NaOH soln.}}{0.250 \text{ mol NaOH}}$ came from the known molarity of the NaOH solution as described above.

Step 4. $0.110 \times \dfrac{1 \text{ L NaOH soln.}}{0.250} = 0.44 \text{ L NaOH soln.} = 0.440 \text{ L NaOH soln.}$

The calculator answer of 0.44 was rounded by adding a trailing zero so the answer had three significant figures to match the three significant figures in 0.110 and 0.250. The number 1 is an exact counting number.

b. The known quantity is 0.150 mol HCl, and the unit of the unknown quantity is the volume of NaOH solution which we will express as L NaOH soln.

Step 1. 0.150 mol HCl

Step 2. 0.150 mol HCl $= $ L NaOH soln.

Step 3. 0.150 ~~mol HCl~~ $\times \dfrac{1 \text{ ~~mol NaOH~~}}{1 \text{ ~~mol HCl~~}} \times \dfrac{1 \text{ L NaOH soln.}}{0.200 \text{ ~~mol NaOH~~}} = $ L NaOH soln.

Note that two factors were needed to cancel the unit of the known quantity and generate the unit of the unknown quantity. The first factor came from statement 2 for the reaction:

$$\text{NaOH(aq)} + \text{HCl(aq)} \rightarrow \text{NaCl(aq)} + \text{H}_2\text{O}(\ell)$$

Statement 2:   1 mol NaOH + 1 mol HCl → 1 mol NaCl + 1 mole H_2O

The second factor came from the 0.200 M concentration of the NaOH solution. Two factors were possible: $\dfrac{0.200 \text{ mol NaOH}}{1 \text{ L NaOH soln.}}$ and $\dfrac{1 \text{ L NaOH soln.}}{0.200 \text{ mol NaOH}}$. The second factor was used because it cancelled the *mol NaOH* unit and generated the *L NaOH soln.* unit.

Step 4. $0.150 \times \dfrac{1}{1} \times \dfrac{1 \text{ L NaOH soln.}}{0.200} = 0.75 \text{ L NaOH soln.} = 0.750 \text{ L NaOH soln.}$

Once again, a zero was added to the calculator answer of 0.75 to give an answer with three significant figures to match the three in 0.150 and 0.200. Both of the 1 numbers are exact counting numbers.

c. The known quantity is 25.0 mL of a 0.225 M HCl soln., and the unit of the unknown is the volume of a 0.185 M NaOH soln. which we will express as mL NaOH soln. to match the mL unit of the known quantity.

Step 1. 25.0 mL HCl soln.

Step 2. 25.0 mL HCl soln. $= $ mL NaOH soln.

Step 3. 25.0 ~~mL HCl soln.~~ $\times \dfrac{0.225 \text{ ~~mol HCl~~}}{1000 \text{ ~~mL HCl soln.~~}}$

$\times \dfrac{1 \text{ ~~mol NaOH~~}}{1 \text{ ~~mol HCl~~}} \times \dfrac{1000 \text{ mL NaOH soln.}}{0.185 \text{ ~~mol NaOH~~}} = $ mL NaOH soln.

Note that three factors were needed. The first came from the molarity of the HCl solution with the solution volume expressed as 1000 mL rather than 1 L. The second factor came from statement 2 for the reaction as was done in part b. The third factor came from the molarity of the NaOH solution with the solution volume expressed as 1000 mL rather than 1 L.

Step 4. $25.0 \times \dfrac{0.225}{1000} \times \dfrac{1}{1} \times \dfrac{1000 \text{ mL NaOH soln.}}{0.185} = 30.4 \text{ mL NaOH soln.}$

7.7 Solution Properties

Learning Objective

7. Do calculations based on the colligative solution properties of boiling point, freezing point, and osmotic pressure.

The experiment represented in ◗ Figure 7.10 reveals an interesting property of some solutions. As shown by the figure, the circuit is completed and the light goes on only when the solution conducts electricity. Experiments of this type demonstrate that conductive solutions are formed when soluble ionic materials (Section 4.3), such as sodium chloride, or highly polar covalent materials (Section 4.9), such as hydrogen chloride, dissolve in water. Solutes that form conductive water solutions are called **electrolytes,** whereas solutes that form nonconductive solutions are called **nonelectrolytes.** Conductive solutions result when the dissolved solute dissociates, or breaks apart, to form ions, as shown in the following equations representing the dissociation into ions of polar covalent hydrogen chloride and ionic calcium nitrate:

$$HCl(aq) \rightarrow H^+(aq) + Cl^-(aq) \tag{7.10}$$

$$Ca(NO_3)_2(aq) \rightarrow Ca^{2+}(aq) + 2NO_3^-(aq) \tag{7.11}$$

electrolyte A solute that when dissolved in water forms a solution that conducts electricity.

nonelectrolyte A solute that when dissolved in water forms a solution that does not conduct electricity.

① A solution of a strong electrolyte conducts electricity well.

② A solution of a weak electrolyte conducts electricity poorly.

③ A solution of a nonelectrolyte does not conduct electricity.

Figure 7.10 Electrical conductivity of solutions.

Some electrolytes, such as HCl and $Ca(NO_3)_2$, dissociate essentially completely in solution and are called *strong electrolytes.* They form strongly conducting solutions. Other electrolytes, such as acetic acid (the acid found in vinegar), dissociate only slightly and form weakly conductive solutions. These solutes are classified as *weak electrolytes.* The solutes in nonconductive solutions dissolve in the solvent but remain in the form of uncharged molecules.

Besides electrical conductivity (or the lack of it), all solutions have properties that depend only on the concentration of solute particles present and not on the actual identity of the solute. Thus, these properties, called **colligative properties,** would be identical for water solutions containing 1 mol of sugar or 1 mol of alcohol per liter.

Three closely related colligative properties are vapor pressure, boiling point, and freezing point. Experiments demonstrate that the vapor pressure of water (solvent) above a solution is lower than the vapor pressure of pure water (see ◗ Figure 7.11). This behavior causes the boiling point of solutions to be higher than the boiling point of the pure solvent used in the solutions, and the freezing point to be lower (see ◗ Table 7.5).

colligative property A solution property that depends only on the concentration of solute particles in solution.

Figure 7.11 The vapor pressure of water above a solution is lower than the vapor pressure of pure water. Remember that pressure is a force per unit area. Think of this force as "pushing" water molecules in the vapor state, and explain the process shown in the photos.

1 Initially, equal volumes of pure water and copper sulfate solution (blue) are placed under a glass dome that prevents water vapor from escaping to the outside.

2 After some time has passed, the volume of pure water has decreased, while that of the copper sulfate solution has increased.

TABLE 7.5 A Comparison of Colligative Properties of Pure Solvent and Solutions

Property	Pure Solvent	Solution
Vapor pressure	Normal	Lower than pure solvent
Boiling point	Normal	Higher than pure solvent
Freezing point	Normal	Lower than pure solvent

The boiling and freezing point differences between pure solvent and solutions can be calculated by using an equation of the general form

$$\Delta T = nkm \tag{7.12}$$

In this equation, m is the solution concentration expressed as a molality, a unit we have not discussed. For very dilute solutions, the molality m and molarity M are essentially equal, and M can be used in Equation 7.12 instead of m. This approximation will be used for calculating colligative properties of solutions in this book. The symbol ΔT is the boiling point or freezing point difference between pure solvent and solution. The specific equations used to calculate ΔT for boiling and freezing points are

$$\Delta T_b = nK_b M \tag{7.13}$$

$$\Delta T_f = nK_f M \tag{7.14}$$

The subscripts b and f refer to boiling or freezing, and K_b and K_f are constants characteristic of the solvent used in the solution (remember, to this point we have focused on water as the solvent). Values for K_b and K_f are given in ▶ Table 7.6 for a number of solvents. In Equations 7.13 and 7.14, M is the molarity of solute in solution, and n is the number of moles of solute particles put into solution when 1 mol of solute dissolves.

TABLE 7.6 Boiling and Freezing Point Constants for Various Solvents

Solvent	Normal Boiling Point (°C)	K_b (°C/M)	Normal Freezing Point (°C)	K_f (°C/M)
Benzene	80.1	2.53	5.5	4.90
Camphor			174.0	40.0
Carbon tetrachloride	76.8	5.03		
Chloroform	61.2	3.63		
Cyclohexane	81.0	2.79	6.5	20.0
Ethyl alcohol	78.5	1.22		
Water	100.0	0.52	0.0	1.86

Example 7.9

Calculate the boiling and freezing points of the following solutions:

a. 171 g of sugar ($C_{12}H_{22}O_{11}$) is dissolved in enough water to give 1.00 L of solution.
b. 13.4 g of NH_4Cl is dissolved in water to form 500 mL of solution.

Solution

In each case, Equations 7.13 and 7.14 are used to calculate the difference between the normal boiling and freezing point of water and the solution. To use these equations, K_b and K_f are obtained from Table 7.6; the solution molarity, M, is calculated; and n is determined.

a. To find the boiling point, calculate solution molarity.

$$(171 \text{ g } C_{12}H_{22}O_{11})\left(\frac{1 \text{ mol } C_{12}H_{22}O_{11}}{342.0 \text{ g } C_{12}H_{22}O_{11}}\right) = 0.500 \text{ mol } C_{12}H_{22}O_{11}$$

$$M = \frac{\text{moles of solute}}{\text{liters of solution}} = \frac{0.500 \text{ mol}}{1.00 \text{ L}} = 0.500 \frac{\text{mol}}{\text{L}}$$

Determine n: Because sugar does not dissociate on dissolving, $n = 1$. Therefore,

$$\Delta T_b = nK_bM = (1)(0.52°C/M)(0.500 \text{ M}) = 0.26°C$$

Because boiling points are higher in solutions, we add ΔT_b to the normal boiling point of water:

$$\text{solution boiling point} = 100.00°C + 0.26°C = 100.26°C$$

To find the freezing point, calculate ΔT_f:

$$\Delta T_f = nK_fM = (1)(1.86°C/M)(0.500 \text{ M}) = 0.93°C$$

Because freezing points are lower in solutions, we subtract ΔT_f from the normal freezing point of water:

$$\text{solution freezing point} = 0.00°C - 0.93°C = -0.93°C$$

b. Similarly,

$$(13.4 \text{ g } NH_4Cl)\left(\frac{1 \text{ mol } NH_4Cl}{53.5 \text{ g } NH_4Cl}\right) = 0.250 \text{ mol } NH_4Cl$$

$$M = \frac{\text{moles of solute}}{\text{liters of solution}} = \frac{0.250 \text{ mol}}{0.500 \text{ L}} = 0.500 \frac{\text{mol}}{\text{L}}$$

Because NH$_4$Cl dissociates in water, 1 mol of solute gives 2 mol of particles (ions):

$$NH_4Cl(aq) \rightarrow NH_4^+(aq) + Cl^-(aq)$$

Thus, we conclude that $n = 2$. Therefore,

$$\Delta T_b = nK_bM = (2)(0.52°C/M)(0.500\ M) = 0.52°C$$
$$\text{solution boiling point} = 100.00°C + 0.52°C = 100.52°C$$

$$\Delta T_f = nK_fM = (2)(1.86°C/M)(0.500\ M) = 1.86°C$$
$$\text{solution freezing point} = 0.00°C - 1.86°C = -1.86°C$$

Example 7.9 demonstrates the influence of solute dissociation on colligative properties. Even though the solutions in parts a and b have the same molarity, the NH$_4$Cl dissociates and produces twice as many solute particles in solution. Hence, it causes twice as much change in the colligative properties.

Osmotic pressure, another important colligative property, can be illustrated by some hypothetical experiments. Consider ❱ Figure 7.12, where a sugar solution is separated from pure water by a barrier Ⓐ. The barrier is removed, but the mixture is not stirred Ⓑ. After a day or so, the barrier is replaced, with the results shown in Ⓒ. These results are not surprising; the sugar has diffused throughout the mixture uniformly. It would have been very surprising if the process had not taken place.

Barrier
Pure water
Sugar solution

Ⓐ   Ⓑ   Ⓒ

Figure 7.12 Diffusion eliminates concentration gradients.

Now consider the experiment shown in ❱ Figure 7.13, in which two solutions similar to those used before are separated by a semipermeable membrane, which has pores large enough to allow small molecules such as water to pass through but small enough to prevent passage by larger molecules or hydrated ions. In Figure 7.13, the membrane allows water molecules to pass but not sugar molecules.

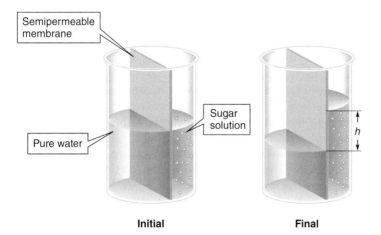

Semipermeable membrane
Sugar solution
Pure water
h

Initial   Final

Figure 7.13 Osmosis.

In this experiment, a concentration difference between the liquids has been created, but diffusion is prevented from taking place as it did before. Water molecules move through the membrane in both directions, but sugar molecules cannot move into the pure water to

Knowing where to start or what to do first is a critical part of working any math-type problem. This might be a factor if molarity calculations are difficult for you. First, of course, you must identify a problem as a molarity problem. This can be done by looking for a key word or phrase (molar, molarity, or moles/liter) or the abbreviation M. Second, remember that you have a formula for molarity

$$M = \frac{\text{moles of solute}}{\text{liters of solution}}$$

and the problem might be treated as a formula-type problem. Look for the given numbers and their units, and see if they match the units of the formula. If they do, put them into the formula and do the calculations. For example, if you are asked to calculate the molarity of a solution that contains 0.0856 mol NaCl dissolved in enough water to give 0.100 L of solution, you could put the numbers and their units directly into the formula:

$$M = \frac{0.0856 \text{ mol NaCl}}{0.100 \text{ L solution}} = 0.856 \text{ mol NaCl/L solution}$$

However, if you must calculate the molarity of a solution that contains 5.00 g of NaCl in enough water to give 100 mL of solution, the units of the numbers do not match those of the formula. The factor-unit method can be used to convert each quantity into the units needed by the formula

$$5.00 \text{ g NaCl} \times \frac{1 \text{ mol NaCl}}{58.4 \text{ g NaCl}} = 0.0856 \text{ mol NaCl}$$

$$100 \text{ mL solution} \times \frac{1 \text{ L solution}}{1000 \text{ mL solution}} = 0.100 \text{ L solution}$$

These numbers and units can then be put into the formula

$$M = \frac{0.0856 \text{ mol NaCl}}{0.100 \text{ L solution}} = 0.856 \text{ mol NaCl/L solution}$$

Alternatively, the problem can be solved by remembering the units of the answer (molarity would have the units mol NaCl/L solution), noting the numbers and units of the given quantities (5.00 g NaCl/100 mL solution), and using the factor-unit method to convert the units of the given quantity to those of the answer:

Step 1. $\dfrac{5.00 \text{ g NaCl}}{100 \text{ mL solution}}$

Step 2. $\dfrac{5.00 \text{ g NaCl}}{100 \text{ mL solution}} = \dfrac{\text{mol NaCl}}{\text{L solution}}$

Step 3. $\dfrac{5.00 \text{ g NaCl}}{100 \text{ mL solution}} \times \dfrac{1 \text{ mol NaCl}}{58.4 \text{ g NaCl}}$

$$\times \frac{1000 \text{ mL solution}}{1 \text{ L solution}} = \frac{\text{mol NaCl}}{\text{L solution}}$$

Step 4. $\dfrac{(5.00)(1 \text{ mol NaCl})(1000)}{(100)(58.4)(1 \text{ L solution})} = 0.856 \dfrac{\text{mol NaCl}}{\text{L solution}}$

The final answer, which has the correct units for a molarity, has been rounded to three significant figures.

---

osmotic pressure The hydrostatic pressure required to prevent the net flow of solvent through a semipermeable membrane into a solution.

osmosis The process in which solvent flows through a semipermeable membrane into a solution.

equalize the concentration. As a result, the net flow of water through the membrane is into the sugar solution. This has the effect of increasing the volume of the sugar solution and decreasing its concentration to a value closer to that of the pure water. The movement of water creates a difference in liquid levels *(h)*, which causes hydrostatic pressure against the membrane. This pressure increases until it becomes high enough to balance the tendency for net water to flow through the membrane into the sugar solution. From that time on, the flow of water in both directions through the membrane is equal, and the volume of liquid on each side of the membrane no longer changes. The hydrostatic pressure required to prevent the net flow of water through a semipermeable membrane into a solution is called the **osmotic pressure** of the solution. The process in which solvent molecules move through semipermeable membranes is called **osmosis.** Although our experiment doesn't illustrate it, solvent will also flow osmotically through semipermeable membranes separating solutions of different concentrations. The net flow of solvent is always from the more dilute solution into the more concentrated (see ▶ Figure 7.14).

The osmotic pressure that will develop across a semipermeable membrane separating pure solvent from a solution of molarity M is given by Equation 7.15, which is similar to the ideal gas law given earlier (Section 6.8):

$$\pi = nMRT \tag{7.15}$$

In this equation, π is the osmotic pressure in units that are the same as the pressure units used in the ideal gas constant *R*. M is the solution molarity, *T* is the temperature in kelvins, and *n*, as before, is the number of moles of solute particles obtained when 1 mol of solute dissolves. Scientists in biological and medical fields often call the product of *n* and M the **osmolarity** of the solution.

osmolarity The product of *n* and M in the equation π = *n*MRT.

▶ Example 7.10

Calculate the osmolarity and the osmotic pressure that would develop across a semipermeable membrane if the solutions of Example 7.9 were separated from pure water by the membrane. Assume a solution temperature of 27°C and use an *R* value of 62.4 L torr/K mol.

Solution

a. The molarity of the sugar solution was found to be 0.500 mol/L. Because sugar does not dissociate, *n* is equal to 1, and the osmolarity *n*M is also equal to 0.500 mol/L.

$$\pi = nMRT = (1)\left(0.500\ \frac{mol}{L}\right)\left(\frac{62.4\ L\ torr}{K\ mol}\right)(300\ K)$$

$$= 9.36 \times 10^3\ torr$$

Thus, this solution would develop a pressure sufficient to support a column of mercury 9.36×10^3 mm high. This is equal to about 12.3 standard atmospheres of pressure (see Table 6.3 for pressure units).

b. The molarity of this solution was also found to be 0.500. But, because this solute dissociates into two ions, *n* = 2. This makes the osmolarity equal to (2)(0.500 mol/L) = 1.00 mol/L:

$$\pi = nMRT = (2)\left(0.500\ \frac{mol}{L}\right)\left(\frac{62.4\ L\ torr}{K\ mol}\right)(300\ K)$$

$$= 1.87 \times 10^4\ torr,\ or\ about\ 24.6\ standard\ atmospheres$$

▶ **LEARNING CHECK 7.8** Calculate the boiling point, freezing point, and osmotic pressure of the following solutions. Assume that the osmotic pressure is measured at 27°C.

a. A 0.100 M solution of $CaCl_2$ in water. The $CaCl_2$ is a strong electrolyte.
b. A 0.100 M solution of ethylene glycol in water. Ethylene glycol does not dissociate in solution.

Figure 7.14 Osmosis through carrot membranes. The tissue of a dried shriveled carrot (left) acts as a semipermeable osmotic membrane when placed in water. After some time, the osmotic flow of water into the carrot rehydrates the carrot (right).

7.8 Colloids

Learning Objective

8. Describe the characteristics of colloids.

Like solutions, **colloids** (or colloidal dispersions) are homogeneous mixtures of two or more components in which there is more of one component than of the others. In solutions, the terms *solvent* and *solute* are used for the components, but in colloids, the terms **dispersing medium** (for solvent) and **dispersed phase** (for solute) are used.

Solute particles (in solutions) and dispersed phase particles (in colloids) cannot be seen and do not settle under the influence of gravity. However, solute particles do not scatter or reflect light, whereas dispersed phase particles do. This and other variations in properties result from the principal difference between solutions and colloids, the size of the particles making up the solute or dispersed phase. The dissolved solute in a solution is present in the form of tiny particles (small molecules or ions) that are less than about 10^{-7} cm (0.1 μm) in diameter. The dispersed phase of colloids is made up of much larger particles (very large molecules or small pieces of matter) with diameters of 10^{-7} to 10^{-5} cm (0.1–10 μm). As a result of light scattering, colloids often appear to be cloudy. When a

colloid A homogeneous mixture of two or more substances in which the dispersed substances are present as larger particles than are found in solutions.

dispersing medium The substance present in a colloidal dispersion in the largest amount.

dispersed phase The substance present in a colloidal dispersion in amounts less than the amount of dispersing medium.

beam of light passes through them, they demonstrate the **Tyndall effect** in which the path of the light becomes visible (▶ Figure 7.15).

© Joel Gordon, courtesy of West Publishing Company/CHEMISTRY by Radel & Navidi, 2d ed

Figure 7.15 The Tyndall effect.

Tyndall effect A property of colloids in which the path of a beam of light through the colloid is visible because the light is scattered.

The word *colloidal* means "gluelike," and some colloids, including some glues, fit this description quite well. However, many, including smoke, shaving cream, and cheese, do not. Colloids are usually differentiated according to the states of the dispersing medium and dispersed phase. Some colloid types are listed in ▶ Table 7.7, together with examples and specific names.

TABLE 7.7 Types of Colloids

Colloid Type			
Dispersing Medium	**Dispersed Phase**	**Name**	**Examples**
Gas	Liquid	Aerosol	Fog, aerosol sprays, some air pollutants
Gas	Solid		Smoke, some air pollutants
Liquid	Gas	Foam	Whipped cream, shaving cream
Liquid	Liquid	Emulsion	Milk, mayonnaise
Liquid	Solid	Sol	Paint, ink, gelatin dessert
Solid	Gas	Solid foam	Marshmallow, pumice stone, foam rubber
Solid	Liquid		Butter, cheese
Solid	Solid		Pearls, opals, colored glass, some metal alloys

Sols that become viscous and semisolid are called *gels*. In these colloids, the solid dispersed phase has a very high affinity for the dispersing medium. The gel "sets" by forming a three-dimensional network of solid and dispersing medium. Other examples of gels are fruit jellies and "canned heat" (jellied alcohol).

Much of the interest in colloids is related to their formation or destruction. Ions that are present in the dispersing medium are attracted to colloid particles and stick on their surfaces. The charge (+ or −) of the ions depends on the nature of the colloid, but all colloid particles within a particular system will attract ions of only one charge or the other. In this way, the colloid particles all acquire the same charge and repel each other. This repulsion helps prevent the particles from coalescing into aggregates large enough to settle out.

Tears: Solutions for Many Eye Problems

Tears are usually associated with the act of crying. However, healthy eyes are constantly bathed with tears, which constitute a sophisticated lubrication system made up of mucus, water, and oil. The tears wash away microbes and debris and help heal any damaged eye tissue.

Some individuals suffer from a condition called dry eye syndrome. The classic symptoms of this condition are a burning sensation in the eyes coupled with a scratchy or gritty feeling. Interestingly, constant tearing and watering are also symptomatic of the condition. More women than men commonly suffer from this ailment, with the onset often occurring near the time they go through menopause. This indicates a possible link between the dry eye condition and hormonal changes in the body.

Dry eye syndrome is not a cause of blindness, and often goes away on its own. However, it can seriously interfere with working, reading, driving, and other normal activities of daily life. A number of simple remedies and practices have been shown to help reduce or eliminate the troublesome symptoms. A number of over-the-counter eye drops that help lubricate the eyes and provide needed moisture are available. However, some such products that contain preservatives or materials to eliminate redness can make dry eye syndrome symptoms worse and should be avoided. Lubricating ointments applied at bedtime work well by moistening the eyes throughout the night. Other practices that help include drinking plenty of water to remain hydrated and avoiding hair dryers, electric fans, persistent winds, overheated rooms, and exposure to smoke. Home air humidifiers have also proved to be helpful.

Anyone with eyes that continue to burn, itch, or water frequently in spite of taking appropriate precautions should consult an ophthalmologist or optometrist to determine if an infection or other serious condition is present.

Examples of typical over-the-counter eye drops used by some to combat dry eye syndrome.

In the Cottrel precipitator, colloidal solids are removed from gaseous smokestack wastes before they are released into the atmosphere. The precipitator contains a number of highly charged plates or electrodes. As smoke passes over the charged surfaces, the colloid particles lose their charges. The particles then coalesce into larger particles that settle out and are collected for disposal.

Some colloids are stabilized (prevented from coalescing) by substances known as **emulsifying agents** or **stabilizing agents.** Mayonnaise-like salad dressing is a colloid of oil in water, with compounds from egg yolk acting as the emulsifying agent. These compounds form a coating around the oil droplets, which keeps them separated and suspended in the water.

The cleaning action of soaps and detergents comes from their activity as emulsifying agents. Both soaps and detergents contain long molecules with structures like the following:

emulsifying agent (stabilizing agent) A substance that when added to colloids prevents them from coalescing and settling.

$$H-\underset{\underset{H}{|}}{\overset{\overset{H}{|}}{C}}-\underset{\underset{H}{|}}{\overset{\overset{H}{|}}{C}}-\underset{\underset{H}{|}}{\overset{\overset{H}{|}}{C}}-\underset{\underset{H}{|}}{\overset{\overset{H}{|}}{C}}-\underset{\underset{H}{|}}{\overset{\overset{H}{|}}{C}}-\underset{\underset{H}{|}}{\overset{\overset{H}{|}}{C}}-\underset{\underset{H}{|}}{\overset{\overset{H}{|}}{C}}-\underset{\underset{H}{|}}{\overset{\overset{H}{|}}{C}}-\underset{\underset{H}{|}}{\overset{\overset{H}{|}}{C}}-\underset{\underset{H}{|}}{\overset{\overset{H}{|}}{C}}-\underset{\underset{H}{|}}{\overset{\overset{H}{|}}{C}}-\overset{\overset{O}{\parallel}}{C}-O^-Na^+$$

Soap

$$H-\underset{\underset{H}{|}}{\overset{\overset{H}{|}}{C}}-\underset{\underset{H}{|}}{\overset{\overset{H}{|}}{C}}-\underset{\underset{H}{|}}{\overset{\overset{H}{|}}{C}}-\underset{\underset{H}{|}}{\overset{\overset{H}{|}}{C}}-\underset{\underset{H}{|}}{\overset{\overset{H}{|}}{C}}-\underset{\underset{H}{|}}{\overset{\overset{H}{|}}{C}}-\underset{\underset{H}{|}}{\overset{\overset{H}{|}}{C}}-\underset{\underset{H}{|}}{\overset{\overset{H}{|}}{C}}-\underset{\underset{H}{|}}{\overset{\overset{H}{|}}{C}}-\underset{\underset{H}{|}}{\overset{\overset{H}{|}}{C}}-\underset{\underset{H}{|}}{\overset{\overset{H}{|}}{C}}-O-\overset{\overset{O}{\parallel}}{\underset{\underset{O}{\parallel}}{S}}-O^-Na^+$$

Detergent

When placed in water, soaps and detergents dissociate to form ions, as shown in Equations 7.16 and 7.17, where the long carbon chains are represented by a wavy line.

$$\text{\wavyline}\underset{O}{\overset{O}{\underset{\|}{C}}}-O^-Na^+ \longrightarrow \text{\wavyline}\underset{O}{\overset{O}{\underset{\|}{C}}}-O^- + Na^+ \quad (7.16)$$

$$\text{\wavyline}\ O-\underset{O}{\overset{O}{\underset{\|}{\overset{\|}{S}}}}-O^-Na^+ \longrightarrow \text{\wavyline}\ O-\underset{O}{\overset{O}{\underset{\|}{\overset{\|}{S}}}}-O^- + Na^+ \quad (7.17)$$

Nonpolar oils and greases are not soluble in water, but they are attracted to the uncharged ends of the soap or detergent ions. As a result of this attraction, the soap or detergent forms a charged layer around the oil droplets, which keeps them separated and suspended (see ❱ Figure 7.16). Certain compounds (lecithins) of egg yolk act in much the same way to stabilize the mayonnaise-like dressing discussed earlier.

Figure 7.16 The cleaning action of soaps and detergents.

Soap ions dissolved in water   **Soil particles suspended in water**

7.9 Dialysis

Learning Objective

6. Describe the process of dialysis, and compare it to the process of osmosis.

dialyzing membrane
A semipermeable membrane with pores large enough to allow solvent molecules, other small molecules, and hydrated ions to pass through.

dialysis A process in which solvent molecules, other small molecules, and hydrated ions pass from a solution through a membrane.

Earlier we discussed semipermeable membranes that selectively allow solvent to pass but retain dissolved solutes during osmosis. **Dialyzing membranes** are semipermeable membranes with larger pores than osmotic membranes. They hold back colloid particles and large molecules but allow solvent, hydrated ions, and small molecules to pass through. The passage of these ions and small molecules through such membranes is called **dialysis.** Dialysis can be used to separate small particles from colloids, as shown in ❱ Figure 7.17.

A mixture containing water, ions, small molecules, and colloid particles is placed inside a bag made from a dialyzing membrane. Water flowing around the bag carries away ions and small molecules that pass through the membrane. Water molecules move through the membrane in both directions, but the colloid particles remain inside the bag.

Global Warming and a Cooler Europe

Approximately 70% of the Earth's surface is covered with water in the form of salty solutions called oceans or seas. Massive rivers of moving water flow between these oceans. One of these rivers, called the Gulf Stream, is made up of warm surface water from the South Atlantic Ocean. It flows through the North Atlantic past Europe on its way north toward Iceland. As it passes Europe, the water cools significantly, releasing heat that increases the average temperature of European countries by 5–8°C.

During its journey from the South Atlantic, the concentration of salt in the Gulf Stream water has steadily increased as a result of evaporation of some of the warm water. This increase in salt concentration coupled with the cooling of the water as it moves north causes the density of the Gulf Stream water to increase to the point that, at approximately the latitude of Iceland, it plunges into the depths and becomes a cold river flowing south along the ocean floor. As the cold salty water sinks and flows south, more warmer surface water flows north to replace it. The south-flowing cold water is eventually pushed to the surface, where it warms up and begins to evaporate, thus completing a cycle. This cycle acts as sort of a huge liquid conveyor belt that continues to cause warm surface water to flow past Europe and warm it.

The concern of climatologists is that the atmosphere of the Earth is undoubtedly warming, and the extra heat is melting ice in the Arctic ocean. The fresh water from the melting ice flows into the salty North Atlantic. If this fresh water dilutes the salty water of the Gulf Stream enough, the Gulf Stream water might not become dense enough as it cools in the North Atlantic to sink to the ocean bottom for the return trip south. If this happens, the Gulf Stream could cease to operate, and the heat delivery to Europe would stop. Ironically, according to this possibility, global warming could result in a cooling of Europe, possibly enough to cause a European mini ice age.

An increase in the rate of breakup of glaciers (calving) as they approach the oceans is one indication of global warming.

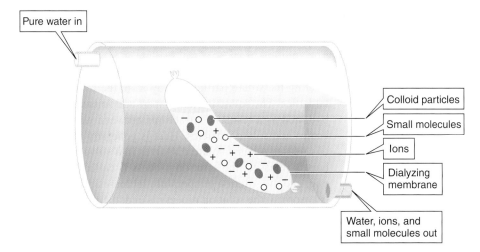

Pure water in

Colloid particles

Small molecules

Ions

Dialyzing membrane

Water, ions, and small molecules out

Figure 7.17 Dialysis. This is one method of dialysis used to purify proteins.

A similar technique is used to clean the blood of people suffering from kidney malfunction. The blood is pumped through tubing made of a dialyzing membrane. The tubing passes through a bath in which impurities collect after passing out of the blood. Blood proteins and other important large molecules remain in the blood.

The U.S. Food and Drug Administration (FDA) does not consider colloidal silver to be safe or effective in treating any disease or condition in humans or animals (FDA, 1997). Ingestion of colloidal silver can cause an irreversible build-up of silver particles in tissues, called *argyria*. Advertisements for colloidal silver may claim that the product will strengthen the immune system or kill bacteria. Some manufactures claim that it treats cancer, HIV-AIDS, or diabetes. These claims are not supported by medical research (NCCAM, 2004).

However, alternative and supportive therapies that are confirmed by reputable research can be helpful. A good example of such a therapy is the use of fish oil supplements to enhance the effects of psychotropic drugs that patients suffering from anxiety or depression might already be taking (Harvard, 2003).

The State has the right to regulate the family in cases of public interest. The U.S. Supreme Court stated in the 1944 case *Prince vs. Massachussetts*, "Acting to guard the general interest in youth's well being, the state as *parens patriae* may restrict the parent's control by requiring school attendance, regulating or prohibiting the child's labor and in many other ways. The right to practice religion freely does not include liberty to expose the community or the child to communicable disease or the latter to ill health or death" (*Prince v. Massachusetts*, 321 U. S. 158, 166). Therefore, a parent does not have the legal right to refuse life-saving medical treatment for a child or treatment for conditions that are a matter of public health. It is, of course, in the child's best interest to keep the family intact, so educating the parents is the best first option before considering removal of the child by the State.

Concept Summary

Physical States of Solutions

Solutions, homogeneous mixtures of a solvent and one or more solutes, can be found in any of the three states of matter: solid, liquid, or gas. The physical state of the solution is often the same as the physical state of the solvent.

Objective 1, Exercise 7.4

Solubility

The amount of solute that will dissolve in a quantity of solvent to form a saturated solution is the solubility of the solute. Solubility depends on similarities in polarities of the solvent and solute; it generally increases with temperature for solid and liquid solutes, but decreases with temperature for gaseous solutes.

Objective 2, Exercises 7.6 and 7.12

The Solution Process

The solution process of solid solute in a liquid solvent can be thought of in terms of the solvent molecules attracting the solute particles away from the solute crystal lattice. The solute is picked apart and solute particles hydrated as a result of attractions between water molecules and solute particles. Heat is generally absorbed or released when a solute dissolves in a solvent. When heat is released, the solution process is called exothermic, and the solution temperature increases. When heat is absorbed, the solution process is endothermic, and the solution cools as solute dissolves.

Objective 3, Exercise 7.16

Solution Concentrations

Relationships between the amount of solute and the amount of solution containing the solute are called concentrations. They may be expressed as molarity, weight/weight percent, weight/volume percent, or volume/volume percent.

Objective 4, Exercises 7.22b, 7.30c, 7.34a, and 7.38c

Solution Preparation

Solutions of specific concentration can be prepared by mixing appropriate amounts of solute and solvent or by diluting a more concentrated solution with solvent.

Objective 5, Exercises 7.46 and 7.48b

Solution Stoichiometry

When solution concentrations are expressed as molarities, the mole concept can be applied to reactions taking place between substances that are solutes in the solutions.

Objective 6, Exercise 7.56

Solution Properties

A number of solution properties differ from the properties of pure solvent. Ionic or highly polar solutes that dissociate in solution result in solutions that conduct electricity. Colligative solution properties depend on the concentration of solute particles in the solution and include vapor pressure, boiling point, freezing point, and osmotic pressure.

Objective 7, Exercises 7.64a & c and 7.74

Colloids

Homogeneous mixtures called colloids, or colloidal dispersions, differ from solutions in terms of the size of the dispersed phase particles. In colloids, the particles are large enough to scatter light and thus show the Tyndall effect. Colloids, like solutions, occur in all three physical states, depending primarily on the physical state of the dispersing medium. In true colloids, the suspension is permanent because dispersed particles acquire similar charges by adsorbing ions from the dispersing medium and repel each other or because emulsifying agents keep the dispersed particles from coalescing.

Objective 8, Exercise 7.82

Dialysis

Dialyzing membranes are semipermeable but with pores large enough to allow solvent molecules, hydrated ions, and small molecules to pass through in a process called dialysis. The process is important in the removal of impurities from the blood and is applied artificially for people suffering from kidney malfunction.

Objective 9, Exercise 7.84

Key Terms and Concepts

Colligative property (7.7)
Colloid (7.8)
Concentration (7.4)
Dialysis (7.9)
Dialyzing membrane (7.9)
Dispersed phase (7.8)
Dispersing medium (7.8)
Dissolving (7.1)
Electrolyte (7.7)
Emulsifying agent (stabilizing agent) (7.8)

Hydrated ion (7.3)
Immiscible (7.2)
Insoluble substance (7.2)
Molarity (M) (7.4)
Nonelectrolyte (7.7)
Osmolarity (7.7)
Osmosis (7.7)
Osmotic pressue (7.7)
Percent (7.4)
Saturated solution (7.2)

Solubility (7.2)
Soluble substance (7.2)
Solute (7.1)
Solution (7.1)
Solvent (7.1)
Supersaturated solution (7.2)
Tyndall effect (7.8)
Volume/volume percent (7.4)
Weight/volume percent (7.4)
Weight/weight percent (7.4)

Key Equations

1. Solution concentration in terms of molarity (Section 7.4):	$M = \dfrac{\text{moles of solute}}{\text{liters of solution}}$	Equation 7.5
2. Solution concentration in terms of weight/weight % (Section 7.4):	$\%(w/w) = \dfrac{\text{solute mass}}{\text{solution mass}} \times 100$	Equation 7.6
3. Solution concentration in terms of weight/volume % (Section 7.4):	$\%(w/v) = \dfrac{\text{grams of solute}}{\text{milliliters of solution}} \times 100$	Equation 7.7
4. Solution concentration in terms of volume/volume % (Section 7.4):	$\%(v/v) = \dfrac{\text{solute volume}}{\text{solution volume}} \times 100$	Equation 7.8
5. Dilution of concentrated solution to make less-concentrated solution (Section 7.5):	$(C_c)(V_c) = (C_d)(V_d)$	Equation 7.9
6. Boiling point elevation of a solution (Section 7.7):	$\Delta T_b = nK_bM$	Equation 7.13
7. Freezing point depression of a solution (Section 7.7):	$\Delta T_f = nK_fM$	Equation 7.14
8. Osmotic pressure of a solution (Section 7.7):	$\pi = nMRT$	Equation 7.15

Exercises

Even-numbered exercises are answered in Appendix B.

Blue-numbered exercises are more challenging.

Physical States of Solutions (Section 7.1)

7.1 Many solutions are found in the home. Some are listed below, with the composition as printed on the label. When no percentage is indicated, components are usually given in order of decreasing amount. When water is present, it is often not mentioned on the label or it is included in the inert ingredients. Identify the solvent and solutes of the following solutions:

   a. Antiseptic mouthwash: alcohol 25%, thymol, eucalyptol, methyl salicylate, menthol, benzoic acid, boric acid

   b. Paregoric: alcohol 45%, opium 0.4%

   c. Baby oil: mineral oil, lanolin (there happens to be no water in this solution—why?)

   d. Distilled vinegar: acetic acid 5%

7.2 Many solutions are found in the home. Some are listed below, with the composition as printed on the label. When no percentage is indicated, components are usually given in order of decreasing amount. When water is present, it is often not mentioned on the label or it is included in the inert ingredients. Identify the solvent and solutes of the following solutions:

   a. Liquid laundry bleach: sodium hypochlorite 5.25%, inert ingredients 94.75%

   b. Rubbing alcohol: isopropyl alcohol 70%

   c. Hydrogen peroxide: 3% hydrogen peroxide

   d. Aftershave: SD alcohol, water, glycerin, fragrance, menthol, benzophenone-1, coloring

7.3 Classify the following as being a solution or not a solution. Explain your reasons when you classify one as *not* a solution. For the ones classified as solutions, identify the solvent and solute(s).

   a. Maple syrup

   b. Milk

   c. Eyedrops

   d. Tomato juice

   e. Tap water

7.4 Classify the following as being a solution or not a solution. Explain your reasons when you classify one as *not* a solution. For the ones classified as solutions, identify the solvent and solute(s).

Unless otherwise noted all art on this page © Cengage Learning 2014

a. Foggy air

b. Tears

c. Freshly squeezed orange juice

d. Strained tea

e. Creamy hand lotion

Solubility (Section 7.2)

7.5 Use the term *soluble, insoluble,* or *immiscible* to describe the behavior of the following pairs of substances when they are shaken together:

a. 25 mL of water and 1 g of salt—the resulting mixture is clear and colorless.

b. 25 mL of water and 1 g of solid silver chloride—the resulting mixture is cloudy and solid settles out.

c. 25 mL of water and 5 mL of mineral oil—the resulting mixture is cloudy and gradually separates into two layers.

7.6 Use the term *soluble, insoluble,* or *immiscible* to describe the behavior of the following pairs of substances when they are shaken together:

a. 25 mL of cooking oil and 25 mL of vinegar—the resulting mixture is cloudy and gradually separates into two layers.

b. 25 mL of water and 10 mL of rubbing alcohol—the resulting mixture is clear and colorless.

c. 25 mL of chloroform and 1 g of roofing tar—the resulting mixture is clear but dark brown in color.

7.7 Define the term *miscible.* It is *not* defined in the text.

7.8 Classify the following solutions as unsaturated, saturated, or supersaturated:

a. A solution to which a small piece of solute is added, and it dissolves.

b. A solution to which a small piece of solute is added, and much more solute comes out of solution.

c. The final solution resulting from the process in part b.

7.9 Suppose you put 35.8 g of ammonium sulfate into a flask and add 100 g of water at 0°C. After stirring to dissolve as much solute as possible, will you have a saturated or unsaturated solution? Explain your answer. See Table 7.2.

7.10 Suppose you have a saturated solution that is at room temperature. Discuss how it could be changed into a supersaturated solution without using any additional solute.

7.11 Classify each of the following solutes into the approximate solubility categories of Table 7.3. The numbers in parentheses are the grams of solute that will dissolve in 100 g of water at the temperature indicated.

a. boric acid, H_3BO_3 (6.35 g at 30°C)

b. calcium hydroxide, $Ca(OH)_2$ (5.35 g at 30°C)

c. antimony(III) sulfide, Sb_2S_3 (1.75×10^{-4} g at 18°C)

d. copper(II) chloride, $CuCl_2$ (70.6 g at 0°C)

e. iron(II) bromide, $FeBr_2$ (109 g at 10°C)

7.12 Classify each of the following solutes into the approximate solubility categories of Table 7.3. The numbers in parentheses are the grams of solute that will dissolve in 100 g of water at the temperature indicated.

a. barium nitrate, $Ba(NO_3)_2$ (8.7 g at 20°C)

b. aluminum oxide, Al_2O_3 (9.8×10^{-5} g at 29°C)

c. calcium sulfate, $CaSO_4$ (0.21 g at 30°C)

d. manganese chloride, $MnCl_2$ (72.3 g at 25°C)

e. lead bromide, $PbBr_2$ (0.46 g at 0°C)

The Solution Process (Section 7.3)

7.13 What is the difference between a nonhydrated ion and a hydrated ion? Draw a sketch using the Cl^- ion to help illustrate your answer.

7.14 Suppose you had a sample of white crystalline solid that was a mixture of calcium carbonate ($CaCO_3$) and calcium chloride ($CaCl_2$). Describe how you could treat the sample to isolate one of the solids in the pure state. Which solid would it be?

7.15 Ground-up limestone ($CaCO_3$) is used as a gentle abrasive in some powdered cleansers. Why is this a better choice than ground-up soda ash (Na_2CO_3)?

7.16 Indicate which of the following substances (with geometries as given) would be soluble in water (a polar solvent) and in benzene (a nonpolar solvent):

a.

(tetrahedral)

b. Ne

c.

$$H \underset{\underset{H}{|}}{\overset{\diagup N \diagdown}{}} H$$

(triangular-based pyramid)

d.

$$F \overset{\displaystyle |}{\underset{\diagup \quad \diagdown}{B}} F$$

(flat triangle)

7.17 Indicate which of the following substances (with geometries as given) would be soluble in water (a polar solvent) and in benzene (a nonpolar solvent):

a. $H - S \diagdown H$

b. HCl

c. $$H \diagup \overset{O - O}{} \diagdown H$$

d. N≡N

7.18 Freons are compounds formerly used in a variety of ways. Explain why Freon-114 was useful as a degreasing agent. The molecular structure is

7.19 Suppose you put a piece of a solid into a beaker that contains water and stir the mixture briefly. You find that the solid does not immediately dissolve completely. Describe three things you might do to try to get the solid to dissolve.

Solution Concentrations (Section 7.4)

7.20 Calculate the molarity of the following solutions:

 a. 1.25 L of solution contains 0.455 mol of solute.

 b. 250 mL of solution contains 0.215 mol of solute.

 c. 0.175 mol of solute is put into a container and enough distilled water is added to give 100 mL of solution.

7.21 Calculate the molarity of the following solutions:

 a. 2.00 L of solution that contains 0.860 mol of solute.

 b. 500 mL of solution that contains 0.304 mol of solute.

 c. 0.115 mol of solute is put into a container and enough distilled water is added to give 250 mL of solution.

7.22 Calculate the molarity of the following solutions:

 a. A sample of solid KBr weighing 11.9 g is put in enough distilled water to give 200 mL of solution.

 b. A 14.2-g sample of solid Na_2SO_4 is dissolved in enough water to give 500 mL of solution.

 c. A 10.0-mL sample of solution is evaporated to dryness and leaves 0.29 g of solid residue that is identified as Li_2SO_4.

7.23 Calculate the molarity of the following solutions:

 a. A sample of solid Na_2SO_4 weighing 0.150 g is dissolved in enough water to make 10.0 mL of solution.

 b. A 6.50-g sample of glucose ($C_6H_{12}O_6$) is dissolved in enough water to give 150 mL of solution.

 c. A 43.5-g sample of K_2SO_4 is dissolved in a quantity of water, and the solution is stirred well. A 25.0-mL sample of the resulting solution is evaporated to dryness and leaves behind 2.18 g of solid K_2SO_4.

7.24 Calculate:

 a. How many moles of solute is contained in 1.75 L of 0.215 M solution?

 b. How many moles of solute is contained in 250 mL of 0.300 M solution?

 c. How many mL of 0.350 M solution contains 0.200 mol of solute?

7.25 Calculate:

 a. How many moles of solute is contained in 2.50 L of 0.350 M solution?

 b. How many moles of solute is contained in 250 mL of 0.750 M solution?

 c. How many mL of 0.415 M solution contains 0.250 mol of solute?

7.26 Calculate:

 a. How many grams of solid would be left behind if 25.0 mL of 0.400 M NaCl solution was evaporated to dryness?

 b. How many liters of 0.255 M HCl solution is needed to provide 0.0400 mol of HCl?

 c. How many mL of 0.500 M $NaNO_3$ solution contains 50.0 g of solute?

7.27 Calculate:

 a. How many grams of solid $AgNO_3$ will be needed to prepare 200 mL of a 0.200 M solution?

 b. How many grams of vitamin C ($C_6H_8O_6$) would be contained in 25.0 mL of a 1.00 M solution?

 c. How many moles of HCl is contained in 250 mL of a 6.0 M solution?

7.28 Calculate the concentration in %(w/w) of the following solutions. Assume water has a density of 1.00 g/mL.

 a. 5.3 g of sugar and 100 mL of water

 b. 5.3 g of any solute and 100 mL of water

 c. 5.3 g of any solute and 100 g of any solvent

7.29 Calculate the concentration in %(w/w) of the following solutions. Assume water has a density of 1.00 g/mL.

 a. 7.5 g of table salt and 100 mL of water

 b. 7.5 g of any solute and 100 mL of water

 c. 7.5 g of any solute and 100 g of any solvent

7.30 Calculate the concentration in %(w/w) of the following solutions. Assume water has a density of 1.00 g/mL.

 a. 15.0 g of salt is dissolved in 200 mL of water.

 b. 0.150 mol of solid NH_4Cl is dissolved in 150 mL of water.

 c. 75.0 g of solid is dissolved in 500 mL of water.

 d. 20.0 mL of ethyl alcohol (density = 0.789 g/mL) is mixed with 15.0 mL of water.

7.31 Calculate the concentration in %(w/w) of the following solutions. Assume water has a density of 1.00 g/mL.

 a. 5.20 g of $CaCl_2$ is dissolved in 125 mL of water.

 b. 0.200 mol of solid KBr is dissolved in 200 mL of water.

 c. 50.0 g of solid is dissolved in 250 mL of water.

 d. 10.0 mL of ethyl alcohol (density = 0.789 g/mL) is mixed with 10.0 mL of ethylene glycol (density = 1.11 g/mL).

7.32 Calculate the concentration in %(w/w) of the following solutions:

 a. 15.0 g of solute is dissolved in enough water to give 100 mL of solution. The density of the resulting solution is 1.20 g/mL.

 b. A 10.0-mL solution sample with a density of 1.15 g/mL leaves 1.54 g of solid residue when evaporated.

 c. A 50.0-g sample of solution on evaporation leaves a 3.12-g residue of $MgCl_2$.

7.33 Calculate the concentration in %(w/w) of the following solutions:

 a. 424 g of solute is dissolved in enough water to give 1.00 L of solution. The density of the resulting solution is 1.18 g/mL.

 b. A 50.0-mL solution sample with a density of 0.898 g/mL leaves 12.6 g of solid residue when evaporated.

 c. A 25.0-g sample of solution on evaporation leaves a 2.32-g residue of NH_4Cl.

Even-numbered exercises answered in Appendix B.
248 Blue-numbered exercises are more challenging.

Unless otherwise noted all art on this page © Cengage Learning 2014

7.34 Calculate the concentration in %(v/v) of the following solutions:

   a. 150 mL of solution contains 12.0 mL of alcohol.

   b. 150 mL of solution contains 12.0 mL of any soluble liquid solute.

   c. 6.0 fluid ounces of oil is added to 2.0 gallons (256 fluid ounces) of gasoline.

   d. A solution of alcohol and water is separated by distillation. A 150-mL solution sample gives 51.3 mL of alcohol.

7.35 Calculate the concentration in %(v/v) of the following solutions:

   a. 250 mL of solution contains 15.0 mL of acetone.

   b. 250 mL of solution contains 15.0 mL of any soluble liquid solute.

   c. 1.0 quart of acetic acid is put into a 5-gallon container, and enough water is added to fill the container.

   d. A solution of acetone and water is separated by distillation. A 300-mL sample gives 109 mL of acetone.

7.36 Consider the blood volume of an adult to be 5.0 L. A blood alcohol level of 0.50% (v/v) can cause a coma. What volume of pure ethyl alcohol, if consumed in one long drink and assumed to be absorbed completely into the blood, would result in this critical blood alcohol level?

7.37 The blood serum acetone level for a person is determined to be 1.8 mg of acetone per 100 mL of serum. Express this concentration as %(v/v) if liquid acetone has a density of 0.79 g/mL.

7.38 Calculate the concentration in %(w/v) of the following solutions:

   a. 150 mL of solution contains 7.50 g of dissolved solid Na_2SO_4.

   b. 150 mL of solution contains 7.50 g of any dissolved solid solute.

   c. 350 mL of solution contains 30.7 g of dissolved solid solute.

7.39 Calculate the concentration in %(w/v) of the following solutions:

   a. 28.0 g of solute is dissolved in 200 mL of water to give a solution with a density of 1.10 g/mL.

   b. A 25.0-mL solution sample on evaporation leaves a solid residue of 0.38 g.

   c. On analysis for total protein, a blood serum sample of 15.0 mL is found to contain 1.02 g of total protein.

7.40 A saturated solution of KBr in water is formed at 20.0°C. Consult Figure 7.3 and calculate the concentration of the solution in %(w/w).

7.41 Assume the density of the solution prepared in Exercise 7.40 is 1.18 g/mL and express the concentration in %(w/v).

Solution Preparation (Section 7.5)

7.42 Explain how you would prepare the following solutions using pure solute and water. Assume water has a density of 1.00 g/mL.

   a. 200 mL of 0.150 M Na_2SO_4 solution

   b. 250 mL of 0.250 M $Zn(NO_3)_2$ solution

   c. 150 g of 2.25% (w/w) NaCl solution

   d. 125 mL of 0.75% (w/v) KCl solution

7.43 Explain how you would prepare the following solutions using pure solute and water. Assume water has a density of 1.00 g/mL.

   a. 500 mL of a 2.00 M NaOH solution

   b. 250 mL of a 40.0% (v/v) alcohol solution (C_2H_5OH)

   c. 100 mL of a 10.0% (w/v) glycerol solution. Glycerol is a liquid with a density of 1.26 g/mL. Describe two ways to measure out the amount of glycerol needed.

   d. Approximately 100 mL of a normal saline solution, 0.89% (w/w) NaCl

7.44 A solution is prepared by mixing 45.0 g of water and 15.0 g of ethyl alcohol. The resulting solution has a density of 0.952 g/mL. Express the solution concentration in %(w/w) ethyl alcohol and %(w/v) ethyl alcohol.

7.45 Calculate the following:

   a. The number of moles of NaI in 50.0 mL of 0.250 M solution

   b. The number of grams of KBr in 120 mL of 0.500 M solution

   c. The number of grams of NaCl in 20.0 mL of 2.00% (w/v) NaCl solution

   d. The number of milliliters of alcohol in 250 mL of 15.0% (v/v) solution

7.46 Calculate the following:

   a. The number of grams of Li_2CO_3 in 250 mL of 1.75 M Li_2CO_3 solution

   b. The number of moles of NH_3 in 200 mL of 3.50 M NH_3 solution

   c. The number of mL of alcohol in 250 mL of 12.5% (v/v) solution

   d. The number of grams of $CaCl_2$ in 50.0 mL of 4.20% (w/v) $CaCl_2$ solution

7.47 Explain how you would prepare the following dilute solutions from the more concentrated ones:

   a. 200 mL of 0.500 M HCl from a 6.00 M HCl solution

   b. 50 mL of 2.00 M H_2SO_4 from a 6.00 M H_2SO_4 solution

   c. 100 mL of normal saline solution, 0.89% (w/v) NaCl, from a 5.0% (w/v) NaCl solution

   d. 250 mL of 5.00% (v/v) acetone from 20.5% (v/v) acetone

7.48 Explain how you would prepare the following dilute solutions from the more concentrated ones:

   a. 5.00 L of 6.00 M H_2SO_4 from a 18.0 M H_2SO_4 solution

   b. 250 mL of 0.500 M $CaCl_2$ from a 3.00 M $CaCl_2$ solution

   c. 200 mL of 1.50% (w/v) KBr from a 10.0% (w/v) KBr solution

   d. 500 mL of 10.0% (v/v) alcohol from 50.0% (v/v) alcohol

7.49 What is the molarity of the solution prepared by diluting 25.0 mL of 0.412 M $Mg(NO_3)_2$ solution to each of the following final volumes?

   a. 40.0 mL

   b. 100 mL

   c. 1.10 L

   d. 350 mL

7.50 What is the molarity of the solution prepared by diluting 100 mL of 0.225 M KBr solution to each of the following volumes?

   a. 2.00 L

   b. 225 mL

   c. 600 mL

   d. 850 mL

Solution Stoichiometry (Section 7.6)

7.51 How many milliliters of 6.00 M HCl solution would be needed to react exactly with 20.0 g of pure solid NaOH?

$$HCl(aq) + NaOH(s) \rightarrow NaCl(aq) + H_2O(\ell)$$

7.52 How many grams of solid Na_2CO_3 will exactly react with 250 mL of a 1.25 M HCl solution?

$$Na_2CO_3(s) + 2HCl(aq) \rightarrow 2NaCl(aq) + CO_2(g) + H_2O(\ell)$$

7.53 How many milliliters of 0.250 M HCl would be needed to react exactly with 10.5 g of solid $NaHCO_3$?

$$NaHCO_3(s) + HCl(aq) \rightarrow NaCl(aq) + CO_2(g) + H_2O(\ell)$$

7.54 How many milliliters of 0.250 M $AgNO_3$ solution will exactly react with 25.0 mL of a 0.200 M NaCl solution?

$$NaCl(aq) + AgNO_3(aq) \rightarrow NaNO_3(aq) + AgCl(s)$$

7.55 How many milliliters of 0.115 M Na_2S solution will exactly react with 35.0 mL of a 0.150 M $AgNO_3$ solution?

$$2AgNO_3(aq) + Na_2S(aq) \rightarrow Ag_2S(s) + 2NaNO_3(aq)$$

7.56 How many milliliters of 0.225 M NH_3 solution will exactly react with 30.0 mL of 0.190 M H_2SO_4 solution?

$$2NH_3(aq) + H_2SO_4(aq) \rightarrow (NH_4)_2SO_4(aq)$$

7.57 How many milliliters of 0.124 M NaOH solution will exactly react with 35.0 mL of 0.210 M H_3PO_4 solution?

$$3NaOH(aq) + H_3PO_4(aq) \rightarrow Na_3PO_4(aq) + 3H_2O(\ell)$$

7.58 How many milliliters of 0.135 M NaOH solution will exactly react with 30.0 mL of 0.125 M HCl solution?

$$NaOH(aq) + HCl(aq) \rightarrow NaCl(aq) + H_2O(\ell)$$

7.59 How many milliliters of 0.124 M NaOH solution will exactly react with 25.0 mL of a 0.210 M H_2SO_4 solution?

$$2NaOH(aq) + H_2SO_4(aq) \rightarrow Na_2SO_4(aq) + 2H_2O(\ell)$$

7.60 Stomach acid is essentially 0.10 M HCl. An active ingredient found in a number of popular antacids is calcium carbonate, $CaCO_3$. Calculate the number of grams of $CaCO_3$ needed to exactly react with 250 mL of stomach acid.

$$CaCO_3(s) + 2HCl(aq) \rightarrow CO_2(g) + CaCl_2(aq) + H_2O(\ell)$$

7.61 An ingredient found in some antacids is magnesium hydroxide, $Mg(OH)_2$. Calculate the number of grams of $Mg(OH)_2$ needed to exactly react with 250 mL of stomach acid (see Exercise 7.60).

$$Mg(OH)_2(s) + 2HCl(aq) \rightarrow MgCl_2(aq) + 2H_2O(\ell)$$

Solution Properties (Section 7.7)

7.62 Before it is frozen, ice cream is essentially a solution of sugar, flavorings, etc., dissolved in water. Use the idea of colligative solution properties and explain why a mixture of ice (and water) and salt is used to freeze homemade ice cream. Why won't just ice work?

7.63 If you look at the labels of automotive products used to prevent radiator freezing (antifreeze) and radiator boiling, you will find the same ingredient listed, ethylene glycol. Use the idea of colligative properties to explain how the same material can prevent an automobile cooling system from freezing and boiling.

7.64 Calculate the boiling and freezing points of water solutions that are 1.50 M in the following solutes:

   a. KCl, a strong electrolyte

   b. glycerol, a nonelectrolyte

   c. $(NH_4)_2SO_4$, a strong electrolyte

   d. $Al(NO_3)_3$, a strong electrolyte

7.65 Calculate the boiling and freezing points of water solutions that are 1.25 M in the following solutes:

   a. KBr, a strong electrolyte

   b. ethylene glycol, a nonelectrolyte

   c. $(NH_4)_2CO_3$, a strong electrolyte

   d. $Al_2(SO_4)_3$, a strong electrolyte

7.66 Calculate the boiling and freezing points of the following solutions. Water is the solvent unless otherwise indicated.

   a. A 0.750 M solution of urea, a nonelectrolyte

   b. A 0.150 M solution of $CaCl_2$, a strong electrolyte

   c. A solution containing 120 g of ethylene glycol $(C_2H_6O_2)$ per 400 mL of solution

7.67 Calculate the boiling and freezing points of the following solutions. Water is the solvent, unless otherwise indicated.

   a. A solution containing 50.0 g of H_2SO_4, a strong electrolyte (both Hs dissociate), per 250 mL

   b. A solution containing 200 g of table sugar $(C_{12}H_{22}O_{11})$, a nonelectrolyte, per 250 mL

   c. A solution containing 75.0 g of octanoic acid $(C_8H_{16}O_2)$, a nonelectrolyte, in enough benzene to give 250 mL of solution

7.68 Calculate the osmolarity for the following solutions:

   a. A 0.20 M solution of glycerol, a nonelectrolyte

   b. A 0.20 M solution of $(NH_4)_2SO_4$, a strong electrolyte

   c. A solution containing 33.7 g of LiCl (a strong electrolyte) per 500 mL

7.69 Calculate the osmolarity for the following solutions:

   a. A 0.25 M solution of KCl, a strong electrolyte

   b. A solution containing 15.0 g of urea (CH_4N_2O), a nonelectrolyte, per 500 mL

   c. A solution containing 50.0 mL of ethylene glycol $(C_2H_6O_2)$, a nonelectrolyte with a density of 1.11 g/mL, per 250 mL

NOTE: In Exercises 7.70–7.79, assume the temperature is 25.0°C, and express your answer in torr, mmHg, and atm.

7.70 Calculate the osmotic pressure of any solution with an osmolarity of 0.250.

7.71 Calculate the osmotic pressure of any solution with an osmolarity of 0.200.

7.72 Calculate the osmotic pressure of a 0.200 M solution of Na_2SO_4, a strong electrolyte.

7.73 Calculate the osmotic pressure of a 0.300 M solution of methanol, a nonelectrolyte.

7.74 Calculate the osmotic pressure of a solution that contains 95.0 g of the nonelectrolyte urea, CH_4N_2O, per 500 mL of solution.

7.75 Calculate the osmotic pressure of a solution that contains 1.50 mol of $CaCl_2$ in 1500 mL.

7.76 Calculate the osmotic pressure of a solution that has a freezing point of 20.35°C.

7.77 Calculate the osmotic pressure of a solution that is 0.122 M in solute and has a boiling point 0.19°C above that of pure water.

7.78 Calculate the osmotic pressure of a solution that contains 5.30 g of NaCl and 8.20 g of KCl per 750 mL.

7.79 Calculate the osmotic pressure of a solution that contains 245.0 g of ethylene glycol ($C_2H_6O_2$), a nonelectrolyte, per liter.

7.80 Suppose an osmotic membrane separates a 5.00% sugar solution from a 10.0% sugar solution. In which direction will water flow? Which solution will become diluted as osmosis takes place?

Colloids (Section 7.8)

7.81 Explain how the following behave in a colloidal suspension: dispersing medium, dispersed phase, and colloid emulsifying agent.

7.82 Explain why detergents or soaps are needed if water is to be used as a solvent for cleaning clothes and dishes.

Dialysis (Section 7.9)

7.83 Suppose you have a bag made of a membrane like that in Figure 7.17. Inside the bag is a solution containing water and dissolved small molecules. Describe the behavior of the system when the bag functions as an osmotic membrane and when it functions as a dialysis membrane.

7.84 Each of the following mixtures was placed in a dialysis bag like the one shown in Figure 7.17. The bag was immersed in pure water. Which substances will pass through the bag into the surrounding water?

   a. NaCl solution and starch (colloid)

   b. Urea (small organic molecule), and starch (colloid)

   c. Albumin (colloid), KCl solution, and glucose solution (small organic molecule)

Additional Exercises

7.85 When 5.0 mL of water and 5.0 mL of rubbing alcohol are mixed together, the volume of the resulting solution is 9.7 mL. Explain in terms of molecules why the final solution volume is not 10.0 mL.

7.86 A student made a 7.00% (w/v) solution by dissolving 7.00 grams of sodium chloride (salt) in an appropriate volume of water. Assume the density of the water was 1.00 g/mL, and that there was no increase in volume when the salt was added to the water. Calculate the density of the resulting solution in g/mL.

7.87 Ethylene glycol is mixed with water and used as an antifreeze in automobile cooling systems. If a 50% (v/v) solution is made using pure ethylene glycol and pure water, it will have a freezing point of about 265°C. What will happen to the freezing point of the mixture if ethylene glycol is added until the concentration is 70% (v/v) ethylene glycol? Explain your answer.

7.88 How many mL of 1.50 M HCl solution contains enough HCl to react completely with 0.500 g of zinc metal? The reaction is $Zn(s) + 2HCl(aq) \rightarrow ZnCl_2(aq) + H_2(g)$.

7.89 At 20°C ethyl alcohol has a vapor pressure of 43.9 torr, and the vapor pressure of water at the same temperature is 17.5 torr. Suppose a 50% (v/v) solution of water and ethyl alcohol was brought to a boil and the first vapors given off were collected and condensed to a liquid. How would the percentage of alcohol in the condensed liquid compare to the percentage of alcohol in the original 50% solution (greater than, less than, equal to)? Explain your reasoning.

Allied Health Exam Connection

The following questions are from these sources:

1. *Nursing School Entrance Exam* © 2005, Learning Express, LLC.

2. *McGraw-Hill's Nursing School Entrance Exams* by Thomas A. Evangelist, Tamara B. Orr and Judy Unrein © 2009, The McGraw-Hill Companies, Inc.

3. *NSEE Nursing School Entrance Exams,* 3rd Edition © 2009, Kaplan Publishing.

4. *Cliffs Test Prep: Nursing School Entrance Exams* by Fred N. Grayson © 2004, Wiley Publishing, Inc.

5. *Peterson's Master the Nursing School and Allied Health Entrance Exams,* 18th Edition by Marion F. Gooding © 2008, Peterson's, a Nelnet Company.

7.90 If a salt is added to water, which of the following is likely to occur?

   a. The boiling point will increase and the freezing point will decrease.

   b. The boiling point will increase and the freezing point will increase.

   c. The boiling point will decrease and the freezing point will decrease.

   d. The boiling point will decrease and the freezing point will increase.

7.91 A cell is in a solution in which the concentration of solutes is higher inside the cell than outside the cell. The cell will likely:

   a. swell up and possibly burst

   b. shrivel and shrink

   c. maintain its size

   d. grow a cell wall for support

7.92 What will happen if a semipermeable membrane is placed between two different concentrations of a NaCl solution?

   a. The solute will move toward the higher concentration.

   b. The solute will move toward the lower concentration.

   c. The solvent will move toward the higher concentration.

   d. The solvent will move toward the lower concentration.

7.93 Given a sample of $C_6H_{12}O_6$(aq), which of the following is true?

  a. The glucose is the solvent and water is the solute.

  b. The glucose is the solvent and water is the solvent.

  c. The glucose is the solute and water is the solute.

  d. The glucose is the solute and water is the solvent.

7.94 Which one of the following compounds is a nonelectrolyte when dissolved in water?

  a. KOH

  b. NH_3

  c. NaBr

  d. $CaCl_2$

7.95 An example of a strong electrolyte is:

  a. sugar

  b. calcium chloride

  c. glycerin

  d. boric acid

7.96 A solution that contains all the solute it can normally dissolve at a given temperature must be:

  a. concentrated

  b. supersaturated

  c. saturated

  d. unsaturated

7.97 Oil and water are immiscible (do not mix) because:

  a. oil is polar and water is polar

  b. oil is nonpolar and water is polar

  c. water is nonpolar and oil is polar

  d. water is nonpolar and oil is nonpolar

7.98 Cells that contain more dissolved salts and sugars than the surrounding solution are called:

  a. isotonic

  b. hypertonic

  c. hypotonic

  d. osmosis

7.99 How many grams of NaOH would be needed to make 250 ml of a 0.200 M solution? (molecular weight of NaOH = 40.0)

  a. 8.00 g

  b. 4.00 g

  c. 2.00 g

  d. 2.50 g

7.100 Ice can be melted most effectively by _____ if 1 mole is used.

  a. sucrose

  b. calcium chloride

  c. sodium chloride

  d. methanol

7.101 Jack has 100 mL of a 12 molar solution of sulfuric acid. How much of it should he put into a graduated cylinder to make 20 mL of a 1.2 molar solution?

  a. 1 mL

  b. 2 mL

  c. 10 mL

  d. 12 mL

7.102 How many grams of sugar are needed to make 500 mL of a 5% (weight/volume) solution of sugar?

  a. 20

  b. 25

  c. 50

  d. 10

7.103 As water is evaporated from a solution, the concentration of the solute in the solution will:

  a. increase

  b. decrease

  c. remain the same

7.104 In a dilute solution of sodium chloride in water, the sodium chloride is the:

  a. solvent

  b. solute

  c. precipitate

  d. reactant

7.105 A salt solution has a molarity of 1.5 M. How many moles of this salt are present in 2.0 L of this solution?

  a. 1.5 moles

  b. 2.0 moles

  c. 3.0 moles

  d. 0.75 moles

7.106 If 58.5 g of NaCl (1 mole of NaCl) are dissolved in enough water to make 0.500 L of solution, what is the molarity of this solution?

  a. 2.0 M

  b. 11.7 M

  c. 1.0 M

  d. The answer cannot be determined from the information above.

7.107 To prepare 100 mL of a 0.20 M NaCl solution from a stock solution of 1.00 M NaCl, you should mix:

  a. 20 mL of stock solution with 80 mL of water

  b. 40 mL of stock solution with 60 mL of water

  c. 20 mL of stock solution with 100 mL of water

  d. 25 mL of stock solution with 75 mL of water

7.108 If a 2.0 M solution is diluted to 0.5 M, and the final volume is 100 ml, what was the original volume?

  a. 400 mL

  b. 200 mL

c. 50 mL

d. 25 mL

7.109 The number of moles of NaCl in 250 mL of a 0.300 M solution of NaCl is:

a. 0.0750

b. 0.150

c. 0.250

d. 1.15

7.110 One liter of solution is made by dissolving 29.2 g of NaCl in water. What is the molarity of the solution?

a. 0.5 M

b. 2.0 M

c. 1.3 M

d. 0.82 M

7.111 If a red blood cell is placed in sea water, it will be in what kind of solution?

a. isotonic

b. hypotonic

c. hypertonic

d. facilitated diffusion

7.112 When placed in distilled water, a human red blood cell:

a. shrivels up

b. neither shrinks nor swells

c. takes up more salts to balance all concentrations

d. swells to a larger size

7.113 The movement of substances from a lesser concentration to a higher concentration is called:

a. osmosis

b. diffusion

c. active transport

d. pinocytosis

7.114 Primary intermolecular interactions between a K cation and H_2O molecules are:

a. hydrogen bonds

b. dipole-dipole interactions

c. ion-dipole interactions

d. London forces

7.115 All of the following are colligative properties of solutions EXCEPT:

a. vapor pressure

b. osmotic pressure

c. density

d. boiling point elevation

Chemistry for Thought

7.116 When a patient has blood cleansed by hemodialysis, the blood is circulated through dialysis tubing submerged in a bath that contains the following solutes in water: 0.6% NaCl, 0.04% KCl, 0.2% $NaHCO_3$, and 0.72% glucose (all percentages are w/v). Suggest one or more reasons why the dialysis tubing is not submerged in pure water.

7.117 Can the terms *saturated* and *supersaturated* be used to describe solutions made of liquids that are soluble in all proportions? Explain.

7.118 Refer to Figure 7.3 and propose a reason why fish sometimes die when the temperature of the water in which they live increases.

7.119 Small souvenir salt-covered objects are made by forming the object out of wire mesh and suspending the mesh object in a container of water from a salt lake such as the Dead Sea or Great Salt Lake. As the water evaporates, the wire mesh becomes coated with salt crystals. Describe this process using the key terms introduced in Section 7.2.

7.120 Refer to Figure 7.6 and answer the question. How would the solubility of sugar compare in equal amounts of hot and iced tea?

7.121 Refer to Figure 7.11 and explain the process as requested. Draw simple diagrams showing the initial appearance and appearance after some time for a similar experiment in which the two liquids are 0.20 M copper sulfate solution and 2.0 M copper sulfate solution. Explain your reasoning.

7.122 Strips of fresh meat can be preserved by drying. In one process, the strips are coated with table salt and exposed to the air. Use a process discussed in this chapter and one discussed in Chapter 6 to explain how the drying takes place.

8 Reaction Rates and Equilibrium

© Evgeniy Ayupov/Shutterstock.com

Case Study

Joy wanted to get the kids out of the house on this beautiful end-of-summer day. Stopping at the gas station on the way to the park, she bought corn dogs. Her toddler, Morgan, was eating his corn dog in the stroller while Joy pushed her daughter on the swing. Suddenly, Morgan screamed. Several yellow jackets were flying around him. Apparently, the insects were fond of corn dogs as well! Morgan had been stung on the lip. Joy scraped off the stinger and comforted her son, but almost immediately his face began to swell. They went home and Joy gave Morgan some liquid antihistamine. Half of his face was swollen and his eye was nearly shut and so Joy took Morgan to the clinic. The physician's assistant (PA) examined Morgan and said that Joy had done the right thing, but that her son seemed to be sensitive to yellow jacket stings. The PA added that it would be wise to have a fast-acting treatment on hand because stings can be dangerous to sensitive individuals. He wrote a prescription for an epi-pen (epinephrine injection) and some cortisone cream.

When is it appropriate to use an epinephrine injection? How can someone distinguish between a mild allergic reaction and a dangerous one?

Follow-up to this Case Study appears at the end the chapter before the Concept Summary.

Learning Objectives

When you have completed your study of this chapter, you should be able to:

1 Use the concepts of energy and entropy to predict the spontaneity of processes and reactions. **(Section 8.1)**

2 Calculate reaction rates from experimental data. **(Section 8.2)**

3 Use the concept of molecular collisions to explain reaction characteristics. **(Section 8.3)**

4 Represent and interpret the energy relationships for reactions by using energy diagrams. **(Section 8.4)**

5 Explain how factors such as reactant concentrations, temperature, and catalysts influence reaction rates. **(Section 8.5)**

6 Relate experimental observations to the establishment of equilibrium. **(Section 8.6)**

7 Write equilibrium expressions based on reaction equations, and do calculations based on equilibrium expressions. **(Section 8.7)**

8 Use Le Châtelier's principle to predict the influence of changes in concentration and reaction temperature on the position of equilibrium for a reaction. **(Section 8.8)**

According to calculations, carbon in the form of a diamond will spontaneously change into graphite. Owners of diamonds seem unconcerned by this fact; they know from experience that their investments in gems are secure. In reality, the change does take place, but so slowly that it is not detectable over many human lifetimes. The low rate of the reaction makes the difference.

The element barium is very poisonous when it is taken into the body as barium ions (Ba^{2+}). When barium sulfate ($BaSO_4$) dissolves, Ba^{2+} and SO_4^{2-} ions are formed. However, suspensions of solid $BaSO_4$ are routinely swallowed by patients undergoing diagnostic X-ray photography of the stomach and intestinal tract. The patients are not affected because very little $BaSO_4$ dissolves in the body. This lack of solubility is an equilibrium property of $BaSO_4$.

8.1 Spontaneous and Nonspontaneous Processes

Learning Objective

1. Use the concepts of energy and entropy to predict the spontaneity of processes and reactions.

---

spontaneous process A process that takes place naturally with no apparent cause or stimulus.

Years ago, tourists in Salt Lake City visited "gravity hill," where an optical illusion made a stream of water appear to flow uphill. The fascination with the scene came from the apparent violation of natural laws—everyone knew that water does not flow uphill. Processes that take place naturally with no apparent cause or stimulus are called **spontaneous processes.** Nonspontaneous processes take place only as the result of some cause or stimulus. For example, imagine you are in the positions depicted in ❱ Figure 8.1 and want to move a boulder.

Figure 8.1 The problem of moving a boulder.

A   Spontaneous process

B   Spontaneous process once started

C   Nonspontaneous process

In Ⓐ, the boulder will roll down the hill as soon as you release it; the process begins and takes place spontaneously. In Ⓑ, you must push the boulder a little to get it over the hump, but once started, the boulder spontaneously rolls downhill. Situation Ⓒ is very different. You must continually push on the boulder, or it will not move up the hill; at no time can you stop pushing and expect the boulder to continue moving up. The process is nonspontaneous; it takes place only because of the continuous application of an external stimulus.

exergonic process A process that gives up energy as it takes place.

endergonic process A process that gains or accepts energy as it takes place.

As the boulder rolls downhill, it gives up energy; it moves from a state of high potential energy to a lower one. As you push the boulder uphill, it gains energy (from you). Processes that give up energy are called **exergonic** (energy out), whereas those that gain energy are called **endergonic** (energy in). Very often, energy changes in chemical processes involve heat. Those changes that do are referred to as either exothermic (heat out) or endothermic (heat in), two terms you have encountered before (Sections 5.8 and 6.11).

entropy A measurement or indication of the disorder or randomness of a system. The more disorderly a system is, the higher its entropy.

Many spontaneous processes give up energy. A piece of wood, once ignited, spontaneously burns and liberates energy as heat and light. At normal room temperature, steam spontaneously condenses to water and releases heat. However, some spontaneous processes take place and either give up no energy or actually gain energy. These spontaneous processes are always accompanied by another change called an entropy increase. **Entropy** describes the disorder or mixed-up character (randomness) of a system (see ❱ Figure 8.2). Thus, an entropy increase accompanies all spontaneous processes in which energy remains constant or is gained. A drop of ink placed in a glass of water will eventually become uniformly distributed throughout the water, even though the water is not stirred. No energy change accompanies the diffusion of the ink, but the distribution of ink throughout the water is a more disorderly (higher-entropy) state than when the ink is all together in a single drop. Ice melts spontaneously at 20°C and in the process absorbs heat. The process takes place because the random distribution of water molecules in the liquid is a higher-entropy state than the orderly molecular arrangement found in crystalline ice.

Energy and entropy changes influence the spontaneity of processes in several ways:

1. A process will always be spontaneous if the energy decreases and the entropy increases. An example of such a process is the burning of a piece of wood in which heat is given up and the gaseous products of combustion are distributed throughout the surroundings, providing the entropy increase.
2. When a spontaneous process is accompanied by an energy increase, a large entropy increase must also occur. Thus, when ice spontaneously melts at 20°C, the increase in entropy is large enough to compensate for the increase in energy that also takes place.
3. A spontaneous process accompanied by an entropy decrease must also be accompanied by a compensating energy decrease. Thus, when water spontaneously freezes at 0°C, the energy loss compensates for the entropy decrease that occurs as the water molecules assume the well-ordered arrangement in ice.

It is useful to think of energy and entropy changes in terms of the directions favoring spontaneity and the directions not favoring spontaneity. It is apparent from the preceding discussion that spontaneity is favored by an energy decrease and/or an entropy increase. Processes in which one of these factors changes in a nonspontaneous direction will be spontaneous only if the other factor changes in a spontaneous direction by a large enough amount to compensate for the nonspontaneous change. The influences listed under 2 and 3 above are examples of this.

Substances that do not undergo spontaneous changes are said to be **stable.** However, stability depends on the surrounding conditions, and a change in those conditions might cause a nonspontaneous process to become spontaneous. Ice is a stable solid at -10°C and 760 torr pressure but spontaneously melts to a liquid at 5°C and 760 torr. Wood is stable at room temperature but spontaneously burns when its temperature equals or exceeds the ignition temperature. In the following discussions, the surrounding conditions are assumed to be those normally encountered. Otherwise, the differences are specified.

stable substance A substance that does not undergo spontaneous changes under the surrounding conditions.

8.2 Reaction Rates

Learning Objective

2. Calculate reaction rates from experimental data.

The speed of a reaction is called the **reaction rate.** It is determined experimentally as the change in concentration of a reactant or product divided by the time required for the change to occur. This average rate is represented for changes in product concentration by Equation 8.1:

reaction rate The speed of a reaction.

$$\text{Rate} = \frac{\Delta C}{\Delta t} = \frac{C_t - C_0}{\Delta t} \tag{8.1}$$

where the delta symbol (Δ) stands for change. C_t and C_0 are the concentrations of a product at the end and beginning, respectively, of the measured time change, Δt. The time can be measured in any convenient unit.

▶ Example 8.1

The gases NO_2 and CO react as follows:

$$NO_2(g) + CO(g) \rightarrow NO(g) + CO_2(g)$$

Calculate the average rate of the reaction if pure NO_2 and CO are mixed and after 50.0 seconds the concentration of CO_2 is found to be 0.0160 mol/L.

Solution

Because the reaction was started by mixing pure NO_2 and CO, the initial concentration C_0 of CO_2 was 0.000 mol/L. After 50.0 seconds, the concentration C_t of CO_2 was 0.0160 mol/L.

$$\text{Rate} = \frac{C_t - C_0}{\Delta t} = \frac{0.0160 \text{ mol/L} - 0.000 \text{ mol/L}}{50.0 \text{ s}} = 3.2 \times 10^{-4} \frac{\text{mol/L}}{\text{s}}$$

The answer is read 3.2×10^{-4} mole per liter per second. It means that during the 50.0-second time interval, an average of 3.2×10^{-4} mol CO_2 was formed in each liter of reacting mixture each second.

▶ LEARNING CHECK 8.1 The Ce^{4+} and Fe^{2+} ions react in solution as follows:

$$Ce^{4+}(aq) + Fe^{2+}(aq) \rightarrow Ce^{3+}(aq) + Fe^{3+}(aq)$$

Solutions of Ce^{4+} and Fe^{2+} are mixed and allowed to react. After 75.0 seconds, the concentration of Ce^{3+} is found to be 1.50×10^{-5} mol/L. Calculate the average rate of the reaction.

8.3 Molecular Collisions

Learning Objective

3. Use the concept of molecular collisions to explain reaction characteristics.

Chemical equations such as those in Example 8.1 and Learning Check 8.1 are useful in identifying reactants and products and in representing the stoichiometry of a reaction. However, they indicate nothing about how a reaction takes place—how reactants get together (or break apart) to form products. The explanation of how a reaction occurs is called a **reaction mechanism.** Reaction mechanisms are not discussed much in this book, but most are based on the following assumptions:

reaction mechanism A detailed explanation of how a reaction actually takes place.

1. Reactant particles must collide with one another in order for a reaction to occur.
2. Particles must collide with at least a certain minimum total amount of energy if the collision is to result in a reaction.
3. In some cases, colliding reactants must be oriented in a specific way if a reaction is to occur.

The validity of the first assumption is fairly obvious. Two molecules cannot react with each other if they are far apart. In order to break bonds, exchange atoms, and form new bonds, they must come in contact. There are, however, some exceptions, such as decompositions in which molecules break into fragments and processes in which molecules react by an internal rearrangement of their atoms.

Time-Release Medications

Most of us have experienced the inconvenience of having to take repeated doses of a medication at specific intervals; this dosing maintains an effective level of the medication in the body throughout the day. In 1961, the first commercial attempt was made to overcome this inconvenience when the decongestant Contac® became available in the form of time-release capsules. The approach used to time the release was simple. Each capsule contained many tiny beads of medication. Each bead was coated with a slow-dissolving polymer, but the thickness of the polymer coating was not the same on all beads. Beads with a thin coating dissolved rapidly, while those with a thicker coating dissolved more slowly. By including an appropriate mix of beads with coatings of different thickness in each capsule, a gradual, steady release of medication over a specific time period was achieved.

This same method for time release is still used in some products, but other methods have also been developed. The modern technology is used not only to provide a steady release of medication but also to control the point of release in the digestive tract. Some medications are known to irritate and damage the stomach lining or to be destroyed by the acidic environment of the stomach. An *enteric coating* that is stable in acid but that dissolves in the more basic environment of the small intestine is used to allow tablets or caplets of such materials to pass

through the stomach and release their medication in the intestines.

In a pharmacy, products in the following categories can be found with enteric coatings or in time-release forms that claim to provide effective levels of medication for periods ranging from 6 to 24 hours: analgesics (pain relievers), allergy treatments, cold/flu remedies, laxatives, sleeping aids, gas-relief medications, appetite-control materials, diuretics (water pills), and tablets to prevent sleeping.

A wide variety of products is available in time-release form.

In general, reactions take place more rapidly in the gaseous or liquid state than in the solid state. This observation verifies the first assumption because molecules of gases and liquids move about freely and can undergo many more collisions than can the rigidly held molecules of solids. Reactions involving solids usually take place only on the solid surface and therefore involve only a small fraction of the total molecules present in the solid. As the reaction proceeds and the products dissolve, diffuse, or fall from the surface, fresh solid is exposed. In this way, the reaction proceeds into the solid. The rusting of iron is an example of such a process.

If collisions were the only factor, however, most gaseous and liquid state reactions would take place almost instantaneously if every collision resulted in a reaction. Such high reaction rates are not observed, a fact that brings us to assumptions 2 and 3.

One of the ways to speed up a chemical reaction is to add energy in the form of heat. The added heat increases both the average speed (kinetic energy) and the internal energy of the molecules. **Internal energy** is the energy associated with molecular vibrations. An increase in internal energy increases the amplitude of molecular vibrations and, if large enough, breaks bonds (see ▶ Figure 8.3). Internal energy is also increased by the conversion of some kinetic energy into internal energy during collisions. When heat is added, both the kinetic and the internal energy of molecules is increased. This increases the frequency and speed of collisions and the chances that a collision will cause a sufficient increase in internal energy to break bonds and bring about a reaction (see ▶ Figure 8.4).

In some reaction mixtures, the average total energy of the molecules is too low at the prevailing temperature for a reaction to proceed at a detectable rate; the reaction mixture is stable. However, some reactions can be started by providing **activation energy.** Once the reaction is started, enough energy is released to activate other molecules

internal energy The energy associated with vibrations within molecules.

activation energy Energy needed to start some spontaneous processes. Once started, the processes continue without further stimulus or energy from an outside source.

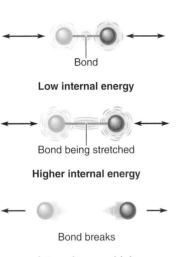

Low internal energy

Bond

Higher internal energy

Bond being stretched

Internal energy high enough to break bonds

Bond breaks

Figure 8.3 The internal energy of molecules.

Figure 8.4 The reaction that makes a Cyalume™ light stick glow takes place more rapidly in hot water (left) than in ice water (right). The light sticks outside the container are at room temperature. Which pair of sticks would glow for a longer time? Explain your reasoning.

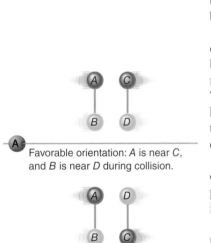

A: Favorable orientation: *A* is near *C*, and *B* is near *D* during collision.

B: Unfavorable orientation: *A* is not near *C*, and *B* is not near *D* during collision.

C: Unfavorable orientation: *B* is near *D*, but *A* and *C* are far removed during collision.

Figure 8.5 Molecular orientations during collisions.

and keep the reaction going at a good rate. The energy needed to get the boulder over the small hump in Figure 8.1B is a kind of activation energy. In many chemical reactions, activation energy causes bonds in reactant molecules to break. When the broken bonds react to form the new bonds of the products, energy is released that can cause bonds in more reactant molecules to break and the reaction to continue. The striking of a kitchen match is a good example. Activation energy is provided by rubbing the match head against a rough surface. Once started, however, the match continues to burn spontaneously.

A number of gases are routinely used in hospitals. Some of these form very flammable or even potentially explosive mixtures. Cyclopropane, a formerly used anesthetic, will burn vigorously in the presence of the oxygen in air. However, a mixture of the two will not react unless activation energy is provided in the form of an open flame or a spark. This is the reason that extreme precautions are taken to avoid open flames and sparks in a hospital operating room. Even sparks from static electricity can set off such gaseous mixtures, so special materials are used in clothing, floor coverings, and so on to prevent static electricity from building up.

Oxygen gas does not burn, but it reacts with anything that is combustible. Oxygen is often used in hospitals in concentrations much higher than the 20% found in air. Thus, precautions must be taken to prevent fire or sparks in any areas where oxygen gas is used in high concentrations.

Orientation effects are related to which side or end of a particle actually hits another particle during a collision. Orientation effects are unimportant in many reactions. For example, the orientation of silver ions (Ag^+) and chloride ions (Cl^-) toward each other during a collision has no effect on the rate of forming AgCl:

$$Ag^+(aq) + Cl^-(aq) \rightarrow AgCl(s) \tag{8.2}$$

The reason is that Ag^+ and Cl^- are both essentially spherical charged particles. However, collision orientation may be important in reactions that involve nonspherical molecules (assumption 3). Consider the following hypothetical reaction:

$$A - B + C - D \rightarrow A - C + B - D \tag{8.3}$$

It is clear that the collision orientation of $A-B$ and $C-D$ shown in ▶ Figure 8.5 **A** is more favorable to reaction than the orientations shown in **B** and **C**.

8.4 Energy Diagrams

Learning Objective

4. Represent and interpret the energy relationships for reactions by using energy diagrams.

Energy relationships for reactions can be represented by energy diagrams like the one in ◗ Figure 8.6. Notice the similarity to the earlier example of rolling boulders.

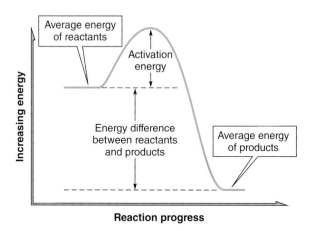

Figure 8.6 A typical energy diagram for chemical reactions.

The energy diagrams for most reactions look generally alike, but there are some differences. Typical diagrams for exothermic (exergonic) and endothermic (endergonic) reactions are given in ◗ Figure 8.7. It is clear that the products of exothermic reactions have lower energy than the reactants and that the products of endothermic reactions have higher energy than the reactants. When exothermic reactions occur, the energy difference between reactants and products is released. This energy often appears as heat, so heat is given up to the surroundings. When endothermic reactions occur, the energy (heat) added to the products is absorbed from the surroundings. Also, different reactions generally have different activation energies, a fact easily represented by energy diagrams.

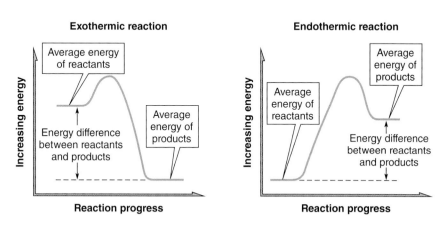

Figure 8.7 Energy diagrams for exothermic and endothermic reactions.

White phosphorus, a nonmetallic element, spontaneously bursts into flame if left in a rather warm room (34°C):

$$4P(s) + 5O_2(g) \rightarrow P_4O_{10}(s) \tag{8.4}$$

Sulfur, another nonmetallic element, will also burn, but it does not ignite until heated to about 232°C:

$$S(s) + O_2(g) \rightarrow SO_2(g) \tag{8.5}$$

The different activation energies for these two exothermic reactions are represented in ▶ Figure 8.8.

Figure 8.8 Differences in activation energies.

It should now be clear why we stressed the notion that some substances that are "stable" at normal living conditions undergo spontaneous changes at other conditions. It is simply that a higher temperature would provide the necessary activation energy. In a room hotter than 232°C, for example, both white phosphorus and sulfur would be "unstable" in the presence of oxygen.

8.5 Factors That Influence Reaction Rates

Learning Objective

5. Explain how factors such as reactant concentrations, temperature, and catalysts influence reaction rates.

Reaction rates are influenced by a number of different factors. Four factors that affect the rates of all reactions are:

1. The nature of the reactants.
2. The concentration of the reactants.
3. The temperature of the reactants.
4. The presence of catalysts.

The formation of insoluble AgCl by mixing together solutions containing Ag^+ and Cl^- ions is represented by Equation 8.2, given earlier. The white solid AgCl forms the instant the two solutions are mixed. This behavior is typical of reactions involving ionic reactants. The high reaction rate results from the attraction of the charged reactants to each other. In contrast, reactions that require covalent bonds to be broken or formed often proceed slowly. The production of water gas, a mixture of hydrogen (H_2) and carbon monoxide (CO), is a reaction involving covalent bonds:

$$H_2O(g) + C(s) \xrightarrow{\text{Heat}} CO(g) + H_2(g) \tag{8.6}$$

In addition, other structural characteristics of reactants, such as bond polarity or molecular size, may also be important factors in reaction rates.

The influence of reactant concentration on reaction rates can be illustrated by using the concept of molecular collisions. Suppose a reaction takes place between the hypothetical molecules A and B, which are mixed in a 1:1 ratio. The proposed reaction is

$$A + B \rightarrow \text{products} \tag{8.7}$$

Collisions with the capability to cause a reaction to occur are called **effective collisions.** Only collisions between *A* and *B* molecules can be classified as effective, since collisions between two *A* molecules or between two *B* molecules cannot possibly yield products. Imagine the reaction is begun with two *A* molecules and two *B* molecules as shown in ❱ Figure 8.9A. The example is simplified by looking only at the collisions of a single *A* molecule. Initially, two of every three collisions of the *A* molecule will be effective Ⓐ. On doubling the number (concentration) of *B* molecules, the number of effective collisions also doubles, and four of every five collisions is effective Ⓑ.

When larger numbers of molecules are used, the results approach more closely those actually observed experimentally for simple chemical reactions. Imagine a mixture containing 1000 *A* molecules and 1000 *B* molecules. On the average, 1000 out of every 1999 collisions of a single *A* molecule will involve a *B* molecule. (Each *A* molecule has almost a 50–50 chance of bumping into a *B* molecule.) This gives essentially a 1:1 ratio of effective to noneffective collisions. When the number of *B* molecules is doubled to 2000, the ratio becomes 2:1, because 2000 of every 2999 collisions will be effective. Therefore, the reaction rate should double when the concentration of one reactant is doubled. This result has been verified in numerous chemical reactions. Thus, higher concentrations produce a larger number of effective collisions in a given period of time, and this increases the reaction rate.

Gas-phase reactions are easily visualized this way, but reacting liquids and solids must be looked at differently. A large piece of solid contains a large number of molecules, but, as described before, only those on the surface can react with the molecules of another substance. For this reason, the total amount of solid in a sample is not as important as the surface area of solid in contact with other reactants. The effective concentration of a solid therefore depends on its surface area and state of division. A 100-pound sack of flour is difficult to burn when it is in a single pile. The same 100 pounds, dispersed in the air as a fine dust, burns very rapidly, and a dust explosion results (see ❱ Figure 8.10). The effective concentrations of reacting liquids must be thought of in a similar way unless the reactants are completely miscible.

effective collision A collision that causes a reaction to occur between the colliding molecules.

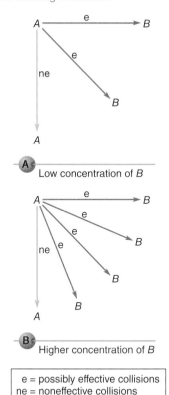

e = possibly effective collisions
ne = noneffective collisions

Figure 8.9 The effect of concentration on reaction rates.

1 A compact pile of lycopodium powder burns sluggishly.

2 A similar sample of lycopodium powder in a blowpipe.

3 The blown powder burns instantly with a bright flash.

Figure 8.10 Surface area influences reaction rates of solids. What is the other reactant in this reaction? How finely divided is it?

The effect of reactant temperature on reaction rates can also be explained by using the molecular collision concept. As was noted earlier, an increase in the temperature of a system increases the average kinetic energy and internal energy of the reacting molecules. The increased molecular speed (kinetic energy) causes more collisions to take place in a given time. Also, because the kinetic and internal energies of the colliding molecules are greater, a larger fraction of the collisions will be effective because they will provide the necessary activation energy.

catalyst A substance that changes (usually increases) reaction rates without being used up in the reaction.

inhibitor A substance that decreases reaction rates.

homogeneous catalyst A catalytic substance that is distributed uniformly throughout the reaction mixture.

heterogeneous or surface catalyst A catalytic substance normally used in the form of a solid with a large surface area on which reactions take place.

Figure 8.11 The effect of catalysts on activation energy.

As a rough rule of thumb, it has been found that the rate of a chemical reaction doubles for every 10°C increase in temperature. The chemical reactions of cooking take place faster in a pressure cooker because of a higher cooking temperature (Section 6.13). Also, cooling or freezing is used to slow the chemical reactions involved in the spoiling of food, souring of milk, and ripening of fruit.

Catalysts are substances that change reaction rates without being used up in the reactions. Usually the term *catalyst* is used to describe substances that speed up reactions. Substances that slow reactions are known as **inhibitors.** Catalysts are used in a number of different forms. Those dispersed uniformly throughout a reaction mixture in the form of individual ions or molecules are called **homogeneous catalysts. Heterogeneous** or **surface catalysts** are used in the form of solids; usually they have large surface areas on which reactions take place readily.

Catalysts enhance a reaction rate by providing an alternate reaction pathway that requires less activation energy than the normal pathway. This effect is represented by ❱ Figure 8.11. According to some theories, a catalyst provides the lower-energy pathway by entering into a reaction and forming an intermediate structure, which then breaks up to produce the final products and regenerate the catalyst:

$$\text{Uncatalyzed:} \qquad A + B \longrightarrow \text{products} \qquad (8.8)$$

$$\text{Catalyzed:} \quad A + B + \text{catalyst} \longrightarrow \underset{\substack{| \quad | \\ \text{catalyst}}}{A - B} \longrightarrow \text{products} + \text{catalyst} \qquad (8.9)$$

Intermediate structure

According to another proposed mechanism, solid catalysts provide a surface to which reactant molecules become attached with a particular orientation. Reactants attached to these surfaces are sufficiently close to one another and oriented favorably enough to allow the reaction to take place. The products of the reaction then leave the surface and make the attachment sites available for catalyzing other reactant molecules.

8.6 **Chemical Equilibrium**

Learning Objective

6. Relate experimental observations to the establishment of equilibrium.

So far, the focus has been only on the reactants of a reaction. However, all reactions can (in principle) go in both directions, and the products can be looked on as reactants. Suppose equal amounts of gaseous H_2 and I_2 are placed in a closed container and allowed to react and form HI. Initially, no HI is present, so the only possible reaction is

$$H_2(g) + I_2(g) \rightarrow 2HI(g) \qquad (8.10)$$

However, after a short time, some HI molecules are produced. They can collide with one another in a way that causes the reverse reaction to occur:

$$2HI(g) \rightarrow H_2(g) + I_2(g) \qquad (8.11)$$

The low concentration of HI makes this reaction slow at first, but as the concentration increases, so does the reaction rate. The rate of Reaction 8.10 decreases as the concentrations of H_2 and I_2 decrease. Eventually, the concentrations of H_2, I_2, and HI in the reaction mixture reach levels at which the rates of the forward (Equation 8.10) and reverse (Equation 8.11) reactions are equal. From that time on, the concentrations of H_2, I_2, and HI

Hypothermia: Surviving the Big Chill

The chemical reactions that maintain life in a healthy, human body take place at or near the normal body temperature of 98.6°F, or 37.0°C. These reactions, like all chemical reactions, slow down as the temperature decreases. If a person's temperature decreases to a level 2°C (4°F) lower than this normal value, the person is said to be suffering from hypothermia. Death is a significant possibility if hypothermia persists for more than a few hours, and anyone whose body temperature drops lower than 32°C (90°F) has a 17–33% chance of dying.

Hypothermia occurs when the amount of heat lost by the body to the surroundings exceeds the amount of heat generated by the body as a result of such things as exercise (muscular movement) or exothermic chemical reactions associated with metabolism. The body loses heat to the surroundings by several processes, including radiation, conduction, evaporation, and convection. Even though it is not visible to the naked eye, heat is radiated from uncovered parts of the warm body in the form of infrared light. Because it is constantly supplied with a large amount of warm blood, an uncovered head loses large amounts of heat to the surroundings by this mechanism. Conduction is the flow of heat from a warm object, such as an ungloved hand, that touches a cooler object, such as a cold piece of wood or metal. The evaporation of perspiration is a well-known natural way for the body to cool itself in warm weather. However, the evaporation of water from clothing that has accidentally become wet in very cold weather can result in severe and dangerous heat loss from the body. Convection is a term used to describe the circulation or movement of liquids or gases. In still air, a thin layer of warm air forms near the surface of the body and creates a warm, surrounding shell. But if a breeze or wind is blowing, or the body is moving as during running, the air passing over the body removes this warm layer, and its contained heat is lost to the surroundings. This cooling by convection is the basis for windchill charts that show a lower effective temperature in the winter when the wind is blowing than when the wind is not blowing, even though the actual air temperature is the same.

Heat can be lost from the body by various combinations of these four processes. The act of exhaling air on a cold day is an example in which all four are involved. Inhaled cold, dry air is heated in the warm moist environment of the lungs by a combination of radiation, conduction, convection and evapora-

tion. When this heated, moist air is exhaled, a significant amount of heat is lost to the surrounding cold air.

Anyone can suffer hypothermia and even death if exposed to extreme cold long enough without adequate protection. However, the elderly are especially susceptible to hypothermia, even in less extreme conditions. There are two primary reasons for this. First, the aging body becomes progressively less able to respond to cold surroundings and maintain an even temperature. Second, the body mechanism that normally detects a drop in body temperature gradually loses its sensitivity as the body ages. As a result of these factors, some elderly individuals suffering from hypothermia do not realize they are dangerously cold, and might even die without warning.

The symptoms of hypothermia that should be watched for, especially in the elderly, are lack of interest, indifference, despondency, drowsiness, mental confusion, and pallor. Prevention, of course, is the best treatment, but if hypothermia is suspected, sudden warming should not be attempted. The individual should be warmed up slowly, and professional medical help should be obtained quickly, especially if the victim is elderly.

The metallic coated wrap around this victim minimizes further cooling by preventing water evaporation from his wet clothing.

in the mixture remain constant, since both reactions take place at the same rate and each substance is being produced as fast as it is used up. When the forward and reverse reaction rates are equal, the reaction is in a **state of equilibrium,** and the concentrations are called **equilibrium concentrations.**

The behavior of reaction rates and reactant concentrations for both the forward and reverse reactions is shown graphically in ▶ Figure 8.12.

state of equilibrium A condition in a reaction system when the rates of the forward and reverse reactions are equal.

equilibrium concentrations The unchanging concentrations of reactants and products in a reaction system that is in a state of equilibrium.

Figure 8.12 Variation of reaction rates and reactant concentrations as equilibrium is established (T_e is the time needed to reach equilibrium).

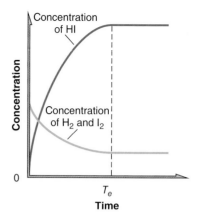

Instead of writing separate equations for both the forward and reverse reactions, the usual practice is to represent reversibility by double arrows. Thus, the equation for the reaction between H_2 and I_2 is written

$$H_2(g) + I_2(g) \rightleftarrows 2HI(g) \tag{8.12}$$

8.7 The Position of Equilibrium

Learning Objective

7. Write equilibrium expressions based on reaction equations, and do calculations based on equilibrium expressions.

position of equilibrium An indication of the relative amounts of reactants and products present at equilibrium.

The **position of equilibrium** for a reaction indicates the relative amounts of reactants and products present at equilibrium. When the position is described as being far to the right, it means that at equilibrium the concentration of products is much higher than the concentration of reactants. A position far to the left means the concentration of reactants is much higher than that of products.

The position of equilibrium can be represented numerically by using the concept of an equilibrium constant. Any reaction that establishes an equilibrium can be represented by a general equation:

$$aA + bB + \cdots \rightleftarrows wW + xX + \cdots \tag{8.13}$$

In this equation, the capital letters stand for substances such as the H_2, I_2, and HI of Equation 8.12, and the lowercase letters are the coefficients in the balanced equation such as the 1, 1, and 2 of Equation 8.12. The dots in Equation 8.13 indicate that any number of reactants and products can be involved in the reaction, but we will limit ourselves to just the four shown.

As we implied earlier, a reaction at equilibrium can be recognized because the concentrations of reactants and products remain constant. That is, they do not change as time passes. For a reaction at equilibrium, the following equation is valid:

$$K = \frac{[W]^w[X]^x}{[A]^a[B]^b} \tag{8.14}$$

equilibrium expression An equation relating the equilibrium constant and reactant and product equilibrium concentrations.

equilibrium constant A numerical relationship between reactant and product concentrations in a reaction at equilibrium.

In this **equilibrium expression,** the brackets [ ] represent molar concentrations of the reactants (*A* and *B*) and the products (*W* and *X*). The *K* is a constant called the **equilibrium constant,** and the powers on each bracket are the coefficients from the balanced equation for the reaction. According to this equation, the product of equilibrium concentrations of products (raised to appropriate powers) divided by the product of the equilibrium concentration of reactants (also raised to appropriate powers) gives a number that does not change with time (the equilibrium constant). The reason *K* does not change with time is that at equilibrium, the concentrations used to calculate *K* do not change with time. We could get a different constant by dividing the reactant concentrations by the product concentrations, but it is the accepted practice to calculate K values as we have shown in Equation 8.14.

▶ Example 8.2

Write equilibrium expressions for the reactions represented by the following equations:

a. $H_2(g) + I_2(g) \rightleftarrows 2HI(g)$
b. $2SO_2(g) + O_2(g) \rightleftarrows 2SO_3(g)$

Solution

a. The concentration of the product HI goes on top and is raised to the power 2. The concentrations of the reactants H_2 and I_2 go on the bottom, and each is raised to the power 1, which is not written but understood:

$$K = \frac{[HI]^2}{[H_2][I_2]}$$

b. The product is SO_3, and the power is 2. The reactants are SO_2 (power = 2) and O_2 (power = 1).

$$K = \frac{[SO_3]^2}{[SO_2]^2[O_2]}$$

▶ **LEARNING CHECK 8.2** Write equilibrium expressions for the reactions represented by the following equations:

a. $N_2O_4(g) \rightleftarrows 2NO_2(g)$
b. $3NO(g) \rightleftarrows N_2O(g) + NO_2(g)$

For a reaction with an equilibrium position far to the right, the product concentrations $[W]$ and $[X]$ will be much higher than the reactant concentrations $[A]$ and $[B]$. According to Equation 8.14, such a situation should result in a large K value. For reactions with equilibrium positions far to the left, a similar line of reasoning leads to the conclusion that K values should be small. Thus, the value of the equilibrium constant gives an indication of the position of equilibrium for a reaction.

▶ Example 8.3

For each of the following equations, the equilibrium molarity concentration is given below each reactant and product. Calculate K for each reaction and comment on the position of equilibrium.

a. $\quad 2NO_2(g) \quad \rightleftarrows \quad N_2O_4(g)$
$\quad\quad$ 8.23 × 10^{-3} M $\quad$ 1.46 × 10^{-2} M at 25°C

b. $CH_4(g) \quad + \quad H_2O(g) \rightleftarrows \quad CO(g) \quad + \quad 3H_2(g)$
$\quad$ 0.200 M $\quad\quad$ 0.150 M $\quad\quad$ 1.37 × 10^{-2} M $\quad\quad$ 4.11 × 10^{-2} M at 900 K

Solution

a. $K = \dfrac{[N_2O_4]}{[NO_2]^2} = \dfrac{(1.46 \times 10^{-2}\ \text{mol/L})}{(8.23 \times 10^{-3}\ \text{mol/L})^2} = \dfrac{2.16 \times 10^2}{\text{mol/L}}$

We see that K is fairly large, which tells us the equilibrium position is toward the right, or product, side of the reaction. The peculiar sounding "per mole per liter" unit is the result of the way concentration terms are arranged in the equilibrium expression. The unit will not be the same for the K of all reactions. These units are not ordinarily shown, a practice we will follow for the remainder of this chapter.

b. $K = \dfrac{[CO][H_2]^3}{[CH_4][H_2O]}$

$= \dfrac{(1.37 \times 10^{-2} \text{ mol/L})(4.11 \times 10^{-2} \text{ mol/L})^3}{(0.200 \text{ mol/L})(0.150 \text{ mol/L})}$

$= 3.17 \times 10^{-5}$

The small value of K indicates the equilibrium position is toward the left.

▶ **LEARNING CHECK 8.3** The molar concentrations are given below the reactants and products in the equations for the following reactions. Calculate K and comment on the position of each equilibrium.

a. $\quad Br_2(g) \quad + \quad I_2(g) \quad \rightleftarrows \quad 2IBr(g)$
$\quad 1.50 \times 10^{-1} \text{ M} \quad 5.00 \times 10^{-2} \text{ M} \quad 1.96 \times 10^{-2} \text{ M at } 25°C$

b. $\quad N_2(g) \quad + \quad 3H_2(g) \quad \rightleftarrows \quad 2NH_3(g)$
$\quad 9.23 \times 10^{-3} \text{ M} \quad 2.77 \times 10^{-2} \text{ M} \quad 9.00 \text{ M at } 25°C$

Example 8.3 illustrates that equilibrium constants for different reactions may be small or large. In fact, the two reactions used are by no means even close to the extremes encountered for K values. Some are so small (such as 1.1×10^{-36}) that for all practical purposes no products are present at equilibrium. Others are so large (such as 1.2×10^{40}) that the reaction can be considered to go completely to products. For reactions with K values between 10^{-3} and 10^3, the position of equilibrium is not extremely favorable to either side, and significant concentrations of both reactants and products can be detected in the equilibrium mixtures.

You might also have noticed in Example 8.3 that the equilibrium concentrations were given for specific temperatures. The reason for this is that K values are constant for a reaction as long as the temperature remains constant. However, K values will change as the temperature is changed. This is a demonstration of the effect of temperature on the position of equilibrium described in the next section.

8.8 Factors That Influence Equilibrium Position

Learning Objective

8. Use Le Châtelier's principle to predict the influence of changes in concentration and reaction temperature on the position of equilibrium for reactions.

Le Châtelier's principle The position of an equilibrium shifts in response to changes made in factors of the equilibrium.

A number of factors can change the position of an established equilibrium. The influence of such factors can be predicted by using a concept known as Le Châtelier's principle, in honor of its originator. According to **Le Châtelier's principle,** when a change is made in any factor of an established equilibrium, the position of equilibrium will shift in a direction that will minimize or oppose the change. The factors we will be most concerned with are concentrations of reactants and products, and reaction temperature.

The effect of concentration changes can be illustrated by the reaction of H_2 with I_2 (Equation 8.12):

$$H_2(g) + I_2(g) \rightleftarrows 2HI(g)$$

Suppose an equilibrium mixture of H_2, I_2, and HI is formed. According to the molecular collision concept, favorable collisions are occurring between H_2 and I_2 molecules to form HI molecules, and favorable collisions are also taking place between HI molecules that cause them to form H_2 and I_2 molecules. Now, suppose some additional I_2 is added to the equilibrium mixture. The chances for favorable collisions between H_2 and I_2 molecules are increased; the rate of formation of HI is increased, and more HI is formed than disappears.

The True Value of Platinum and Gold

Metallic gold has been known since prehistoric times and used through much of human history. This is likely the result of the resistance of gold to react with other materials and undergo such processes as corrosion. This property also explains why gold occurs in nature primarily in the form of small metallic grains or nuggets, or as veins of the metal in rocks. Skillfully worked gold jewelry has been found in good condition in royal graves known to be at least 5000 years old.

Historically, gold has been regarded as the ultimate symbol of political power and wealth. The mythical king Midas could turn any material into gold simply by touching it. The biblical king Solomon built a temple and had all the inside surfaces overlaid with gold. Today, the highest achievements in athletics and other competitions are often recognized by the awarding of gold medals.

Does gold live up to this symbolic reputation in terms of its actual price when bought and sold in the marketplace? In the past, its market value remained quite high compared to other precious metals such as silver and platinum. However, by 2000, the price of platinum had increased more rapidly than gold and they both sold for about $400 per ounce. The price of platinum continued to increase more rapidly than gold and by 2005, platinum sold for $850 per ounce compared to gold's price of $425 per ounce.

The value of platinum overtaking the value of gold is a classic case of economics and the concept of supply and demand. During each of the years between 2000 and 2005, worldwide demand for platinum exceeded the supply. As a result, the market price of platinum increased. But why did the demand for platinum increase at a greater rate than the demand for gold? The answer becomes evident when the pattern of use for each metal is analyzed.

About 75% of the gold produced worldwide is used to make jewelry. Much smaller amounts are used in other ways such as dental work, electronic devices, and glass making. In dental work, alloys of gold are used to make crowns, fillings and bridges. The electronics industry uses electroplated gold to prevent corrosion of electrical connections, printed circuit boards and other copper components. Colloidal gold added to molten glass gives the finished glass a red-to-purple color, and metallic gold applied as a thin film to window glass reflects heat from sunlight that falls on the windows.

Like gold, the largest use of platinum is for jewelry making, but this accounts for only 50% of the worldwide production. Another 30% is used to make catalytic converters for the exhaust system of automobiles, and the remaining 20% goes into other industrial applications. It has been estimated that one of every five purchased products contains some platinum or requires platinum for its production.

Most of the non-jewelry applications of platinum utilize its ability to act as a catalyst in chemical reactions. Catalytic converters are devices attached to the end of the exhaust system of internal combustion engines. At the high temperature of combustion in these engines, hydrocarbon-containing fuels such as gasoline are mixed with air (O_2 and N_2 gases) and, ideally, burned to form carbon dioxide gas (CO_2) and water vapor (H_2O). However, the ideal reactions do not take place completely during combustion, and small amounts of undesirable products are also formed, including carbon monoxide (CO) and nitrogen oxide (NO). These undesirable products and a small amount of unburned fuel are then included in the exhaust given off by the engine. When they are exposed to sunlight, these air pollutants are converted into an unhealthy mixture of other substances called smog. The role of the catalytic converter is to catalyze the conversion of the CO, NO, and unburned hydrocarbons into CO_2, H_2O and nitrogen gas, N_2. The use of catalytic converters on exhaust systems has had a significant impact on air quality in the cities of the world. In 1960, the exhaust from an average automobile contained 100 grams of pollutants for each mile driven. A modern automobile equipped with a catalytic converter puts out only about 2 grams of pollutants per mile. The catalytic property of platinum is also employed in numerous industrial processes such as the production of nitric acid, a very important industrial chemical.

The use of platinum and gold as components of jewelry or other decorative objects has changed little during the last 50 years. However, the increased industrial use of platinum as a catalyst during that time is the factor that accounts for its increased demand and accompanying price increase. Thus, the true value of gold or platinum, as measured by importance to society, is not so much in their use to adorn ourselves and impress others, but more in how they are used in practical, beneficial ways.

The production of jewelry accounts for the largest use of gold in the world today.

Platinum metal in the form of a woven fabric functions as a catalyst in this industrial equipment.

The rate of reaction of HI to give H_2 and I_2 increases as the concentration of HI increases, and eventually a new equilibrium position and a new set of equilibrium concentrations will be established. The new equilibrium mixture will contain more HI than the original mixture, but the amounts of the reactants will also be different such that the equilibrium constant remains unchanged. The original equilibrium of the reaction has been shifted toward the right.

Le Châtelier's principle can also be used to predict the influence of temperature on an equilibrium by treating heat as a product or reactant. Consider the following equation for a hypothetical exothermic reaction:

$$A + B \rightleftarrows \text{products} + \text{heat} \qquad (8.15)$$

If the temperature is increased (by adding more heat, which appears on the right side of the equation), the equilibrium shifts to the left in an attempt to use up the added heat. In the new equilibrium position, the concentrations of A and B will be higher, whereas the chemical product concentrations will be lower than those in the original equilibrium. In this case the value of the equilibrium constant is changed by the change in temperature (see ▶ Figure 8.13).

Figure 8.13 The effect of temperature on the position of equilibrium for the reaction $N_2O_4 \rightleftarrows 2NO_2$. On which side of the equation will heat appear? Is the reaction exothermic or endothermic?

① A sealed tube containing an equilibrium mixture of red-brown NO_2 and colorless N_2O_4 is cooled in an ice bath.

② The same sealed tube is heated in a hot water bath.

© 1995 Richard Megna/Fundamental Photographs

© 1995 Richard Megna/Fundamental Photographs

▶ **Example 8.4**

Using Le Châtelier's principle, answer the following questions:

a. Ammonia is made from hydrogen and atmospheric nitrogen:

$$N_2(g) + 3H_2(g) \rightleftarrows 2NH_3(g) + \text{heat}$$

What effect will cooling have on an equilibrium mixture?

b. What effect will removing H_2 have on the equilibrium mixture described in part a?

c. Consider the following equation for a reaction that takes place in solution:

$$\underset{\substack{\text{light} \\ \text{brown}}}{Fe^{3+}(aq)} + \underset{\text{colorless}}{6SCN^-(aq)} \rightleftarrows \underset{\text{dark red}}{Fe(SCN)_6^{3-}(aq)}$$

What effect will the addition of colorless NaSCN solid have on the color of the solution?

Solution

a. Heat is removed by cooling. Heat is replenished when the equilibrium shifts to the right. Thus, more NH_3 will be present at equilibrium in the cooler mixture.

b. The equilibrium will shift to the left in an attempt to replenish the H_2. At the new equilibrium position, less NH_3 will be present.

c. When solid NaSCN is added, it will dissolve and form ions:

$$NaSCN(s) \rightarrow Na^+(aq) + SCN^-(aq)$$

The SCN^- concentration in the solution will therefore be increased. In an attempt to use up the added SCN^-, the equilibrium will shift to the right. This shift generates more of the dark red $Fe(SCN)_6^{3-}$. Therefore, the new position of equilibrium will be characterized by a darker red color.

▶ **LEARNING CHECK 8.4** Use Le Châtelier's principle and answer the following:

a. A saturated solution of ammonium nitrate is formed as follows. The solution process is endothermic:

$$heat + NH_4NO_3(s) \rightleftharpoons NH_4^+(aq) + NO_3^-(aq)$$

Which way will the equilibrium shift when heat is added? What does this shift mean in terms of the solubility of NH_4NO_3 at the higher temperature?

b. Sulfur trioxide gas is formed when oxygen and sulfur dioxide gas react:

$$2SO_2(g) + O_2(g) \rightleftharpoons 2SO_3(g)$$

Which way would the equilibrium shift if O_2 were removed from the system? How would the new equilibrium concentration of SO_3 compare with the earlier equilibrium concentration?

Catalysts cannot change the position of an equilibrium. This fact becomes clear when you remember that a catalyst functions by lowering the activation energy for a reaction. A lowering of the energy barrier for the forward reaction also lowers the barrier for the reverse reaction (see ▶ Figure 8.14). Hence, a catalyst speeds up both the forward and reverse reactions and cannot change the position of equilibrium. However, the lowered activation energy allows equilibrium to be established more quickly than if the catalyst were absent.

Figure 8.14 The influence of catalysts on forward and reverse activation energies.

Le Châtelier's principle is extremely important in laboratory work and in the chemical industry when the goal is to obtain the maximum amount of product from a reaction. In such cases, increasing reactant concentrations, or adjusting the reaction temperature or pressure to shift the equilibrium position to the product side of the reaction, is a common practice.

To help you understand Le Châtelier's principle, it is often useful to observe familiar situations and events and interpret them in terms of shifting equilibria. For example, when you stand in an upright position, you are in equilibrium with the force of gravity. If a friend leans against you, your equilibrium is upset, and you will lose your balance unless you respond by leaning toward your friend such that the forces on you are once again balanced.

Children playing on a seesaw provide a similar example of forces in equilibrium. If one child slides toward the center of the seesaw, the force acting on that side is reduced, the equilibrium is upset, and the other child will be firmly on the ground. However, if the second child also shifts toward the center, the forces on each side can be made equal again, and play can resume.

The preceding examples involved equilibrium between forces, but we are not limited to only those types of examples. Imagine you are enjoying a warm shower when someone turns on a nearby hot water faucet. Suddenly, your shower turns cold. You respond to this stress by adjusting the shower controls (after shrieking loudly) to let in more hot water in an attempt to restore the pleasant temperature you were enjoying. (However, be aware of what will happen when the other faucet is turned off.)

Be mindful of your surroundings and activities, and you will see many more examples that will remind you of responses to stress that are attempts to restore equilibrium to a previous situation.

Case Study Follow-up

Allergic reactions are common and range from mild to severe. *Anaphylaxis* is the proper name for a severe allergic reaction. Anaphylaxis is a whole-body reaction to an allergen, and it can be life threatening. Anaphylaxis can occur in response to any allergen, including insect bites and stings, certain foods, and some medicines. Death can occur in as little as 15 minutes—anaphylaxis is a true medical emergency. It is appropriate to call 911 immediately if a person is showing the following symptoms after exposure to an allergen: wheezing, difficulty breathing, palpitations, slurred speech, or difficulty swallowing. Other allergy symptoms that warrant medical attention but might not be an immediate emergency are vomiting, diarrhea, hives, nasal congestion, and cough (NIH, 2010).

Epinephrine is the treatment of choice for severe allergic reactions. It acts rapidly and has a short duration. Any time a severe allergic reaction is suspected, an epinephrine injection is appropriate for individuals with a prescription for it or if administered by qualified medical personnel. Whenever someone uses an epinephrine injection, he or she should seek additional medical care immediately (FDA, 2008). Anyone who experiences a moderate to severe allergic reaction should consult his or her healthcare provider about carrying an epinephrine injector.

Concept Summary

Spontaneous and Nonspontaneous Processes

Spontaneous processes take place naturally with no apparent cause or stimulus. Process spontaneity depends on the energy and entropy changes that accompany the process. Energy decreases and entropy increases favor spontaneity. However, a nonspontaneous change in one of these factors can be compensated for by a large spontaneous change in the other to cause processes to be spontaneous.

Objective 1, Exercise 8.6

Reaction Rates

The speed of a reaction is called a reaction rate, which can be determined by measuring how fast reactants are used up or products are formed.

Objective 2, Exercise 8.14

Molecular Collisions

Explanations of how reactions take place are called reaction mechanisms. Most mechanisms are based on three assumptions: (1) Molecules must collide with one another, (2) the collision must involve a certain minimum of energy, and (3) some colliding molecules must be oriented in a specific way during collision in order to react.

Objective 3, Exercise 8.20

Energy Diagrams

Energy relationships for reactions can be represented by energy diagrams, in which energy is plotted versus the reaction progress. The concepts of exothermic and endothermic reactions and activation energy are clearly represented by such diagrams.

Objective 4, Exercise 8.26

Factors That Influence Reaction Rates

Four factors affect the rates of all reactions: (1) the nature of the reactants, (2) reactant concentrations, (3) reactant temperature, and (4) the presence of catalysts.

Objective 5, Exercise 8.30

Chemical Equilibrium

Reactions are in equilibrium when the rate of the forward reaction is equal to the rate of the reverse reaction. Equilibrium is emphasized in equations for reactions by writing double arrows pointing in both directions between reactants and products.

Objective 6, Exercise 8.38

The Position of Equilibrium

The relative amounts of reactants and products present in a system at equilibrium define the position of equilibrium. The equilibrium position is toward the right when a large amount of product is present and toward the left when a large amount of reactant is present. The position is indicated by the value of the equilibrium constant.

Objective 7, Exercises 8.40 and 8.46

Factors That Influence Equilibrium Position

Factors known to influence the position of equilibrium include changes in amount of reactants and/or products and changes in temperature. The influence of such factors can be predicted by using Le Châtelier's principle. Catalysts cannot change the position of equilibrium.

Objective 8, Exercise 8.52

Key Terms and Concepts

Activation energy (8.3)
Catalyst (8.5)
Effective collision (8.5)
Endergonic process (8.1)
Entropy (8.1)
Equilibrium concentrations (8.6)
Equilibrium constant (8.7)

Equilibrium expression (8.7)
Exergonic process (8.1)
Heterogeneous (surface) catalyst (8.5)
Homogeneous catalyst (8.5)
Inhibitor (8.5)
Internal energy (8.3)
Le Châtelier's principle (8.8)

Position of equilibrium (8.7)
Reaction mechanism (8.3)
Reaction rate (8.2)
Spontaneous process (8.1)
Stable substance (8.1)
State of equilibrium (8.6)

Key Equations

1. Calculation of reaction rate (Section 8.2):	$\text{Rate} = \dfrac{\Delta C}{\Delta t} = \dfrac{C_t - C_0}{\Delta t}$	Equation 8.1
2. Equilibrium expression for general reaction (Section 8.7):	$aA + bB + \cdots \rightleftarrows wW + xX + \cdots$	Equation 8.13
	$K = \dfrac{[W]^w[X]^x}{[A]^a[B]^b}$	Equation 8.14

Exercises

Even-numbered exercises are answered in Appendix B.

Blue-numbered exercises are more challenging.

Spontaneous and Nonspontaneous Processes (Section 8.1)

8.1 Classify the following processes as spontaneous or nonspontaneous. Explain your answers in terms of whether energy must be continually supplied to keep the process going.

  a. Water is decomposed into hydrogen and oxygen gas by passing electricity through the liquid.

  b. An explosive detonates after being struck by a falling rock.

  c. A coating of magnesium oxide forms on a clean piece of magnesium exposed to air.

  d. A light bulb emits light when an electric current is passed through it.

  e. A cube of sugar dissolves in a cup of hot coffee.

8.2 Classify the following processes as spontaneous or nonspontaneous. Explain your answers in terms of whether energy must be continually supplied to keep the process going.

  a. The space shuttle leaves its pad and goes into orbit.

  b. The fuel in a booster rocket of the space shuttle burns.

  c. Water boils at 100°C and 1 atm pressure.

  d. Water temperature increases to 100°C at 1 atm pressure.

  e. Your bedroom becomes orderly.

8.3 Classify the following processes as exergonic or endergonic. Explain your answers.

  a. Any combustion process

  b. Perspiration evaporating from the skin

  c. Melted lead solidifying

  d. An explosive detonating

  e. An automobile being pushed up a slight hill (from point of view of the automobile)

8.4 Classify the following processes as exergonic or endergonic. Explain your answers.

  a. An automobile being pushed up a slight hill (from point of view of the one pushing)

  b. Ice melting (from point of view of the ice)

  c. Ice melting (from point of view of surroundings of the ice)

  d. Steam condensing to liquid water (from point of view of the steam)

  e. Steam condensing to liquid water (from point of view of surroundings of the steam)

8.5 Describe the energy and entropy changes that occur in the following processes, and indicate whether the processes are spontaneous under the conditions stated:

  a. Lumber becomes a house

  b. A seed grows into a tree

  c. On a hot day, water evaporates from a lake

8.6 Describe the energy and entropy changes that occur in the following processes, and indicate whether the processes are spontaneous under the conditions stated:

  a. On a cold day, water freezes.

  b. A container of water at 40°C cools to room temperature.

  c. The odor from an open bottle of perfume spreads throughout a room.

8.7 Pick the example with the highest entropy from each of the following sets. Explain your answers.

  a. Solid ice, liquid water, or steam

  b. Leaves on a tree, fallen leaves blown about on the ground, fallen leaves raked and placed in a basket

  c. A stack of sheets of paper, a wastebasket containing sheets of paper, a wastebasket containing torn and crumpled sheets of paper

  d. A 0.10 M sugar solution, a 1.0 M sugar solution, a 10.0 M sugar solution

  e. A banquet table set for dinner, a banquet table during dinner, a banquet table immediately after dinner

8.8 Pick the example with the highest entropy from each of the following sets. Explain your answers.

  a. Two opposing football teams just before the ball is snapped, two opposing football teams 1 second after the ball is snapped, two opposing football teams when the whistle is blown, ending the play

  b. A 10% copper/gold alloy, a 2% copper/gold alloy, pure gold

  c. A purse on which the strap just broke, a purse just hitting the ground, a purse on the ground with contents scattered

  d. Coins in a piggy bank, coins in piles containing the same type of coins, coins in stacks of the same type of coins

  e. A dozen loose pearls in a box, a dozen pearls randomly strung on a string, a dozen pearls strung on a string in order of decreasing size

8.9 You probably know that, on exposure to air, silver tarnishes and iron rusts; but gold, stainless steel, and chromium do not change. Explain these facts, using the concept of stability.

Reaction Rates (Section 8.2)

8.10 Classify the following processes according to their rates as very slow, slow, or fast:

  a. The cooking of a pizza in a hot oven

  b. The change in leaf color from green to red or orange in the autumn

  c. The drying of a wet swimsuit hung out in the sun

  d. The burning of a match after it has been ignited

  e. The growing of grass during a warm summer

8.11 Classify the following processes according to their rates as very slow, slow, or fast:

  a. The melting of butter put into a hot pan

  b. The ripening of a piece of fruit stored at room temperature

  c. The cooking of a raw potato in a hot oven

  d. The melting of an ice cube in a glass of cool water

  e. The combustion of gasoline in the engine of a car

8.12 Describe the observations or measurements that could be made to allow you to follow the rate of the following processes:

  a. The melting of a block of ice

  b. The setting (hardening) of concrete

  c. The burning of a candle

8.13 Describe the observations or measurements that could be made to allow you to follow the rate of the following processes:

  a. The diffusion of ink from a drop placed in a pan of quiet, undisturbed water

  b. The loss of water from a pan of boiling water

  c. The growth of a corn plant

8.14 Consider the following hypothetical reaction:

$$A + B \rightarrow C$$

Calculate the average rate of the reaction on the basis of the following information:

  a. Pure A and B are mixed, and after 12.0 minutes the measured concentration of C is 0.396 mol/L.

  b. Pure A, B, and C are mixed together at equal concentrations of 0.300 M. After 8.00 minutes, the concentration of C is found to be 0.455 M.

8.15 Consider the following reaction:

$$A + B \rightarrow C$$

Calculate the average rate of the reaction on the basis of the following information:

  a. Pure A, B, and C are mixed together at concentrations of $A = B = 0.400$ M, $C = 0.150$ M. After 9.00 minutes, the concentration of C is 0.418 M.

  b. Pure A and B are mixed together at the same concentration of 0.361 M. After 6.00 minutes, the concentration of A is found to be 0.048 M.

8.16 A reaction generates chlorine gas (Cl_2) as a product. The reactants are mixed and sealed in a 250-mL container. After 30.0 minutes, 2.97×10^{-2} mol of Cl_2 has been generated. Calculate the average rate of the reaction.

8.17 A reaction generates hydrogen gas (H_2) as a product. The reactants are mixed in a sealed 250-mL vessel. After 15.0 minutes, 3.91×10^{-2} mol H_2 has been generated. Calculate the average rate of the reaction.

8.18 Ammonium and nitrite ions react in solution to form nitrogen gas:

$$NH_4^+(aq) + NO_2^-(aq) \rightarrow N_2(g) + 2H_2O(\ell)$$

A reaction is run, and the liberated N_2 gas is collected in a previously evacuated 500-mL container. After the reaction has gone on for 750 seconds, the pressure of N_2 in the 500-mL container is 2.77×10^{-2} atm, and the temperature of the N_2 is 25.0°C. Use the ideal gas law (Equation 6.9) to calculate the number of moles of N_2 liberated. Then calculate the average rate of the reaction.

8.19 Suppose a small lake is contaminated with an insecticide that decomposes with time. An analysis done in June shows the decomposition product concentration to be 7.8×10^{-4} mol/L. An analysis done 35 days later shows the concentration of decomposition product to be 9.9×10^{-4} mol/L. Assume the lake volume remains constant and calculate the average rate of decomposition of the insecticide.

Molecular Collisions (Section 8.3)

8.20 In each of the following, which reaction mechanism assumption is apparently being violated? Explain your answers.

    a. A reaction takes place more rapidly when the concentration of reactants is decreased.

    b. A reaction takes place more rapidly when the reaction mixture is cooled.

    c. The reaction rate of $A + B \rightarrow A\!-\!B$ increases as the concentration of A is increased but does not change as the concentration of B is increased.

8.21 Which reaction mechanism assumptions are unimportant in describing simple ionic reactions between cations and anions? Why?

8.22 Describe two ways by which an increase in temperature increases a reaction rate.

Energy Diagrams (Section 8.4)

8.23 Sketch energy diagrams to represent each of the following. Label the diagrams completely and tell how they are similar to each other and how they are different.

    a. Exothermic (exergonic) reaction with activation energy

    b. Exothermic (exergonic) reaction without activation energy

8.24 Sketch energy diagrams to represent each of the following. Label the diagrams completely and tell how they are similar to each other and how they are different.

    a. Endothermic (endergonic) reaction with activation energy

    b. Endothermic (endergonic) reaction without activation energy

8.25 Use energy diagrams to compare catalyzed and uncatalyzed reactions.

8.26 One reaction occurs at room temperature and liberates 500 kJ/mol of reactant. Another reaction does not take place until the reaction mixture is heated to 150°C. However, it also liberates 500 kJ/mol of reactant. Draw an energy diagram for each reaction and indicate the similarities and differences between the two.

Factors That Influence Reaction Rates (Section 8.5)

8.27 The following reactions are proposed. Make a rough estimate of the rate of each one—rapid, slow, won't react. Explain each answer.

    a. $H_2O(\ell) + H^+(aq) \rightarrow H_3O^+(aq)$

    b. $H_3O^+(aq) + H^+(aq) \rightarrow H_4O^{2+}(aq)$

    c. $3H_2(g) + N_2(g) \rightarrow 2NH_3(g)$

    d. $Ba^{2+}(aq) + SO_4^{2-}(aq) \rightarrow BaSO_4(s)$

8.28 The following reactions are proposed. Make a rough estimate of the rate of each one—rapid, slow, won't react. Explain each answer.

    a. $2I^-(aq) + Pb^{2+}(aq) \rightarrow PbI_2(s)$

    b. $Br^-(aq) + I^-(aq) \rightarrow BrI^{2-}(aq)$

    c. $NH_3(g) + HCl(g) \rightarrow NH_4Cl(s)$

    d. $CaO(s) + CO_2(g) \rightarrow CaCO_3(s)$

8.29 Which reaction mechanism assumption is most important in explaining the following? Why?

    a. The influence of concentration on reaction rates

    b. The influence of catalysts on reaction rates

8.30 Suppose you are running a reaction and you want to speed it up. Describe three things you might try to do this.

8.31 A reaction is started by mixing reactants. As time passes, the rate decreases. Explain this behavior that is characteristic of most reactions.

8.32 A reaction is run at 20°C and takes 30 minutes to go to completion. How long would it take to complete the same reaction at a temperature of 40°C?

8.33 What factor is more important than simply the amount of solid reactant present in determining the rate of a reaction? Explain.

8.34 Describe two ways catalysts might speed up a reaction.

Chemical Equilibrium (Section 8.6)

8.35 Describe the establishment of equilibrium in a system represented by a shopper walking up the "down" escalator.

8.36 Describe the observation or measurement result that would indicate when each of the following had reached equilibrium:

    a. $\quad H_2 \quad + \quad I_2 \quad \rightleftarrows \quad 2HI$
         colorless gas   violet gas   colorless gas

    b. solid sugar + water $\rightleftarrows$ sugar solution

    c. $\quad N_2 \quad + \quad 2O_2 \quad \rightleftarrows \quad 2NO_2$
         colorless gas   colorless gas   red-brown gas

8.37 Describe the observation or measurement result that would indicate when each of the following had reached equilibrium:

    a. $\quad 2CO \quad + \quad O_2 \quad \rightleftarrows \quad 2CO_2$
         colorless gas   colorless gas   colorless gas
                                     (apply Dalton's law)

    b. $\quad LiOH \quad + \quad CO_2 \quad \rightleftarrows \quad LiHCO_3$
         colorless solid   colorless gas   colorless solid

    c. paycheck $\rightarrow$ checking account $\rightarrow$ checks to pay bills

8.38 Colorless hydrogen gas (H_2) and red-brown colored bromine gas (Br_2) react to form colorless HBr gas. Describe how the concentrations of H_2, Br_2, and HBr would change (increase or decrease) as equilibrium was established in a sealed container that initially contained only H_2 and Br_2. What observation would indicate that equilibrium had been established?

8.39 Colorless N_2O_4 gas decomposes to form red-brown colored NO_2 gas. Describe how the concentrations of N_2O_4 and NO_2 would change (increase or decrease) as equilibrium was established in a sealed container that was initially filled with N_2O_4. What observation would indicate that equilibrium had been established?

The Position of Equilibrium (Section 8.7)

8.40 Write an equilibrium expression for each of the following gaseous reactions:

   a. $2CO + O_2 \rightleftarrows 2CO_2$

   b. $N_2O_4 \rightleftarrows 2NO_2$

   c. $2C_2H_6 + 7O_2 \rightleftarrows 4CO_2 + 6H_2O$

   d. $2NOCl \rightleftarrows 2NO + Cl_2$

   e. $2Cl_2O_5 \rightleftarrows O_2 + 4ClO_2$

8.41 Write an equilibrium expression for each of the following gaseous reactions:

   a. $H_2 + Br_2 \rightleftarrows 2HBr$

   b. $2H_2S + 3O_2 \rightleftarrows 2H_2O + 2SO_2$

   c. $3NO_2 \rightleftarrows N_2O_5 + NO$

   d. $4NH_3 + 3O_2 \rightleftarrows 2N_2 + 6H_2O$

   e. $2NO + 2H_2 \rightleftarrows N_2 + 2H_2O$

8.42 The following equilibria are established in water solutions. Write an equilibrium expression for each reaction.

   a. $Fe^{3+} + 6CN^- \rightleftarrows Fe(CN)_6^{3-}$

   b. $Ag^+ + 2NH_3 \rightleftarrows Ag(NH_3)_2^+$

   a. $Au^{3+} + 4Cl^- \rightleftarrows AuCl_4^-$

8.43 The following equilibria are established in water solutions. Write an equilibrium expression for each reaction.

   a. $Ni^{2+} + 6NH_3 \rightleftarrows Ni(NH_3)_6^{2+}$

   b. $Sn^{2+} + 2Fe^{3+} \rightleftarrows Sn^{4+} + 2Fe^{2+}$

   c. $F_2 + 2Cl^- \rightleftarrows 2F^- + Cl_2$

8.44 Write a reaction equation that corresponds to each of the following equilibrium expressions:

   a. $K = \dfrac{[CO_2][H_2O]^2}{[CH_4][O_2]^2}$

   b. $K = \dfrac{[CH_4][H_2O]}{[H_2]^3[CO]}$

   c. $K = \dfrac{[O_2]^3}{[O_3]^2}$

   d. $K = \dfrac{[NH_3]^4[O_2]^7}{[NO_2]^4[H_2O]^6}$

8.45 Write an equation that corresponds to each of the following equilibrium expressions:

   a. $K = \dfrac{[PH_3][F_2]^3}{[HF]^3[PF_3]}$

   b. $K = \dfrac{[O_2]^7[NH_3]^4}{[NO_2]^4[H_2O]^6}$

   c. $K = \dfrac{[O_2][ClO_2]^4}{[Cl_2O_5]^2}$

   d. $K = \dfrac{[N_2][H_2O]^2}{[NO]^2[H_2]^2}$

8.46 A sample of gaseous BrCl is allowed to decompose in a closed container at 25°C:

$$2BrCl(g) \rightleftarrows Br_2(g) + Cl_2(g)$$

When the reaction reaches equilibrium, the following concentrations are measured: [BrCl] = 0.38 M, [Cl_2] = 0.26 M, [Br_2] = 0.26 M. Evaluate the equilibrium constant for the reaction at 25°C.

8.47 At 600°C, gaseous CO and Cl_2 are mixed together in a closed container. At the instant they are mixed, their concentrations are CO = 0.79 mol/L and Cl_2 = 0.69 mol/L. After equilibrium is established, their concentrations are CO = 0.25 mol/L and Cl_2 = 0.15 mol/L. Evaluate the equilibrium constant for the reaction

$$CO(g) + Cl_2(g) \rightleftarrows COCl_2(g)$$

8.48 A mixture of the gases NOCl, Cl_2, and NO is allowed to reach equilibrium at 25°C. The measured equilibrium concentrations are [NO] = 0.92 mol/L, [NOCl] = 1.31 mol/L, and [Cl_2] = 0.20 mol/L. What is the value of the equilibrium constant at 25°C for the reaction

$$2NOCl(g) \rightleftarrows 2NO(g) + Cl_2(g)$$

8.49 Consider the following equilibrium constants. Describe how you would expect the equilibrium concentrations of reactants and products to compare with each other (larger than, smaller than, etc.) for each case.

   a. $K = 2.1 \times 10^{-6}$

   b. $K = 0.15$

   c. $K = 1.2 \times 10^8$

   d. $K = 0.00036$

8.50 Consider the following equilibrium constants. Describe how you would expect the equilibrium concentrations of reactants and products to compare with each other (larger than, smaller than, etc.) for each case.

   a. $K = 5.9$

   b. $K = 3.3 \times 10^6$

   c. $K = 2.7 \times 10^{-4}$

   d. $K = 0.0000558$

Factors That Influence Equilibrium Position (Section 8.8)

8.51 Use Le Châtelier's principle to predict the direction of equilibrium shift in the following equilibria when the indicated stress is applied:

   a. $2A + B + \text{heat} \rightleftarrows D$; some B is removed.

   b. $2A + B + \text{heat} \rightleftarrows C + D$; the system is heated.

   c. $N_2O_4 \rightleftarrows 2NO_2$; some NO_2 is added.

8.52 Use Le Châtelier's principle to predict the direction of equilibrium shift in the following equilibria when the indicated stress is applied:

   a. $Ag^+(aq) + Cl^-(aq) \rightleftarrows AgCl(s)$; some Ag^+ is removed.

   b. $2HI(g) + \text{heat} \rightleftarrows H_2(g) + I_2(g)$; the system is heated.

   c. $6Cu(s) + N_2(g) + \text{heat} \rightleftarrows 2Cu_3N(s)$; the system is cooled and some N_2 is removed.

8.53 Using Le Châtelier's principle, predict the direction of equilibrium shift and the changes that will be observed (color, amount of precipitate, etc.) in the following equilibria when the indicated stress is applied:

   a. $\text{heat} + Co^{2+}(aq) + 4Cl^-(aq) \rightleftarrows CoCl_4^{2-}(aq)$;
   pink   colorless   blue
   the equilibrium mixture is cooled.

   b. $\text{heat} + Co^{2+}(aq) + 4Cl^-(aq) \rightleftarrows CoCl_4^{2-}(aq)$;
   pink   colorless   blue
   Cl^- is added to the equilibrium mixture.

   c. $Fe^{3+}(aq) + 6SCN^-(aq) \rightleftarrows Fe(SCN)_6^{3-}(aq)$;
   brown   colorless   red
   Fe^{3+} is added to the equilibrium mixture.

   d. $Pb^{2+}(aq) + 2Cl^-(aq) \rightleftarrows PbCl_2(s) + \text{heat}$;
   colorless colorless white solid
   Cl^- is added to the equilibrium mixture.

   e. $C_2H_4 \quad + \quad I_2 \quad \rightleftarrows \quad C_2H_4I_2 \quad + \quad \text{heat}$;
   colorless gas   violet gas   colorless gas
   a catalyst is added to the equilibrium mixture.

8.54 Using Le Châtelier's principle, predict the direction of equilibrium shift and the changes that will be observed (color, amount of precipitate, etc.) in the following equilibria when the indicated stress is applied:

   a. $Cu^{2+}(aq) + 4NH_3(aq) \rightleftarrows Cu(NH_3)_4^{2+}(aq)$;
   blue   colorless   dark purple
   some NH_3 is added to the equilibrium mixture.

   b. $Pb^{2+}(aq) + 2Cl^-(aq) \rightleftarrows PbCl_2(s) + \text{heat}$;
   colorless colorless white solid
   the equilibrium mixture is cooled.

   c. $C_2H_4 \quad + \quad I_2 \quad \rightleftarrows \quad C_2H_4I_2 \quad + \quad \text{heat}$;
   colorless gas   violet gas   colorless gas
   some $C_2H_4I_2$ is removed from the equilibrium mixture.

   d. $C_2H_4 \quad + \quad I_2 \quad \rightleftarrows \quad C_2H_4I_2 \quad + \quad \text{heat}$;
   colorless gas   violet gas   colorless gas
   the equilibrium mixture is cooled.

   e. $\text{heat} \quad + \quad 4NO_2 \quad + \quad 6H_2O \rightleftarrows \quad 7O_2 \quad + \quad 4NH_3$;
           brown   colorless colorless colorless
           gas   gas   gas   gas
   a catalyst is added, and NH_3 is added to the equilibrium mixture.

8.55 Tell what will happen to each equilibrium concentration in the following when the indicated stress is applied and a new equilibrium position is established:

   a. $H^+(aq) + HCO_3^-(aq) \rightleftarrows H_2O(\ell) + CO_2(g)$;
   HCO_3^- is added.

   b. $CO_2(g) + H_2O(\ell) \rightleftarrows H_2CO_3(aq) + \text{heat}$;
   CO_2 is removed.

   c. $CO_2(g) + H_2O(\ell) \rightleftarrows H_2CO_3(aq) + \text{heat}$;
   the system is cooled.

8.56 Tell what will happen to each equilibrium concentration in the following when the indicated stress is applied and a new equilibrium position is established:

   a. $2N_2O(g) + 3O_2(g) \rightleftarrows 4NO_2(g) + \text{heat}$; O_2 is added.

   b. $2SO_3(g) + \text{heat} \rightleftarrows 2SO_2(g) + O_2(g)$; the system is heated.

   c. $2CO(g) + O_2(g) \rightleftarrows 2CO_2(g) + \text{heat}$; the system is cooled.

8.57 The gaseous reaction $2HBr(g) \rightleftarrows H_2(g) + Br_2(g)$ is endothermic. Tell which direction the equilibrium will shift for each of the following:

   a. Some H_2 is removed.

   b. The temperature is decreased.

   c. Some Br_2 is added.

   d. A catalyst is added.

   e. Some HBr is added.

   f. The temperature is decreased, and some HBr is removed.

8.58 The gaseous reaction $N_2(g) + 3H_2(g) \rightleftarrows 2NH_3(g)$ is exothermic. Tell which direction the equilibrium will shift for each of the following:

   a. Some N_2 is added.

   b. The temperature is increased.

   c. Some NH_3 is removed.

   d. Some H_2 is removed.

   e. A catalyst is added.

   f. The temperature is increased, and some H_2 is removed.

Additional Exercises

8.59 Assume the following reactions take place under identical conditions of concentration and temperature. Which reaction would you expect to take place faster? Explain your answer.

   a. $H_2(g) + F_2(g) \rightarrow 2HF(g)$

   b. $H_2(g) + I_2(g) \rightarrow 2HI(g)$

8.60 Gases A and B react as follows: $A(g) + B(g) \rightarrow C(s)$. Suppose gases A and B are mixed and used to fill a balloon. How could you increase the concentration of A and B, and speed up the reaction?

8.61 Bacteria found in lakes use dissolved oxygen gas to metabolize organic contaminants and use them as a food source, thus removing the contaminants from the lakes. When lake temperatures increase significantly, the rate of this chemical decontamination decreases. Remember your study of gases and explain this apparent violation of the factors influencing reaction rates studied in this chapter.

8.62 In Section 8.1 of this chapter, three criteria that determine the spontaneity of a process are given. Review these criteria, and determine which one is involved in each of the following spontaneous phase changes:

  a. Evaporation of a liquid

  b. Condensation of a gas to a liquid

  c. Sublimation of a solid to a gas

  d. Liquefaction of a gas to a liquid

  e. Crystallization of a liquid to a solid

Allied Health Exam Connection

The following questions are from these sources:

1. *Nursing School Entrance Exam* © 2005, Learning Express, LLC.

2. *McGraw-Hill's Nursing School Entrance Exams* by Thomas A. Evangelist, Tamara B. Orr and Judy Unrein © 2009, The McGraw-Hill Companies, Inc.

3. *NSEE Nursing School Entrance Exams,* 3rd Edition © 2009, Kaplan Publishing.

4. *Cliffs Test Prep: Nursing School Entrance Exams* by Fred N. Grayson © 2004, Wiley Publishing, Inc.

5. *Peterson's Master the Nursing School and Allied Health Entrance Exams,* 18th Edition by Marion F. Gooding © 2008, Peterson's, a Nelnet Company.

8.63 A reaction takes place between an acid and 0.5 grams of solid magnesium ribbon. Another reaction takes place between an acid and 0.5 grams of powdered magnesium. Which statement is true?

  a. The powdered magnesium reacts faster because the activation energy has been lowered.

  b. The magnesium strip reacts faster because it has a higher concentration of magnesium.

  c. The powdered magnesium reacts faster because it has a greater surface area.

  d. The magnesium strip reacts faster because it will create a higher temperature once the reaction starts.

8.64 If the reaction: $A + B \rightarrow C + D$ is designated as first order, the rate depends on:

  a. the concentration of only one reactant

  b. the concentration of each reactant

  c. no specific concentration

  d. the temperature only

8.65 Reaction kinetics deals with:

  a. equilibrium position

  b. reaction rates

  c. molecular reactant size

  d. none of the above

8.66 A book is held six feet above the floor and then dropped. Which statement is true?

  a. The potential energy of the book is converted to kinetic energy.

  b. The potential energy of the book is destroyed.

  c. Kinetic energy is created.

  d. The total energy of the system will not be conserved.

8.67 When a crane at a building site lifts a beam to its top height, what type of energy is created?

  a. kinetic energy

  b. potential energy

  c. chemical energy

  d. electrical energy

8.68 Stored energy is referred to as:

  a. activation energy

  b. kinetic energy

  c. potential energy

  d. electrical energy

8.69 In exergonic reactions, the energy is:

  a. used

  b. stored

  c. released

  d. lost

8.70 Which is an example of an exothermic change?

  a. sublimation

  b. condensation

  c. melting

  d. evaporation

8.71 Which of the following reactions releases heat energy?

  a. double replacement

  b. decomposition

  c. endothermic

  d. exothermic

8.72 Which of the following is the best example of potential energy changing to kinetic energy?

  a. Pushing a rock off a cliff

  b. Sitting in a rocking chair

  c. Observing a bird fly

  d. Standing on a table

8.73 Which is NOT an example of an endothermic change?

  a. melting

  b. sublimation

  c. freezing

  d. evaporation

8.74 Which of the following processes is endothermic?

  a. ice melting

  b. a piece of paper burning

  c. a bomb exploding

  d. an organism's metabolism producing a certain amount of heat

8.75 Which sentence best describes the following reaction?

$$2H_2(g) + O_2(g) \rightarrow 2H_2O(\ell) + \text{heat}$$

  a. It is an endothermic reaction.

  b. It is an exothermic double replacement reaction.

  c. It is a synthesis reaction that is also exothermic.

  d. It is a decomposition reaction that is also endothermic.

8.76 By which of the following mechanisms does a catalyst operate?

  a. It decreases the activation energy barrier for a reaction.

  b. It serves as a reactant and is consumed.

  c. It increases the temperature of a reaction.

  d. It increases the concentration of reactants.

8.77 Which of the following is NOT true of reversible chemical reactions?

  a. A chemical reaction is never complete.

  b. The products of the reaction also react to reform the original reactants.

  c. When the reaction is finished, both reactants and products are present in equal amounts.

  d. The reaction can result in an equilibrium.

8.78 Given the reaction:

$$2CO(g) + O_2(g) = 2CO_2(g)$$

When there is an increase in pressure to the system one would expect:

  a. an increase in the amount of carbon dioxide

  b. an increase in the amount of carbon monoxide and oxygen

  c. a decrease in the amount of carbon dioxide

  d. no change in the system

8.79 The following reaction is exothermic: $AgNO_3 + NaCl \leftrightarrow AgCl + NaNO_3$. How will the equilibrium be changed if the temperature is increased?

  a. Equilibrium will shift to the right.

  b. Equilibrium will shift to the left.

  c. The reaction will not proceed.

  d. Equilibrium will not change.

8.80 Consider the reaction $N_2(g) + 3H_2(g) \rightarrow 2NH_3(g) + \text{heat}$. Indicate the incorrect statement.

  a. An increase in temperature will shift the equilibrium to the right.

  b. An increase in pressure applied will shift the equilibrium to the right.

  c. The addition of ammonia will shift the equilibrium to the left.

  d. The addition of H_2 will shift the equilibrium to the right.

8.81 In a chemical reaction, [A] and [B] combine to form [C] and [D], as expressed by the reaction [A] + [B] = [C] + [D]. Select the statement that best describes the equilibrium condition.

  a. The reaction is shifted to the right.

  b. The concentrations of reactants and products are constant.

  c. The reaction is shifted to the left.

  d. The concentration of products is greater than the concentration of reactants.

8.82 What is the effect of the addition of a catalyst to a reaction in equilibrium?

  a. The reaction favors the formation of the products.

  b. The reaction favors the formation of the reactants.

  c. There is no change in composition of the reaction.

  d. The rate of the reaction slows.

8.83 For the reaction: $H_2(g) + Br_2(g) \rightarrow 2HBr(g)$, the reaction can be driven to the left by:

  a. increasing the pressure

  b. increasing the hydrogen

  c. increasing hydrogen bromide

  d. decreasing hydrogen bromide

8.84 What is the equilibrium constant K_c for the following equation?

$$2SO_2(g) + O_2(g) \rightleftarrows 2SO_3(g)$$

  a. $K_c = \dfrac{[SO_3]}{[SO_2][O_2]}$

  b. $K_c = \dfrac{[SO_2][O_2]}{[SO_3]}$

  c. $K_c = \dfrac{[SO_2]^2[O_2]}{[SO_3]^2}$

  d. $K_c = \dfrac{[SO_3]^2}{[SO_2]^2[O_2]}$

Chemistry for Thought

8.85 Refer to Figure 8.4 and answer the question. How would the total energy released as light for each light stick compare when they have both stopped glowing?

8.86 A mixture of the gases NOCl, NO, and Cl_2 is allowed to come to equilibrium at 400°C in a 1.50-L reaction container. Analysis shows the following number of moles of each substance to be present at equilibrium: NOCl = 1.80 mol, NO = 0.70 mol, and Cl_2 = 0.35 mol. Calculate the value of the equilibrium constant for the reaction at 400°C. The equation for the reaction is

$$2NOCl(g) \rightleftarrows 2NO(g) + Cl_2(g)$$

8.87 The equilibrium constant for the reaction $PCl_5 \rightleftarrows PCl_3 + Cl_2$ is 0.0245 at 250°C. What molar concentration of PCl_5 would be present at equilibrium if the concentrations of PCl_3 and Cl_2 were both 0.250 M?

8.88 At 448°C, the equilibrium constant for the reaction $H_2 + I_2 \rightleftarrows 2HI$ is 50.5. What concentration of I_2 would be found in an equilibrium mixture in which the concentrations of HI and H_2 were 0.500 M and 0.050 M, respectively?

8.89 Refer to Figure 8.10 and answer the questions. Would you expect a crushed antacid tablet (like Alka-Seltzer) to dissolve faster or slower than a whole tablet? Explain.

8.90 Refer to Figure 8.13 and answer the questions. Would the presence of a catalyst in the tube influence the equilibrium concentrations of the two gases? Explain.

8.91 In the blood, both oxygen (O_2) and poisonous carbon monoxide (CO) can react with hemoglobin (Hb). The respective compounds are HbO_2 and HbCO. When both gases are present in the blood, the following equilibrium is established.

$$HbCO + O_2 \rightleftarrows HbO_2 + CO$$

Use Le Châtelier's principle to explain why pure oxygen is often administered to victims of CO poisoning.

8.92 Use the concept of reaction rates to explain why no smoking is allowed in hospital areas where patients are being administered oxygen gas.

8.93 Suppose you have two identical unopened bottles of carbonated beverage. The contents of both bottles appear to be perfectly clear. You loosen the cap of one of the bottles and hear a hiss as gas escapes, and at the same time gas bubbles appear in the liquid. The liquid in the unopened bottle still appears to be perfectly clear. Explain these observations using the concept of equilibrium and Le Châtelier's principle. Remember, a carbonated beverage contains carbon dioxide gas dissolved in a liquid under pressure.

8.94 Someone once suggested that it is impossible to unscramble a scrambled egg. Describe an unscrambled and a scrambled egg in terms of the concept of entropy.

Case Study

Dallas grew up on a farm. He learned to work hard and to just tough it out when he didn't feel well. After his 50th birthday, he started having frequent bouts of heartburn. No matter how hard he tried, the heartburn pain could not be ignored. He treated it by drinking baking soda in water, which helped a little, but not much. Then he tried liquid antacids and tablets, but his daughter was concerned because he seemed to be eating them like candy, and she convinced him to see a doctor. The doctor diagnosed Dallas with gastro-esophageal reflux disease (GERD). The doctor explained to Dallas that the valve between his stomach and esophagus did not close tightly and that stomach acid was entering his esophagus and causing his pain. The doctor prescribed a proton-pump inhibitor to reduce the amount of acid produced and recommended that Dallas avoid certain foods. She also ordered an endoscopy to check for possible damage to the esophagus.

Can too many over-the-counter antacids be harmful? How can heartburn symptoms be distinguished from heart attack? What foods aggravate heartburn?

Follow-up to this Case Study appears at the end the chapter before the Concept Summary.

Learning Objectives

When you have completed your study of this chapter, you should be able to:

1 Write reaction equations that illustrate Arrhenius acid–base behavior. **(Section 9.1)**

2 Write reaction equations that illustrate Brønsted acid–base behavior, and identify Brønsted acids and bases from written reaction equations. **(Section 9.2)**

3 Name common acids. **(Section 9.3)**

4 Do calculations using the concept of the self-ionizaton of water. **(Section 9.4)**

5 Do calculations using the pH concept. **(Section 9.5)**

6 Write reaction equations that illustrate the characteristic reactions of acids. **(Section 9.6)**

7 Write reaction equations that represent neutralization reactions between acids and bases. **(Section 9.7)**

8 Write reaction equations that illustrate various ways to prepare salts, and do calculations using the concept of an equivalent of salt. **(Section 9.8)**

9 Demonstrate an understanding of the words *weak* and *strong* as applied to acids and bases. **(Section 9.9)**

10 Demonstrate an understanding of the titration technique used to analyze acids and bases. **(Section 9.10)**

11 Do calculations related to the analysis of acids and bases by titration. **(Section 9.11)**

12 Explain the concept of salt hydrolysis, and write equations to illustrate the concept. **(Section 9.12)**

13 Explain how buffers work, and write equations to illustrate their action. **(Section 9.13)**

A cids, bases, and salts are among the most common and important solutes found in solutions. Until late in the 19th century, these substances were characterized by such properties as taste and color changes induced in certain dyes. Acids taste sour; bases, bitter; and salts, salty. Litmus, a dye, is red in the presence of acids and blue in the presence of bases. These and other observations led to the correct conclusions that acids and bases are chemical opposites, and that salts are produced when acids and bases react with each other. Today, acids and bases are defined in more precise ways that are useful when studying their characteristics.

9.1 The Arrhenius Theory

Learning Objective

1. Write reaction equations that illustrate Arrhenius acid–base behavior.

In 1887, Swedish chemist Svante Arrhenius proposed a theory dealing with electrolytic dissociation. He defined **acids** as substances that dissociate when dissolved in water and produce hydrogen ions (H^+). Similarly, **bases** are substances that dissociate and release hydroxide ions (OH^-) into the solution. Hydrogen chloride (HCl) and sodium

Arrhenius acid Any substance that provides H^+ ions when dissolved in water.

Arrhenius base Any substance that provides OH^- ions when dissolved in water.

hydroxide (NaOH) are examples of an Arrhenius acid and base, respectively. They dissociate in water as follows:

$$HCl(aq) \rightarrow H^+(aq) + Cl^-(aq) \tag{9.1}$$

$$NaOH(aq) \rightarrow Na^+(aq) + OH^-(aq) \tag{9.2}$$

Note that the hydrogen ion is a bare proton, the nucleus of a hydrogen atom.

9.2 The Brønsted Theory

Learning Objective

2. Write reaction equations that illustrate Brønsted acid–base behavior, and identify Brønsted acids and bases from written reaction equations.

Arrhenius did not know that free hydrogen ions cannot exist in water. They covalently bond with water molecules to form hydronium ions (H_3O^+). The water molecules provide both electrons used to form the covalent bond:

$$H^+ + :\ddot{O}-H \longrightarrow \left[H-\ddot{O}-H\right]^+ \tag{9.3}$$
$$\qquad\qquad |\qquad\qquad\qquad\quad |$$
$$\qquad\qquad H \qquad\qquad\qquad\ H$$

new bond

Or, more simply,

$$H^+(aq) + H_2O(\ell) \rightarrow H_3O^+(aq) \tag{9.4}$$

In 1923, Johannes Brønsted in Denmark and Thomas Lowry in England proposed an acid–base theory that took into account this behavior of hydrogen ions. They defined an **acid** as any hydrogen-containing substance that donates a proton (hydrogen ion) to another substance and a **base** as any substance that accepts a proton.

In conformity with this theory, the acidic behavior of covalently bonded HCl molecules in water is written

$$HCl(aq) + H_2O(\ell) \rightleftharpoons H_3O^+(aq) + Cl^-(aq) \tag{9.5}$$

The HCl behaves as a Brønsted acid by donating a proton to a water molecule. The water molecule, by accepting the proton, behaves as a base.

The double arrows of unequal length in Equation 9.5 indicate that the reaction is reversible with the equilibrium lying far to the right. In actual water solutions, essentially 100% of the dissolved HCl is in the ionic form at equilibrium. Remember, both the forward and the reverse reactions are taking place at equilibrium (Section 8.6). When the reverse reaction occurs, hydronium ions donate protons to chloride ions to form HCl and H_2O molecules. Thus, H_3O^+ behaves as a Brønsted acid, and Cl^- behaves as a Brønsted base. We see from this discussion that when a substance like HCl behaves as a Brønsted acid by donating a proton, the species that remains (Cl^-) is a Brønsted base. The Cl^- is called the **conjugate base** of HCl. Every Brønsted acid and the base formed when it donates a proton is called a **conjugate acid–base pair.** Thus, in Equation 9.5, we see that HCl and Cl^- form a conjugate acid–base pair, as do H_3O^+ and H_2O for the reverse reaction. Notice that the acid and base in a conjugate acid–base pair differ only by a proton, H^+.

Brønsted acid Any hydrogen-containing substance that is capable of donating a proton (H^+) to another substance.

Brønsted base Any substance capable of accepting a proton from another substance.

conjugate base The species remaining when a Brønsted acid donates a proton.

conjugate acid–base pair A Brønsted acid and its conjugate base.

▶ Example 9.1

Identify all Brønsted acids, bases, and acid–base conjugate pairs in the reactions represented by the following equations:

a. $HNO_3(aq) + H_2O(\ell) \rightleftharpoons H_3O^+(aq) + NO_3^-(aq)$
b. $HClO_4(aq) + H_2O(\ell) \rightleftharpoons H_3O^+(aq) + ClO_4^-(aq)$
c. $NH_3(aq) + H_2O(\ell) \rightleftharpoons OH^-(aq) + NH_4^+(aq)$

Solution

a. Nitric acid, HNO_3, behaves as a Brønsted acid by donating a proton to H_2O, a base, in the forward reaction. In the reverse reaction, H_3O^+, the conjugate acid of H_2O, donates a proton to NO_3^-, the conjugate base of HNO_3. In summary, the Brønsted acids are HNO_3 and H_3O^+, the Brønsted bases are H_2O and NO_3^-, and the conjugate acid–base pairs are HNO_3/NO_3^- and H_3O^+/H_2O:

$$HNO_3(aq) + H_2O(\ell) \rightleftharpoons H_3O^+(aq) + NO_3^-(aq)$$

acid   base   acid   base

conjugate pair

conjugate pair

b. Similarly, perchloric acid ($HClO_4$) is a Brønsted acid, and H_2O is a Brønsted base (forward reaction). Also, H_3O^+ is a Brønsted acid, and the perchlorate ion (ClO_4^-) is a Brønsted base (reverse reaction). The conjugate acid–base pairs are $HClO_4/ClO_4^-$ and H_3O^+/H_2O:

$$HClO_4(aq) + H_2O(\ell) \rightleftharpoons H_3O^+(aq) + ClO_4^-(aq)$$

acid   base   acid   base

conjugate pair

conjugate pair

c. In this reaction, water donates a proton instead of accepting one. Therefore, H_2O is a Brønsted acid, and ammonia (NH_3) is a Brønsted base (forward reaction). The ammonium ion (NH_4^+) is an acid, and the hydroxide ion (OH^-) is a base (reverse reaction). Note that an Arrhenius base was a substance that released the OH^-, whereas according to Brønsted, the OH^- is a base. The conjugate acid–base pairs are H_2O/OH^- and NH_4^+/NH_3:

$$NH_3(aq) + H_2O(\ell) \rightleftharpoons OH^-(aq) + NH_4^+(aq)$$

base   acid   base   acid

conjugate pair

conjugate pair

▶ **LEARNING CHECK 9.1** Identify all Brønsted acids, bases, and acid–base conjugate pairs in the reactions represented by the following equations:

a. $HC_2H_3O_2(aq) + H_2O(\ell) \rightleftharpoons H_3O^+(aq) + C_2H_3O_2^-(aq)$
b. $NO_2^-(aq) + H_2O(\ell) \rightleftharpoons HNO_2(aq) + OH^-(aq)$
c. $HS^-(aq) + H_2O(\ell) \rightleftharpoons H_3O^+(aq) + S^{2-}(aq)$

9.3 Naming Acids

Learning Objective

3. Name common acids.

In Section 4.4, the rules for naming binary ionic compounds were given. The rules for naming binary covalent compounds and ionic compounds that contain polyatomic ions were discussed in Section 4.10. We now conclude our discussion of inorganic nomenclature by giving the rules used to name hydrogen-containing compounds that behave as acids. Examples of the two types of compounds that behave as acids were given in Section 9.2. Acids of the first type, represented by HCl, are compounds in which hydrogen is covalently bonded to a nonmetal. In acids of the second type, hydrogen is covalently bonded to a polyatomic ion. An example of the second type is HNO_3.

You probably recognized that HCl is a binary covalent compound that can be named by the rules given earlier in Section 4.10. According to those rules, HCl should be named hydrogen chloride. In fact, that is the correct name for the compound HCl that has not been dissolved in water and is represented in reaction equations by the notation HCl(g). Such compounds that have not been dissolved in water are said to be anhydrous (without water). However, when the gas is dissolved in water and represented in equations by the notation HCl(aq), it behaves as an acid and is given another name. The following rules are used to name acidic water solutions of such compounds:

1. The word *hydrogen* in the anhydrous compound name is dropped.
2. The prefix *hydro-* is attached to the stem of the name of the nonmetal that is combined with hydrogen.
3. The suffix *-ide* on the stem of the name of the nonmetal that is combined with hydrogen is replaced with the suffix *-ic*.
4. The word *acid* is added to the end of the name as a separate word.

▶ Example 9.2

Determine the name that would be given to each of the following binary covalent compounds in the anhydrous form and in the form of water solutions:

a. HCl (stomach acid)
b. H_2S (a gas produced when some sulfur-containing foods such as eggs decay)

Solution

a. The name of the anhydrous compound was given above as hydrogen chloride. The name of the water solution is obtained by dropping *hydrogen* from the anhydrous compound name and adding the prefix *hydro-* to the stem *chlor*. The *-ide* suffix on the *chlor* stem is replaced by the suffix *-ic* to give the name *hydrochloric*. The word *acid* is added, giving the final name *hydrochloric acid* for the water solution.

b. According to the rules of Section 4.10, the anhydrous compound name is hydrogen sulfide. The first two steps in obtaining the name of the water solution are to drop *hydrogen* and to add the prefix *hydro-* to the stem *sulf*. However, in acids involving sulfur as the nonmetal combined with hydrogen, the stem *sulf* is replaced by the entire name *sulfur* for pronunciation reasons. The next steps involve dropping the suffix *-ide*, adding the suffix *-ic*, and adding the word *acid*. The resulting name of the water solution is *hydrosulfuric acid*.

▶ **LEARNING CHECK 9.2** Determine the name that would be given to each of the following binary covalent compounds in the anhydrous form and in the form of water solutions:

a. HI
b. HBr

Acids of the second type, in which hydrogen is covalently bonded to a polyatomic ion, have the same name in the anhydrous form and in the form of water solutions. The names for these acids are based on the name of the polyatomic ion to which the hydrogen is bonded. The rules are as follows:

1. All hydrogens that are written as the first part of the formula of the acid are removed. The hydrogens are removed in the form of H^+ ions.
2. The polyatomic ion that remains after the H^+ ions are removed is named by referring to sources such as Table 4.7.
3. When the remaining polyatomic ion has a name ending in the suffix *-ate*, the suffix is replaced by the suffix *-ic*, and the word *acid* is added.
4. When the remaining polyatomic ion has a name ending in the suffix *-ite*, the suffix is replaced by the suffix *-ous*, and the word *acid* is added.
5. If the polyatomic ion contains sulfur or phosphorus, the stems *-sulf* or *-phosph* that remain in Steps 3 or 4, when the suffixes *-ate* or *-ite* are replaced, are expanded for pronunciation reasons to *-sulfur* and *-phosphor* before the *-ic* or *-ous* suffixes are added.

Compounds derived from the acid H_3PO_4 serve numerous important functions in the body, including the control of acidity in urine and body cells and the storage of energy in the form of ATP. Name this important acid.

Solution

The removal of the three H^+ ions leaves behind the PO_4^{3-} polyatomic ion. This ion is named the phosphate ion in Table 4.7. According to Rules 1–4, the *-ate* suffix is replaced by the *-ic* suffix to give the name *phosphic acid*. However, Rule 5 must be used for this phosphorus-containing acid. The stem is expanded to *phosphor* to give the final name, *phosphoric acid*.

▶ **LEARNING CHECK 9.3** The acid H_2CO_3 is involved in many processes in the body, including the removal of CO_2 gas produced by cellular metabolism and the control of the acidity of various body fluids. Name this important acid.

9.4 The Self-Ionization of Water

Learning Objective

4. Do calculations using the concept of the self-ionization of water.

In Examples 9.1a and b, water behaved as a Brønsted base. In Example 9.1c, it was a Brønsted acid. But what happens when only pure water is present? The answer is that water behaves as both an acid and a base and undergoes a self- or auto-ionization. The equation representing this self-ionization is

$$H_2O(\ell) + H_2O(\ell) \rightleftarrows H_3O^+(aq) + OH^-(aq) \tag{9.6}$$

or

$$\begin{array}{ccccc}
\overset{..}{O}-H & & \overset{..}{O}-H & & \left[H-\overset{..}{O}-H \right]^+ \\
| & + & | & \longrightarrow & | & + & \overset{..}{O}-H^- \\
H & & H & & H
\end{array}$$

The transfer of a proton from one water molecule (the acid) to another (the base) causes one H_3O^+ and one OH^- to form. Therefore, in pure water the concentrations of H_3O^+ and OH^- must be equal. At 25°C these concentrations are 10^{-7} mol/L (M). Thus, the equilibrium position is far to the left, as indicated by the arrows in Equation 9.6. Unless noted otherwise, all concentrations and related terms in this chapter are given at 25°C.

The term **neutral** is used to describe any water solution in which the concentrations of H_3O^+ and OH^- are equal. Thus, pure water is neutral because each liter of pure water contains 10^{-7} mol H_3O^+ and 10^{-7} mol OH^-, at equilibrium. Although all water solutions are not necessarily neutral, it is true that in any solution that contains water, the product of the molar concentrations of H_3O^+ and OH^- is a constant. This becomes apparent by writing the equilibrium expression for Reaction 9.6:

neutral A term used to describe any water solution in which the concentrations of H_3O^+ and OH^- are equal. Also, a water solution with pH = 7.

$$K = \frac{[H_3O^+][OH^-]}{[H_2O][H_2O]} = \frac{[H_3O^+][OH^-]}{[H_2O]^2} \tag{9.7}$$

This expression contains the square of the concentration of water in the denominator. However, only a tiny amount of water actually reacts to form H_3O^+ and OH^- (10^{-7} mol/L), so the concentration of water remains essentially constant. Rearrangement of Equation 9.7 gives

$$K[H_2O]^2 = [H_3O^+][OH^-] \tag{9.8}$$

This equation may be written as

$$K[H_2O]^2 = K_w = [H_3O^+][OH^-] \qquad (9.9)$$

ion product of water The equilibrium constant for the dissociation of pure water into H_3O^+ and OH^-.

where K_w is a new constant called the **ion product of water.** It is a constant because it is equal to the product of two constants, K and $[H_2O]^2$. At 25°C, K_w can be evaluated from the measured values of $[H_3O^+]$ and $[OH^-]$ in pure water.

$$\begin{aligned} K_w &= [H_3O^+][OH^-] = (1.0 \times 10^{-7}\,\text{mol/L})(1.0 \times 10^{-7}\,\text{mol/L}) \\ &= 1.0 \times 10^{-14}\,(\text{mol/L})^2 \end{aligned} \qquad (9.10)$$

acidic solution A solution in which the concentration of H_3O^+ is greater than the concentration of OH^-. Also, a solution in which pH is less than 7.

basic or alkaline solution A solution in which the concentration of OH^- is greater than the concentration of H_3O^+. Also, a solution in which pH is greater than 7.

Equation 9.10 is valid not only for pure water but for any solution in which water is the solvent. Notice that we are including units for equilibrium constants in this discussion. This makes the calculation of concentrations easier to follow and understand.

A solution is classified as **acidic** when the concentration of H_3O^+ is greater than the concentration of OH^-. In a **basic** or **alkaline solution,** the concentration of OH^- is greater than that of H_3O^+. However, the product of the molar concentrations of H_3O^+ and OH^- will be $1.0 \times 10^{-14}\,(\text{mol/L})^2$ in either case. Many acidic and basic materials are found in the home (see) Figure 9.1).

Figure 9.1 Acidic and basic materials are common in the home. Look at the photos, and note a single category to which most of the acidic materials belong, and one to which most of the basic materials belong.

© Cengage Learning/West

© Cengage Learning/West

1 Acidic materials found in the home.

2 Basic or alkaline materials found in the home.

▶ Example 9.4

Classify each of the following solutions as acidic, basic, or neutral. Calculate the molar concentration of the ion whose concentration is not given.

a. $[H_3O^+] = 1.0 \times 10^{-4}$ mol/L
b. $[OH^-] = 1.0 \times 10^{-9}$ mol/L
c. $[OH^-] = 1.0 \times 10^{-6}$ mol/L

Solution

a. Because $[H_3O^+] = 1.0 \times 10^{-4}$ mol/L, a rearrangement of Equation 9.10 gives

$$[OH^-] = \frac{1.0 \times 10^{-14}\,(\text{mol/L})^2}{[H_3O^+]} = \frac{1.0 \times 10^{-14}\,(\text{mol/L})^2}{1.0 \times 10^{-4}\,\text{mol/L}}$$

$$= 1.0 \times 10.0^{-10}\,\text{mol/L}$$

Thus,

$$[H_3O^+] = 1.0 \times 10^{-4}\,\text{mol/L} \quad \text{and}$$

$$[OH^-] = 1.0 \times 10^{-10}\,\text{mol/L}$$

The solution is acidic because the $[H_3O^+]$ is greater than the $[OH^-]$; 1.0×10^{-4} is greater than 1.0×10^{-10}.

b. Similarly, $[OH^-] = 1.0 \times 10^{-9}$ mol/L; therefore,

$$[H_3O^+] = \frac{1.0 \times 10^{-14}\,(\text{mol/L})^2}{[OH^-]} = \frac{1.0 \times 10^{-14}\,(\text{mol/L})^2}{1.0 \times 10^{-9}\,\text{mol/L}}$$

$$= 1.0 \times 10.0^{-5}\,\text{mol/L}$$

This solution is also acidic because 1.0×10^{-5} is greater than 1.0×10^{-9}.

c. $[OH^-] = 1.0 \times 10^{-6}$ mol/L; therefore,

$$[H_3O^+] = \frac{1.0 \times 10^{-14}\,(\text{mol/L})^2}{[OH^-]} = \frac{1.0 \times 10^{-14}\,(\text{mol/L})^2}{1.0 \times 10^{-6}\,\text{mol/L}}$$

$$= 1.0 \times 10.0^{-8}\,\text{mol/L}$$

This solution is basic because the OH^- concentration (1.0×10^{-6} mol/L) is greater than the H_3O^+ concentration (1.0×10^{-8} mol/L).

▶ **LEARNING CHECK 9.4** Classify each of the following solutions as acidic, basic, or neutral. Calculate the molarity of the ion whose concentration is not given.

a. $[OH^-] = 1.0 \times 10^{-5}$ mol/L
b. $[H_3O^+] = 1.0 \times 10^{-9}$ mol/L
c. $[H_3O^+] = 1.0 \times 10^{-2}$ mol/L

9.5 The pH Concept

Learning Objective

5. Do calculations using the pH concept.

In Section 9.4, you learned that the concentration of H_3O^+ in pure water is 1.0×10^{-7} M. Chemists, technologists, and other laboratory personnel routinely work with solutions in which the H_3O^+ concentration may be anywhere from 10 to 10^{-14} M. Because of the inconvenience of working with numbers that extend over such a wide range, chemists long ago adopted a shortcut notation known as the pH. Mathematically, the **pH** is defined in terms of the negative logarithm (log) of $[H_3O^+]$ by the following two equations, where we now introduce the common practice of substituting H^+ for H_3O^+ to simplify equations:

$$pH = -\log[H^+] \qquad (9.11)$$

$$[H^+] = 1 \times 10^{-pH} \qquad (9.12)$$

Thus, pH is simply the negative of the exponent used to express the hydrogen-ion concentration in moles per liter.

pH The negative logarithm of the molar concentration of H^+ (H_3O^+) in a solution.

▶ **Example 9.5**

Express the following concentrations in terms of pH:

a. $[H^+] = 1 \times 10^{-5}$ mol/L
b. $[OH^-] = 1 \times 10^{-9}$ mol/L
c. $[H^+] = 1 \times 10^{-7}$ mol/L
d. $[H^+] = 1 \times 10^{-11}$ mol/L

Solution

a. pH is the negative of the exponent used to express $[H^+]$. Therefore, pH $= -(-5) = 5.0$
b. Here $[OH^-]$ is given, and $[H^+]$ must be calculated. We remember Equation 9.10 and get

$$[H^+] = \frac{1.0 \times 10^{-14}\,(mol/L)^2}{[OH^-]} = \frac{1.0 \times 10^{-14}\,(mol/L)^2}{1 \times 10^{-9}\,mol/L} = 1 \times 10^{-5}\,mol/L$$

Therefore, pH $= -(-5) = 5.0$
c. This pH $= -(-7) = 7.0$. $[H^+] = 1 \times 10^{-7}$ mol/L and corresponds to pure water and neutral solutions. Thus, a pH of 7 represents neutrality.
d. pH $= -(-11) = 11.0$

▶ **LEARNING CHECK 9.5** Express the following concentrations in terms of pH.

a. $[H^+] = 1 \times 10^{-14}$ mol/L
b. $[OH^-] = 1.0$ mol/L
c. $[OH^-] = 1 \times 10^{-8}$ mol/L

The pH value of 5.0 obtained in Example 9.5a corresponds to a solution in which the H^+ concentration is greater than the OH^- concentration. Thus, the solution is acidic. Any solution with a pH less than 7 is classified as acidic. Any solution with a pH greater than 7 is classified as basic or alkaline. The pH values of some familiar solutions are given in ▶ Table 9.1.

TABLE 9.1 Relationships Between $[H^+]$, $[OH^-]$, and pH

	$[H^+]$	$[OH^-]$	pH	Examples (solids are dissolved in water)
	10^0	10^{-14}	0	HCl (1 mol/L)
	10^{-1}	10^{-13}	1	Gastric juice
	10^{-2}	10^{-12}	2	Lemon juice
	10^{-3}	10^{-11}	3	Vinegar, carbonated drink Aspirin
	10^{-4}	10^{-10}	4	Orange juice Apple juice
	10^{-5}	10^{-9}	5	Black coffee
	10^{-6}	10^{-8}	6	Normal urine (average value) Milk, liquid dishwashing detergent
Neutral	10^{-7}	10^{-7}	7	Saliva, pure water Blood
	10^{-8}	10^{-6}	8	Soap (not synthetic detergent) Baking soda Phosphate-containing detergent
	10^{-9}	10^{-5}	9	Milk of magnesia Powdered household cleanser
	10^{-10}	10^{-4}	10	Phosphate-free detergent
	10^{-11}	10^{-3}	11	Household ammonia Liquid household cleaner
	10^{-12}	10^{-2}	12	
	10^{-13}	10^{-1}	13	NaOH (0.1 mol/L)
	10^{-14}	10^0	14	NaOH (1 mol/L)

▶ Example 9.6

Determine the H^+ and OH^- molar concentrations that correspond to the following pH values:

a. pH = 9.0
b. pH = 3.0
c. pH = 11.0

Solution

In each case, the relationships $[H^+] = 1 \times 10^{-pH}$ and $K_w = [H^+][OH^-] = 1.0 \times 10^{-14}$ (mol/L)^2 can be used. Note that we have substituted $[H^+]$ for $[H_3O^+]$ in Equation 9.10.

a. Because pH = 9.0, $[H^+] = 1 \times 10^{-pH} = 1 \times 10^{-9}$ mol/L

$$[OH^-] = \frac{K_w}{[H^+]} = \frac{1.0 \times 10^{-14}\,(mol/L)^2}{1 \times 10^{-9}\,mol/L} = 1 \times 10^{-5}\,mol/L$$

b. Because pH = 3.0, $[H^+] = 1 \times 10^{-pH} = 1 \times 10^{-3}$ mol/L

$$[OH^-] = \frac{K_w}{[H^+]} = \frac{1.0 \times 10^{-14}\,(mol/L)^2}{1 \times 10^{-3}\,mol/L} = 1 \times 10^{-11}\,mol/L$$

c. Because pH = 11.0, $[H^+] = 1 \times 10^{-pH} = 1 \times 10^{-11}$ mol/L

$$[OH^-] = \frac{K_w}{[H^+]} = \frac{1.0 \times 10^{-14}\,(mol/L)^2}{1 \times 10^{-11}\,mol/L} = 1 \times 10^{-3}\,mol/L$$

▶ **LEARNING CHECK 9.6** Determine the $[H^+]$ and $[OH^-]$ values that correspond to the following pH values:

a. pH = 10.0
b. pH = 4.0
c. pH = 5.0

It is apparent from Table 9.1 that not all solutions have pH values that are neat whole numbers. For example, the pH of vinegar is about 3.3. How do we deal with such numbers? The H^+ concentration of vinegar could be written $1 \times 10^{-3.3}$, but it is not convenient to work with exponents that are not whole numbers. When exact values are not needed, the pH, $[H^+]$, and so on can be expressed as a range. Thus, the H^+ concentration of vinegar is between 1×10^{-3} and 1×10^{-4} mol/L. Similarly, a solution with $[H^+] = 2 \times 10^{-5}$ mol/L has a pH between 4 and 5.

When more exact values are needed, we must work with logarithms. This is most easily done by using a hand calculator. A hydrogen-ion concentration is converted to pH by taking the logarithm and changing its sign. ▶ Table 9.2 gives the steps of a typical calculator procedure (what button is pushed, etc.) and a typical calculator readout or display for the conversion of $[H^+] = 3.6 \times 10^{-4}$ mol/L into pH. The pH from Table 9.2 would be recorded as 3.44.

▶ **LEARNING CHECK 9.7** Convert the following $[H^+]$ values into pH:

a. $[H^+] = 4.2 \times 10^{-5}$ mol/L
b. $[H^+] = 8.1 \times 10^{-9}$ mol/L

TABLE 9.2 Calculating pH from Molarity with a Calculator

Step	Calculator Procedure	Calculator Display
1. Enter 3.6	Press buttons 3, ., 6	3.6
2. Enter 10^{-4}	Press button that activates exponential mode (EE, Exp, etc.)	3.6^{00}
	Press 4	3.6^{04}
	Press change-sign button ($\pm$, etc.)	3.6^{-04}
3. Take logarithm	Press log button (log, etc.)	-3.4437
4. Change sign	Press change-sign button ($\pm$, etc.)	3.4437

Calculators can also be used to convert pH values into corresponding molar concentrations. The steps are given in ▶ Table 9.3 for the conversion of a pH value of 5.92 into $[H^+]$. As shown in Table 9.3, a pH of 5.92 corresponds to a $[H^+]$ of 1.2×10^{-6} mol/L. Note that the number of figures to the right of the decimal in a pH value should be the same as the number of significant figures in the $[H^+]$ value. Thus, the two figures to the right of the decimal in pH = 5.92 is reflected in the two significant figures in $[H^+] = 1.2 \times 10^{-6}$ mol/L.

TABLE 9.3 Calculating Molarity from pH with a Calculator

Step	Calculator Procedure	Calculator Display
1. Enter 5.92	Press 5, ., 9, 2	5.92
2. Change sign	Press change-sign button ($\pm$, etc.)	-5.92
3. Take antilog	Press antilog or 10^x button, or (more commonly) press inv. or 2nd function button and then log button	.0000012 or 1.2^{-06}

▶ **LEARNING CHECK 9.8** Convert the following pH values into molar concentrations of H^+:

a. pH = 2.75
b. pH = 8.33

9.6 Properties of Acids

Learning Objective

6. Write reaction equations that illustrate the characteristic reactions of acids.

Acids and bases are used so often in most laboratories that stock solutions are kept readily available at each work space. The common solutions, their concentrations, and label designations are given in ▶ Table 9.4.

▶ **Example 9.7**

Describe how you would make 250 mL of 1 M HNO_3 by using a dilute HNO_3 stock solution.

Solution

According to Table 9.4, the dilute HNO_3 stock solution is 6 M. The volume of 6 M HNO_3 needed to make 250 mL of 1 M HNO_3 is obtained by using Equation 7.9. It must be remembered that this equation is useful only for dilution problems such as this one. The equation is not useful for calculations involving the volume of one solution that will react with a specific volume of a second solution (Sections 9.10 and 9.11).

$$C_c V_c = C_d V_d$$

$$(6 \text{ M}) V_c = (1 \text{ M})(250 \text{ mL})$$

$$V_c = \frac{(1 \text{ M})(250 \text{ mL})}{6 \text{ M}} = 41.7 \text{ mL}$$

Because the molarity is given using only one significant figure, the solution can be made without too much attention to volumetric flasks and the like. Thus, 42 mL of 6 M HNO_3 is measured with a graduated cylinder, and this is added to 208 mL of distilled water that has also been measured with a graduated cylinder.

▶ **LEARNING CHECK 9.9** Describe how you would prepare 500 mL of 3.0 M aqueous ammonia using a concentrated NH_3 stock solution (Table 9.4).

TABLE 9.4 Common Laboratory Acids and Bases

Name	Formula	Label Concentration	Molarity
Acids			
Acetic acid	$HC_2H_3O_2$	Glacial	18
Acetic acid	$HC_2H_3O_2$	Dilute	6
Hydrochloric acid	HCl	Concentrated	12
Hydrochloric acid	HCl	Dilute	6
Nitric acid	HNO_3	Concentrated	16
Nitric acid	HNO_3	Dilute	6
Sulfuric acid	H_2SO_4	Concentrated	18
Sulfuric acid	H_2SO_4	Dilute	3
Bases			
Aqueous ammonia[a]	NH_3	Concentrated	15
Aqueous ammonia	NH_3	None usually given	6
Sodium hydroxide	NaOH	None usually given	6

[a]Often erroneously called ammonium hydroxide and given the formula NH_4OH.

Different acids have different properties that make some more practical than others for specific uses. However, all acids have certain properties in common. Two of these were mentioned earlier—all acids taste sour and produce H_3O^+ ions when dissolved in water. In addition, all acids undergo characteristic double-replacement reactions with solid oxides, hydroxides, carbonates, and bicarbonates (see ▶ Figure 9.2).

Figure 9.2 The reaction of hydrochloric acid with three natural forms of CaCo₃. Left to right: limestone, marble, and eggshell.

$$2HCl(aq) + CuO(s) \rightarrow CuCl_2(aq) + H_2O(\ell) \tag{9.13}$$
$$\text{copper}$$
$$\text{oxide}$$

$$2HCl(aq) + Ca(OH)_2(s) \rightarrow CaCl_2(aq) + 2H_2O(\ell) \tag{9.14}$$
$$\text{calcium}$$
$$\text{hydroxide}$$

$$2HCl(aq) + CaCO_3(s) \rightarrow CaCl_2(aq) + CO_2(g) + H_2O(\ell) \tag{9.15}$$
$$\text{calcium}$$
$$\text{carbonate}$$

$$2HCl(aq) + Sr(HCO_3)_2(s) \rightarrow SrCl_2(aq) + 2CO_2(g) + 2H_2O(\ell) \tag{9.16}$$
$$\text{strontium}$$
$$\text{bicarbonate}$$

Notice that the preceding reactions are written using molecular equations (Section 5.7).

Reactions involving ionic substances can also be written as total ionic equations or net ionic equations. Equation 9.13 is written in total ionic form in Equation 9.17.

$$\underbrace{2H^+(aq) + 2Cl^-(aq)}_{2HCl} + CuO(s) \longrightarrow \underbrace{Cu^{2+}(aq) + 2Cl^-(aq)}_{CuCl_2} + H_2O(\ell) \tag{9.17}$$

We see that the chloride ions are spectator ions, so a net ionic equation can be written as shown in Equation 9.18.

$$2H^+(aq) + CuO(s) \rightarrow Cu^{2+}(aq) + H_2O(\ell) \tag{9.18}$$

The general nature of the reaction is emphasized in the net ionic form because the H^+ could come from any acid.

Remember that correctly written molecular equations must have their atoms balanced (Section 5.1). Total ionic and net ionic equations must also have their atoms balanced, but in addition the total charges on each side of the equation must balance. In Equation 9.18, for example, the two H^+ ions provide two positive charges on the left, which are balanced by the two positive charges of Cu^{2+} on the right.

▶ Example 9.8

Write Equations 9.14 and 9.15 in total ionic and net ionic forms.

Solution

In Equation 9.14, HCl and $CaCl_2$ are soluble and ionizable.

Total ionic: $2H^+(aq) + 2Cl^-(aq) + Ca(OH)_2(s) \rightarrow Ca^{2+}(aq) + 2Cl^-(aq) + 2H_2O(\ell)$

The Cl^- is a spectator ion.

$\qquad$ Net ionic: $2H^+(aq) + Ca(OH)_2(s) \rightarrow Ca^{2+}(aq) + 2H_2O(\ell)$

In Equation 9.15, HCl and $CaCl_2$ are soluble and ionizable.

Total ionic: $2H^+(aq) + 2Cl^-(aq) + CaCO_3(s) \rightarrow Ca^{2+}(aq) + 2Cl^-(aq) + CO_2(g) + H_2O(\ell)$

Again, Cl^- is a spectator ion.

$\qquad$ Net ionic: $2H^+(aq) + CaCO_3(s) \rightarrow Ca^{2+}(aq) + CO_2(g) + H_2O(\ell)$

▶ **LEARNING CHECK 9.10** Write Equation 9.16 in total and net ionic forms. Consider HCl and $SrCl_2$ as soluble and ionizable.

Another property of acids is their ability to react with (and dissolve) certain metals to yield hydrogen gas. This is a redox reaction (Section 5.3), as evidenced by the change in oxidation number of hydrogen as the reaction takes place. In compounds such as acids, hydrogen has an oxidation number of $+1$. In hydrogen gas, the oxidation number is 0 (Section 5.3). Thus, hydrogen is reduced during the reaction, and, as shown by Equations 9.19 and 9.20, the metal is oxidized. The ability to reduce hydrogen ions to hydrogen gas is not the same for all metals. Some are such strong reducing agents that they can react with hydrogen ions of very low concentrations such as that found in water. Others are so weak as reducing agents that they cannot react with H^+ at the high concentration found in concentrated acids. These tendencies are represented by the **activity series** of metals shown in ▶ Table 9.5. The higher a metal is in the series, the more active it is as a reducing agent. Some typical reactions are given in Equations 9.19 and 9.20 (see ▶ Figure 9.3).

activity series A tabular representation of the tendencies of metals to react with H^+.

TABLE 9.5 The Activity Series of the Metals

Metal	Symbol	Comments
Potassium	K	React violently with cold water
Sodium	Na	
Calcium	Ca	Reacts slowly with cold water
Magnesium	Mg	React very slowly with steam, but quite rapidly in higher H^+ concentrations
Aluminum	Al	
Zinc	Zn	
Chromium	Cr	
Iron	Fe	React in moderately high H^+ concentrations
Nickel	Ni	
Tin	Sn	
Lead	Pb	
Copper	Cu	Do not react with H^+
Mercury	Hg	
Silver	Ag	
Platinum	Pt	
Gold	Au	

Figure 9.3 Metals vary in their ability to reduce hydrogen ions (H^+) to hydrogen gas (H_2). This difference is apparent when copper, iron, zinc, and magnesium (left to right) are put into hydrochloric acid solution (HCl) of the same molarity. Are these results consistent with Table 9.5? Explain.

$$\text{Molecular equation: } Zn(s) + 2HCl(aq) \rightarrow ZnCl_2(aq) + H_2(g) \tag{9.19}$$

$$\text{Net ionic equation: } Zn(s) + 2H^+(aq) \rightarrow Zn^{2+}(aq) + H_2(g)$$

$$\text{Molecular equation: } 2K(s) + 2H_2O(\ell) \rightarrow 2KOH(aq) + H_2(g) \tag{9.20}$$

$$\text{Net ionic equation: } 2K(s) + 2H_2O(\ell) \rightarrow 2K^+(aq) + 2OH^-(aq) + H_2(g)$$

▶ **LEARNING CHECK 9.11** Write molecular, total ionic, and net ionic equations to represent the following reactions:

a. Calcium (Ca) with cold water (NOTE: $Ca(OH)_2$ is not soluble in water.)
b. Mg with H_2SO_4 (NOTE: $MgSO_4$ is soluble and ionizable in water.)

9.7 Properties of Bases

Learning Objective

7. Write reaction equations that represent neutralization reactions between acids and bases.

neutralization reaction A reaction in which an acid and base react completely, leaving a solution that contains only a salt and water.

Solutions containing bases feel soapy or slippery and change the color of litmus from red to blue. Equation 9.14 illustrates their most characteristic chemical property—they react readily with acids. In most of the earliest acid–base reactions studied, the complete reaction of an acid with a base produced a neutral solution. For this reason, such reactions were often called **neutralization reactions** (see below). It is now known that many "neutralization" reactions do not produce neutral solutions (Section 9.12). However, the name for the reactions is still used. More than 20 billion pounds of sodium hydroxide (NaOH) is produced and used in the United States each year. This useful crystalline solid is quite soluble in water and dissociates to form basic solutions.

$$NaOH(s) + H_2O(\ell) \rightarrow Na^+(aq) + OH^-(aq) \tag{9.21}$$

The neutralization reaction between NaOH and HCl is

Molecular equation: $HCl(aq) + NaOH(aq) \rightarrow NaCl(aq) + H_2O(\ell)$ (9.22)

Total ionic equation: $H^+(aq) + Cl^-(aq) + Na^+(aq) + OH^-(aq) \rightarrow$ (9.23)
$$Na^+(aq) + Cl^-(aq) + H_2O(\ell)$$

Net ionic equation: $H^+(aq) + OH^-(aq) \rightarrow H_2O(\ell)$ (9.24)

The molecular equation (9.22) illustrates the following statement, which is a common definition of neutralization: During a neutralization reaction, an acid and a base combine to form a salt and water. (More is said about salts in the next section.) The net ionic form of the equation (9.24) emphasizes the general nature of neutralization reactions: H^+ ions (from any source) react with OH^- (from any source) to form water.

Bases also react with fats and oils and convert them into smaller, soluble molecules. For this reason, most household cleaning products contain basic substances. For example, lye (NaOH) is the active ingredient in numerous drain cleaners, and many liquid household cleaners contain ammonia.

> ▶ **LEARNING CHECK 9.12** Write molecular, total ionic, and net ionic equations to represent neutralization reactions between the following acids and bases:
>
> **a.** HNO_3 and NaOH
> **b.** H_2SO_4 and KOH

9.8 Salts

Learning Objective

8. Write reaction equations that illustrate various ways to prepare salts, and do calculations using the concept of an equivalent of salt.

At room temperature, **salts** are solid ionic compounds that contain the **cation** (positive ion) of a base and the **anion** (negative ion) of an acid. Thus, ordinary table salt (NaCl) contains Na^+, the cation of NaOH, and Cl^-, the anion of HCl (look again at Equation 9.22). Similarly, $CuSO_4$ is a salt containing the cation of $Cu(OH)_2$ and the anion of H_2SO_4. You must be careful to think of the term *salt* in a general way, and not as representing only table salt (NaCl).

Some acids and bases are not stable enough to be isolated even though their salts are. For example, carbonic acid (H_2CO_3) cannot be isolated in the pure state. When it forms in water, it promptly decomposes:

$$H_2CO_3(aq) \rightleftharpoons H_2O(\ell) + CO_2(g)$$ (9.25)

Despite this characteristic, salts of carbonic acid, such as Na_2CO_3 and $NaHCO_3$, are quite stable.

It is not necessary to identify the parent acid and base in order to write correct salt formulas or names. Just remember that the cation of a salt can be any positive ion except H^+, and it will usually be a simple metal ion or NH_4^+. The salt anion can be any negative ion except OH^-. Most of the polyatomic anions you will use were given earlier in Table 4.7. The rules for naming salts were given in Sections 4.4 and 4.10.

As previously observed, salts dissolved in solution are dissociated into ions. The salt can be recovered by evaporating away the water solvent. When this is done carefully, some salts retain specific numbers of water molecules as part of the solid crystalline structure. Such salts are called **hydrates,** and the retained water is called the **water of hydration.** Most hydrates lose all or part of the water of hydration when they are heated to moderate or high temperatures. A number of useful hydrates are given in ▶ Table 9.6.

salt A solid crystalline ionic compound at room temperature that contains the cation of a base and the anion of an acid.

cation A positively charged ion.

anion A negatively charged ion.

hydrate A salt that contains specific numbers of water molecules as part of the solid crystalline structure.

water of hydration Water retained as part of the solid crystalline structure of some salts.

TABLE 9.6 Some Useful and Common Hydrates

Formula	Chemical Name	Common Name	Uses
$CaSO_4 \cdot H_2O$	Calcium sulfate monohydrate	Plaster of Paris	Plaster, casts, molds
$CaSO_4 \cdot 2H_2O$	Calcium sulfate dihydrate	Gypsum	Casts, molds, wallboard
$MgSO_4 \cdot 7H_2O$	Magnesium sulfate heptahydrate	Epsom salts	Cathartic
$Na_2B_4O_7 \cdot 10H_2O$	Sodium tetraborate decahydrate	Borax	Laundry
$Na_2CO_3 \cdot 10H_2O$	Sodium carbonate decahydrate	Washing soda	Water softener
$Na_3PO_4 \cdot 12H_2O$	Sodium phosphate dodecahydrate	Trisodium phosphate (TSP)	Water softener
$Na_2SO_4 \cdot 10H_2O$	Sodium sulfate decahydrate	Glauber's salt	Cathartic
$Na_2S_2O_3 \cdot 5H_2O$	Sodium thiosulfate pentahydrate	Hypo	Film photography

Many salts occur in nature, and some are used as industrial raw materials. Examples are sodium chloride, NaCl (a source of Cl_2 and NaOH); calcium carbonate or limestone, $CaCO_3$ (a source of cement and building stone); and calcium phosphate or rock phosphate, $Ca_3(PO_4)_2$ (a source of fertilizer). In the laboratory, salts can be prepared by reacting a solution of an appropriate acid with a metal, a metal oxide, a metal hydroxide, a metal carbonate, or a metal bicarbonate. These reactions, given earlier as examples of acid properties (Equations 9.19, 9.13, 9.14, 9.15, 9.16), are given below in a general form:

$$\text{acid} + \text{metal} \rightarrow \text{salt} + H_2 \tag{9.26}$$

$$\text{acid} + \text{metal oxide} \rightarrow \text{salt} + H_2O \tag{9.27}$$

$$\text{acid} + \text{metal hydroxide} \rightarrow \text{salt} + H_2O \tag{9.28}$$

$$\text{acid} + \text{metal carbonate} \rightarrow \text{salt} + H_2O + CO_2 \tag{9.29}$$

$$\text{acid} + \text{metal bicarbonate} \rightarrow \text{salt} + H_2O + CO_2 \tag{9.30}$$

▶ Example 9.9

Write equations to represent the preparation of $Mg(NO_3)_2$, using Reactions 9.26 through 9.30. Use molecular equations to emphasize the salt formation.

Solution

In each case, a dilute solution of nitric acid (HNO_3) is reacted with magnesium metal or the appropriate magnesium compound:

$$2HNO_3(aq) + Mg(s) \rightarrow Mg(NO_3)_2(aq) + H_2(g)$$

$$2HNO_3(aq) + MgO(s) \rightarrow Mg(NO_3)_2(aq) + H_2O(\ell)$$

$$2HNO_3(aq) + Mg(OH)_2(s) \rightarrow Mg(NO_3)_2(aq) + 2H_2O(\ell)$$

$$2HNO_3(aq) + MgCO_3(s) \rightarrow Mg(NO_3)_2(aq) + H_2O(\ell) + CO_2(g)$$

$$2HNO_3(aq) + Mg(HCO_3)_2(s) \rightarrow Mg(NO_3)_2(aq) + 2H_2O(\ell) + 2CO_2(g)$$

▶ **LEARNING CHECK 9.13** Write balanced equations to represent the preparation of $AlCl_3$, using Reactions 9.26 through 9.30. Use molecular equations to emphasize the salt formation.

We have expressed the amounts of materials involved in chemical reactions primarily in terms of mass (grams) or number of moles. However, in certain applications, it is important to express the amount of salt in a solution in terms of the amount of electrical charge represented by the ions of the salt. This is especially true in medical applications, where the levels and balance of electrolytes in various body fluids are extremely important.

A unit that expresses the amount of ionic electrical charge for salts is the equivalent. One **equivalent of salt** is the amount that will produce 1 mol of positive (or negative) charges when dissolved and dissociated. To determine the amount of salt that represents 1 equivalent (eq), you must know what ions are produced when the salt dissociates. For example, potassium chloride (KCl), an electrolyte often administered to patients following surgery, dissociates as follows: $KCl(aq) \rightarrow K^+(aq) + Cl^-(aq)$. Thus, 1 mol, or 74.6 g, of solid KCl provides 1 mol of positive charges (1 mol of K^+ ions) when it dissolves and dissociates. Thus, 1 eq of KCl is equal to 1 mol, or 74.6 g, of KCl.

equivalent of salt The amount that will produce 1 mol of positive electrical charge when dissolved and dissociated.

▶ Example 9.10

Determine the number of equivalents and milliequivalents (meq) of salt contained in the following:

a. 0.050 mol KCl
b. 0.050 mol $CaCl_2$

Solution

a. As shown above, 1 mol KCl = 1 eq KCl. Therefore, 0.050 mol KCl = 0.050 eq KCl. Also, because 1 eq = 1000 meq,

$$0.050 \text{ eq} \times \frac{1000 \text{ meq}}{1 \text{ eq}} = 50 \text{ meq}$$

b. The dissociation reaction is

$$CaCl_2(aq) \rightarrow Ca^{2+}(aq) + 2Cl^-(aq)$$

Thus, we see that 1 mol $CaCl_2$ produces 1 mol Ca^{2+} or 2 mol of positive charges. Thus, 1 mol $CaCl_2$ = 2 eq $CaCl_2$. Therefore,

$$0.050 \text{ mol } CaCl_2 \times \frac{2 \text{ eq } CaCl_2}{1 \text{ mol } CaCl_2} = 0.10 \text{ eq } CaCl_2$$

Also, because 1 eq = 1000 meq,

$$0.10 \text{ eq} \times \frac{1000 \text{ meq}}{1 \text{ eq}} = 1.0 \times 10^2 \text{ meq}$$

Notice that we could have focused on the negative charges in either part above and still arrived at the same answers. In part a, 0.050 mol KCl produces 0.050 mol Cl^-, or 0.050 mol of negative charge. Similarly, in part b, 0.050 mol $CaCl_2$ produces 2×0.050, or 0.10 mol, Cl^- ions, or 0.10 mol of negative charge. We have arbitrarily chosen to use the positive charges, but either will work. Just remember that you do not count both the negative and positive charges for a salt.

▶ LEARNING CHECK 9.14 Determine the number of equivalents and milliequivalents in each of the following:

a. 0.10 mol NaCl
b. 0.10 mol $Mg(NO_3)_2$

▶ Example 9.11

A sample of blood serum contains 0.139 equivalents of Na^+ ion per liter of serum. Assume the Na^+ comes from dissolved NaCl, and calculate the number of equivalents, number of moles and number of grams of NaCl in 250 mL of the serum.

Solution

The Na^+ ion has a single charge, so we may write the following relationships:

$$1.00 \text{ mol NaCl} = 1.00 \text{ mol } Na^+ = 1.00 \text{ eq. } Na^+ = 1.00 \text{ eq NaCl}$$

These relationships provide numerous factors that can be used in the factor-unit method to solve problems. Two that will be used to solve these problems are:

$$\frac{1.00 \text{ eq NaCl}}{1.00 \text{ eq } Na^+} \quad \text{and} \quad \frac{1.00 \text{ mol NaCl}}{1.00 \text{ eq NaCl}}$$

The known quantity is 250 mL of serum. This is converted into equivalents of Na^+ by using the given concentration of $\dfrac{0.139 \text{ eq } Na^+}{\text{L serum}}$ as the factor.

$$(0.250 \text{ L serum}) \times \frac{0.139 \text{ eq } Na^+}{1.00 \text{ L serum}} = 0.0347 \text{ eq } Na^+$$

This number of equivalents of Na^+ will now be converted into the quantities asked for by using the two factors given above, and a factor obtained from the formula weight for NaCl of 58.44 u.

$$(0.0347 \text{ eq } Na^+) \times \frac{1.00 \text{ eq NaCl}}{1.00 \text{ eq } Na^+} = 0.0347 \text{ eq NaCl}$$

$$(0.0347 \text{ eq NaCl}) \times \frac{1.00 \text{ mol NaCl}}{1.00 \text{ eq NaCl}} = 0.0347 \text{ mol NaCl}$$

$$(0.0347 \text{ mol NaCl}) \times \frac{58.44 \text{ g NaCl}}{1.00 \text{ mol NaCl}} = 2.03 \text{ g NaCl}$$

▶ **LEARNING CHECK 9.15** A sample of blood serum contains 0.103 equivalents of Cl^- ions per liter. Assume the Cl^- comes from dissolved NaCl and calculate the number of equivalents, number of moles, and number of grams of NaCl in 250 mL of serum.

9.9 The Strengths of Acids and Bases

Learning Objective

9. Demonstrate an understanding of the words *weak* and *strong* as applied to acids and bases.

strong acids and strong bases Acids and bases that dissociate (ionize) completely when dissolved to form a solution.

weak (or moderately weak) acids and bases Acids and bases that dissociate (ionize) less than completely when dissolved to form a solution.

When salts dissolve in water, they generally dissociate completely, but this is not true for all acids and bases. The acids and bases that do dissociate almost completely are classified as **strong acids** and **strong bases** (they are also strong electrolytes). Those that dissociate to a much smaller extent are called **weak** or **moderately weak,** depending on the degree of dissociation (they are also weak or moderately weak electrolytes). Examples of strong and weak acids are given in ▶ Table 9.7.

Carbon Dioxide in the Oceans

Carbon dioxide, CO_2, is generally recognized as an atmospheric greenhouse gas that contributes to global warming. Concerns are now being raised about the potential adverse impact of the dissolved CO_2 on the oceans of the world.

Measurements of CO_2 levels in the atmosphere begun in 1958 coupled with studies of air bubbles trapped in old glacial ice indicate that the concentration of CO_2 in the atmosphere remained essentially constant for several thousand years. However, in about 1800 the concentration began to increase rapidly as a result of the industrial revolution and the accompanying increase in the burning of fossil fuels to provide the energy needed to power machinery.

It is now estimated that about one-third of the CO_2 produced by burning fossil fuels in our highly industrialized, modern world ends up dissolved in the oceans of the world. This might seem to be a positive development because it reduces the amount of CO_2 contributing to global warming, but the results of continuing studies raise questions about the possible negative effects of this dissolved CO_2 on ocean ecosystems.

When CO_2 dissolves in water, it reacts with the water to form carbonic acid (Reaction 1).

$$1. \quad CO_2(g) + H_2O(\ell) \rightleftarrows H_2CO_3(aq)$$

The carbonic acid dissociates into ions in two steps (Reactions 2 and 3).

$$2. \quad H_2CO_3(aq) \rightleftarrows H^+(aq) + HCO_3^-(aq)$$

$$3. \quad HCO_3^-(aq) \rightleftarrows H^+(aq) + CO_3^{2-}(aq)$$

Reaction 3 occurs only to a slight extent. The net effect of these two dissociation reactions is to increase the concentration of H^+ ions and decrease the pH of the oceans from their normal range of $8.0 – 8.3$. According to one prediction, the pH range could decrease to $7.6 – 7.9$ by the year 2100 if current trends continue.

The increase in H^+ concentration has the potential to be very harmful to some forms of marine life, especially those forms that use carbonate ions, CO_3^{2-}, to build their shells or other hard body parts out of calcium carbonate, $CaCO_3$. At first, this doesn't seem to make sense because, according to Reaction 3, the dissolved CO_2 actually forms CO_3^{2-} ions. However, as mentioned earlier, very little of the dissolved CO_2 actually undergoes Reaction 3. In fact, the extra H^+ ions added to the water from Reaction 2 actually react with CO_3^{2-} ions that are normally found in seawater, and convert them to HCO_3^- ions in a reaction that is the reverse of Reaction 3. This process thus reduces the concentration of CO_3^{2-} ions to levels lower than that normally found in seawater.

Scientists estimate that this mechanism could reduce the concentration of CO_3^{2-} ions in seawater to one-half its current value during the next 100 years if current trends continue. It is feared that such reductions will interfere with the growth and survival of carbonate-using organisms ranging in size from tiny phytoplankton that serve as major food sources for fish and marine mammals, to corals, the animals responsible for building huge coral reefs, the largest biological structures in the world.

Scientific research results such as these give us another reason to pause, seriously consider the negative impact of some of our modern practices on the world around us, and maybe even change some of those practices.

A Coral reef and a close-up of coral.

Some Common Strong and Weak Acids

Name	Formula	% Dissociation[a]	K_a	Classification
Hydrochloric acid	HCl	100	Very large	Strong
Hydrobromic acid	HBr	100	Very large	Strong
Nitric acid	HNO_3	100	Very large	Strong
Sulfuric acid	H_2SO_4	100	Very large	Strong
Phosphoric acid	H_3PO_4	28	7.5×10^{-3}	Moderately weak
Sulfurous acid[b]	H_2SO_3	34	1.5×10^{-2}	Moderately weak
Acetic acid	$HC_2H_3O_2$	1.3	1.8×10^{-5}	Weak
Boric acid	H_3BO_3	0.01	7.3×10^{-10}	Weak
Carbonic acid[b]	H_2CO_3	0.2	4.3×10^{-7}	Weak
Nitrous acid[b]	HNO_2	6.7	4.6×10^{-4}	Weak

[a]Based on dissociation of one proton in 0.1 M solution at 25°C.

[b]Unstable acid.

A 0.10 M solution of hydrochloric acid could be prepared by dissolving 0.10 mol (3.7 g) of HCl gas in enough water to give 1.0 L of solution. According to Table 9.7, 100% of the dissolved gas would dissociate into H^+ and Cl^-. Thus, the concentration of H^+ in a 0.10 M HCl solution is 0.10 mol/L, and the pH is 1.00.

The strength of acids and bases is shown quantitatively by the value of the equilibrium constant for the dissociation reaction in water solutions. Equation 9.31 represents the dissociation of a general acid, HB, in water, where B^- represents the conjugate base of the acid:

$$HB(aq) + H_2O(\ell) \rightleftarrows H_3O^+(aq) + B^-(aq) \tag{9.31}$$

The equilibrium expression for this reaction is

$$K = \frac{[H_3O^+][B^-]}{[HB][H_2O]} \tag{9.32}$$

In Equation 9.32, the brackets, again, represent molar concentrations of the materials in the solution. Only a tiny amount of the water in the solution actually enters into the reaction, so the concentration of water is considered to be constant (see Section 9.4). We can then write Equation 9.33, where K_a is a new constant called the **acid dissociation constant:**

acid dissociation constant The equilibrium constant for the dissociation of an acid.

$$K[H_2O] = K_a = \frac{[H_3O^+][B^-]}{[HB]} \tag{9.33}$$

In Equation 9.33, $[H_3O^+]$ and $[B^-]$ are, respectively, the equilibrium concentrations of the hydronium ion and the anion conjugate base that is characteristic of the acid. The [HB] represents the concentration of that part of the dissolved acid that remains undissociated in the equilibrium mixture. In solutions of strong acids, $[H_3O^+]$ and $[B^-]$ values are quite large, while [HB] has a value near 0, so K_a is quite large. In weak acids, [HB] has larger values, while $[H_3O^+]$ and $[B^-]$ are smaller, so K_a is smaller. Thus, the larger a K_a value, the stronger the acid it represents. This is illustrated by the K_a values given in Table 9.7. If we simplify by substituting H^+ for H_3O^+, introduced in Section 9.5, we obtain

$$K_a = \frac{[H^+][B^-]}{[HB]} \tag{9.34}$$

It is important to remember that the terms *weak* and *strong* apply to the extent of dissociation and not to the concentration of an acid or base. For example, gastric juice (0.05% HCl) is a dilute (not weak) solution of a strong acid.

▶ Example 9.12

Write dissociation reactions and expressions for K_a for each of the following weak acids:

a. Hydrocyanic acid (HCN)
b. Phosphoric acid (H_3PO_4) (1st H only)
c. Dihydrogen phosphate ion ($H_2PO_4^-$) (1st H only)

Solution

In each case, H^+ has been substituted for H_3O^+.

a. $HCN \rightleftharpoons H^+ + CN^-$; $K_a = \dfrac{[H^+][CN^-]}{[HCN]}$

b. $H_3PO_4 \rightleftharpoons H^+ + H_2PO_4^-$; $K_a = \dfrac{[H^+][H_2PO_4^-]}{[H_3PO_4]}$

c. $H_2PO_4^- \rightleftharpoons H^+ + HPO_4^{2-}$; $K_a = \dfrac{[H^+][HPO_4^{2-}]}{[H_2PO_4^-]}$

▶ LEARNING CHECK 9.16 Write dissociation reactions and K_a expressions for the following weak acids:

a. Hydrogen phosphate ion (HPO_4^{2-})
b. Nitrous acid (HNO_2)
c. Hydrofluoric acid (HF)

Acid behavior is linked to the loss of protons. Thus, acids must contain hydrogen atoms that can be removed to form H^+. **Monoprotic acids** can lose only one proton per molecule, whereas **diprotic** and **triprotic** acids can lose two and three, respectively. For example, HCl is monoprotic, H_2SO_4 is diprotic, and H_3PO_4 is triprotic. Di- and triprotic acids dissociate in steps, as shown for H_2SO_4 in Equations 9.35 and 9.36:

monoprotic acid An acid that gives up only one proton (H^+) per molecule when dissolved.

diprotic acid An acid that gives up two protons (H^+) per molecule when dissolved.

$$H_2SO_4(aq) \rightleftharpoons H^+(aq) + HSO_4^-(aq) \tag{9.35}$$

$$HSO_4^-(aq) \rightleftharpoons H^+(aq) + SO_4^{2-}(aq) \tag{9.36}$$

triprotic acid An acid that gives up three protons (H^+) per molecule when dissolved.

The second proton is not as easily removed as the first because it must be pulled away from a negatively charged particle, HSO_4^-. Accordingly, HSO_4^- is a weaker acid than H_2SO_4.

The number of ionizable hydrogens cannot always be determined from the molecular formula for an acid. For example, acetic acid ($HC_2H_3O_2$) is monoprotic even though the molecule contains four hydrogen atoms. The dissociation of acetic acid is represented by Equation 9.37, where structural formulas are used to emphasize the different H atoms in the molecule:

$$
\begin{array}{ccccc}
& H & O & & \\
& | & \| & & \\
H-C-C & - & O-H \\
& | & & \\
& H & &
\end{array}
\rightleftharpoons
H^+ +
\begin{array}{ccc}
H & O & \\
| & \| & \\
H-C-C & - & O^- \\
| & & \\
H & &
\end{array}
\tag{9.37}
$$

Only the hydrogen bound to the oxygen is ionizable. Those hydrogens bound to C are too tightly held to be removed. ▶ Table 9.8 contains other examples, with the ionizable hydrogens shown in color.

TABLE 9.8 Examples of Monoprotic, Diprotic, and Triprotic Acids

Name	Formula	Structural Formula	Classification
Butyric acid	$HC_4H_7O_2$		Monoprotic
Carbonic acid	H_2CO_3		Diprotic
Formic acid	$HCHO_2$		Monoprotic
Nitric acid	HNO_3		Monoprotic
Phosphoric acid	H_3PO_4		Triprotic
Phosphorous acid	H_3PO_3		Diprotic

We have focused our attention on the strength of acids, using the extent of dissociation as a basis. However, all acid dissociations are reversible to some degree, and in the reverse reactions, anions produced by the forward reaction behave as Brønsted bases. What can be said about the strength of these bases? Because Brønsted acid–base behavior is really just competition for protons, we can answer this question by looking again at the simplified form of the equation for dissociation of a general acid:

$$HB(aq) \rightleftharpoons H^+(aq) + B^-(aq) \tag{9.38}$$

We see from this equation that the strength of HB as an acid depends on how tightly the conjugate base B^- holds onto the proton. If HB is a strong acid, the conjugate base holds onto the proton only weakly. If HB is a weak acid, the conjugate base holds on more strongly, depending on the strength of HB. Thus, we have answered our earlier question. If HB is a strong acid, the H^+ is held only weakly by the B^-. We can conclude then that B^- is not strongly attracted to protons—it is a weak base. Conversely, if HB is a weak acid, the H^+ is held tightly by the B^-, and we conclude that B^- is more strongly attracted to protons—it is a stronger base than the B^- from a strong acid.

In general, the conjugate base anions produced by the dissociation of strong Brønsted acids are weak Brønsted bases. The conjugate base anions of weak acids are stronger bases, with their strengths dependent on the strength of the parent acid.

Writing Reactions of Acids

As you study this chapter, you will acquire a knowledge of the characteristic reactions of acids and the ability to write balanced equations for the reactions. In the Key Equations section at the end of the chapter, five characteristic reactions are summarized in item 3. One useful way to remember the reactions is to learn them as general word equations, such as

"An acid plus a metal gives a salt plus hydrogen gas"

rather than as specific equations such as $H_2SO_4(aq) + Zn(s) \rightarrow ZnSO_4(aq) + H_2(g)$. If you learn the general word equations and you recognize the starting materials for a reaction (such as an acid and a metal), you will know what the products will be (a salt and H_2 gas).

A second approach is to remember that all five of the general reactions of acids are either single-replacement

$$\text{acid} + \text{metal} \rightarrow \text{salt} + H_2$$

or double-replacement (the remaining four acid reactions in the Key Equations section). The formulas of the products of double-replacement reactions can be predicted by simply breaking each reactant into its positive and negative parts and recombining the parts in the other possible way (the positive part of one reactant with the negative part of the other reactant). For example, let's determine the products and the balanced equation for a reaction between H_2SO_4 and KOH. If you remember that H_2SO_4 is an acid and KOH is a base, the general word equation says the products will be a salt and water:

$$H_2SO_4(aq) + KOH(aq) \rightarrow \text{salt}(aq) + H_2O(\ell)$$

H_2SO_4 breaks apart to give H^+ and SO_4^{2-}; KOH breaks apart to give K^+ and OH^-. If we combine the positive part of the acid (H^+) with the negative part of the base (OH^-), we get the water (H_2O). A similar combination of the positive part

of the base (K^+) with the negative part of the acid (SO_4^{2-}) gives the salt. We must remember that the total charges of the combined parts must add up to 0. Thus, two K^+ will combine with one SO_4^{2-} to give the salt K_2SO_4. The balanced equation is

$$H_2SO_4(aq) + 2KOH(aq) \rightarrow K_2SO_4(aq) + 2H_2O(\ell)$$

As another example, consider a reaction between HBr and KOH. KBr breaks into H^+ and Br^-. KOH breaks into K^+ and OH^-. Now, switch the parts and recombine:

The products are KBr and H_2O, and the balanced equation is

$$HBr(aq) + KOH(aq) \rightarrow KBr(aq) + H_2O(\ell)$$

For a final example, let's try a reaction between HBr and $NaHCO_3$. The word equation predicts that the products should be a salt, water, and carbon dioxide gas. HBr breaks into H^+ and Br^-; $NaHCO_3$ breaks into Na^+ and HCO_3^-. Now, switch the parts and recombine:

The products are NaBr and H_2CO_3. However, H_2CO_3 is not stable; it decomposes to give H_2O and CO_2, the products predicted earlier. The balanced equation is

$$HBr(aq) + NaHCO_3(aq) \rightarrow NaBr(aq) + H_2O(\ell) + CO_2(g)$$

Ammonia (NH_3) is the weak base most often encountered in addition to the anions of strong acids. The dissociation reaction of gaseous NH_3 in water, given earlier in Example 9.1, is

$$NH_3(aq) + H_2O(\ell) \rightleftharpoons NH_4^+(aq) + OH^-(aq) \tag{9.39}$$

The most common strong bases are the hydroxides of group IA(1) metals (NaOH, KOH, etc.) and the hydroxides of group IIA(2) metals ($Mg(OH)_2$, $Ca(OH)_2$, etc.).

▶ **Example 9.13**

Classify each of the following pairs by identifying the stronger of the pair according to the indicated behavior. Information from Tables 9.7 and 9.8 may be used.

a. H_3PO_4 and $H_2PO_4^-$ (acid)
b. $H_2PO_4^-$ and HPO_4^{2-} (base)
c. HNO_3 and HNO_2 (acid)

CHEMISTRY AND YOUR HEALTH 9.1

Do You Have Acid Reflux Disease?

The medical name for acid reflux disease is *gastroesophageal reflux disease,* which is often abbreviated and referred to as GERD. The disease is often mistaken for occasional heartburn and treated with over-the-counter remedies. GERD is the result of a malfunctioning muscle (the lower esophageal sphincter [LES] muscle) located at the bottom of the esophagus, just above the stomach. When operating normally, this muscle relaxes and opens to allow food to pass from the esophagus down into the stomach, then contracts to close the opening and prevent the acidic contents of the stomach from backing up into the esophagus.

When the muscle relaxes at inappropriate times, the acidic stomach contents get into the esophagus and cause the burning chest pain called heartburn. However, when this occurs repeatedly and frequently, the acidic stomach contents can also erode the lining of the esophagus. GERD is a complex condition with many degrees of severity, ranging from only frequent heartburn symptoms to erosive esophagitis, in which the esophagus can suffer different degrees of damage. In extreme cases of erosive esophagitis, ulcers develop in the esophagus and lead to esophageal bleeding that, if persistent and undetected, can lead to iron deficiency and anemia as well as extreme pain and weight loss. In some cases, severe GERD can lead to other serious medical conditions that require hospitalization and even surgery to correct.

Doctors often recommend lifestyle and dietary changes for most GERD patients, including the avoidance of foods and beverages that weaken the LES muscle. These foods include chocolate, peppermint, fatty foods, coffee, and alcoholic beverages. The use of foods and beverages that can irritate a damaged esophageal lining, such as acidic fruits and juices, pepper, and tomato products is also discouraged. GERD symptoms in overweight individuals often diminish when some weight is lost. Smokers who quit also

generally gain some relief. Prescription medications that reduce the amount of acid in the stomach are also available. Two types of medication are available. Both types, called H2 blockers and proton (acid) pump inhibitors, decrease the amount of acid secreted into the stomach, but by different mechanisms.

The main symptom that may indicate the presence of GERD is frequent, persistent heartburn that occurs two or more times a week. Other symptoms include difficulty in swallowing and frequent belching and regurgitation. Less common symptoms that occur in some people resemble respiratory conditions and include a persistent sore throat, wheezing, chronic coughing, and hoarseness. Individuals who suspect they might be suffering from some degree of GERD should consult a physician to determine the extent of the disease and proper treatment.

GERD has many degrees of severity.

Solution

a. H_3PO_4 is stronger as an acid. In di- and triprotic acids, each proton in the removal sequence is harder to remove, so the corresponding acid is weaker.

b. The stronger acid produces the weaker anion base. Thus, $H_2PO_4^-$, the anion of the stronger acid H_3PO_4, would be a weaker base than HPO_4^{2-}, the anion of the weaker acid $H_2PO_4^-$. So HPO_4^{2-} is a stronger base than $H_2PO_4^-$.

c. HNO_3 is the stronger acid according to Table 9.7. Generally, when related acids (same atoms, etc.) are compared for strength, the one containing more oxygen atoms will be the stronger.

▶ **LEARNING CHECK 9.17** Classify each of the following according to strength for the indicated behavior. If more than two are compared, list them with the strongest at the top and the weakest at the bottom. Use Tables 9.7 and 9.8 as needed.

a. $HClO$, $HClO_3$, $HClO_2$ (acid)
b. NO_2^- and NO_3^- (base)
c. $HC_2H_3O_2$ and $C_2H_3O_2^-$ (acid)

Unless otherwise noted all art on this page © Cengage Learning 2014

9.10 Analyzing Acids and Bases

Learning Objective

10. Demonstrate an understanding of the titration technique used to analyze acids and bases.

The analysis of solutions for the total amount of acid or base they contain is a regular activity in many laboratories. The total amount of acid in a solution is indicated by its capacity to neutralize a base. The pH is related to the acidity or concentration of H^+ in solution, while the capacity to neutralize a base depends on the total amount of H^+ available. For example, a 0.10 M acetic acid solution has an H^+ concentration of about 1.3×10^{-3} M (pH = 2.89). However, 1 L of the solution can neutralize 0.10 mol of OH^-, not just 1.3×10^{-3} mol. The reason is that the dissociation equilibrium of acetic acid is

$$HC_2H_3O_2(aq) \rightleftarrows H^+(aq) + C_2H_3O_2^-(aq) \qquad (9.40)$$

As OH^- is added, H^+ reacts to form water (see Equation 9.24). The removal of H^+ causes the equilibrium of Reaction 9.40 to shift right in accordance with Le Châtelier's principle. The continued addition of OH^- and removal of H^+ will eventually cause all of the acetic acid molecules to dissociate and react.

A common procedure often used to analyze acids and bases is called **titration** (❱ Figure 9.4). Suppose the total acidity of an unknown acid solution needs to be determined. A known volume of the acidic solution is first measured out by drawing it up

titration An analytical procedure in which one solution (often a base) of known concentration is slowly added to a measured volume of an unknown solution (often an acid). The volume of the added solution is measured with a buret.

Figure 9.4 Titration.

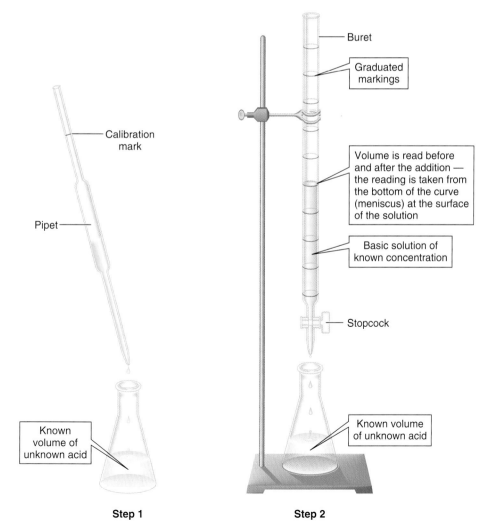

Buret

Graduated markings

Calibration mark

Volume is read before and after the addition — the reading is taken from the bottom of the curve (meniscus) at the surface of the solution

Basic solution of known concentration

Pipet

Stopcock

Known volume of unknown acid

Known volume of unknown acid

Step 1

Step 2

equivalence point of a titration The point at which the unknown solution has exactly reacted with the known solution. Neither is in excess.

end point of a titration The point at which the titration is stopped on the basis of an indicator color change or pH meter reading.

to the calibration mark of a pipet. This solution is placed in a container (Step 1). A basic solution of known concentration (a standard solution) is added to the acid solution in the container until the **equivalence point** is reached. This is the point where the unknown acid is completely reacted with base. The volume of base needed to reach the equivalence point is obtained from the buret readings.

To successfully complete a titration, the point at which the reaction is completed must somehow be detected. One way to do this is to add an indicator to the solution being titrated. An indicator is an organic compound that changes to different colors depending on the pH of its surroundings. An indicator is selected that will change color at a pH as close as possible to the pH the solution will have at the equivalence point. If the acid and base are both strong, the pH at that point will be 7. However, for reasons discussed later, the pH is not always 7 at the equivalence point. The point at which the indicator changes color and the titration is stopped is called the titration **endpoint.** ❱ Figure 9.5 gives the colors shown by a number of indicators at various pH values.

Figure 9.5 Indicators change color with changes in pH (the numbers across the bottom). Methyl red (top row) goes from red at low pH to yellow at higher pH. Bromthymol blue (middle row) goes from yellow to blue, and phenolphthalein (bottom row) goes from colorless to red. Which of the three indicators would be the best to use to differentiate between the solutions with pH values of 4 and 6?

Indicator papers (litmus paper, pH paper, etc.) are often used to make routine pH measurements. These papers are impregnated with one or more indicators that change to a variety of colors depending on the pH. However, indicators may not be practical under certain conditions. For example, there might not be any indicator that changes color close enough to the equivalence point, or the solution being titrated might be so highly colored that an indicator color change cannot be detected. Under such circumstances, a pH meter can be used. The electrodes of the meter are placed in the solution being titrated, and the pH is read directly (see ❱ Figure 9.6). The titration is continued until the meter reading matches the pH of the equivalence point.

At the beginning, the pH meter gives the pH of the acid solution being titrated.

 Partway through the titration, the pH meter reading is of a solution of un-reacted acid and the salt produced by the reaction.

 At the end of the titration, the pH meter gives the pH of the salt solution formed by the complete reaction of acid with base.

Figure 9.6 An acid–base titration using a pH meter to detect the equivalence point.

9.11 Titration Calculations

Learning Objective

11. Do calculations related to the analysis of acids and bases by titration.

Equations 9.41 and 9.42 represent the reactions that occur when solutions of nitric and sulfuric acid are titrated with sodium hydroxide:

$$\text{HNO}_3(\text{aq}) + \text{NaOH}(\text{aq}) \rightarrow \text{H}_2\text{O}(\ell) + \text{NaNO}_3(\text{aq}) \tag{9.41}$$

$$\text{H}_2\text{SO}_4(\text{aq}) + 2\text{NaOH}(\text{aq}) \rightarrow 2\text{H}_2\text{O}(\ell) + \text{Na}_2\text{SO}_4(\text{aq}) \tag{9.42}$$

One mole of HNO_3 requires 1 mol of NaOH for a complete reaction, but 1 mol of H_2SO_4 requires 2 mol of NaOH. Because these are solution reactions, stoichiometric calculations can be done using the methods described in Section 7.6.

▶ Example 9.14

Calculate the molarity of the HNO_3 and H_2SO_4 solutions involved in the following titrations:

a. A 25.0-mL sample of an HNO_3 solution requires the addition of 16.3 mL of 0.200 M NaOH to reach the equivalence point.
b. A 25.0-mL sample of an H_2SO_4 solution requires the addition of 32.6 mL of 0.200 M NaOH to reach the equivalence point.

Solution
In each case we know the volume of acid solution reacted. If we also knew the number of moles of acid in the volume of reacted solution, we could calculate the solution molarity by using Equation 7.5:

$$M = \frac{\text{moles of solute}}{\text{liters of solution}}$$

Therefore, our task will be to calculate the number of moles of acid reacted in each case. The methods described in Section 7.6 will be used. We will begin with the volume of NaOH solution used in each titration, and convert that volume to moles of acid using the factor-unit method.

a. Nitric acid

Step 1. 0.0163 L NaOH solution
Step 2. 0.0163 L NaOH solution $= \text{mol HNO}_3$

Step 3. $0.0163 \text{ L NaOH solution} \times \dfrac{0.200 \text{ mol NaOH}}{1 \text{ L NaOH solution}} \times \dfrac{1 \text{ mol HNO}_3}{1 \text{ mol NaOH}}$

$$= \text{mol HNO}_3$$

The factor $\dfrac{0.200 \text{ mol NaOH}}{1 \text{ L NaOH solution}}$ came from the molarity of the NaOH solution, and the factor $\dfrac{1 \text{ mol HNO}_3}{1 \text{ mol NaOH}}$ came from the 1 to 1 reaction ratio of HNO_3 with NaOH.

Step 4. $0.0163 \times \dfrac{0.200}{1} \times \dfrac{1 \text{ mol HNO}_3}{1} = 0.00326 \text{ mol HNO}_3$

This result tells us that the 25.0 mL titrated sample of HNO_3 contained 0.00326 moles of HNO_3. After converting the HNO_3 volume to liters, the molarity can be calculated:

$$M = \frac{\text{moles of HNO}_3}{\text{liters of HNO}_3 \text{ solution}} = \frac{0.00326 \text{ mol HNO}_3}{0.0250 \text{ L HNO}_3 \text{ solution}} = \frac{0.130 \text{ mol HNO}_3}{\text{L HNO}_3 \text{ solution}}$$

$$= 0.130 \text{ M HNO}_3$$

b. Sulfuric acid

This calculation follows the same pattern as the HNO_3 calculation, except for the fact that the reaction ratio of NaOH and H_2SO_4 is 2 mol of NaOH reacts with 1 mol of H_2SO_4. This gives a factor of $\dfrac{1 \text{ mol H}_2SO_4}{2 \text{ mol NaOH}}$ that will be used in step 3.

Step 1. 0.0326 L NaOH solution
Step 2. 0.0326 L NaOH solution $= \text{mol H}_2SO_4$
Step 3. $0.0326 \text{ L NaOH solution} \times \dfrac{0.200 \text{ mol NaOH}}{1 \text{ L NaOH solution}} \times \dfrac{1 \text{ mol H}_2SO_4}{2 \text{ mol NaOH}}$

$$= \text{mol H}_2SO_4$$

Step 4. $0.0326 \times \dfrac{0.200}{1} \times \dfrac{1 \text{ mol H}_2SO_4}{2} = 0.00326 \text{ mol H}_2SO_4$

The molarity of the solution is calculated as before:

$$M = \frac{\text{moles of H}_2SO_4}{\text{liters of H}_2SO_4 \text{ solution}} = \frac{0.0326 \text{ mol H}_2SO_4}{0.0250 \text{ L H}_2SO_4 \text{ solution}} = \frac{0.130 \text{ mol H}_2SO_4}{\text{L H}_2SO_4 \text{ solution}}$$

$$= 0.130 \text{ M H}_2SO_4$$

Thus, we see that the two acid solutions have the same molar concentrations. Why did the second require twice the volume of NaOH solution for the titration?

▶ **LEARNING CHECK 9.18** Calculate the molarity of a solution of phosphoric acid (H_3PO_4) if a 25.0-mL sample of acid solution requires 14.1 mL of a 0.250 M NaOH solution to titrate to the equivalence point. The equation for the reaction is

$$H_3PO_4(aq) + 3NaOH(aq) \rightarrow 3H_2O(aq) + Na_3PO_4(aq)$$

9.12 Hydrolysis Reactions of Salts

Learning Objective

12. Explain the concept of salt hydrolysis, and write equations to illustrate the concept.

hydrolysis reaction Any reaction with water. For salts it is a reaction of the acidic cation and/or basic anion of the salt with water.

In general, a **hydrolysis reaction** is a reaction with water. Many types of hydrolysis reactions are known, but at this point we will discuss only one, the hydrolysis of salts. In Section 9.10, we said that the pH at the equivalence point is not 7 for all acid–base titrations. However, we pointed out earlier that the only products of acid–base reactions are water and a salt. Therefore, it seems reasonable to conclude that some salts and water must interact (a hydrolysis reaction) and cause the solution pH to differ from that of pure water.

Suppose three solutions are prepared by dissolving equal molar amounts of sodium acetate, ammonium chloride, and sodium chloride in identical volumes of pure water. Measurement of pH shows that the sodium acetate solution is alkaline (pH higher than 7), the ammonium chloride solution is acidic (pH lower than 7), and the sodium chloride solution is neutral (pH = 7). The alkaline sodium acetate solution must contain more OH^- than H^+ ions. However, the only source of both ions in a solution of salt and water is the dissociation of water, so both should be present in equal amounts. The equation for the water dissociation is

$$H_2O(\ell) \rightleftarrows H^+(aq) + OH^-(aq) \tag{9.43}$$

The OH^- excess would result if something removed H^+ from this equilibrium, causing it to shift right and produce more OH^-. This is exactly what happens. The acetate anion is the conjugate base of a weak acid (acetic acid), and it is a strong enough Brønsted base to react with the H^+ ions of the above equilibrium:

$$C_2H_3O_2^-(aq) + H^+(aq) \rightleftarrows HC_2H_3O_2(aq) \tag{9.44}$$

The overall result of this reaction can be seen by adding Equations 9.43 and 9.44:

$$H_2O(\ell) + C_2H_3O_2^-(aq) + H^+(aq) \rightleftarrows H^+(aq) + OH^-(aq) + HC_2H_3O_2(aq) \tag{9.45}$$

Net reaction: $H_2O(\ell) + C_2H_3O_2^-(aq) \rightleftarrows OH^-(aq) + HC_2H_3O_2^-(aq)$ (9.46)

Now it can be seen that the excess OH^- comes from a reaction between the acetate ion and water. But what about the Na^+ ion that is also in the solution? Does it react with water? The Na^+ cation is the conjugate acid of the strong base NaOH. As a result, Na^+ is an extremely weak Brønsted acid and makes no contribution to the pH of the solution.

In general, salts influence the pH of water solutions as follows: (1) Salts containing an anion of a weak acid and a cation of a strong base will form alkaline water solutions. (2) Salts containing an anion of a strong acid and a cation of a weak base will form acidic solutions. (3) Salts containing an anion and a cation from equal-strength acids and bases (both weak or both strong) will form neutral solutions (see ▶ Figure 9.7).

Figure 9.7 Samples of pure water (left) and sodium acetate dissolved in water (right) behave differently when phenolphthalein indicator is added. The acetate ion hydrolyzes in water to form a basic solution that turns phenolphthalein to a pink color. Why didn't the Na^+ ion from the sodium acetate hydrolyze?

The acidic ammonium chloride solution is an example of the second category of salts, where the NH_4^+ cation is the conjugate acid of the weak base NH_3, and the Cl^- anion is the conjugate base of the strong acid HCl. The hydrolysis reaction is given in Equation 9.47, where H_3O^+ is used to emphasize the reaction with water:

$$NH_4^+(aq) + H_2O(\ell) \rightleftarrows H_3O^+(aq) + NH_3(aq) \tag{9.47}$$

The neutral sodium chloride solution is an example of the third category. Both NaOH and HCl are strong, and an NaCl solution is neutral.

▶ **LEARNING CHECK 9.19** Predict the relative pH value (higher than 7, lower than 7, etc.) for water solutions of the following salts. Table 9.7 may be useful.

  a. Sodium nitrate ($NaNO_3$)
  b. Sodium nitrite ($NaNO_2$)
  c. Potassium borate (K_3BO_3)

9.13 Buffers

Learning Objective

13. Explain how buffers work, and write equations to illustrate their action.

buffer A solution with the ability to resist changing pH when acids (H^+) or bases (OH^-) are added.

Buffers are solutions that have the ability to resist changes in pH when acids or bases are added (see ▶ Figure 9.8). Most buffers consist of a pair of compounds, one with the ability to react with H^+ and the other with the ability to react with OH^-. An example is a mixture of acetic acid and its salt, sodium acetate. The acetate ion from sodium acetate is the conjugate base of acetic acid and reacts with any added acid (H^+ ions):

$$C_2H_3O_2^-(aq) + H^+(aq) \rightleftarrows HC_2H_3O_2(aq) \tag{9.48}$$

Any added base (OH^- ions) reacts with nonionized acetic acid molecules:

$$HC_2H_3O_2(aq) + OH^-(aq) \rightleftarrows C_2H_3O_2^-(aq) + H_2O(\ell) \tag{9.49}$$

buffer capacity The amount of acid (H^+) or base (OH^-) that can be absorbed by a buffer without causing a significant change in pH.

In this way, the buffer solution resists changes in pH. The amount of H^+ or OH^- that a buffer system can absorb without allowing significant pH changes to occur is called its **buffer capacity.**

① The solution on the left is not buffered; the one on the right is; universal indicator has been added to each solution.

② Sodium hydroxide has been added to each solution.

③ Hydrochloric acid has been added to two fresh samples that originally looked like the first pair of samples.

Figure 9.8 Buffered solutions resist changing pH when bases and acids are added. What observation of this experiment indicates that the pH of the buffered solutions did not change significantly when base or acid was added?

Buffered solutions are found extensively in the body, where they serve to protect us from the disastrous effects of large deviations in pH. An important buffer system in blood is composed of carbonic acid (H_2CO_3) and bicarbonate salts such as $NaHCO_3$. The unstable carbonic acid results when dissolved CO_2 reacts with water in the blood:

$$H_2O(\ell) + CO_2(aq) \rightleftarrows H_2CO_3(aq) \tag{9.50}$$

Added H^+ is neutralized by reacting with HCO_3^- from the dissolved bicarbonate salts:

$$HCO_3^-(aq) + H^+(aq) \rightleftarrows H_2CO_3(aq) \tag{9.51}$$

The carbonic acid protects against added OH^-:

$$H_2CO_3(aq) + OH^-(aq) \rightleftarrows HCO_3^-(aq) + H_2O(\ell) \qquad (9.52)$$

If large amounts of H^+ or OH^- are added to a buffer, the buffer capacity can be exceeded, the buffer system is overwhelmed, and the pH changes. For example, if large amounts of H^+ were added to the bicarbonate ion–carbonic acid buffer, Reaction 9.51 would take place until the HCO_3^- was depleted. The pH would then drop as additional H^+ ions were added. In blood, the concentration of HCO_3^- is ten times the concentration of H_2CO_3. Thus, this buffer has a greater capacity against added acid than against bases. This is consistent with the normal functions of the body that cause larger amounts of acidic than basic substances to enter the blood.

The pH of buffers can be calculated by using a form of Equation 9.34 that was given earlier and is repeated here:

$$K_a = \frac{[H^+][B^-]}{[HB]}$$

This equation can be rearranged to give

$$[H^+] = K_a \frac{[HB]}{[B^-]} \qquad (9.53)$$

Application of the logarithm concept to Equation 9.53 gives

$$pH = pK_a + \log \frac{[B^-]}{[HB]} \qquad (9.54)$$

In this equation, $pH = -\log[H^+]$ (Equation 9.11), and **pKa** $= -\log K_a$. Equation 9.54 is known as the **Henderson–Hasselbalch equation;** it is often used by biologists, biochemists, and others who frequently work with buffers. We see from Equation 9.54 that when the concentrations of a weak acid and its conjugate base (anion) are equal in a solution, the pH of the solution is equal to pK_a. When it is desired to produce a buffer with a pH different from the exact pK_a, Equation 9.54 indicates it can be done. An acid with a pK_a near the desired pH is selected, and the ratio of the concentrations of conjugate base (anion of the acid) and acid is adjusted to give the desired pH. When a weak acid and its salt (the source of the acid's conjugate base anion) are mixed in solution, you can generally assume that the amount of acid that dissociates is small and may be neglected. This means that the buffer concentrations of the acid and anion (the salt of acid) are "equal" to the made-up concentrations. ▶ Table 9.9 lists some weak acids, together with values of K_a and pK_a.

pKa The negative logarithm of K_a.

Henderson–Hasselbalch equation A relationship between the pH of a buffer, pK_a, and the concentrations of acid and salt in the buffer.

▶ Example 9.15

a. Calculate the pH of a buffer solution that contains 0.10 mol acetic acid (CH_3COOH) and 0.10 mol sodium acetate (CH_3COONa) per liter.
b. What is the pH of a buffer in which the concentration of NaH_2PO_4 is 0.10 M and that of Na_2HPO_4 is 0.50 M?
c. A buffer system consisting of the ions $H_2PO_4^-$ and HPO_4^{2-} helps control the pH of urine. What value of the ratio $[HPO_4^{2-}]/[H_2PO_4^-]$ would be required to maintain a pH of 6.00, the average value for normal urine?

Solution

a. The acetic acid concentration is 0.10 M and that of its conjugate base, the acetate ion from the sodium acetate salt, is also 0.10 M. The pK_a for acetic acid is 4.74:

$$pH = pK_a + \log \frac{[CH_3COO^-]}{[CH_3COOH]} = 4.74 + \log \left(\frac{0.10 \text{ mol/L}}{0.10 \text{ mol/L}} \right)$$

$$= 4.74 + \log(1) = 4.74 + 0 = 4.74$$

TABLE 9.9 K_a and pK_a Values for Selected Weak Acids

Name	Formula	K_a	pK_a
Acetic acid[a]	CH_3COOH	1.8×10^{-5}	4.74
Ammonium ion	NH_4^+	5.6×10^{-10}	9.25
Boric acid	H_3BO_3	7.3×10^{-10}	9.14
Dihydrogen borate ion	$H_2BO_3^-$	1.8×10^{-13}	12.75
Hydrogen borate ion	HBO_3^{2-}	1.6×10^{-14}	13.80
Carbonic acid	H_2CO_3	4.3×10^{-7}	6.37
Bicarbonate ion	HCO_3^-	5.6×10^{-11}	10.25
Citric acid[a]	$C_3H_4(OH)(COOH)_3$	8.4×10^{-4}	3.08
Dihydrogen citrate ion[a]	$C_3H_4(OH)(COOH)_2COO^-$	1.8×10^{-5}	4.74
Hydrogen citrate ion[a]	$C_3H_4(OH)(COOH)(COO)_2^{2-}$	4.0×10^{-6}	5.40
Formic acid[a]	$HCOOH$	1.8×10^{-4}	3.74
Lactic acid[a]	$C_2H_4(OH)COOH$	1.4×10^{-4}	3.85
Nitrous acid	HNO_2	4.6×10^{-4}	3.33
Phosphoric acid	H_3PO_4	7.5×10^{-3}	2.12
Dihydrogen phosphate ion	$H_2PO_4^-$	6.2×10^{-8}	7.21
Hydrogen phosphate ion	HPO_4^{2-}	2.2×10^{-13}	12.66
Sulfurous acid	H_2SO_3	1.5×10^{-2}	1.82
Bisulfite ion	HSO_3^-	1.0×10^{-7}	7.00

[a]The hydrogen that ionizes in organic acids and ions is a part of a carboxylic acid group, represented by COOH.

We see that, as mentioned earlier, when the acid and conjugate base concentrations are equal, pH = pK_a.

b. In this case, both compounds are salts that produce the $H_2PO_4^-$ and HPO_4^{2-} ions. In such situations, the ion with more hydrogens will be the acid, and the one with fewer will be the anion or conjugate base. Therefore, the acid is $H_2PO_4^-$ at a concentration of 0.10 M, and the conjugate base or anion is HPO_4^{2-} at a concentration of 0.50 M. The pK_a for $H_2PO_4^-$ is 7.21:

$$pH = pK_a + \log \frac{[HPO_4^{2-}]}{[H_2PO_4^-]} = 7.21 + \log \left(\frac{0.50 \text{ mol/L}}{0.10 \text{ mol/L}} \right)$$

$$= 7.21 + \log(5.00) = 7.21 + 0.70 = 7.91$$

c. As in part (b), the acid is $H_2PO_4^-$ and the conjugate base is HPO_4^{2-}. The desired pH is 6.00. Substitution into Equation 9.54 gives

$$pH = pK_a + \log \frac{[HPO_4^{2-}]}{[H_2PO_4^-]}$$

$$6.00 = 7.21 + \log \frac{[HPO_4^{2-}]}{[H_2PO_4^-]}$$

$$6.00 - 7.21 = \log \frac{[HPO_4^{2-}]}{[H_2PO_4^-]}$$

$$-1.21 = \log \frac{[HPO_4^{2-}]}{[H_2PO_4^-]}$$

DR. MELINA JAMPOLIS

Dr. Melina Jampolis is a San Francisco, California, physician nutrition specialist—a specialty practiced by only about 200 physicians in the United States. She focuses exclusively on nutrition for weight loss and disease prevention and treatment.

Ask an Expert 9.1

Q: Does diet play a role in peptic ulcer disease?

A: Peptic ulcer disease (PUD), which includes ulcer formation in the stomach or the duodenum (the first portion of the small intestine), affects a significant portion of the U.S. population and is the leading reason for gastrointestinal surgery. Since PUD affects your digestive tract, it makes sense that what you eat on a daily basis should impact the development of ulcer disease. However, direct associations between what many people believe to be ulcer-causing components of the diet and the development of PUD are surprisingly limited and inconsistent. Large observational studies have not established a direct link between the intake of alcohol, caffeine, or spicy foods and ulcer formation. This is true despite the fact that alcohol and caffeine are known to stimulate gastric acid secretion, which in turn makes ulcer symptoms worse and interferes with ulcer healing. Research over the past few decades has found that 80–90% of peptic ulcers are actually caused by a bacterium called *Helicobacter pylori* (*H. pylori*). So while diet might play a role, particularly in terms of ulcer healing, *H. pylori* infection is the primary cause of PUD.

Milk, which many people still mistakenly believe protects against or even helps to treat PUD, might actually increase its severity by causing an increase in acid secretion and masking ulcer symptoms. Masked symptoms allow the disease to progress without treatment. On the other hand, fermented milk products, including yogurt and cheese, have been associated with a reduction in PUD, possibly due to their antimicrobial properties. Research shows that probiotic strains, including *Lactobacillus* and *Bifidobacterium bifidum*, inhibit the growth of the *H. pylori*. These probiotic strains have also been shown to improve the effectiveness of treatment for *H. pylori* infection.

In the past, fiber consumption was discouraged for patients with PUD. Current research suggests that soluble fiber might actually play a protective role. In a large prospective study, those with the highest intakes of soluble fiber, particularly fiber from legumes, including beans, tofu, peanut butter, and nuts, had a 50% lower incidence of duodenal ulcer compared to those consuming the least amount of soluble fiber. Vegetable fiber and vitamin A intake also appear to exert a protective effect. Conversely, higher intakes of refined grains (which are lower in fiber) and refined sugar have been associated with a higher risk of duodenal ulcer disease.

Polyphenols, plant-based nutrients that play a role in decreasing the risk of numerous diseases, also appear to play an important role in ulcer prevention. Animal studies have shown that several different polyphenols help prevent PUD by increasing the formation of gut-protecting mucus; decreasing acid secretion; inhibiting the growth of *H. pylori*; and

A. Peptic Ulcer Disease
B. Fermented dairy products are good for PUD
C. Ginger is good for PUD
D. Fennel is good for PUD
E. Milk may increase the severity of PUD
F. Patients with PUD should avoid refined sugars

© Veronika Burmeister/Visuals Unlimited/Corbis © tacar/Shutterstock.com © Reika/Shutterstock.com © Franco Deriu/Shutterstock.com © Jiri Hera /Shutterstock.com © Olga Danylenko /Shutterstock.com

© Michael Roud Photography

Ask an Expert 9.1 (continued)

acting as powerful antioxidants, which help protect the lining of the gut. Promising research in rats has shown a potential beneficial effect of numerous polyphenol-rich plant compounds, including ginger, fennel, and olive leaf extract.

Smoking has been strongly associated with the incidence of PUD through numerous mechanisms. On the other hand, physical activity has been shown to decrease the risk of duodenal ulcer. So, the adoption of a healthy lifestyle, including a diet containing a variety of fruits and vegetables, plenty of soluble fiber, and fermented dairy products with limited added sugar; not smoking; and exercising regularly can play a significant role in the prevention of ulcer disease.

At this point, we know the value of the log of the desired ratio. To get the ratio, we must evaluate the antilog of -1.21. Use Step 3 given in Table 9.3 to get 0.062 as the value of the ratio. This result says the ratio $[HPO_4^{2-}]/[H_2PO_4^-]$ must be 0.062. In other words, the HPO_4^{2-} concentration must be only about 0.06 times the value of the $H_2PO_4^-$ concentration. Any concentrations giving this ratio would be satisfactory. For example, if $[H_2PO_4^-] = 0.50$ M, then $[HPO_4^{2-}]$ would have to equal 0.062×0.5 M, or 0.031 M.

▶ LEARNING CHECK 9.20

a. What is the pH of a buffer solution in which formic acid (HCOOH) and sodium formate (HCOONa) are both at a concentration of 0.22 M?

b. What is the pH of a buffer solution that is 0.25 M in sulfurous acid (H_2SO_3) and 0.10 M in sodium bisulfite ($NaHSO_3$)?

c. A buffer system in the blood consists of the bicarbonate ion, HCO_3^-, and carbonic acid, H_2CO_3. What value of the ratio $[HCO_3^-]/[H_2CO_3]$ would be required to maintain blood at the average normal pH of 7.40?

Case Study Follow-up

Over-the-counter medications should be taken as directed. Some antacids contain large amounts of sodium, which might be problematic for someone on a sodium-restricted diet, and calcium-containing antacids might exacerbate existing kidney problems. Antacids should only be used for occasional relief. If a person is taking antacids on a daily basis for extended periods of time, a different medication, such as a proton-pump inhibitor, should be considered. Frequent long-lasting heartburn may indicate gastroesophageal reflux disease (GERD) (Mayo Clinic 2011).

Symptoms of heartburn include a burning sensation in the chest area occurring after a meal or while lying down, or a sour taste in the mouth from stomach acid. Heart attack symptoms might be mistaken for heartburn. However, if chest pain occurs during exercise; is accompanied by shortness of breath, nausea, or dizziness; and is radiating to the arm, neck, shoulder, or jaw, heart attack is a possibility—and emergency help should be sought immediately (Mayo Clinic, 2011).

Heartburn may be initiated or made worse by certain foods. Someone suffering from heartburn should avoid fatty foods, tomato sauce, chocolate, caffeine, and spicy foods. Other factors that aggravate heartburn are excess body weight, tight clothing, and lying down too soon after a meal. Maintaining a healthy weight and eating frequent small meals instead of infrequent large meals, especially before bedtime, will help control heartburn.

Concept Summary

The Arrhenius Theory

Svante Arrhenius defined acids as substances that dissociate in water to provide hydrogen ions (H^+), and bases as substances that dissociate in water to provide hydroxide ions, (OH^-).

Objective 1, Exercise 9.2

The Brønsted Theory

Johannes Brønsted and Thomas Lowry proposed a theory in which acids are defined as any hydrogen-containing substances capable of donating protons to other substances. Bases are substances that accept and form covalent bonds with protons. When a substance behaves as a Brønsted acid by donating a proton, the substance becomes a conjugate base.

Objective 2, Exercises 9.6 and 9.10

Naming Acids

Two types of acids are named differently. Water solutions of binary covalent compounds containing hydrogen and a nonmetal are named following the pattern hydro(stem)ic acid, where (stem) is the stem of the name of the nonmetal bonded to hydrogen. Acids in which hydrogen is bonded to polyatomic ions have names based on the name of the polyatomic ion to which hydrogen is bonded.

Objective 3, Exercise 9.22

The Self-Ionization of Water

Water, a hydrogen-containing material, is able to behave as both a Brønsted acid and a Brønsted base. In pure water, a small number of water molecules (10^{-7} mol/L) donate protons to other water molecules.

Objective 4, Exercises 9.28a & b and 9.30a & b

The pH Concept

The pH is the negative logarithm of the molar H^+ concentration of a solution. Solutions with pH values lower than 7 are acidic, those with pH values higher than 7 are basic or alkaline, and those with a pH value of 7 are neutral.

Objective 5, Exercises 9.36 and 9.40

Properties of Acids

All acids have certain characteristic properties that include (1) a sour taste; (2) a reaction with water to produce H_3O^+; (3) reactions with solid metallic oxides, hydroxides, carbonates, and bicarbonates; and (4) a reaction with certain metals to give hydrogen gas.

Objective 6, Exercise 9.50

Properties of Bases

Basic solutions feel soapy or slippery and change the color of litmus from red to blue. Their most characteristic chemical property is a neutralization reaction with acids to produce water and a salt.

Objective 7, Exercise 9.60

Salts

At room temperature, salts are solid crystalline substances that contain the cation of a base and the anion of an acid. Hydrated salts contain specific numbers of water molecules as a part of their crystalline structures. Salts can be prepared by reacting an appropriate acid with one of a number of other materials.

Objective 8, Exercises 9.68 and 9.74

The Strengths of Acids and Bases

Acids and bases that dissociate completely when dissolved in solution are called strong, and those that do not are called weak or moderately weak, depending on the degree of dissociation they undergo. Acid strength is indicated by the value of K_a, the acid dissociation constant. In general, polyprotic acids become weaker as an acid during each successive dissociation reaction. The Brønsted base produced by the dissociation of an acid has a strength opposite that of the acid. The acid strengths of cations produced by the dissociation of bases follow a similar pattern.

Objective 9, Exercise 9.86

Analyzing Acids and Bases

The neutralization reaction of acids and bases is used in a process called titration to analyze acids and bases. During a typical titration, a base solution of known concentration is added slowly to an acid solution of unknown concentration. The titration is stopped at the endpoint when a color change occurs in an indicator. The volumes of acid and base required are used to calculate the acid concentration.

Objective 10, Exercise 9.92

Titration Calculations

Titrations are used to determine the total amount of acid or base in solutions. Data collected from titrations are treated like other stoichiometric data in calculations.

Objective 11, Exercises 9.98 and 9.100a

Hydrolysis Reactions of Salts

The cation and anion of a salt may have the same or different Brønsted acid and base strengths. When they are the same, solutions of the salt have a pH of 7. When they are different, salt solutions have H^+ and OH^- concentrations that are unequal and pH values different than 7. The reactions of salt ions with water that cause these results are called hydrolysis reactions.

Objective 12, Exercise 9.108

Buffers

Solutions with the ability to maintain essentially constant pH values when acid (H^+) or base (OH^-) are added are called buffers. All buffers have a limit to the amount of acid or base they can absorb without changing pH. This limit is called the buffer capacity.

Objective 13, Exercise 9.116

Key Terms and Concepts

Acid dissociation constant (9.9)
Acidic solution (9.4)
Activity series (9.6)
Anion (9.8)
Arrhenius acid (9.1)
Arrhenius base (9.1)
Basic or alkaline solution (9.4)
Brønsted acid (9.2)
Brønsted base (9.2)
Buffer (9.13)
Buffer capacity (9.13)
Cation (9.8)

Conjugate acid–base pair (9.2)
Conjugate base (9.2)
Diprotic acid (9.9)
Endpoint of a titration (9.10)
Equivalence point of a titration (9.10)
Equivalent of salt (9.8)
Henderson–Hasselbalch equation (9.13)
Hydrate (9.8)
Hydrolysis reaction (9.12)
Ion product of water (9.4)
Monoprotic acid (9.9)
Neutral (9.4)

Neutralization reaction (9.7)
pH (9.5)
pK_a (9.13)
Salt (9.8)
Strong acids and strong bases (9.9)
Titration (9.10)
Triprotic acid (9.9)
Water of hydration (9.8)
Weak (or moderately weak) acids and bases (9.9)

Key Equations

1.	Relationship between [H^+] and [OH^-] in water solutions (Section 9.4):	$K_w = [H_3O^+][OH^-] = 1.0 \times 10^{-14}\ (mol/L)^2$	Equation 9.10
2.	Relationships between pH and [H^+] (Section 9.5):	$pH = -\log[H^+]$ $[H^+] = 1 \times 10^{-pH}$	Equation 9.11 Equation 9.12
3.	General reactions of acids (Section 9.8):	acid + metal → salt + H_2 acid + metal oxide → salt + H_2O acid + metal hydroxide → salt + H_2O acid + metal carbonate → salt + H_2O + CO_2 acid + metal bicarbonate → salt + H_2O + CO_2	Equation 9.26 Equation 9.27 Equation 9.28 Equation 9.29 Equation 9.30
4.	Dissociation constant for acids (Section 9.9):	$K_a = \dfrac{[H^+][B^-]}{[HB]}$	Equation 9.34
5.	Calculation of buffer pH, Henderson–Hasselbalch equation (Section 9.13):	$pH = pK_a + \log \dfrac{[B^-]}{[HB]}$	Equation 9.54

Exercises

Even-numbered exercises are answered in Appendix B.

Blue-numbered exercises are more challenging.

The Arrhenius Theory (Section 9.1)

9.1 Write the dissociation equations for the following that emphasize their behavior as Arrhenius acids:

  a. HI

  b. HBrO

  c. HCN

  d. $HClO_2$

9.2 Write the dissociation equations for the following that emphasize their behavior as Arrhenius acids:

  a. $HBrO_2$

  b. HS^-

  c. HBr

  d. $HC_2H_3O_2$ (only the 1st listed H dissociates)

9.3 Each of the following produces a basic solution when dissolved in water. Identify those that behave as Arrhenius bases and write dissociation equations to illustrate that behavior.

  a. CsOH

  b. CH_3NH_2

  c. NH_3

  d. $Ca(OH)_2$

9.4 Each of the following produces a basic solution when dissolved in water. Identify those that behave as Arrhenius bases and write dissociation equations to illustrate that behavior.

  a. LiOH

  b. $C_2H_5NH_2$

  c. $Sr(OH)_2$

  d. $NaNH_2$

The Brønsted Theory (Section 9.2)

9.5 Identify each Brønsted acid and base in the following equations. Note that the reactions are assumed to be reversible.

  a. $HBr(aq) + H_2O(\ell) \rightleftarrows H_3O^+(aq) + Br^-(aq)$

  b. $H_2O(\ell) + N_3^-(aq) \rightleftarrows HN_3(aq) + OH^-(aq)$

  c. $H_2S(aq) + H_2O(\ell) \rightleftarrows H_3O^+(aq) + HS^-(aq)$

  d. $SO_3^{2-}(aq) + H_2O(\ell) \rightleftarrows HSO_3^-(aq) + OH^-(aq)$

  e. $HCN(aq) + H_2O(\ell) \rightleftarrows H_3O^+(aq) + CN^-(aq)$

9.6 Identify each Brønsted acid and base in the following equations. Note that the reactions are assumed to be reversible.

  a. $HC_2O_4^-(aq) + H_2O(\ell) \rightleftarrows H_3O^+(aq) + C_2O_4^{2-}(aq)$

  b. $HNO_2(aq) + H_2O(\ell) \rightleftarrows H_3O^+(aq) + NO_2^-(aq)$

  c. $PO_4^{3-}(aq) + H_2O(\ell) \rightleftarrows HPO_4^{2-}(aq) + OH^-(aq)$

  d. $H_2SO_3(aq) + H_2O(\ell) \rightleftarrows HSO_3^-(aq) + H_3O^+(\ell)$

  e. $F^-(aq) + H_2O(\ell) \rightleftarrows HF(aq) + OH^-(aq)$

9.7 Identify each conjugate acid–base pair in the equations of Exercise 9.5.

9.8 Identify each conjugate acid–base pair in the equations of Exercise 9.6.

9.9 Write equations to represent the Brønsted acid behavior for each of the following acids in water solution. Remember to represent the reactions as being reversible.

  a. HI

  b. HBrO

  c. HCN

  d. HSe^-

9.10 Write equations to represent the Brønsted acid behavior for each of the following acids in water solution. Remember to represent the reactions as being reversible.

  a. HF

  b. $HClO_3$

  c. HClO

  d. HS^-

9.11 Write a formula for the conjugate base formed when each of the following behaves as a Brønsted acid:

  a. HSO_3^-

  b. HPO_4^{2-}

  c. $HClO_3$

  d. $CH_3NH_3^+$

  e. $H_2C_2O_4$

9.12 Write a formula for the conjugate base formed when each of the following behaves as a Brønsted acid:

  a. HSO_4^-

  b. $CH_3NH_3^+$

  c. $HClO_4$

  d. NH_4^+

  e. HCl

9.13 Write a formula for the conjugate acid formed when each of the following behaves as a Brønsted base:

  a. NH_2^-

  b. CO_3^{2-}

  c. OH^-

  d. $(CH_3)_2NH$

  e. NO_2^-

9.14 Write a formula for the conjugate acid formed when each of the following behaves as a Brønsted base:

  a. HCO_3^-

  b. S^{2-}

  c. HS^-

  d. $HC_2O_4^-$

  e. $HN_2O_2^-$

9.15 The following reactions illustrate Brønsted acid–base behavior. Complete each equation.

a. $HI(aq) + ? \rightarrow H_3O^+(aq) + I^-(aq)$

b. $NH_3(\ell) + ? \rightarrow NH_4^+ + NH_2^-$

c. $H_2C_2O_4(aq) + H_2O(\ell) \rightarrow ? + HC_2O_4^-(aq)$

d. $H_2N_2O_2(aq) + H_2O(\ell) \rightarrow H_3O^+(aq) + ?$

e. $? + H_2O(\ell) \rightarrow H_3O^+(aq) + CO_3^{2-}(aq)$

9.16 The following reactions illustrate Brønsted acid–base behavior. Complete each equation.

a. $H_2AsO_4^-(aq) + ? \rightarrow NH_4^+(aq) + HAsO_4^{2-}(aq)$

b. $C_6H_5NH_2(aq) + ? \rightarrow C_6H_5NH_3^+(aq) + OH^-(aq)$

c. $S^{2-}(aq) + H_2O(\ell) \rightarrow ? + OH^-(aq)$

d. $(CH_3)_2NH(aq) + ? \rightarrow (CH_3)_2NH_2^+(aq) + Br^-(aq)$

e. $H_2PO_4^-(aq) + CH_3NH_2(aq) \rightarrow HPO_4^{2-}(aq) + ?$

9.17 Write equations to illustrate the acid–base reaction when each of the following pairs of Brønsted acids and bases are combined:

Acid	Base
a. HOCl	H_2O
b. $HClO_4$	NH_3
c. H_2O	NH_2^-
d. H_2O	OCl^-
e. $HC_2O_4^-$	H_2O

9.18 Write equations to illustrate the acid–base reaction of each of the following pairs of Brønsted acids and bases:

Acid	Base
a. HS^-	NH_3
b. H_2O	ClO_3^-
c. H_2O	NH_2^-
d. HBO_3^{2-}	H_2O
e. HNO_2	NH_3

Naming Acids (Section 9.3)

9.19 A water solution of HF gas is used to etch glass. Name the water solution as an acid.

9.20 Hydrogen cyanide, HCN, behaves in a water solution very much like the binary covalent compounds of hydrogen, but it liberates the cyanide ion, CN^-. Name the acidic water solution by following the rules for binary covalent compounds of hydrogen.

9.21 Name the following acids. Refer to Table 4.7 as needed.

a. $H_2Se(aq)$

b. $HClO_3$

c. H_2SO_4

d. HNO_3

9.22 Name the following acids. Refer to Table 4.7 as needed.

a. $H_2Te(aq)$

b. $HClO$

c. H_2SO_3

d. HNO_2

9.23 The acid $H_3C_6H_5O_7$ forms the citrate ion, $C_6H_5O_7^{3-}$, when all three hydrogens are removed. This acid is involved in an important energy-storing process in the body. Name the acid.

9.24 The acid $H_2C_4H_4O_4$ forms the succinate ion, $C_4H_4O_4^{2-}$, when both hydrogens are removed. This acid is involved in the same energy-storing process as the acid of Exercise 9.23. Name $H_2C_4H_4O_4$ as an acid.

9.25 Refer to Table 4.7, and write the formula for chromic acid.

9.26 Refer to Table 4.7, and write the formula for carbonic acid.

The Self-Ionization of Water (Section 9.4)

9.27 Calculate the molar concentration of OH^- in water solutions with the following H_3O^+ molar concentrations:

a. 1.0×10^{-7}

b. 3.2×10^{-3}

c. 4.7×10^{-11}

d. 1.2

e. 0.043

9.28 Calculate the molar concentration of OH^- in water solutions with the following H_3O^+ molar concentrations:

a. 0.044

b. 1.3×10^{-4}

c. 0.0087

d. 7.9×10^{-10}

e. 3.3×10^{-2}

9.29 Calculate the molar concentration of H_3O^+ in water solutions with the following OH^- molar concentrations:

a. 1.0×10^{-7}

b. 5.2×10^{-4}

c. 9.9×10^{-10}

d. 0.092

e. 3.7

9.30 Calculate the molar concentration of H_3O^+ in water solutions with the following OH^- molar concentrations:

a. 6.9×10^{-5}

b. 0.074

c. 4.9

d. 1.7×10^{-3}

e. 9.2×10^{-9}

9.31 Classify the solutions represented in Exercises 9.27 and 9.29 as acidic, basic, or neutral.

9.32 Classify the solutions represented in Exercises 9.28 and 9.30 as acidic, basic, or neutral.

Even-numbered exercises answered in Appendix B.

Unless otherwise noted all art on this page © Cengage Learning 2014

The pH Concept (Section 9.5)

9.33 Classify solutions with the following characteristics as acidic, basic, or neutral:

  a. pH = 10

  b. pH = 4

  c. pH = 7.3

  d. pH = 6

9.34 Classify solutions with the following characteristics as acidic, basic, or neutral.

  a. pH = 4.7

  b. pH = 8.8

  c. pH = 1.5

  d. pH = 11

  e. pH = 5.9

9.35 Determine the pH of water solutions with the following characteristics. Classify each solution as acidic, basic, or neutral.

  a. $[H^+] = 1.0 \times 10^{-5}$

  b. $[OH^-] = 6.0 \times 10^{-3}$

  c. $[H^+] = [OH^-]$

  d. $[H^+] = 9.0 \times 10^{-4}$

  e. $[OH^-] = 3.0 \times 10^{-9}$

9.36 Determine the pH of water solutions with the following characteristics. Classify each solution as acidic, basic, or neutral.

  a. $[H^+] = 4.1 \times 10^{-9}$

  b. $[OH^-] = 9.4 \times 10^{-4}$

  c. $[OH^-] = 10[H^+]$

  d. $[H^+] = 2.3 \times 10^{-2}$

  e. $[OH^-] = 5.1 \times 10^{-10}$

9.37 Determine the pH of water solutions with the following characteristics. Classify each solution as acidic, basic, or neutral.

  a. $[H^+] = 3.7 \times 10^{-4}$

  b. $[H^+] = 7.4 \times 10^{-8}$

  c. $[H^+] = 1.9 \times 10^{-10}$

  d. $[OH^-] = 1.3 \times 10^{-1}$

  e. $[OH^-] = 6.8 \times 10^{-7}$

9.38 Determine the pH of water solutions with the following characteristics. Classify each solution as acidic, basic, or neutral.

  a. $[H^+] = 2.2 \times 10^{-3}$

  b. $[H^+] = 3.9 \times 10^{-12}$

  c. $[H^+] = 7.5 \times 10^{-6}$

  d. $[OH^-] = 2.5 \times 10^{-4}$

  e. $[OH^-] = 8.6 \times 10^{-10}$

9.39 Determine the $[H^+]$ value for solutions with the following characteristics:

  a. pH = 4.45

  b. pH = 13.12

  c. pH = 7.73

9.40 Determine the $[H^+]$ value for solutions with the following characteristics:

  a. pH = 9.27

  b. pH = 2.55

  c. pH = 5.42

9.41 Convert the following pH values into both $[H^+]$ and $[OH^-]$ values:

  a. pH = 8.00

  b. pH = 6.15

  c. pH = 1.30

9.42 Convert the following pH values into both $[H^+]$ and $[OH^-]$ values:

  a. pH = 3.95

  b. pH = 4.00

  c. pH = 11.86

9.43 The pH values listed in Table 9.1 are generally the average values for the listed materials. Most natural materials, such as body fluids and fruit juices, have pH values that cover a range for different samples. Some measured pH values for specific body fluid samples are given below. Convert each one to $[H^+]$, and classify the fluid as acidic, basic, or neutral.

  a. Blood, pH = 7.41

  b. Gastric juice, pH = 1.60

  c. Urine, pH = 5.93

  d. Saliva, pH = 6.85

  e. Pancreatic juice, pH = 7.85

9.44 The pH values listed in Table 9.1 are generally the average values for the listed materials. Most natural materials, such as body fluids and fruit juices, have pH values that cover a range for different samples. Some measured pH values for specific body fluid samples are given below. Convert each one to $[H^+]$, and classify the fluid as acidic, basic, or neutral.

  a. Bile, pH = 8.05

  b. Vaginal fluid, pH = 3.93

  c. Semen, pH = 7.38

  d. Cerebrospinal fluid, pH = 7.40

  e. Perspiration, pH = 6.23

9.45 The pH values of specific samples of food items are listed below. Convert each value to $[H^+]$, and classify the sample as acidic, basic, or neutral.

  a. Milk, pH = 6.39

  b. Coffee, pH = 5.10

  c. Orange juice, pH = 4.07

  d. Vinegar, pH = 2.65

9.46 The pH values of specific samples of food items are listed below. Convert each value to $[H^+]$, and classify the sample as acidic, basic, or neutral.

  a. Soft drink, pH = 2.91

  b. Tomato juice, pH = 4.11

  c. Lemon juice, pH = 2.32

  d. Grapefruit juice, pH = 3.07

Properties of Acids (Section 9.6)

9.47 Using the information in Table 9.4, describe how you would prepare each of the following solutions.

    a. About 750 mL of 0.5 M H_2SO_4 from dilute sulfuric acid

    b. About 200 mL of 0.1 M NaOH from stock sodium hydroxide solution

    c. About 1.0 L of 1 M acetic acid from glacial acetic acid

9.48 Use the information in Table 9.4 and describe how you would prepare each of the following solutions.

    a. About 2 L of 3.0 M HNO_3 from dilute nitric acid solution

    b. About 500 mL of 1.5 M aqueous ammonia from concentrated aqueous ammonia solution

    c. About 5 L of 0.2 M HCl from concentrated hydrochloric acid solution

9.49 Write balanced molecular equations to illustrate the following characteristic reactions of acids, using nitric acid (HNO_3).

    a. Reaction with water to form hydronium ions

    b. Reaction with the solid oxide, CaO

    c. Reaction with the solid hydroxide, $Mg(OH)_2$

    d. Reaction with the solid carbonate, $CuCO_3$

    e. Reaction with the solid bicarbonate, $KHCO_3$

    f. Reaction with Mg metal

9.50 Write balanced molecular equations to illustrate the following characteristic reactions of acids, using sulfuric acid (H_2SO_4).

    a. Reaction with water to form hydronium ions

    b. Reaction with the solid oxide, CaO

    c. Reaction with the solid hydroxide, $Mg(OH)_2$

    d. Reaction with the solid carbonate, $CuCO_3$

    e. Reaction with the solid bicarbonate, $KHCO_3$

    f. Reaction with Mg metal

9.51 Write each molecular equation of Exercise 9.49 in total ionic and net ionic form. Use Table 7.4 to decide which products will be soluble.

9.52 Write each molecular equation of Exercise 9.50 in total ionic and net ionic form. Use Table 7.4 to decide which products will be soluble.

9.53 Write balanced molecular equations to illustrate five different reactions that could be used to prepare $BaCl_2$ from hydrochloric acid (HCl) and other appropriate substances.

9.54 Write balanced molecular equations to illustrate five different reactions that could be used to prepare $MgCl_2$ from hydrochloric acid (HCl) and other appropriate substances.

9.55 Write balanced molecular, total ionic, and net ionic equations to illustrate each of the following reactions. All the metals form 2+ ions.

    a. Zinc with H_2SO_4

    b. Magnesium with HCl

    c. Calcium with $HC_2H_3O_2$

9.56 Write balanced molecular, total ionic, and net ionic equations to illustrate each of the following reactions. All the metals form 2+ ions.

    a. Tin with H_2SO_3

    b. Magnesium with H_3PO_4

    c. Calcium with HBr

Properties of Bases (Section 9.7)

9.57 Write balanced molecular, total ionic, and net ionic equations to represent neutralization reactions between RbOH and the following acids. Use all H's possible for each acid.

    a. HCl

    b. HNO_3

    c. H_2SO_4

9.58 Write balanced molecular, total ionic, and net ionic equations to represent the neutralization reactions between RbOH and the following acids. Use all H's possible for each acid.

    a. HBr

    b. H_2SO_3

    c. H_3PO_3 (only two H's react)

9.59 Some polyprotic acids can form more than one salt, depending on the number of H's that react with base. Write balanced molecular, total ionic, and net ionic equations to represent the following neutralization reactions between KOH and

    a. H_2SO_4 (only react one H)

    b. H_2SO_4 (react both H's)

    c. H_3PO_4 (only react one H)

9.60 Some polyprotic acids can form more than one salt, depending on the number of H's that react with base. Write balanced molecular, total ionic, and net ionic equations to represent the following neutralization reactions between KOH and

    a. H_3PO_4 (react two H's)

    b. H_3PO_4 (react three H's)

    c. $H_2C_2O_4$ (react one H)

Salts (Section 9.8)

9.61 Identify with ionic formulas the cations and anions of the following salts:

    a. LiCl

    b. $Cu(NO_3)_2$

    c. $SrSO_4$

    d. K_3PO_4

    e. K_2HPO_4

    f. $CaCO_3$

9.62 Identify with ionic formulas the cations and anions of the following salts:

    a. $CuCl_2$

    b. $(NH_4)_2SO_4$

    c. Li_3PO_4

    d. $MgCO_3$

    e. $Ca(C_2H_3O_2)_2$

    f. KNO_3

Even-numbered exercises answered in Appendix B.

9.63 Identify with formulas the acid and base from which the anion and cation of each salt in Exercise 9.61 was derived. Pay special attention to salts derived from polyprotic acids and be sure to list the acid formula with all H's.

9.64 Identify with formulas the acid and base from which the anion and cation of each salt in Exercise 9.62 was derived. Pay special attention to salts derived from polyprotic acids and be sure to list the acid formula with all H's.

9.65 Calculate the mass of water that would be released if the water of hydration were completely driven off 1.0 mol of (a) plaster of Paris and (b) gypsum (see Table 9.6). How would the products of these reactions compare?

9.66 Calculate the mass of water that would be released if the water of hydration were completely driven off 1.0 mol of (a) Epsom salts and (b) borax (see Table 9.6). How would the products of these reactions compare?

9.67 Write formulas for the acid and indicated solid that could be used to prepare each of the following salts:

   a. $CuCl_2$ (solid is an oxide)

   b. $MgSO_4$ (solid is a carbonate)

   c. LiBr (solid is a hydroxide)

9.68 Write formulas for the acid and indicated solid that could be used to prepare each of the following salts:

   a. $Mg(NO_3)_2$ (solid is a carbonate)

   b. $CaCl_2$ (solid is an oxide)

   c. Rb_2SO_4 (solid is a bicarbonate)

9.69 Write balanced molecular equations to illustrate each salt preparation described in Exercise 9.67.

9.70 Write balanced molecular equations to illustrate each salt preparation described in Exercise 9.68.

9.71 Determine the number of moles of each of the following salts that would equal 1 eq of salt:

   a. KNO_3

   b. Li_2CO_3

   c. $SrCl_2$

9.72 Determine the number of moles of each of the following salts that would equal 1 eq of salt:

   a. $CaCl_2$

   b. $Cu(NO_3)_2$

   c. $CrCl_3$

9.73 Determine the number of equivalents and milliequivalents in each of the following:

   a. 0.10 mol KI

   b. 0.25 mol $MgCl_2$

   c. 4.73×10^{-2} mol $AgNO_3$

9.74 Determine the number of equivalents and milliequivalents in each of the following:

   a. 0.22 mol $ZnCl_2$

   b. 0.45 mol CsCl

   c. 3.12×10^{-2} mol $Fe(NO_3)_2$

9.75 Determine the number of equivalents and milliequivalents in 5.00 g of each of the following salts. Include any waters of

hydration given in the salt formula when you calculate salt formula weights.

   a. NaCl

   b. $NaNO_3$

   c. Na_3PO_4

   d. $MgSO_4 \cdot 7H_2O$

9.76 Determine the number of equivalents and milliequivalents in 5.00 g of each of the following salts. Include any waters of hydration given in the salt formula when you calculate salt formula weights.

   a. $Na_2CO_3 \cdot 10H_2O$

   b. $CuSO_4 \cdot 5H_2O$

   c. Li_2CO_3

   d. NaH_2PO_4

9.77 A sample of intracellular fluid contains 45.1 meq/L of Mg^{2+} ion. Assume the Mg^{2+} comes from dissolved $MgCl_2$, and calculate the number of moles and number of grams of $MgCl_2$ that would be found in 250 mL of the intracellular fluid.

9.78 A sample of intracellular fluid contains 133 meq/L of K^+ ion. Assume the K^+ comes from dissolved K_2SO_4, and calculate the number of moles and number of grams of K_2SO_4 that would be found in 150 mL of the intracellular fluid.

The Strengths of Acids and Bases (Section 9.9)

9.79 Illustrate the difference between weak, moderately strong, and strong acids by writing dissociation reactions for the hypothetical acid HB, using arrows of various lengths.

9.80 The K_a values have been determined for four acids and are listed below. Arrange the acids in order of increasing acid strength (weakest first, strongest last).

$$acid\ A\ (K_a = 5.6 \times 10^{-5})$$
$$acid\ B\ (K_a = 1.8 \times 10^{-5})$$
$$acid\ C\ (K_a = 1.3 \times 10^{-4})$$
$$acid\ D\ (K_a = 1.1 \times 10^{-3})$$

9.81 Arrange the four acids classified as weak in Table 9.7 in order of increasing strength (weakest first, strongest last).

9.82 K_a values for four weak acids are given below:

$$acid\ A\ (K_a = 2.6 \times 10^{-4})$$
$$acid\ B\ (K_a = 3.7 \times 10^{-5})$$
$$acid\ C\ (K_a = 5.8 \times 10^{-4})$$
$$acid\ D\ (K_a = 1.5 \times 10^{-3})$$

   a. Arrange the four acids in order of increasing acid strength (weakest first, strongest last).

   b. Arrange the conjugate bases of the acids (identify as base A, etc.) in order of increasing base strength (weakest base first, strongest last).

9.83 Write dissociation reactions and K_a expressions for the following weak acids:

   a. Hypobromous acid, HBrO

   b. Sulfurous acid, H_2SO_3 (1st H only)

   c. Hydrogen sulfite ion, HSO_3^-

d. Hydroselenic acid, H_2Se (1st H only)

e. Arsenic acid, H_3AsO_4 (1st H only)

9.84 Write dissociation reactions and K_a expressions for the following weak acids:

a. Nitrous acid, HNO_2

b. Hydrogen carbonate ion, HCO_3^-

c. Dihydrogen phosphate ion, $H_2PO_4^-$ (1st H only)

d. Hydrogen sulfide ion, HS^-

e. Hypobromous acid, $HBrO$

9.85 Equal molar solutions are made of three monoprotic acids HA, HB, and HC. The pHs of the solutions are, respectively, 4.82, 3.16, and 5.47. Rank the acids in order of increasing acid strength and explain your reasoning.

9.86 If someone asked you for a weak acid solution, which of the following would you provide according to definitions in this chapter?

a. 0.05 M HCl

b. 20% acetic acid

If the individual really wanted the other solution, what term should have been used instead of *weak*?

9.87 Arsenic acid (H_3AsO_4) is a moderately weak triprotic acid. Write equations showing its stepwise dissociation. Which of the three anions formed in these reactions will be the strongest Brønsted base? Which will be the weakest Brønsted base? Explain your answers.

Analyzing Acids and Bases (Section 9.10)

9.88 Explain the purpose of doing a titration.

9.89 Describe the difference between the information obtained by measuring the pH of an acid solution and by titrating the solution with base.

9.90 Suppose a student is going to titrate an acidic solution with a base and just picks an indicator at random. Under what circumstances will (a) the endpoint and equivalence point be the same? (b) The endpoint and equivalence point be different?

9.91 Determine the number of moles of NaOH that could be neutralized by each of the following:

a. 1.00 L of 0.200 M HCl

b. 500 mL of 0.150 M HNO_3

9.92 Determine the number of moles of NaOH that could be neutralized by each of the following:

a. 250 mL of 0.400 M HBr

b. 750 mL of 0.300 M $HClO_4$

Titration Calculations (Section 9.11)

9.93 Write a balanced molecular equation to represent the neutralization reaction between NaOH and each of the following acids. React all of the acid H's.

a. Molybdic acid, H_2MoO_4

b. Permanganic acid, $HMnO_4$

c. Phosphoric acid, H_3PO_4

9.94 Write a balanced molecular equation to represent the neutralization reaction between NaOH and each of the following acids. React all of the acid H's.

a. Chromic acid, H_2CrO_4

b. Chloric acid, $HClO_3$

c. Arsenous acid, H_3AsO_3

9.95 Write a balanced molecular equation to represent the neutralization reaction between HCl and each of the following bases:

a. $Cd(OH)_2$

b. $Cr(OH)_3$

c. $Fe(OH)_2$

9.96 Write a balanced molecular equation to represent the neutralization reaction between HCl and the following bases:

a. $Sr(OH)_2$

b. $Ni(OH)_3$

c. $Zn(OH)_2$

9.97 A 25.00-mL sample of gastric juice is titrated with a 0.0210 M NaOH solution. The titration to the equivalence point requires 29.8 mL of NaOH solution. If the equation for the reaction is

$$HCl(aq) + NaOH(aq) \rightarrow NaCl(aq) + H_2O(\ell)$$

what is the molarity of HCl in the gastric juice?

9.98 A 25.00-mL sample of $H_2C_2O_4$ solution required 43.88mL of 0.1891 M NaOH solution to titrate it to the equivalence point. Calculate the molarity of the $H_2C_2O_4$ solution.

9.99 A 20.00-mL sample of each of the following acid solutions is to be titrated to the equivalence point using a 0.120 M NaOH solution. Determine the number of milliliters of NaOH solution that will be needed for each acid sample.

a. 0.180 M HCl

b. 0.180 M H_2SO_4

c. 0.100 M HCl

d. 10.00 g of H_3PO_4 in 250 mL of solution

e. 0.150 mol H_2MoO_4 in 500 mL of solution

f. 0.215 mol H_2MoO_4 in 700 mL of solution

9.100 A 20.00-mL sample of each of the following acid solutions is to be titrated to the equivalence point using a 0.120 M NaOH solution. Determine the number of milliliters of NaOH solution that will be needed for each acid sample.

a. 0.200 M $HClO_4$

b. 0.125 M H_2SO_4

c. 0.150 M $H_4P_2O_6$

d. 0.120 mol H_3PO_4 in 500 mL of solution

e. 6.25 g of H_2SO_4 in 250 mL of solution

f. 0.500 mol $HClO_3$ in 1.00 L of solution

9.101 The following acid solutions were titrated to the equivalence point with the base listed. Use the titration data to calculate the molarity of each acid solution.

   a. 25.00 mL of HI solution required 27.15 mL of a 0.250 M NaOH solution.

   b. 20.00 mL of H_2SO_4 solution required 11.12 mL of a 0.109 M KOH solution.

   c. 25.00 mL of gastric juice (HCl) required 18.40 mL of a 0.0250 M NaOH solution.

9.102 The following acid solutions were titrated to the equivalence point with the base listed. Use the titration data to calculate the molarity of each acid solution.

   a. 5.00 mL of dilute H_2SO_4 required 29.88 mL of a 1.17 M NaOH solution.

   b. 10.00 mL of vinegar (acetic acid) required 35.62 mL of a 0.250 M KOH solution.

   c. 10.00 mL of muriatic acid (HCl) used to clean brick and cement required 20.63 mL of a 6.00 M NaOH solution.

9.103 A 20.00-mL sample of diprotic oxalic acid ($H_2C_2O_4$) solution is titrated with a 0.250 M NaOH solution. A total of 27.86 mL of NaOH is required. Calculate:

   a. The number of moles of oxalic acid in the 20.00-mL sample.

   b. The molarity of the oxalic acid solution.

   c. The number of grams of oxalic acid in the 20.00-mL sample.

9.104 A sample of monoprotic benzoic acid weighing 0.5823 g is dissolved in about 25 mL of water. The solution is titrated to the equivalence point using 0.1021 M NaOH. The volume of base required is 46.75 mL. Calculate the molecular weight of the solid acid.

Hydrolysis Reactions of Salts (Section 9.12)

9.105 A solution of solid NH_4Cl dissolved in pure water is acidic (the pH is less than 7). Explain.

9.106 A solution of solid Na_2CO_3 dissolved in pure water has a pH significantly higher than the pH of pure water. Explain why.

9.107 Predict the relative pH (greater than 7, less than 7, etc.) for water solutions of the following salts. Table 9.9 may be useful. For each solution in which the pH is greater or less than 7, explain why and write a net ionic equation to justify your answer.

   a. Potassium sulfite, K_2SO_3

   b. Lithium nitrite, $LiNO_2$

   c. Sodium carbonate, Na_2CO_3

   d. Methylammonium chloride, CH_3NH_3Cl (CH_3NH_2 is a weak base)

9.108 Predict the relative pH (greater than 7, less than 7, etc.) for water solutions of the following salts. Table 9.9 may be useful. For each solution in which the pH is greater or less than 7, explain why and write a net ionic equation to justify your answer.

   a. Sodium hypochlorite, NaOCl (HOCl is a weak acid)

   b. Sodium formate, $NaCHO_2$

   c. Potassium nitrate, KNO_3

   d. Sodium phosphate, Na_3PO_4

9.109 A chemist has 20.00-mL samples of 0.100 M acid A and 0.100 M acid B in separate flasks. Both acids are monoprotic. Unfortunately, the flasks were not labeled, so the chemist doesn't know which sample is in which flask. But fortunately, it is known that acid A is strong and acid B is weak. Before thinking about the problem, the chemist adds 20.00 mL of 0.100 M NaOH solution to each flask. Explain how the chemist could use a pH meter (or pH paper) to determine which flask originally contained which acid.

9.110 Explain why the hydrolysis of salts makes it necessary to have available in a laboratory more than one acid–base indicator for use in titrations.

9.111 How would the pH values of equal molar solutions of the following salts compare (highest, lowest, etc.)? NaH_2PO_4, Na_2HPO_4, and Na_3PO_4.

Buffers (Section 9.13)

9.112 Write equations similar to Equations 9.48 and 9.49 of the text to illustrate how a mixture of sodium hydrogen phosphate (Na_2HPO_4) and sodium dihydrogen phosphate (NaH_2PO_4) could function as a buffer when dissolved in water. Remember that phosphoric acid (H_3PO_4) ionizes in three steps.

9.113 Could a mixture of ammonia (NH_3), a weak base, and ammonium chloride (NH_4Cl) behave as a buffer when dissolved in water? Use reaction equations to justify your answer.

9.114 Some illnesses lead to a condition of excess acid (acidosis) in the body fluids. An accepted treatment is to inject solutions containing bicarbonate ions (HCO_3^-) directly into the bloodstream. Write an equation to show how this treatment would help combat the acidosis.

9.115 Calculate the pH of a buffer made by dissolving 1 mol formic acid (HCOOH) and 1 mol sodium formate (HCOONa) in 1 L of solution (see Table 9.9).

9.116 a. Calculate the pH of a buffer that is 0.1 M in lactic acid ($C_2H_4(OH)COOH$) and 0.1 M in sodium lactate, $C_2H_4(OH)COONa$.

   b. What is the pH of a buffer that is 1 M in lactic acid and 1 M in sodium lactate?

   c. What is the difference between the buffers described in parts a and b?

9.117 Which of the following weak acids and its conjugate base would you use to make a buffer with a pH of 5.00? Explain your reasons: acetic acid, nitrous acid, or formic acid.

9.118 Calculate the pH of buffers that contain the acid and conjugate base concentrations listed below.

   a. $[CH_3COOH] = 0.40$ M, $[CH_3COO^-] = 0.25$ M

   b. $[H_2PO_4^-] = 0.10$ M, $[HPO_4^{2-}] = 0.40$ M

   c. $[HSO_3^-] = 1.50$ M, $[SO_3^{2-}] = 0.20$ M

9.119 Calculate the pH of buffers that contain the acid and conjugate base concentrations listed below.

   a. $[HPO_4^{2-}] = 0.33$ M, $[PO_4^{3-}] = 0.52$ M

   b. $[HNO_2] = 0.029$ M, $[NO_2^-] = 0.065$ M

   c. $[HCO_3^-] = 0.50$ M, $[CO_3^{2-}] = 0.15$ M

9.120 What ratio of concentrations of NaH_2PO_4 and Na_2HPO_4 in solution would give a buffer with pH $= 7.65$?

9.121 A citric acid–citrate buffer has a pH of 3.20. You want to increase the pH to a value of 3.35. Would you add citric acid or sodium citrate to the solution? Explain.

Additional Exercises

9.122 Consider the following reaction equation:

$$Cl^-(aq) + H_3O^+(aq) \rightarrow HCl(aq) + H_2O(\ell)$$

Draw Lewis structures (electron-dot formulas) for all the substances in the reaction equation, identify each reactant as a Brønsted acid or base, and then propose a definition for an acid and a base based on the ability of the substance to accept or donate a pair of electrons to form a covalent bond.

9.123 Consider the following dissociation reaction of a weak acid, HA:

$$HA(aq) + H_2O(\ell) \rightleftarrows H_3O^+(aq) + A^-(aq)$$

It was determined that 2.63% of the acid in a 0.150 M solution of HA in water was dissociated. Calculate the pH of the 0.150 M solution.

9.124 The value of K_w for water varies with temperature. At 25°C, $K_w = 1.0 \times 10^{-14}$. At 50°C, $K_w = 5.5 \times 10^{-14}$. Calculate the pH of water at 50°C.

9.125 When sodium metal, Na, reacts with water, hydrogen gas, H_2, and sodium hydroxide, NaOH, in solution are produced. Write a balanced equation for the reaction, and explain how a basic solution is produced.

Allied Health Exam Connection

The following questions are from these sources:

1. *Nursing School Entrance Exam* © 2005, Learning Express, LLC.

2. *McGraw-Hill's Nursing School Entrance Exams* by Thomas A. Evangelist, Tamara B. Orr and Judy Unrein © 2009, The McGraw-Hill Companies, Inc.

3. *NSEE Nursing School Entrance Exams,* 3rd Edition © 2009, Kaplan Publishing.

4. *Cliffs Test Prep: Nursing School Entrance Exams* by Fred N. Grayson © 2004, Wiley Publishing, Inc.

5. *Peterson's Master the Nursing School and Allied Health Entrance Exams,* 18th Edition by Marion F. Gooding © 2008, Peterson's, a Nelnet Company.

9.126 An acid is a substance that dissociates in water into one or more _____ ions and one or more _____.

   a. hydrogen . . . anions

   b. hydrogen . . . cations

   c. hydroxide . . . anions

   d. hydroxide . . . cations

9.127 A base is a substance that dissociates in water into one or more _____ ions and one or more _____.

   a. hydrogen . . . anions

   b. hydrogen . . . cations

   c. hydroxide . . . anions

   d. hydroxide . . . cations

9.128 Which is true of alkaline solutions?

   a. More H^+ ion than OH^- ion

   b. Same amount of H^+ ion + OH^- ion

   c. More OH^- ion than H^+ ion

   d. None of the above

9.129 Which of the following is NOT an acid/conjugate base pair?

   a. HCN/CN^-

   b. H_2CO_3/OH^-

   c. H_2SO_4/HSO_4^-

   d. $H_3PO_4/H_2PO_4^-$

9.130 What is the formula of the hydronium ion?

   a. H^+

   b. NH_4^+

   c. H_3O^+

   d. H_2O^+

9.131 Which of the following substances has a pH closest to 7?

   a. ammonia

   b. blood

   c. lemon juice

   d. vinegar

9.132 Dissolving H_2SO_4 in water creates an acid solution by increasing the:

   a. sulfate ions

   b. water ions

   c. hydrogen ions

   d. oxygen ions

9.133 When a solution has a pH of 7, it is:

   a. a strong base

   b. a strong acid

   c. a weak base

   d. neutral

9.134 A common detergent has a pH of 11.0, so the detergent is:

   a. neutral

   b. acidic

   c. alkaline

   d. none of the above

9.135 In a 0.001 M solution of HCl, the pH is:

   a. 2

   b. -3

   c. 1

   d. 3

9.136 The pH of a blood sample is 7.40 at room temperature. The pOH is therefore:

   a. 6.60

   b. 7.40

   c. 6×10^{-6}

   d. 4×10^{-7}

9.137 As the concentration of hydrogen ions in a solution decreases:

   a. the pH numerically decreases

   b. the pH numerically increases

c. the product of the concentrations $[H^+] \times [OH^-]$ comes closer to 1×10^{-14}

d. the solution becomes more acidic

9.138 Blood with a pH of 7.4 indicates that the blood sample is:

a. strongly acidic

b. strongly basic

c. weakly acidic

d. weakly basic

9.139 Atmospheric moisture (H_2O) combines with oxides of carbon, nitrogen, and sulfur (CO_2, NO_3 and SO_2) to produce:

a. alkaline precipitation

b. acid rain

c. alcohol

d. none of the above

9.140 Which of the following substances is most likely to taste sour?

a. NaOH

b. NaCl

c. NH_3

d. $HC_2H_3O_2$

9.141 Which of the following compounds would be classified as a salt?

a. Na_2CO_3

b. $Ca(OH)_2$

c. H_2CO_3

d. CH_3OH

9.142 A substance that functions to prevent rapid, drastic changes in the pH of a body fluid by changing strong acids and bases into weak acids and bases is called a(n):

a. salt

b. buffer

c. enzyme

d. coenzyme

9.143 Which one of the following equations represents neutralization?

a. $2Na + Cl_2 \rightarrow 2NaCl$

b. $CO_2 + H_2O \rightarrow H_2CO_3$

c. $HNO_3 + KOH \rightarrow KNO_3 + H_2O$

d. $CaO + H_2O \rightarrow Ca(OH)_2$

9.144 Which reaction below demonstrates a neutralization reaction?

a. $^{232}_{90}Th \rightarrow {}^{228}_{88}Ra + {}^4_2He$

b. $NaCl + H_2O \rightarrow HCl + NaOH$

c. $2HNO_3 + Mg(OH)_2 \rightarrow 2H_2O + Mg(NO_3)_2$

d. $CH_4 + 2O_2 \rightarrow CO_2 + 2H_2O$

9.145 In a titration of 40.0 mL of 0.20 M NaOH with 0.4 M HCl, what will be the final volume of the solution when the sodium hydroxide is completely neutralized?

a. 42 mL

b. 20 mL

c. 60 mL

d. 80 mL

9.146 When titrating 50 mL of 0.2 M HCl, what quantity of 0.5 M NaOH is needed to bring the solution to the equivalence point?

a. 80 mL

b. 40 mL

c. 20 mL

d. 10 mL

Chemistry for Thought

9.147 In an early industrial method, H_2SO_4 was manufactured in lead-lined chambers. Propose an explanation for this.

9.148 A saturated solution of solid $Ca(OH)_2$ in water has a $[OH^-]$ of only 2.50×10^{-2} M, and yet $Ca(OH)_2$ is a strong base. Explain this apparent contradiction.

9.149 Imagine that a solution of weak acid is being titrated with a strong base. Describe the substances present in the solution being titrated when it has been titrated halfway to the equivalence point. How is the pH of this half-titrated solution related to the pK_a for the weak acid?

9.150 Calculate K_a for the following weak acids based on the equilibrium concentrations and pH values given:

a. Benzoic acid, represented by HBz:

pH = 2.61, $[Bz^-] = 2.48 \times 10^{-3}$, [HBz] = 0.0975

b. Abietic acid, represented by HAb:

pH = 4.16, $[Ab^-] = 6.93 \times 10^{-5}$, [HAb] = 0.200

c. Cacodylic acid, represented by HCc:

pH = 3.55, $[Cc^-] = 0.000284$, [HCc] = 0.150

9.151 Bottles of ketchup are routinely left on the counters of cafés, yet the ketchup does not spoil. Why not?

9.152 Refer to Figure 9.2 and answer the question. What implications does this reaction have for the long-term durability of marble structures when exposed to acid rain?

9.153 Refer to Figure 9.3 and answer the question. Do you think the results would be the same if sulfuric acid (H_2SO_4) were substituted for the HCl? Explain.

9.154 Refer to Figure 9.5. Which of the three indicators would be the best to use to differentiate between the solutions with pH values of 4 and 6?

9.155 Refer to Figure 9.6 and explain why the pH reading of the meter might not be 7.00 at the equivalence point in a titration. Would the pH at the equivalence point be greater or less than 7 if hydrochloric acid were titrated with aqueous ammonia?

9.156 Refer to Figure 9.7 and answer the question. List at least two other ions that would behave like the Na^+ ion.

9.157 Refer to Figure 9.8 and answer the question. What color is universal indicator in acidic solutions? In basic solutions? Explain how you arrived at your conclusions.

Case Study

Katie was a third-grade teacher. She loved working with kids and even had three of her own. Katie was diligent about having her annual physical examination. It was hard to make time for it, but she knew it was important. During her most recent exam, she was surprised that the doctor was concerned about her thyroid gland. Kate had heard that premature gray hair was somehow related to thyroid issues, but early gray hair ran in her family, and her relatives seemed to be fine. The doctor referred her for tests, which revealed thyroid cancer. Katie was told that the tumor could be treated with brachytherapy, or implant-radiation therapy and that maintenance thyroid medication would also be necessary. One of the most difficult aspects of the treatment was that for several days, while the radiation was destroying cancer cells, she would not be allowed to be in contact with children because her body would be radioactive. She knew it would be difficult when she could not see her own children and her students, but she was thankful for the life-saving treatment. On the bright side, Katie knew she would have some great lessons for the classroom!

What types of cancer respond well to internal radiation therapy?

Follow-up to this Case Study appears at the end the chapter before the Concept Summary.

Learning Objectives

When you have completed your study of this chapter, you should be able to:

1 Describe and characterize the common forms of radiation emitted during radioactive decay and other nuclear processes. **(Section 10.1)**

2 Write balanced equations for nuclear reactions. **(Section 10.2)**

3 Solve problems using the half-life concept. **(Section 10.3)**

4 Describe the effects of radiation on health. **(Section 10.4)**

5 Describe and compare the units used to measure quantities of radiation. **(Section 10.5)**

6 Describe, with examples, medical uses of radioisotopes. **(Section 10.6)**

7 Describe, with examples, nonmedical uses of radioisotopes. **(Section 10.7)**

8 Show that you understand the concept of induced nuclear reactions. **(Section 10.8)**

9 Describe the differences between nuclear fission and nuclear fusion reactions. **(Section 10.9)**

In Chapters 3 and 4, we learned that the arrangements of electrons around the nuclei of atoms determine the chemical behavior of elements and compounds. We saw that chemical reactions take place between atoms and molecules as electrons rearrange to achieve noble gas configurations for the atoms involved. These noble gas configurations represent a stable arrangement for the electrons. In this chapter, we will see that the achievement of stability is also a driving force that causes certain atomic nuclei to undergo changes that release small subnuclear particles and energy.

10.1 Radioactive Nuclei

Learning Objective

1. Describe and characterize the common forms of radiation emitted during radioactive decay and other nuclear processes.

In 1896 Henri Becquerel, a French physicist, discovered that uranium compounds emitted rays that could expose and fog photographic plates wrapped in lightproof paper. Research conducted since that time has shown that the penetrating rays originate from changes that occur in the nuclei of some atoms. We saw earlier (Section 2.3) that isotopes of the same element differ only in the number of neutrons present in their nuclei. Thus, each of the three isotopes of hydrogen, represented by the symbols ^1_1H, ^2_1H, and ^3_1H, contains one proton in the nucleus and one electron outside the nucleus. However, besides the one nuclear proton, ^2_1H and ^3_1H contain one and two nuclear neutrons, respectively. It has been found that some combinations of neutrons and protons in the nucleus are stable and do not change spontaneously, but some combinations are not stable. It is these unstable nuclei that emit radiation as they become more stable. For example, both ^1_1H and ^2_1H are stable, but ^3_1H emits radiation. Nuclei that emit radiation are said to be **radioactive nuclei.**

radioactive nuclei Nuclei that undergo spontaneous changes and emit energy in the form of radiation.

Henri Becquerel discovered natural radioactivity by chance. Through research he found that radiation was emitted by any compound of uranium. He further found that the intensity of the radiation was unaffected by factors that normally influence the rates of chemical reactions: the temperature, pressure, and type of uranium compound used.

Later studies by other investigators showed that the radiation emitted by uranium, and by other radioactive elements discovered later, could be separated into three types by an electrical or magnetic field (see ▶ Figure 10.1). The three types had different electrical charges: One was positive, one was negative, and one carried no charge. The types of radiation were given names that are still used today: alpha rays (positive), beta rays (negative), and gamma rays (uncharged).

Figure 10.1 The three types of emitted radiation.

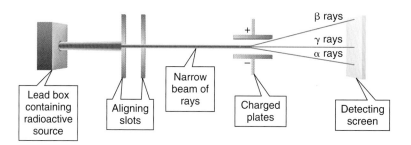

Today it is known that other types of radiation, such as neutrons and positrons, are also emitted by radioactive nuclei. However, alpha, beta, and gamma are the most common. ▶ Table 10.1 summarizes the characteristics of these forms. In this book, we usually use the symbol given first for each form; the symbols in parentheses are alternatives you might find elsewhere. Except for gamma radiation, which is very high-energy radiation somewhat like X-rays, the radiation emitted by radioactive nuclei consists of streams of particles. The emission of radiation by unstable nuclei is often called **radioactive decay.**

radioactive decay A process in which an unstable nucleus changes energy states and in the process emits radiation.

TABLE 10.1 Characteristics of Nuclear Radiation

Type of Radiation	Symbols	Mass Number	Charge	Composition
Alpha	$^4_2\alpha$ (α, ^4_2He, He^{2+})	4	+2	Helium nuclei, 2 protons + 2 neutrons
Beta	$^0_{-1}\beta$ (β, β^-, $^0_{-1}e$, e^-)	0	−1	Electrons produced in nucleus and ejected
Gamma	γ ($^0_0\gamma$)	0	0	Electromagnetic radiation
Neutron	^1_0n (n)	1	0	Neutrons
Positron	$^0_1\beta$ (β^+, ^0_1e, e^+)	0	+1	Positive electrons

We note from Table 10.1 that the particles that make up alpha rays are identical to the nuclei of helium atoms; that is, they are clusters containing two protons and two neutrons. Because they are positively charged, alpha particles are attracted toward the negatively charged plate in Figure 10.1. **Alpha particles** move at a speed of nearly one-tenth the speed of light. They are the most massive particles emitted by radioactive materials, and they have the highest charge. Because of their mass and charge, they collide often with molecules of any matter through which they travel. As a result of these collisions, their energy is quickly dissipated, and they do not travel far. They cannot even penetrate through a few sheets of writing paper or the outer cells of the skin. However, exposure to intense alpha radiation can result in severe burns of the skin.

Beta particles are actually electrons, but they are produced within the nucleus and then emitted. They are not part of the group of electrons moving around the nucleus that is responsible for chemical characteristics of the atoms. Beta particles have a smaller charge than alpha particles and are also 7000 times less massive than alpha particles. They undergo far fewer collisions with molecules of matter through which they pass and therefore are much more penetrating than alpha particles. Because of their negative charge and tiny mass, beta particles are attracted toward the positive plate of Figure 10.1, and are deflected more than alpha particles.

Gamma rays are not streams of particles but rays of electromagnetic radiation similar to X-rays that have higher energies. Like X-rays, gamma radiation is very penetrating and dangerous to living organisms. Adequate shielding from gamma rays is provided only by thick layers of heavy materials like metallic lead or concrete. Because they are not charged, gamma rays are not attracted to either charged plate in Figure 10.1.

alpha particle The particle that makes up alpha rays. It is identical to the helium nucleus and is composed of two protons and two neutrons.

beta particle The particle that makes up beta rays. It is identical to an electron but is produced in the nucleus when a neutron is changed into a proton and an electron.

gamma ray A high-energy ray that is like an X-ray, but with a higher energy.

10.2 Equations for Nuclear Reactions

Learning Objective

2. Write balanced equations for nuclear reactions.

Isotopes of elements that emit nuclear radiation are called **radioisotopes.** In nuclear reactions, a specific isotope of an element may behave differently from another isotope of the same element. Thus, all particles involved in nuclear reactions are designated by a symbol, a mass number (the sum of protons and neutrons in the particle), and an atomic number (or charge for electrons or positrons). The symbolism used was introduced in Section 2.3. It is $_Z^A X$, where X is the symbol for the particle, A is the mass number, and Z is the atomic number or charge. When X represents the nucleus of an isotope of an element, the chemical symbol for the element is generally used.

radioisotope An isotope of an element that emits nuclear radiation.

▶ Example 10.1

Write appropriate symbols for the following particles using the $_Z^A X$ symbolism:

a. A lead-214 nucleus
b. An alpha particle
c. A deuteron, a particle containing one proton and one neutron

Solution

a. Lead-214 is an isotope of lead, so the symbol X is replaced by Pb. The mass number is 214 according to rules learned earlier for designating isotopes, so $A = 214$. The atomic number of lead obtained from the periodic table is 82, so $Z = 82$, and the symbol is $_{82}^{214}\text{Pb}$.

b. An alpha particle is represented by the symbol α. Since it contains 2 protons and 2 neutrons, $A = 4$ and $Z = 2$. The symbol is $^4_2\alpha$, as shown in Table 10.1.

c. A deuteron is not familiar to you, so we will replace its symbol X with D. It contains one proton and one neutron, so $A = 2$ and $Z = 1$. The symbol is ^2_1D. This particle is the nucleus of a deuterium or heavy hydrogen atom and so could also be represented as ^2_1H.

▶ **LEARNING CHECK 10.1** Write appropriate symbols for the following particles using the $^A_Z X$ symbolism:

a. An iodine-131 nucleus
b. A beta particle
c. A hypothetical particle composed of four protons
d. A hypothetical particle composed of four electrons

A nuclear equation is balanced when the sum of the atomic numbers on the left side equals the sum of the atomic numbers on the right side, and the sum of the mass numbers on the left equals the sum of the mass numbers on the right. Thus, we focus on mass numbers and atomic numbers in nuclear reactions. As shown in the following example, a nucleus of an element that undergoes a nuclear reaction by emitting a particle is changed to a nucleus of a different element.

▶ Example 10.2

Uranium-238 nuclei undergo radioactive decay by emitting an alpha particle and changing to the nucleus of another element. Write a balanced nuclear equation for the process.

Solution

Because the uranium-238 nucleus gives up an alpha particle, $^4_2\alpha$, the mass number of the uranium-238 must decrease by 4 and the atomic number must decrease by 2. As a result, the new nucleus, called a **daughter nucleus,** has a mass number of 234 and an atomic number of 90 (remember, the atomic number of uranium can be obtained from a periodic table). The balanced nuclear equation is

$$^{238}_{92}\text{U} \rightarrow\, ^4_2\alpha +\, ^{234}_{90}\text{Th}$$

Note that this equation is balanced because

$$238 = 4 + 234 \qquad \text{and} \qquad 92 = 2 + 90$$

The symbol for the daughter (thorium) was obtained by locating element number 90 in the periodic table.

Processes such as this where one alpha particle is ejected can be represented by the general equation

$$^A_Z X \rightarrow\, ^4_2\alpha +\, ^{A-4}_{Z-2} Y$$

where Y is the symbol of the daughter nucleus.

▶ **LEARNING CHECK 10.2** Thorium, element number 90, exists in a number of isotopic forms. One form, thorium-234, decays by emitting a beta particle. Write a balanced nuclear equation for the process and write a general equation for decays in which one beta particle is ejected.

Because gamma rays have no mass or atomic numbers, they do not enter into the balancing process of nuclear reactions. However, they do represent energy and should be included in balanced equations when they are known to be emitted.

daughter nuclei The new nuclei produced when unstable nuclei undergo radioactive decay.

As described earlier, beta particles are electrons ejected from the nucleus of a radioactive atom, and yet, we know from our discussions of atomic theory in Chapter 2 that atomic nuclei contain only protons and neutrons. Where then do the beta particles originate? According to a simplified theory of nuclear behavior, a beta particle (or electron) is produced in the nucleus when a neutron changes into a proton. The balanced equation for the nuclear reaction is

$$_0^1 n \rightarrow \ _1^1 p + \ _{-1}^0 \beta \qquad (10.1)$$

Note that the criteria for balance in this equation are satisfied; that is, $1 = 1 + 0$ and $0 = 1 + (-1)$. Thus, a nuclear neutron is converted to a proton during beta emission, and the daughter has an atomic number higher by 1 than the decaying nucleus. You found this result if you did Learning Check 10.2 correctly. Similar changes in nuclear particles occur during two other types of nuclear reactions. Some nuclei decay by positron emission. A **positron** is a positively charged electron with the symbol (see Table 10.1). When a nucleus emits a positron, a nuclear proton is changed to a neutron. Positron-emitting isotopes are used in diagnostic PET scans of the brain (see ◗ Figure 10.2).

positron A positively charged electron.

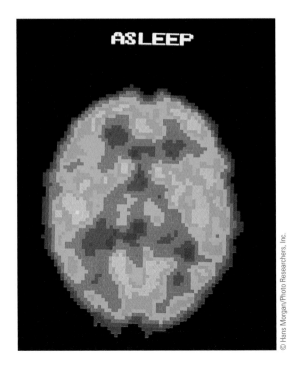

© Hans Morgan/Photo Researchers, Inc.

Figure 10.2 Positron emission tomography (PET) scan shows normal brain activity during sleep. A radioactive isotope that emits positrons is made into a chemical compound that is absorbed by active areas of the brain. The emitted positrons collide with nearby electrons and produce gamma rays that pass through the skull to detectors surrounding the patient's head. A computer uses the detector data to construct the image.

◗ **LEARNING CHECK 10.3** Write a general balanced equation for the nuclear process that occurs during positron emission. How do the atomic numbers of daughter nuclei compare with those of the decaying nucleus?

Certain nuclear changes occur when an electron from outside the nucleus is drawn into the nucleus, where it reacts with a proton and converts it to a neutron. This process, called **electron capture,** is not as common as other decay processes.

◗ **LEARNING CHECK 10.4** Write a general balanced equation for the nuclear process that occurs during electron capture. How do the atomic numbers of daughter nuclei compare with those of the decaying nucleus?

electron capture A mode of decay for some unstable nuclei in which an electron from outside the nucleus is drawn into the nucleus, where it combines with a proton to form a neutron.

▶ Example 10.3

Write a balanced nuclear equation for the decay of each of the following. The mode of decay is indicated in parentheses.

a. $^{14}_{6}C$ (beta emission)

b. $^{122}_{53}I$ (positron emission)

c. $^{55}_{26}Fe$ (electron capture)

d. $^{212}_{84}Po$ (alpha emission)

Solution

In each case, the atomic number of the daughter is obtained by balancing atomic numbers in the equation. The symbol for the daughter is obtained by matching atomic numbers to symbols in the periodic table.

a. $^{14}_{6}C \rightarrow\ ^{0}_{-1}\beta\ +\ ^{14}_{7}N$

b. $^{122}_{53}I \rightarrow\ ^{0}_{1}\beta\ +\ ^{122}_{52}Te$

c. $^{55}_{26}Fe\ +\ ^{0}_{-1}e \rightarrow\ ^{55}_{25}Mn$

d. $^{212}_{84}Po \rightarrow\ ^{4}_{2}\alpha\ +\ ^{208}_{82}Pb$

▶ **LEARNING CHECK 10.5** Write a balanced nuclear equation for the decay of each of the following. The mode of decay is indicated in parentheses.

a. $^{50}_{25}Mn$ (positron emission)

b. $^{54}_{25}Mn$ (electron capture)

c. $^{56}_{25}Mn$ (beta emission)

d. $^{224}_{88}Ra$ (alpha emission)

10.3 Isotope Half-Life

Learning Objective

3. Solve problems using the half-life concept.

half-life The time required for one-half the unstable nuclei in a sample to undergo radioactive decay.

Some radioactive isotopes are more stable than others. The more stable isotopes undergo radioactive decay more slowly than the less stable isotopes. The **half-life** of an isotope is used to indicate stability, and it is equal to the time required for one-half (50%) of the atoms of a sample of the isotope to decay. ▶ Table 10.2 contains examples showing the wide range of half-lives that have been determined.

TABLE 10.2 Examples of Half-Lives

Isotope	Half-Life	Source
$^{238}_{92}U$	4.5×10^9 years	Naturally occurring
$^{40}_{19}K$	1.3×10^9 years	Naturally occurring
$^{226}_{88}Ra$	1600 years	Naturally occurring
$^{14}_{6}C$	5600 years	Naturally occurring
$^{239}_{94}Pu$	24,000 years	Synthetically produced
$^{90}_{38}Sr$	28 years	Synthetically produced
$^{131}_{53}I$	8 days	Synthetically produced
$^{24}_{11}Na$	15 hours	Synthetically produced
$^{15}_{8}O$	2 minutes	Synthetically produced
$^{5}_{3}Li$	10^{-21} seconds	Synthetically produced

▶ Example 10.4

Rubidium-79 decays by positron emission and forms krypton-79, which is a gas. A weighed 100.00-mg sample of solid rubidium-79 was allowed to decay for 42 minutes, then weighed again. Its mass was 25.00 mg. What is the half-life of rubidium-79?

Solution

We assume all the gaseous krypton-79 that was formed escaped into the surrounding air and so was not weighed. In the first half-life, one-half the original 100.00-mg sample of rubidium would have been lost, so the sample would have a mass of 50.00 mg. During the second half-life, this 50.00-mg sample would again lose half its mass and would be reduced to 25.00 mg. Thus, we can conclude that two half-lives passed during the 42 minutes that elapsed between the weighings. Because two half-lives equal 42 minutes, one half-life equals 21 minutes.

▶ **LEARNING CHECK 10.6** Potassium-38 decays by positron emission to argon-38, a gas. Potassium-38 has a half-life of 7.7 minutes. How long will it take a 200.00-mg sample of potassium-38 to be reduced to 25.00 mg?

Because radioisotopes are continuously decaying, any found in nature must belong to one of three categories. They may have very long half-lives and decay very slowly ($^{238}_{92}U$), they may be daughters produced by the decay of long-lived isotopes ($^{226}_{88}Ra$), or they may result from natural processes such as cosmic-ray bombardment of stable nuclei ($^{14}_{6}C$).

The fraction of original radioactive atoms remaining in a sample after a specific time has passed can be determined from the half-life. After one half-life has passed, $\frac{1}{2}$ the original number have decayed, so $\frac{1}{2}$ remain. During the next half-life, $\frac{1}{2}$ the remaining $\frac{1}{2}$ decay, so $\frac{1}{4}$ of the original atoms remain. After three half-lives, $\frac{1}{2} \times \frac{1}{2} \times \frac{1}{2} = \frac{1}{8}$ of the original atoms remain undecayed (see ▶ Figure 10.3).

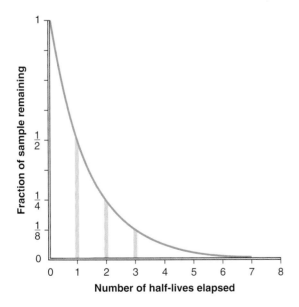

Figure 10.3 A radioactive decay curve (bars give values at 1, 2, and 3 half-lives).

The amount of radiation given off by a sample of radioactive material is proportional to the number of radioactive atoms present in the sample. The amount of radiation given off is called the *activity* of the sample and is measured by devices such as the Geiger–Müller counter described in Section 10.5. Data such as the sample fractions shown in Figure 10.3 are obtained quite easily. For example, suppose a sample gave an initial reading of 120 units of activity, and it took 35 minutes for the reading to drop to 60 units. The reading of 60 units means that the sample contained just half as many radioactive atoms as it contained when the reading was 120 units. Thus, the 35 minutes represents one half-life, and the fraction remaining from the time of the initial measurement is $\frac{1}{2}$.

▶ Example 10.5

The activity (amount of radiation emitted) of a radioactive substance is measured at 9 A.M. At 5 P.M. the activity is found to be only $\frac{1}{16}$ the original (9 A.M.) value. What is the half-life of the radioactive material?

Solution

Because only $\frac{1}{16}$ the original radioactive material is present, four half-lives have passed: $\frac{1}{2} \times \frac{1}{2} \times \frac{1}{2} \times \frac{1}{2} = \frac{1}{16}$. The time between measurements is 8 hours (9 A.M. to 5 P.M.). Therefore, 8 hours is four half-lives, and the half-life is 2 hours.

▶ **LEARNING CHECK 10.7** Iodine-123 is a radioisotope used to diagnose the function of the thyroid gland. It has a half-life of 13.3 hours. What fraction of a diagnostic dose of iodine-123 would be present in a patient 79.8 hours (a little over 3 days) after it was administered?

10.4 The Health Effects of Radiation

Learning Objective

4. Describe the effects of radiation on health.

radical or free radical An electron-deficient particle that is very reactive.

You are probably aware of the fact that radiation is hazardous to living organisms. But why? What does radiation do that is so dangerous? The greatest danger to living organisms results from the ability of high-energy or ionizing radiation to knock electrons out of compounds and generate highly reactive particles called **radicals** or **free radicals** in tissue through which the radiation passes. These electron-deficient radicals are very reactive. They cause reactions to occur among more stable materials in the cells of living organisms. When these reactions involve genetic materials such as genes and chromosomes, the changes might lead to genetic mutations, cancer, or other serious consequences.

acute radiation syndrome The condition associated with and following short-term exposure to intense radiation.

Long-term exposure to low-level radiation is more likely to cause the problems just described than are short exposures to intense radiation. Short-term intense radiation tends to destroy tissue rapidly in the area exposed and can cause symptoms of so-called **acute radiation syndrome.** Some of these symptoms are given in ▶ Table 10.3. It is this rapid destruction of tissue that makes relatively intense radiation a useful tool in the treatment of some cancers.

The health hazards presented by radiation make it imperative that exposures be minimized. This is especially true for individuals with occupations that continually present opportunities for exposure. One approach is to absorb radiation by shielding. As shown in ▶ Table 10.4, alpha and beta rays are the easiest to stop. Gamma radiation and X-rays require the use of very dense materials such as lead to provide protection. If the type of radiation is known, a careful choice of shielding materials can provide effective protection.

TABLE 10.3 The Effects on Humans of Short-Term, Whole-Body Radiation Exposure

Dose (rems)[a]	Effects
0–25	No detectable clinical effects.
25–100	Slight short-term reduction in number of some blood cells, disabling sickness not common.
100–200	Nausea and fatigue, vomiting if dose is greater than 125 rems, longer-term reduction in number of some blood cells.
200–300	Nausea and vomiting first day of exposure, up to a 2-week latent period followed by appetite loss, general malaise, sore throat, pallor, diarrhea, and moderate emaciation. Recovery in about 3 months unless complicated by infection or injury.
300–600	Nausea, vomiting, and diarrhea in first few hours. Up to a 1-week latent period followed by loss of appetite, fever, and general malaise in the second week, followed by hemorrhage, inflammation of mouth and throat, diarrhea, and emaciation. Some deaths in 2–6 weeks. Eventual death for 50% if exposure is above 450 rems; others recover in about 6 months.
600 or more	Nausea, vomiting, and diarrhea in first few hours. Rapid emaciation and death as early as second week. Eventual death of nearly 100%.

[a]The rem, a biological unit of radiation, is defined in Section 10.5.

Source: U.S. Atomic Energy Commission.

TABLE 10.4 Penetrating Abilities of Alpha, Beta, and Gamma Rays

Type of Radiation	Depth of Penetration Into		
	Dry Air	Tissue	Lead
Alpha	4 cm	0.05 mm	0
Beta[a]	6–300 cm	0.06–4 mm	0.005–0.3 mm
Gamma[b]	400 m	50 cm	300 mm

[a]Depth depends on energy of rays.

[b]Depth listed is that necessary to reduce initial intensity by 10%.

Another protection against radiation is distance. Because radiation spreads in all directions from a source, the amount falling on a given area decreases the farther that area is from the source. Imagine a light bulb at the center of a balloon that is being filled slowly with air. The intensity of light on each square centimeter of the balloon will decrease as the surface of the balloon moves away from the light. It has been shown that the intensity is inversely proportional to the square of the distance of the surface from the radiation source. This **inverse square law of radiation** is true only for radiation traveling in a vacuum, but it gives pretty good results when the radiation travels through air. The law is represented by Equation 10.2, where I_x is the intensity at distance d_x and I_y is the intensity at distance d_y:

inverse square law of radiation
A mathematical way of saying that the intensity of radiation is inversely proportional to the square of the distance from the source of the radiation.

$$\frac{I_x}{I_y} = \frac{d_y^2}{d_x^2} \qquad (10.2)$$

The intensity of radiation is 18.5 units at a distance of 50 cm from a source. What is the intensity at a distance of 100 cm?

Solution

Let I_x = 18.5 units and d_x = 50 cm. Then I_y is wanted at d_y = 100 cm. Substitution into Equation 10.2 gives

$$\frac{18.5}{I_y} = \frac{(100 \text{ cm})^2}{(50 \text{ cm})^2}$$

or

$$I_y = \frac{(50 \text{ cm})^2 (18.5)}{(100 \text{ cm})^2} = 4.63 \text{ units}$$

Thus, doubling the distance cut the intensity to one-fourth the initial intensity.

▶ **LEARNING CHECK 10.8** The intensity of a radioactive source is measured at a distance of 25 feet and equals 10.0 units. What is the intensity 5 feet from the source?

10.5 Measurement Units for Radiation

Learning Objective

5. Describe and compare the units used to measure quantities of radiation.

physical unit of radiation A radiation measurement unit indicating the activity of the source of the radiation; for example, the number of nuclear decays per minute.

biological unit of radiation A radiation measurement unit indicating the damage caused by radiation in living tissue.

curie A physical unit of radiation measurement corresponding to 3.7×10^{10} nuclear disintegrations per second.

becquerel A physical unit of radiation measurement corresponding to one nuclear disintegration per second.

roentgen A biological unit of radiation measurement used with X-rays and gamma rays; the quantity of radiation that generates 2.1×10^9 ion pairs per 1 cm^3 of dry air or 1.8×10^{12} ion pairs per 1 g of tissue.

rad A biological unit of radiation measurement corresponding to the transfer of 1×10^{-2} J or 2.4×10^{-3} cal of energy to 1 kg of tissue.

gray A biological unit of radiation measurement corresponding to the transfer of 1 J of energy to 1 kg of tissue.

rem A biological unit of radiation measurement corresponding to the health effect produced by 1 roentgen of gamma or X-rays regardless of the type of radiation involved.

Two methods, physical and biological, are used to describe quantities of radiation. **Physical units** indicate the activity of a source of radiation, typically in terms of the number of nuclei that decay per unit of time. **Biological units** are related to the damage caused by radiation and account for the fact that a given quantity (or number of particles) of one type of radiation does not have the same damaging effect on tissue as the same quantity of another type of radiation.

The curie and the becquerel are physical units. The **curie** is most widely used at the present time and is equal to 3.7×10^{10} nuclear disintegrations per second. One curie (Ci) represents a large unit, so fractions such as the millicurie (mCi), the microcurie (μCi), and the picocurie (pCi) are often used. The **becquerel** (Bq) is a relatively new unit that has achieved little use among U.S. scientists, even though it is smaller and more manageable than the curie. One becquerel is equal to one disintegration per second.

A number of biological units are used, including the roentgen (R), rad (D), gray (Gy), and rem. The **roentgen** is a biological unit used with X-rays or gamma rays. One roentgen is the quantity of radiation that will generate 2.1×10^9 ion pairs in 1 cm^3 of dry air at normal temperature and pressure. The ionization effect of X-rays and gamma rays is greater when the radiation passes through tissue; one roentgen generates about 1.8×10^{12} ion pairs per gram of tissue.

The rad and the gray are biological units used to describe the effects of radiation in terms of the amount of energy transferred from the radiation to the tissue through which it passes. One **rad** transfers 1×10^{-2} J or 2.4×10^{-3} cal of energy to 1 kg of tissue. One **gray,** a larger unit, transfers 1 J per 1 kg of tissue. From a health point of view, the roentgen and rad are essentially equal. One roentgen of X-rays or gamma rays equals 0.96 rad.

The rem is a biological unit devised to account for health differences in various types of radiation. A 1-rad dose of gamma radiation does not produce the same health effects as a 1-rad dose of alpha radiation. An additive unit was needed so that one unit of alpha had the same health effects as one unit of gamma. The devised unit is the rem, which stands for roentgen equivalent in man. One **rem** of any type of radiation has the same health effect as 1 roentgen of gamma rays or X-rays. The

Radiation Exposure in Modern Life

Throughout our existence, we humans have been continually exposed to radiation from natural sources such as alpha, beta, or gamma radiation from naturally occurring radioisotopes, sunlight, and cosmic rays from outer space. The discovery of X-rays in 1895 by Wilhelm Roentgen contributed to this exposure when X-rays were introduced as a diagnostic tool for patients. Today, X-rays are used in numerous medical and dental diagnostic tests and procedures.

In the early 1970s, X-rays were integrated with computers to create a diagnostic tool called *computed tomography*, or CT scans. In this procedure, the intensity of X-rays after they pass through the body is interpreted using a computer, and the computer results are used to create a cross-sectional image of the part of the body through which the x-rays passed.

In 1946, the property called *nuclear magnetic resonance* was discovered. Today, this property is widely used medically in the diagnostic procedure called *magnetic resonance imaging*, or MRI. During an MRI procedure, the patient is placed in a very strong magnetic field and radiation in the form of powerful radio waves is passed through the patient's body. The energy of the radio waves is absorbed by hydrogen atoms in the body and then re-emitted. The re-emitted radio waves are interpreted by a powerful computer program that then produces cross-sectional images of the body.

Airline travelers are increasingly being required to pass through full-body scanners before being allowed to board airliners. In this process, the passengers are exposed to low-energy X-rays. Some individuals have received publicity for refusing to go through the scanners in spite of assurances that the radiation dose from such a scan presents, at most, a very tiny risk. An alternative physical search is usually available for such passengers. According to a recent study, an airline passenger receives a radiation dose from flying for one hour at an altitude of 35,000 feet that is at least 100 times more than the dose received by being scanned before the flight. In another comparison, a report from the British Institute of Radiology found that the radiation dose from a single airport scan was 100,000 times lower than the average total annual dose of radiation a person receives from natural background radiation and medical sources.

However, some radiation experts have taken a very cautious approach to the question of scanner safety. One expert urged researchers to carry out investigations that focused on the way airport scans affected specific groups such as children or individuals with gene mutations. The same expert has proposed that the neck and head areas should not be scanned to avoid increasing the risk of a certain type of skin cancer.

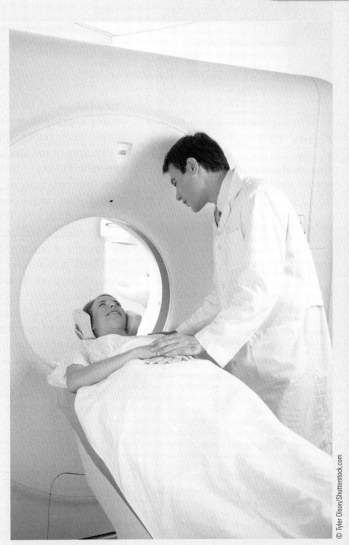

Magnetic resonance imaging (MRI) scans enable medical personnel to view soft tissues in detail.

advantage of such a unit is that individuals exposed to low levels of various types of radiation have to keep track only of the rems absorbed from each type and add them up to get the total health effect. If other units were used, the number of units from each type of radiation would have to be determined and the health effects calculated for each type. The units used for measuring radiation are summarized in ▶ Table 10.5, together with relationships among the units.

TABLE 10.5 Units for Measuring Radiation

Unit	Type of Unit	Relationships Between Units
Curie (Ci)	Physical	$1\ Ci = 3.7 \times 10^{10}\ Bq$
Becquerel (Bq)	Physical	$1\ Bq = 2.7 \times 10^{-11}\ Ci$
Roentgen (R)	Biological	$1\ R = 0.96\ D$ (tissue)
Rad (D)	Biological	$1\ D = 1 \times 10^{-2}\ Gy$
Gray (Gy)	Biological	$1\ Gy = 1 \times 10^{2}\ D$
Rem	Biological	$1\ rem = 1\ R$ X-ray or gamma ray in health effect

Figure 10.4 Film badges provide a convenient way to monitor the total amount of radiation received during a specific time period.

scintillation counter A radiation-detection device operating on the principle that phosphors give off light when struck by radiation.

Geiger–Müller tube A radiation-detection device operating on the principle that ions form when radiation passes through a tube filled with low-pressure gas.

Radiation can be detected in a number of ways. Film badges are commonly used by people working in environments where they might be exposed to radiation. The badges contain photographic film that becomes exposed when subjected to radiation. The extent of exposure of the film increases with the amount of radiation. The film is developed after a specific amount of time, and the degree of exposure indicates the radiation dose that has been absorbed during that time (see ▶ Figure 10.4).

Radiation causes certain substances, called phosphors, to give off visible light. The light appears as brief flashes when radiation strikes a surface coated with a thin layer of phosphors. The number of flashes (called *scintillations*) per unit time is proportional to the amount of radiation striking the surface. Devices employing this approach are called **scintillation counters.**

A third type of detector, called a **Geiger–Müller tube,** is based on the fact that electricity is conducted through a gas that contains charged particles. A Geiger–Müller tube is represented in ▶ Figure 10.5. The tube essentially consists of two electrodes charged to high voltages of opposite sign. The tube is filled with a gas at very low pressure. When radiation passes through the tube, it creates a path of charged particles (ions) by knocking electrons off gas molecules. They conduct a brief pulse of current between the electrodes. This brief current is amplified and triggers a counting circuit to record one "count." The number of counts per unit time is proportional to the amount of radiation passing through the Geiger–Müller tube.

Figure 10.5 A cross section of a detection tube used in Geiger–Müller counters.

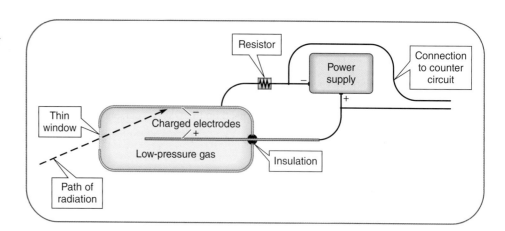

10.6 Medical Uses of Radioisotopes

Learning Objective

6. Describe, with examples, medical uses of radioisotopes.

Chemically, radioisotopes undergo the same reactions as nonradioactive isotopes of the same element. For example, radioactive iodine-123 is absorbed by the thyroid gland and used to produce the hormone thyroxine just as if it were ordinary nonradioactive iodine. This characteristic, together with the fact that the location of radioisotopes in the body can be readily detected, makes them useful for diagnostic and therapeutic medical applications. In diagnostic applications, radioisotopes are used as **tracers** whose progress through the body or localization in specific organs can be followed.

To minimize the risks to patients from exposure to radiation, radioisotopes used as diagnostic tracers should have as many of the following characteristics as possible:

1. Tracers should have short half-lives so they will decay while the diagnosis is being done but will give off as little radiation as possible after the diagnostic procedure is completed.
2. The daughter produced by the decaying radioisotope should be nontoxic and give off little or no radiation of its own. Ideally, it will be stable.
3. The radioisotope should have a long enough half-life to allow it to be prepared and administered conveniently.
4. The radiation given off by the radioisotope should be penetrating gamma rays, if possible, to ensure that they can be detected readily by detectors located outside the body.
5. The radioisotope should have chemical properties that make it possible for the tissue being studied to either concentrate it in diseased areas and form a **hot spot** or essentially reject it from diseased areas to form a **cold spot.**

Iodine-123 is used diagnostically to determine whether a thyroid gland is properly functioning. As mentioned earlier, the thyroid gland concentrates iodine and uses it in the production of thyroxine, an iodine-containing hormone. Iodine is an excellent tracer to use because the thyroid is the only user of iodine in the body. Iodine-123 is a good tracer because it is a gamma-emitter with a half-life of 13.3 hours. Iodine-131 is also radioactive and has a short half-life (8 days), but it emits gamma and beta radiation. The beta radiation cannot penetrate through the tissue to be detected diagnostically and therefore is simply an added radiation risk to the patient. For this reason, iodine-123 is preferred. An overactive thyroid absorbs more iodine and forms a hot spot, whereas an underactive or nonactive gland absorbs less iodine and may show a cold spot on a thyroid scan (see ◗ Figure 10.6).

Radioisotopes administered internally for therapeutic use should ideally have the following characteristics:

1. The radioisotope should emit less penetrating alpha or beta radiation to restrict the extent of damage to the desired tissue.
2. The half-life should be long enough to allow sufficient time for the desired therapy to be accomplished.
3. The decay products should be nontoxic and give off little or no radiation.
4. The target tissue should concentrate the radioisotope to restrict the radiation damage to the target tissue.

Iodine-131 is used therapeutically to treat thyroid cancer and hyperthyroidism. The primary advantage of this radioisotope is the ability of thyroid tissue to absorb it and localize its effects.

Until the mid-1950s, when improvements in the generation of penetrating X-rays made them the therapy of choice, cobalt-60 was widely used to treat cancer. This radioisotope emits beta and gamma radiation and has a half-life of 5.3 years. When used therapeutically

tracer A radioisotope used medically because its progress through the body or localization in specific organs can be followed.

hot spot Tissue in which a radioactive tracer concentrates.

cold spot Tissue from which a radioactive tracer is excluded or rejected.

Figure 10.6 A thyroid scan produced after the administration of a radioactive iodine isotope. Is either a hot spot or a cold spot present?

today, a sample of cobalt-60 is placed in a heavy lead container with a window aimed at the cancerous site. The beam of gamma radiation that exits through the window is focused on a small area of the body where the tumor is located. ❱ Table 10.6 contains other examples of radioisotopes used diagnostically and therapeutically.

TABLE 10.6 Examples of Medically Useful Radioisotopes

Isotope	Emission	Half-Life	Applications
$^{3}_{1}H$	Beta	12.3 years	To measure water content of the body
$^{32}_{15}P$	Beta	14.3 days	Detection of tumors, treatment of a form of leukemia
$^{51}_{24}Cr$	Gamma	27.8 days	Diagnosis: size and shape of spleen, gastrointestinal disorders
$^{59}_{26}Fe$	Beta	45.1 days	Diagnosis: anemia, bone marrow function
$^{60}_{27}Co$	Beta, gamma	5.3 years	Therapy: cancer treatment
$^{67}_{31}Ga$	Gamma	78.1 hours	Diagnosis: various tumors
$^{75}_{34}Se$	Beta	120.4 days	Diagnosis: pancreatic scan
$^{81}_{36}Kr$	Gamma	2.1×10^5 years	Diagnosis: lung ventilation scan
$^{85}_{38}Sr$	Gamma	64 days	Diagnosis: bone scan
$^{99}_{43}Tc$	Gamma	6 hours	Diagnosis: brain, liver, kidney, bone, and heart muscle scans
$^{123}_{53}I$	Gamma	13.3 hours	Diagnosis: thyroid cancer
$^{131}_{53}I$	Beta, gamma	8.1 days	Diagnosis and treatment: thyroid cancer
$^{197}_{80}Hg$	Gamma	65 hours	Diagnosis: kidney scan

10.7 Nonmedical Uses of Radioisotopes

Learning Objective

7. Describe, with examples, nonmedical uses of radioisotopes.

The study of the photosynthetic process in plants has been made much easier because of the use of radioactive carbon-14 as a tracer. The overall photosynthetic process is represented by Equation 10.3:

$$6CO_2(g) + 6H_2O(\ell) \rightarrow C_6H_{12}O_6(aq) + 6O_2(g) \qquad (10.3)$$

However, the conversion of water (H_2O) and carbon dioxide (CO_2) into carbohydrates ($C_6H_{12}O_6$) involves numerous chemical reactions and carbon-containing intermediate compounds. By using CO_2 that contains radioactive carbon-14, scientists have been able to follow the chemical path of CO_2 through the various intermediate compounds.

Interesting uses for radioactive tracers are found in industry. In the petroleum industry, for example, radioisotopes are used in pipelines to indicate the boundary between different products moving through the lines (see ❱ Figure 10.7). In another application, tracer added to fluids moving in pipes makes it easy to detect leaks. Also, the effectiveness of lubricants has been studied by the use of radioactive metal isotopes as components of metal parts; as the metal part wears, the isotope shows up in the lubricant, and the amount of wear can be determined.

Radon: A Chemically Inert Health Risk

Substances that are chemically unreactive are generally not thought to be dangerous to one's health. However, radon, a colorless, odorless, chemically inert noble gas, is considered to be a serious health risk because it is radioactive. It is believed to be the leading cause of lung cancer among nonsmokers. It has been estimated by U.S. officials that up to 8% of the annual lung cancer deaths in the United States can be attributed to indoor exposure to radon-222.

Radon-222 is formed during a multistep process in which naturally occurring uranium-238 decays to stable lead-206. During one step of the process, radium-226, with a half-life of 1600 years, emits alpha and gamma radiation to produce radon-222:

$$^{226}_{88}\text{Ra} \rightarrow {}^{4}_{2}\alpha + {}^{222}_{86}\text{Rn} + \gamma$$

Radon-222, with a half-life of 4 days, is produced in rocks and soil that contain even minute amounts of uranium-238. Because it is a gas, radon migrates readily from the soil into the surrounding air. It seeps into houses and other buildings through openings around pipes, and through cracks in basement floors and walls.

The average level of radon in homes is lower than 1 picocurie/L of air. A picocurie (10^{-12} curie) is equivalent to the radioactive decay of 2 radon nuclei per minute. The Environmental Protection Agency recommends an upper limit of 4 picocuries/L for indoor air, and it has been estimated that 99% of the homes in the United States have levels below this value. However, new studies reported by the World Health Organization have led to recommendations that new upper limits of 2.7 picocuries/L be established. This level is 33% lower than the current upper limit EPA recommendations.

Levels of radiation much higher than 4 picocuries/L have been found in some homes. The highest recorded level of radioactivity from radon in a home was 2700 picocuries/L of air. This was found in a home in Pennsylvania that was built on soil rich in uranium-238. The level of radon in home air can be minimized by sealing the entry points and providing good ventilation.

The hazards associated with breathing radon-222 come not only from the alpha radiation emitted as the radon decays but also from the daughter products of that decay. None of these decay products are gases, but all are radioactive. Because they are not gases, they are not easily exhaled and can become lodged in the lungs, where their radiation can cause serious damage.

It is difficult to decide just how much concern people should have about radon in their homes. Some experts in the field agree with the EPA recommendations, but others question the data on which those recommendations are based. It is quite easy and relatively inexpensive to test for radon in a home, and homes built in locations where the gas has been identified as a problem should be tested.

Radon detectors are available for home use.

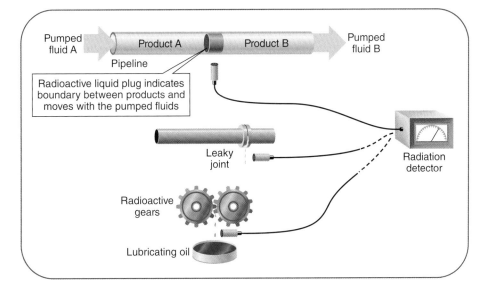

Figure 10.7 Radioactive tracers used in industry.

Pumped fluid A

Product A

Pipeline

Product B

Pumped fluid B

Radioactive liquid plug indicates boundary between products and moves with the pumped fluids

Leaky joint

Radioactive gears

Lubricating oil

Radiation detector

Radioisotopes are also useful in forms other than tracers. The thickness of manufactured materials, such as metal foils or sheets, can be determined continuously by the use of a radioisotope. The amount of radiation that penetrates the material to the detector is related to the thickness of the material.

An interesting application of radioisotopes involves the use of naturally occurring radioactive materials to determine the ages of artifacts and rocks. Perhaps the most widely known example of **radioactive dating** is the use of carbon-14, which is produced in the atmosphere when cosmic-ray neutrons strike nitrogen nuclei. The equation for the reaction is

radioactive dating A process for determining the age of artifacts and rocks, based on the amount and half-life of radioisotopes contained in the object.

$$^{14}_{7}N + ^{1}_{0}n \rightarrow ^{14}_{6}C + ^{1}_{1}p \tag{10.4}$$

The resulting $^{14}_{6}C$ is converted to carbon dioxide $^{14}_{6}CO_2$ through a chemical reaction with oxygen. In this form, the radioactive carbon is absorbed by plants and converted into cellulose or other carbohydrates during photosynthesis. As long as the plant lives, it takes in $^{14}_{6}CO_2$. As a result, an equilibrium is established in which the plant contains the same fraction of carbon-14 as the surrounding air. Thus, as long as the plant lives, a constant fraction of the total carbon present is carbon-14. When the plant dies, $^{14}_{6}CO_2$ intake stops, and the fractional amount of carbon-14 begins to decrease. From a measurement of this amount in, say, a wooden object, and a knowledge of the half-life of carbon-14 (5600 years), it is possible to calculate the age of the object—how long since the tree was cut down to make the object. An object containing only about one-eighth as much carbon-14 as a fresh wood sample from a living tree would be about three half-lives, or 16,800 years old.

This method is limited to objects less than about 50,000 years old because of difficulties in measuring the small amount of carbon-14 present. For this reason, other dating methods have been developed. One method involves potassium-40 and argon-40. The half-life of potassium-40, which absorbs one of its own inner electrons to produce argon-40, is 1.3×10^9 years. By determining the amount of argon-40 in a potassium-containing mineral, it is possible to estimate the age of the mineral. The equation for the reaction is

$$^{40}_{19}K + ^{0}_{-1}e \rightarrow ^{40}_{18}Ar \tag{10.5}$$

10.8 Induced Nuclear Reactions

Learning Objective

8. Show that you understand the concept of induced nuclear reactions.

Before 1934, the study of radioactivity was limited to reactions of the relatively few radioisotopes found in nature. In that year, Irene and Frederic Joliot-Curie, French physical chemists, found that radioactivity could be induced in nonradioactive nuclei by bombarding them with small, subatomic particles. They produced an artificial radioactive isotope, nitrogen-13, by bombarding boron-10 with alpha particles from a natural radioisotope:

$$^{10}_{5}B + ^{4}_{2}\alpha \rightarrow ^{13}_{7}N + ^{1}_{0}n \tag{10.6}$$

Notice that a neutron, $^{1}_{0}n$, is produced in addition to $^{13}_{7}N$. Fifteen years earlier, British physicist Ernest Rutherford had used a similar reaction to produce nonradioactive oxygen-17 from nitrogen-14. A small particle, a proton, was produced in addition to oxygen-17:

$$^{14}_{7}N + ^{4}_{2}\alpha \rightarrow ^{17}_{8}O + ^{1}_{1}p \tag{10.7}$$

A great deal of research has been done since these first experiments were performed, and today it is known that a variety of nuclear reactions can be induced by bombarding either stable or naturally radioactive nuclei with high-energy particles. These artificially induced reactions may lead to one of four results. The first possibility is that the bombarded nucleus may be changed into a different nucleus that is stable. A second possibility is that a new nucleus may be formed that is unstable and undergoes radioactive decay. The other two possibilities are that nuclear fission or nuclear fusion will take place. These are discussed in more detail in Section 10.9. The result of bombardment depends on the kind of nuclei bombarded and the nature and energy of the bombarding particles.

Charged or uncharged particles may be used to bombard nuclei, and each type presents unique problems to overcome. It is believed that a bombarding particle will be drawn into the nucleus of an atom by strong nuclear forces if it can get to within 10^{-12} cm of the nucleus. Neutrons are captured readily if their speeds are reduced enough to allow the nuclear forces to work. Neutrons are slowed by passing them through materials known as **moderators.** Graphite is an example of a neutron moderator. Charged particles present a different problem. If they are negatively charged ($_{-1}^{0}\beta$), they are repelled by the electron cloud around the nucleus; for a reaction to take place they must have sufficient energy to overcome this repulsion. Similarly, positive particles must be energetic enough to overcome repulsion by the positively charged nucleus.

moderators Materials capable of slowing down neutrons that pass through them.

Neutrons for bombardment reactions are obtained from other nuclear reactions and, together with proper moderators, are used as they come from the sources. Charged particles are obtained from radioactive materials or from ionization reactions. For example, alpha particles are produced by an ionization reaction in which the two electrons are removed from helium atoms. These charged particles are generally accelerated to high speeds (and energies) before bombarding the intended target nuclei. Two types of particle accelerators, cyclic and linear, are used. The **cyclotron,** one type of cyclic accelerator, is represented in ◗ Figure 10.8.

cyclotron A cyclic particle accelerator that works by changing electrical polarities as charged particles cross a gap. The particles are kept moving in a spiral path by a strong magnetic field.

Figure 10.8 The cyclotron. Early versions of the cyclotron were about the size of a dinner plate.

The charged particles to be accelerated enter the evacuated chamber at the center of the cyclotron and, because of the magnetic field, move in a circular path toward the gap between the Dees. Just as the particles reach the gap, the electrical charge on the Dees is adjusted so the particles are repelled by the Dee they are leaving and attracted to the other one. The particles then coast inside the Dee until they again reach a gap, at which point the charges are again adjusted to cause acceleration. This process continues, and the

linear accelerator A particle accelerator that works by changing electrical polarities as charged particles cross gaps between segments of a long tube.

particles are accelerated each time they pass through the gap. As the speed and energy of the particles increase, so does the radius of their circular path, until they finally leave the Dees and strike the target.

In a **linear accelerator,** the particles are accelerated through a series of axially aligned, charged tubes located in an evacuated chamber (see ▶ Figure 10.9). Each time a particle passes through a gap and goes from one tube to another, it is accelerated in the same way as particles passing between the Dees of a cyclotron. To make the particle accelerate, the charges on the tubes are reversed at the proper times. The tubes get successively longer to allow the particles the same residence or coasting time as they move toward the end at ever-increasing speeds. After acceleration, the particles exit and strike the target.

Figure 10.9 A linear particle accelerator.

One of the most interesting results of bombardment is the creation of completely new elements. Four of these elements, produced between 1937 and 1941, filled gaps in the periodic table for which no naturally occurring element had been found. These four are technetium (Tc, number 43), promethium (Pm, number 61), astatine (At, number 85), and francium (Fr, number 87). The equations for the reactions for their production are:

$$^{96}_{42}\text{Mo} + {}^{2}_{1}\text{H} \rightarrow {}^{1}_{0}n + {}^{97}_{43}\text{Tc} \text{ (half-life: } 2.6 \times 10^{6} \text{ years)} \tag{10.8}$$

$$^{142}_{60}\text{Nd} + {}^{1}_{0}n \rightarrow {}^{0}_{-1}\beta + {}^{143}_{61}\text{Pm} \text{ (half-life: 265 days)} \tag{10.9}$$

$$^{209}_{83}\text{Bi} + {}^{4}_{2}\alpha \rightarrow 3{}^{1}_{0}n + {}^{210}_{85}\text{At} \text{ (half-life: 8.3 hours)} \tag{10.10}$$

$$^{230}_{90}\text{Th} + {}^{1}_{1}p \rightarrow 2{}^{4}_{2}\alpha + {}^{223}_{87}\text{Fr} \text{ (half-life: 22 minutes)} \tag{10.11}$$

transuranium elements Synthetic elements with atomic numbers greater than that of uranium.

Synthetic elements heavier than uranium, the heaviest naturally occurring element, have also been produced. All these elements, called **transuranium elements,** are radioactive. The first, neptunium (Np, number 93), was produced in 1940; since then at least 21 more (numbers 94 through 114) have been synthesized.

Significant quantities of only a few of these elements have been produced. Plutonium is synthesized in quantities large enough to make it available for use in atomic weapons and reactors. In 1968 the entire world's supply of californium (10^{-4} g) was gathered and made into a target for bombardment by heavy, accelerated particles. These experiments led to the discovery of several elements heavier than californium. Usually, the amounts of the new elements produced are extremely small—sometimes only hundreds of atoms. When the half-life is short, this small amount quickly disappears. Thus, these elements have to be identified not by chemical properties but by the use of instruments that analyze the characteristic radiation emitted by each new element.

▶ Example 10.7

An isotope of gallium (Ga) useful in medical diagnostic work is formed when zinc-66 is bombarded with accelerated protons. Write the equation for this nuclear reaction and determine what isotope of gallium is produced.

Solution

The equation for the reaction is written

$$^{66}_{30}\text{Zn} + ^{1}_{1}p \rightarrow ?$$

where the product is to be determined. By balancing the mass numbers and atomic numbers, we see that the product has a mass number of 67 and an atomic number of 31. Reference to the periodic table shows element number 31 to be gallium, Ga. Therefore, the isotope produced is gallium-67, and the equation for the reaction is:

$$^{66}_{30}\text{Zn} + ^{1}_{1}p \rightarrow ^{67}_{31}\text{Ga}$$

▶ **LEARNING CHECK 10.9** When manganese-55 is bombarded by protons, a neutron is one product. Write the equation for the process and identify the other product.

10.9 Nuclear Energy

Learning Objective

9. Describe the differences between nuclear fission and nuclear fusion.

In 1903, Ernest Rutherford and Frederick Soddy made observations that led them to conclude that the nuclei of all atoms, not just radioactive ones, must contain large quantities of energy. However, the form it took and the means for releasing it remained a mystery. In 1905, Albert Einstein, who was then working as a patent clerk, provided a partial answer. According to his theory of relativity, matter and energy were equivalent and one could be converted into the other. This fact was expressed in his now-famous equation,

$$E = mc^2 \tag{10.12}$$

According to his equation, a small amount of mass, *m,* would, upon conversion, yield a huge amount of energy, *E.* The amount of energy liberated is huge because of the size of the multiplier c^2, the velocity of light squared. The velocity of light is a very large number (3.0×10^8 m/s, or 1.9×10^5 mi/s). The verification of Einstein's theory was not made until 1939, when the work of Otto Hahn, Lise Meitner, Fritz Strassman, and Enrico Fermi led to the discovery of **nuclear fission.** Attempts to produce new heavier transuranium elements by bombarding uranium-235 with neutrons led instead to the production of elements much less massive than uranium-235. The uranium-235 nuclei had undergone fission and split into smaller fragments. In the process, a small amount of matter was converted into energy, just as Einstein had predicted.

nuclear fission A process in which large nuclei split into smaller, approximately equal-sized nuclei when bombarded by neutrons.

Today, we know that uranium-235 is the only naturally occurring isotope that will undergo fission, and a number of different products can result. Equations 10.13–10.16 represent some of the numerous fission reactions that can occur.

$$^{235}_{92}\text{U} + ^{1}_{0}n \begin{cases} \rightarrow ^{135}_{53}\text{I} + ^{97}_{39}\text{Y} + 4\,^{1}_{0}n & (10.13) \\ \rightarrow ^{139}_{56}\text{Ba} + ^{94}_{36}\text{Kr} + 3\,^{1}_{0}n & (10.14) \\ \rightarrow ^{102}_{42}\text{Mo} + ^{132}_{50}\text{Sn} + 2\,^{1}_{0}n & (10.15) \\ \rightarrow ^{139}_{54}\text{Xe} + ^{95}_{38}\text{Sr} + 2\,^{1}_{0}n & (10.16) \end{cases}$$

For every uranium-235 nucleus undergoing fission, more than one neutron is generated, which opens up the possibility for a chain reaction to take place. If only one of the neutrons produced each time reacts with another uranium-235 nucleus, the process will become a **chain reaction** that continues at a constant rate. If more than one of the neutrons

chain reaction A nuclear reaction in which the products of one reaction cause a repeat of the reaction to take place. In the case of uranium fission, neutrons from fission reactions cause other fission reactions to occur.

expanding or branching chain reaction A reaction in which the products of one reaction cause more than one more reaction to occur.

generated per fission reaction produces another reaction, the process becomes an **expanding or branching chain reaction** that leads to an explosion. A constant-rate chain reaction is called a **critical reaction.** A branching chain reaction that will lead to an explosion is called **supercritical** (see ❿ Figure 10.10).

Figure 10.10 Nuclear chain reactions.

critical reaction A constant-rate chain reaction.

supercritical reaction A branching chain reaction.

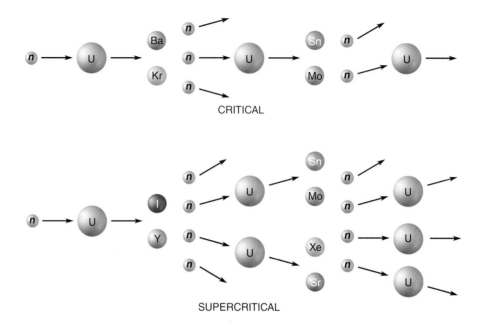

History's first critical chain reaction was carried out in an atomic pile, which was literally a stack or pile of uranium-containing graphite blocks and blocks of pure graphite that served as a neutron moderator (see ❿ Figure 10.11). The pile was constructed in a squash court beneath the stands of an athletic fieldhouse at the University of Chicago, and the experiments were carried out by a team of scientists led by Italian physicist Enrico Fermi. In December 1942, the team observed that nuclear reactions in the pile had become self-sustaining, or critical. The control rods, composed of strong neutron absorbers, were pushed into the pile to stop the reaction.

Figure 10.11 An atomic pile, the world's first nuclear reactor.

Unless otherwise noted all art on this page © Cengage Learning 2014

Once it had been demonstrated that uranium-235 would undergo self-sustaining fission, the construction of an atomic bomb was fairly simple in principle, but it depended on the concept of critical mass. Neutrons released inside a small piece of fissionable material might escape to the outside before they collide with and split another atom. If the majority of neutrons are lost this way, a chain reaction will not take place. However, as the size of the piece of fissionable material is increased, the chance for neutrons to escape without hitting other atoms decreases. The amount of fissionable material needed to cause a critical reaction to occur is the **critical mass.** The amount needed to cause a branching chain reaction and explosion to occur is called a **supercritical mass.**

During World War II, one major difficulty encountered in building a nuclear bomb was the size of the device needed to bring together subcritical masses to form a supercritical mass. Any bomb used in combat would have to fit into a B-29, the largest bomber available at the time, and could not be heavier than the load capacity of the airplane. The configurations of the first two devices actually used are represented in ▶ Figure 10.12. The spherical configuration resulted in a short, wide bomb, while the tubular configuration gave rise to a long, narrow bomb. In each configuration, explosive charges were used to drive subcritical masses together to form a supercritical mass. These nuclear bombs were the first applications of the concepts of subcritical and supercritical mass.

critical mass The minimum amount of fissionable material needed to sustain a critical chain reaction at a constant rate.

supercritical mass The minimum amount of fissionable material that must be present to cause a branching chain reaction to occur.

Figure 10.12 Configurations used in fission-type nuclear bombs.

On August 6, 1945, the United States became the first nation to use a nuclear weapon in combat. This bomb used a tubular configuration. Approximately 70,000 inhabitants of Hiroshima, Japan, were killed in the explosion; an equal number were seriously injured. A few days later a second bomb, with a spherical configuration, was dropped on the city of Nagasaki. The casualty figures were similar, and World War II came to an end. The Nagasaki bomb was made of plutonium, an artificial fissionable material produced in atomic piles by breeding reactions.

Breeder reactions are induced reactions in which the product nuclei undergo fission reactions. The plutonium used in the Nagasaki bomb was made by bombarding uranium-238 with neutrons in the first atomic pile. Uranium-238 is the most abundant isotope of uranium found in nature, but it will not undergo fission, as will the much less

breeder reaction A nuclear reaction in which isotopes that will not undergo spontaneous fission are changed into isotopes that will.

Is Irradiated Food Safe?

In 1993, four deaths and hundreds of illnesses were linked to the eating of undercooked ground beef that was contaminated with *E. coli* bacteria. In 1997 the largest recall of meat products in U.S. history up to that time took place when a producer voluntarily recalled 25 million pounds of ground beef suspected of being contaminated with the same type of bacteria. In 2002, a new meat product recall record was established when 27.4 million pounds of poultry products was recalled because of possible contamination by *Listeria,* a potentially deadly type of bacteria.

It is a well-established fact that treating food by exposing it to radiation destroys disease-causing organisms, kills insects, delays maturation and sprouting of seeds, and slows spoiling. But is irradiated food safe? After years of scientific study, the U.S. government began answering this question with a somewhat-qualified *yes* in 1963, when the Food and Drug Administration (FDA) first allowed radiation-treated wheat and wheat flour to be marketed. Ongoing studies have focused on such areas as the chemical effects of radiation on food, the impact of radiation on nutrient content, and toxicity concerns. Food irradiation for the control of disease-causing organisms has received endorsements and praise from numerous food industry and health organizations.

Since 1963, the list of foods approved for radiation treatment has grown steadily. The effect of the radiation on specific foods is determined by controlling the dose of radiation used. The following table shows the types of foods approved for radiation at this time, along with the approved dose and the desired effect the dose has on the food. The doses are measured in units of kilograys (kGy), which were defined in Table 10.5.

purchase and use them. Some irradiated foods are used unintentionally as a result of the labeling rules. For example, irradiated spices or fresh strawberries sold in a grocery store must have labels that indicate they have been irradiated. However, when such irradiated foods are used as ingredients in other foods, the other foods are not required to have irradiation labels. Also, irradiation labeling does not apply to foods served in restaurants.

Two factors that currently contribute to low consumer acceptance of irradiated food are continuing misconceptions about the effects of radiation on food and the low availability of irradiated foods for sale. Some misconceptions that are difficult to eliminate despite reassuring results from many studies are that irradiation causes food to become radioactive, irradiation produces harmful compounds in food, and irradiation reduces the amounts of nutrients in food. Studies indicate that continued education about the safety of irradiated food will eventually overcome this problem. The availability problem stems primarily from the lack of food-irradiation facilities. However, this factor is rapidly becoming less important as more facilities are being built in response to the acceptance of irradiation as a protection against harmful bacteria in food, especially in meat products. Some large meat-packing companies are adding irradiation facilities to existing packing plants, and including irradiation capabilities in plans for new plants, in order to eliminate the costs of shipping their products away from the packing plants to be irradiated.

It appears that food irradiation is becoming more and more acceptable to consumers, and the acceptance will continue to grow. It is likely that in the future, food irradiation will be considered by most consumers to be a normal practice in providing a safe food supply to the consuming public.

Food	Effect	Dose (kGy)
Spices and dry vegetable seasonings	Decontaminates, controls insects and harmful bacteria	30
Dry or dehydrated enzymes	Controls insects and harmful bacteria	10
All foods	Controls insects	1
Fresh foods	Delays maturation	1
Poultry	Controls harmful bacteria	3
Red meat	Controls spoilage and harmful bacteria	4.5 (fresh) 7 (frozen)

In spite of scientific studies, endorsements from health organizations, and repeated assurances about the safety of irradiated foods, most consumers do not intentionally

The irradiated strawberries on the left have a much longer shelf life than the untreated berries on the right.

abundant uranium-235. However, uranium-238 reacts as follows when placed in a neutron-rich environment.

$$^{238}_{92}\text{U} + ^{1}_{0}n \rightarrow ^{239}_{92}\text{U} \qquad (10.17)$$

$$^{239}_{92}\text{U} \rightarrow ^{239}_{93}\text{Np} + ^{0}_{-1}\beta \qquad (10.18)$$

$$^{239}_{93}\text{Np} \rightarrow ^{239}_{94}\text{Pu} + ^{0}_{-1}\beta \qquad (10.19)$$

The plutonium-239 will undergo fission and, when collected into a supercritical mass, will cause an explosion. Reactions such as 10.17–10.19 are called breeding reactions because a fissionable material is "bred" from a nonfissionable material.

Curiosity about the seemingly unlimited energy supply of the stars (especially the sun) eventually led scientists to an understanding of a second nuclear process that releases vast amounts of energy. In 1920, Sir Arthur Eddington suggested that the energy of the stars is a byproduct of a reaction in which hydrogen is changed into helium. This was followed in 1929 by the concept of **thermonuclear reactions,** nuclear reactions that are started by very high temperatures. According to this idea, at high temperatures the nuclei of very lightweight elements will combine to form nuclei of heavier elements, and in the process some mass is converted into energy. This idea was consistent with Eddington's suggestion that hydrogen is converted to helium in the sun and other stars. It was not until nuclear fission reactions became available that high enough temperatures could be achieved to test the theory of **nuclear fusion.** The theory proved to be correct, and today most nuclear weapons in the world are based on fusion reactions.

thermonuclear reactions Nuclear fusion reactions that require a very high temperature to start.

The fusion reactions responsible for the energy output of the sun are thought to be

$$2^{1}_{1}\text{H} \rightarrow ^{2}_{1}\text{H} + ^{0}_{1}\beta \qquad (10.20)$$

$$^{1}_{1}\text{H} + ^{2}_{1}\text{H} \rightarrow ^{3}_{2}\text{He} + \gamma \qquad (10.21)$$

$$^{3}_{2}\text{He} + ^{3}_{2}\text{He} \rightarrow ^{4}_{2}\text{He} + 2^{1}_{1}\text{H} \qquad (10.22)$$

nuclear fusion A process in which small nuclei combine or fuse to form larger nuclei.

The net overall reaction is

$$4^{1}_{1}\text{H} \rightarrow ^{4}_{2}\text{He} + 2^{0}_{1}\beta + 2\gamma \qquad (10.23)$$

which is what Eddington proposed; hydrogen ($^{1}_{1}\text{H}$) is converted to helium ($^{4}_{2}\text{He}$).

A goal of many of the early developers of nuclear energy was to harness it and make it a virtually unlimited source of useful energy. Today, that goal has been partially achieved. Fission reactors generate a small percentage of the electricity used in the world, and researchers are making progress in their attempts to harness fusion reactions.

The essential components of a fission-powered generating plant are shown in ▶ Figure 10.13. The use of nuclear power plants has created much controversy. The plants release large amounts of waste heat and thus cause thermal pollution of natural waters. In addition, radioactive wastes create disposal problems—where do you dump radioactive

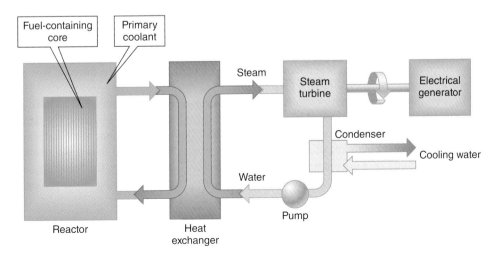

Figure 10.13 Components of a nuclear electricity power plant.

The Do's and Don'ts of Buying Prescription Drugs Online

In our world of modern technology, consumers are relying more and more on the Internet to make purchases. Wedding gifts are ordered online, gift wrapped, and shipped directly to the bride and groom. Movies are downloaded to computers for viewing or are selected online and mailed to our homes, making a trip to the video store unnecessary. Often, both time and money are saved by "shopping around" for the best value in the virtual world without ever leaving the comfort of our home or office. However, remember *caveat emptor*—let the buyer beware. When shopping for prescription drugs, your health and maybe even your life are potentially on the line. Although there are many reputable pharmacies offering online convenience, others are questionable. Examples of the questionable practices followed by such pharmacies include distributing expired or illegal drugs or drugs not approved by the FDA, not checking for drug interactions, and even supplying medications other than those specified by a physician.

If you order your medications online, follow these few simple guidelines. First, consult a professional health care provider before ordering or using any medications. This helps eliminate purchasing medications with false claims or those that have harmful interactions with medications currently being taken. Second, obtain a written prescription for any controlled substances and make sure that the online provider requires a copy of a valid prescription before dispensing such substances. Third, check the standing of the online pharmacy's license with the National Association of Boards of Pharmacy. Approved sites often bear the seal of VIPPS (Verified Internet Pharmacy Practice Sites). Fourth, if your motivation for ordering online is to save money, check the online prices with local pharmacies to see if the online prices are actually less than the local pharmacy prices. Fifth, before you divulge any personal information, check the online provider's policies regarding security and the selling of information to other companies. Sixth, be sure the online pharmacy provides access to a licensed pharmacist as well as contact information such as a phone number and/or physical address for the pharmacy itself. Finally, if the online pharmacy is based out of a country other than the United States, be especially cautious about the quality and legality of the medications, the language on the labels, and the potential differences in the names of the provided medications.

Buying prescriptions online might save money, but precautions should be followed.

materials that will be a health hazard for thousands of years? Electricity consumers and producers have lost some of their confidence in the safety of nuclear power plants as a result of accidents such as the well-known incidents at Three-Mile Island (1979) and Chernobyl (1986). Another problem is the relatively short supply of fissionable uranium-235. Breeder reactors could solve this problem, since they can convert abundant supplies of uranium-238 and thorium-232 into fissionable fuels, but the breeder reactor program in the United States is at a standstill. As a result of these factors, the future of fission power generation is uncertain at the present time.

In principle, fusion reactors can be used as controlled energy sources just as fission reactors are. Fusion reactors would have many advantages over fission reactors. The deuterium fuel (hydrogen-2) is readily available since it is a minor constituent of all water on Earth, no significant amounts of radioactive wastes would be produced during operation, and the possibility of a serious accident is much smaller. Unfortunately, the feasibility of fusion reactors has not yet been established, although scientists are getting closer through continuing research.

Concept Summary

Radioactive Nuclei

Some nuclei are unstable and undergo radioactive decay. The common types of radiation emitted during decay processes are alpha, beta, and gamma, which can be characterized by mass and charge values.

Objective 1, Exercise 10.2

Equations for Nuclear Reactions

Nuclear reactions can be represented by balanced equations in which the focus is on mass number and atomic number balance on each side. Symbols are used in nuclear equations that make it convenient to balance mass and atomic numbers.

Objective 2, Exercise 10.12

Isotope Half-Life

Different radioisotopes generally decay at different rates, which are indicated by half-lives. One half-life is the time required for one-half of the unstable nuclei in a sample to undergo radioactive decay.

Objective 3, Exercise 10.16

The Health Effects of Radiation

Radiation generates free radicals in tissue as it passes through. Radiation is hazardous even at low intensity if there is long-term exposure. Radiation sickness is caused by short-term intense radiation. Those working around radioactive sources can minimize exposure by using shielding or distance as a protection.

Objective 4, Exercise 10.22

Measurement Units for Radiation

Two systems, physical and biological, are used to describe quantities of radiation. Physical units indicate the number of nuclei of radioactive material that decay per unit of time. Biological units are related to the damage caused by radiation in living tissue. The common physical units are the curie and its fractions and the becquerel. Biological units include the roentgen (for gamma and X-rays), the rad, the gray, and the rem.

Objective 5, Exercise 10.24

Concept Summary (continued)

Medical Uses of Radioisotopes

Radioisotopes behave chemically like nonradioisotopes of the same element and can be used diagnostically and therapeutically. Diagnostically, radioisotopes are used as tracers whose movement or localization in the body can be followed. Therapeutic radioisotopes localize in diseased areas of the body, where their radiation can destroy diseased tissue.

Objective 6, Exercise 10.30

Nonmedical Uses of Radioisotopes

Nonmedical uses of radioisotopes include (1) tracers in chemical reactions, (2) boundary markers between liquids in pipelines, (3) tracers to detect leaks, (4) metal-wear indicators, and (5) thickness indicators for foils and sheet metal. An especially interesting application is the determination of the ages of artifacts and rocks.

Objective 7, Exercise 10.36

Induced Nuclear Reactions

The bombardment of nuclei with small high-energy particles causes the bombarded nuclei to (1) change into stable nuclei, (2) change into radioactive nuclei, (3) break into smaller pieces (fission), or (4) form larger nuclei (fusion). Most bombarding particles are accelerated by using cyclic or linear accelerators. Bombardment reactions have been used to produce elements that are not found in nature.

Objective 8, Exercise 10.38

Nuclear Energy

During spontaneous nuclear fission reactions, heavy nuclei split when bombarded by neutrons, and release large amounts of energy. This process was used to produce two atomic bombs whose use ended World War II. A second energy-releasing nuclear process, fusion, is the basis for today's hydrogen bombs. Nuclear fission is in limited use as a source of electrical power; however, this use is controversial. Nuclear fusion has not yet proved feasible as a controlled source of power, but research toward this end continues.

Objective 9, Exercise 10.48

Key Terms and Concepts

Acute radiation syndrome (10.4)
Alpha particle (10.1)
Becquerel (10.5)
Beta particle (10.1)
Biological unit of radiation (10.5)
Breeder reaction (10.9)
Chain reaction (10.9)
Cold spot (10.6)
Critical mass (10.9)
Critical reaction (10.9)
Curie (10.5)
Cyclotron (10.8)
Daughter nuclei (10.2)
Electron capture (10.2)

Expanding or branching chain reaction (10.9)
Gamma ray (10.1)
Geiger–Müller tube (10.5)
Gray (10.5)
Half-life (10.3)
Hot spot (10.6)
Inverse square law of radiation (10.4)
Linear accelerator (10.8)
Moderators (10.8)
Nuclear fission (10.9)
Nuclear fusion (10.9)
Physical unit of radiation (10.5)
Positron (10.2)

Rad (10.5)
Radical or free radical (10.4)
Radioactive dating (10.7)
Radioactive decay (10.1)
Radioactive nuclei (10.1)
Radioisotope (10.2)
Rem (10.5)
Roentgen (10.5)
Scintillation counter (10.5)
Supercritical mass (10.9)
Supercritical reaction (10.9)
Thermonuclear reaction (10.9)
Tracer (10.6)
Transuranium elements (10.8)

Key Equations

1. Symbol for isotopes (Section 10.2):	${}^{A}_{Z}X$	
2. Conversion of a neutron to a proton and electron in the nucleus, leading to beta emission (Section 10.2):	${}^{1}_{0}n \rightarrow {}^{1}_{1}p + {}^{0}_{-1}\beta$	Equation 10.1
3. Variation of radiation intensity with distance (Section 10.4):	$\dfrac{I_x}{I_y} = \dfrac{d_y^2}{d_x^2}$	Equation 10.2
4. Fission of uranium-235 (Section 10.9): See also Equations 10.14, 10.15, and 10.16.	${}^{235}_{92}U + {}^{1}_{0}n \rightarrow {}^{135}_{53}I + {}^{97}_{39}Y + 4{}^{1}_{0}n$	Equation 10.13
5. Hydrogen fusion reaction of the sun (Section 10.9):	$4{}^{1}_{1}H \rightarrow {}^{4}_{2}He + 2{}^{0}_{1}\beta + 2\gamma$	Equation 10.23

Exercises

Even-numbered exercises are answered in Appendix B.

Blue-numbered exercises are more challenging.

Radioactive Nuclei (Section 10.1)

10.1 Define the term *radioactive;* then criticize the following statements:

  a. Beta rays are radioactive.

  b. Radon is a stable radioactive element.

10.2 Group the common nuclear radiations (Table 10.1) into the following categories:

  a. Those with a mass number of 0

  b. Those with a positive charge

  c. Those with a charge of 0

10.3 Group the common nuclear radiations (Table 10.1) into the following categories:

  a. Those with a negative charge

  b. Those with a mass number greater than 0

  c. Those that consist of particles

10.4 Characterize the following nuclear particles in terms of the fundamental particles—protons, neutrons, and electrons:

  a. A beta particle

  b. An alpha particle

  c. A positron

10.5 Discuss how the charge and mass of particles that comprise radiation influence the range or ability of the radiation to penetrate matter.

Equations for Nuclear Reactions (Section 10.2)

10.6 Summarize how the atomic number and mass number of daughter nuclei compare with the original nuclei after:

  a. An alpha particle is emitted.

  b. A beta particle is emitted.

  c. An electron is captured.

  d. A gamma ray is emitted.

  e. A positron is emitted.

10.7 Write appropriate symbols for the following particles using the ${}^{A}_{Z}X$ symbolism:

  a. A tin-117 nucleus

  b. A nucleus of the chromium (Cr) isotope containing 26 neutrons

  c. A nucleus of element number 20 that contains 24 neutrons

10.8 Write appropriate symbols for the following particles using the ${}^{A}_{Z}X$ symbolism:

  a. An aluminum-27 nucleus

  b. A nucleus of element number 38 with a mass number of 90

  c. A nucleus of the krypton isotope that contains 40 neutrons

10.9 Complete the following equations, using appropriate notations and formulas:

  a. ${}^{10}_{4}Be \rightarrow ? + {}^{10}_{5}B$

  b. ${}^{210}_{83}Bi \rightarrow {}^{4}_{2}\alpha + ?$

  c. ${}^{15}_{8}O \rightarrow ? + {}^{15}_{7}N$

  d. ${}^{44}_{22}Ti + {}^{0}_{-1}e \rightarrow ?$

  e. ${}^{8}_{4}Be \rightarrow ? + {}^{4}_{2}He$

  f. ${}^{46}_{23}V \rightarrow ? + {}^{46}_{22}Ti$

10.10 Complete the following equations, using appropriate notations and formulas:

  a. ${}^{234}_{90}Th \rightarrow ? + {}^{234}_{91}Pa$

  b. ${}^{35}_{16}S \rightarrow ? + {}^{0}_{-1}\beta$

  c. $? \rightarrow {}^{4}_{2}\alpha + {}^{244}_{98}Cf$

  d. ${}^{226}_{88}Ra \rightarrow ? + {}^{4}_{2}\alpha$

  e. ${}^{113}_{47}Ag \rightarrow {}^{113}_{48}Cd + ?$

  f. ${}^{41}_{18}Ar + {}^{0}_{-1}e \rightarrow ?$

10.11 Write balanced equations to represent decay reactions of the following isotopes. The decay process or daughter isotope is given in parentheses.

   a. $^{66}_{29}Cu$ (beta emission)

   b. $^{22}_{11}Na$ (positron emission)

   c. $^{19}_{8}O$ (daughter $= ^{19}_{9}F$)

   d. $^{192}_{78}Pt$ (alpha emission)

   e. $^{108}_{50}Sn$ (electron capture)

   f. $^{67}_{28}Ni$ (beta emission)

10.12 Write balanced equations to represent decay reactions of the following isotopes. The decay process or daughter isotope is given in parentheses.

   a. $^{157}_{63}Eu$ (beta emission)

   b. $^{190}_{78}Pt$ (daughter $=$ osmium-186)

   c. $^{138}_{62}Sm$ (electron capture)

   d. $^{188}_{80}Hg$ (daughter $=$ Au-188)

   e. $^{234}_{90}Th$ (beta emission)

   f. $^{218}_{85}At$ (alpha emission)

Isotope Half-Life (Section 10.3)

10.13 What is meant by a half-life?

10.14 Describe half-life in terms of something familiar, such as a cake or cookies or your checking account.

10.15 An isotope of lead, $^{194}_{82}Pb$ has a half-life of 11 minutes. What fraction of the lead-194 atoms in a sample would remain after 44 minutes had elapsed?

10.16 Technetium-99 has a half-life of 6 hours. This isotope is used diagnostically to perform brain scans. A patient is given a 9.0-ng dose. How many nanograms will be present in the patient 24 hours later?

10.17 An archaeologist sometime in the future analyzes the iron used in an old building. The iron contains tiny amounts of nickel-63, with a half-life of 92 years. On the basis of the amount of nickel-63 and its decay products found, it is estimated that about 0.78% (1/128) of the original nickel-63 remains. If the building was constructed in 1980, in what year did the archaeologist make the discovery?

10.18 An archaeologist unearths the remains of a wooden box, analyzes them for the carbon-14 content, and finds that about 12.5% of the carbon-14 initially present remains. Estimate the age of the box. The half-life of carbon-14 is 5600 years.

10.19 Germanium-66 decays by positron emission, with a half-life of 2.5 hours. What mass of germanium-66 remains in a sample after 10.0 hours if the sample originally weighed 50.0 mg?

10.20 A grain sample was found in a cave. The amount of carbon-14 in the sample was 1/16 the amount found in a fresh sample of grain. How old was the grain in the cave? The half-life of carbon-14 is 5600 years.

The Health Effects of Radiation (Section 10.4)

10.21 A source of radiation has an intensity of 130 units at a distance of 10 feet. How far away from the source would you have to be to reduce the intensity to 15 units?

10.22 Compare and contrast the general health effects of long-term exposure to low-level radiation and short-term exposure to intense radiation.

10.23 A radiologist found that in a 30-minute period the dose from a radioactive source was 75 units at a distance of 12.0 m from the source. What would the dose be in the same amount of time at a distance of 2.00 m from the source?

Measurement Units for Radiation (Section 10.5)

10.24 Explain the difference between physical and biological units of radiation and give an example of each type.

10.25 Explain why the rem is the best unit to use when evaluating the radiation received by a person working in an area where exposure to several types of radiation is possible.

10.26 An individual receives a short-term whole-body dose of 3.1 rads of beta radiation. How many roentgens of X-rays would represent the same health hazard?

10.27 Describe how scintillation counters and Geiger–Müller counters detect radiation.

10.28 One Ci corresponds to 3.7×10^{10} nuclear disintegrations per second. How many disintegrations per second would take place in a sample containing 4.6 μ Ci of radioisotope?

Medical Uses of Radioisotopes (Section 10.6)

10.29 Explain what a diagnostic tracer is and list the ideal characteristics one should have.

10.30 Describe the importance of hot and cold spots in diagnostic work using tracers.

10.31 List the ideal characteristics of a radioisotope that is to be administered internally for therapeutic use.

10.32 Chromium-51 is used medically to monitor kidney activity. Chromium-51 decays by electron capture. Write a balanced equation for the decay process and identify the daughter that is produced.

10.33 Gold-198 is a β emitter used to treat leukemia. It has a half-life of 2.7 days. The dosage is 1.0 mCi/kg body weight. How long would it take for a 70-mCi dose to decay to an activity of 2.2 mCi?

Nonmedical Uses of Radioisotopes (Section 10.7)

10.34 A mixture of water (H_2O) and hydrogen peroxide (H_2O_2) will give off oxygen gas when solid manganese dioxide is added as a catalyst. Describe how you could use a tracer to determine if the oxygen comes from the water or the peroxide.

10.35 Suppose you are planning a TV commercial for motor oil. Describe how you would set it up to show that your oil prevents engine wear better than a competing brand.

10.36 Propose a method for measuring the volume of water in an irregularly shaped swimming pool. You have 1 gallon of water that contains a radioisotope with a long half-life, and a Geiger–Müller counter.

10.37 Explain why carbon-14, with a half-life of about 5600 years, is not a good radioisotope to use if you want to determine the age of a coal bed thought to be several million years old.

Induced Nuclear Reactions (Section 10.8)

10.38 Write a balanced equation to represent the synthesis of silicon-27 that takes place when magnesium-24 reacts with an accelerated alpha particle. A neutron is also produced.

10.39 Americium-241 (Am) results when plutonium-239 captures two neutrons. A beta particle is also produced. Write a balanced equation for the process.

10.40 Write a balanced equation to represent the net breeding reaction that occurs when uranium-238 reacts with a neutron to form plutonium-239 and two beta particles.

10.41 Explain why neutrons cannot be accelerated in a cyclotron or linear accelerator.

10.42 Describe the role of a moderator in nuclear reactions involving neutrons.

10.43 To make indium-111 for diagnostic work, silver-109 is bombarded with α particles. What isotope forms when a silver-109 nucleus captures an α particle? Write an equation to represent the process. What two particles must be emitted by the isotope to form indium-111?

10.44 To make gallium-67 for diagnostic work, zinc-66 is bombarded with accelerated protons. Write a balanced equation to represent the process when a zinc-66 nucleus captures a single proton.

Nuclear Energy (Section 10.9)

10.45 Write one balanced equation that illustrates nuclear fission.

10.46 Write one balanced equation that illustrates nuclear fusion.

10.47 Explain why critical mass is an important concept in neutron-induced fission chain reactions.

10.48 Describe the difference between nuclear fission and nuclear fusion processes. Use nuclear equations to illustrate the difference.

10.49 Complete the following equations, which represent two additional ways uranium-235 can undergo nuclear fission.

  a. $^{235}_{92}U + ^{1}_{0}n \rightarrow ^{160}_{62}Sm + ? + 4^{1}_{0}n$

  b. $^{235}_{92}U + ^{1}_{0}n \rightarrow ^{87}_{35}Br + ? + 3^{1}_{0}n$

10.50 Plutonium-238 is used to power batteries for pacemakers. It decays by α emission. Write a balanced equation for the process.

Additional Exercises

10.51 Polonium-210 decays by alpha emission and has a half-life of 138 days. Suppose a sample of Po-210 contains 6.02×10^{20} Po-210 atoms and undergoes decay for 138 days. What was the mass in grams of the original sample? Assume all alpha particles formed by the decay escape, and the daughter nuclei remain in the sample as nonradioactive isotopes; then calculate the mass of the sample in grams after 138 days.

10.52 Propose an explanation for the fact that the fusion reactions occurring in the sun only take place at extremely high temperatures.

10.53 Explain how you would make a 10.0-mL sample of 0.100 M NaI that contained iodine in the form of the $^{123}_{53}I$ isotope.

10.54 Zinc-71 decays by beta emission, and has a half-life of 2.4 minutes. Suppose a 0.200-g sample of Zn-71 was put inside the tube of a Geiger–Müller counter that counted every beta particle emitted by the sample for 2.4 minutes. What would the average reading of the Geiger–Müller counter be in counts per minute during the 2.4 minutes? How many curies is this?

10.55 At one time, nuclear bombs were tested by exploding them above ground. The fallout from such tests contained some Sr-90, a radioactive isotope of strontium. If Sr-90 gets into the food supply, it can become incorporated into the bones of humans and other animals. Explain how this can happen even though strontium is not a normal component of bones.

Allied Health Exam Connection

The following questions are from these sources:

1. *Nursing School Entrance Exam* © 2005, Learning Express, LLC.

2. *McGraw-Hill's Nursing School Entrance Exams* by Thomas A. Evangelist, Tamara B. Orr and Judy Unrein © 2009, The McGraw-Hill Companies, Inc.

3. *NSEE Nursing School Entrance Exams*, 3rd Edition © 2009, Kaplan Publishing.

4. *Cliffs Test Prep: Nursing School Entrance Exams* by Fred N. Grayson © 2004, Wiley Publishing, Inc.

5. *Peterson's Master the Nursing School and Allied Health Entrance Exams*, 18th Edition by Marion F. Gooding © 2008, Peterson's, a Nelnet Company.

10.56 What happens when a radioactive element decays?

  a. The nucleus gives off particles.

  b. The ribosomes give off particles.

  c. The mitochondria give off particles.

  d. The Golgi apparatus gives off particles.

10.57 Which of the following particles/rays is not a type of radioactive emanation?

  a. alpha particle

  b. beta particle

  c. gamma ray

  d. carbon-12

10.58 Which of the following forms of radiation contains the least energy?

  a. radio waves

  b. gamma rays

  c. ultraviolet radiation

  d. visible light

10.59 Which of the following is NOT a form of radioactive decay?

  a. electron capture

  b. beta emission

  c. alpha emission

  d. proton emission

10.60 The least penetrating radiation given off by a radioactive substance consists of:

  a. alpha particles

  b. beta particles

  c. gamma rays

  d. X-rays

10.61 Alpha, beta, and gamma radiation are the three major products of natural radioactive decay. Which of the following is (are) the most penetrating radiation?

a. alpha rays

b. beta rays

c. gamma rays

d. they are all equally penetrating

10.62 The isotope $^{234}_{92}U$ decays by releasing an alpha particle. What is the resulting isotope?

a. $^{234}_{93}Np$

b. $^{233}_{91}Pa$

c. $^{234}_{91}Pa$

d. $^{230}_{90}Th$

10.63 Carbon dating involves the decay of a carbon-14 isotope by beta emission. Which of the following equations describes this decay?

a. $^{14}_{6}C \rightarrow ^{13}_{5}B + ^{1}_{1}H$

b. $^{14}_{6}C \rightarrow ^{14}_{7}N + ^{0}_{-1}\beta$

c. $^{14}_{6}C \rightarrow ^{13}_{5}B + ^{1}_{0}n$

d. $^{14}_{7}N + ^{1}_{0}n \rightarrow ^{14}_{6}C + ^{1}_{1}H$

10.64 What is the name of particle X in the following reaction?

$$^{232}_{90}Th \rightarrow ^{228}_{88}Ra + X$$

a. deuterium

b. gamma radiation

c. beta particle

d. alpha particle

10.65 If element A below gives off an alpha particle, what is the atomic number and mass of the resulting element B? $^{210}_{83}A$

a. $^{210}_{81}B$

b. $^{206}_{81}B$

c. $^{206}_{83}B$

d. $^{204}_{81}B$

10.66 If element B below gives off a beta particle and gamma rays, what is the resulting element? $^{238}_{92}B$

a. $^{238}_{93}B$

b. $^{234}_{90}B$

c. $^{239}_{92}B$

d. $^{239}_{91}B$

10.67 The loss of an alpha particle from the radioactive atom $^{228}_{88}X$ would leave:

a. $^{224}_{86}Rn$

b. $^{222}_{86}Rn$

c. $^{224}_{88}Ra$

d. $^{230}_{90}Th$

10.68 What is the missing product?

$$^{42}_{17}X \rightarrow ^{42}_{18}Y + ?$$

a. $^{4}_{2}He$

b. $^{0}_{0}\gamma$

c. $^{0}_{1}e$

d. $^{0}_{-1}\beta$

10.69 What is the missing product?

$$^{60}_{24}A \rightarrow ^{60}_{24}B + ?$$

a. $^{4}_{2}He$

b. $^{0}_{0}\gamma$

c. $^{0}_{-1}e$

d. $^{0}_{1}\beta$

10.70 The half-life of a certain radioactive isotope is 5 days. How much of a 100 g sample of this radioactive isotope will remain after 10 days?

a. 25

b. 50

c. 75

d. 100

10.71 The radioactive isotope I-131 has a half-life of 8 days. What fraction of a sample of I-131 will remain after 24 days?

a. $\dfrac{1}{4}$

b. $\dfrac{1}{16}$

c. $\dfrac{1}{8}$

d. $\dfrac{1}{2}$

10.72 If 40 g of a radioactive substance naturally decays to 10 g after 16 days, what is the half-life of the substance?

a. 40 days

b. 8 days

c. 16 days

d. 10 days

10.73 A dating technique involves electron capture by the potassium-40 isotope according to the following equation: $^{40}_{19}K + ^{0}_{-1}e \rightarrow ^{40}_{18}Ar$. If the half-life is 1.2×10^9 years, how long does it take for only 10 g to remain of the original 40 g of potassium-40 in a rock sample?

a. 1.2×10^9 years

b. 0.6×10^9 years

c. 2.4×10^9 years

d. 1.8×10^9 years

10.74 If the half-life of a certain isotope is 5 years, what fraction of a sample of that isotope will remain after 15 years?

a. $\dfrac{1}{5}$

b. $\dfrac{1}{8}$

c. $\dfrac{1}{15}$

d. $\dfrac{1}{20}$

10.75 The half-life of a given element is 70 years. How long will it take 5.0 g of this element to be reduced to 1.25 g?

a. 70 years

b. 140 years

c. 210 years

d. 35 years

10.76 I^{131} has a half-life of eight days. A 100-milligram sample of this radioactive element would decay to what amount after eight days?

a. 50 milligrams

b. 40 milligrams

c. 30 milligrams

d. 20 milligrams

10.77 If a radioactive element with a half-life ($t^{1/2}$) of 100 years has 31.5 kg remaining after 400 years of decay, the amount in the original sample was close to:

a. 2,500 kg

b. 500 kg

c. 50 kg

d. 5,000 kg

10.78 Carbon-14 has a half-life of 5.73×10^3 years. If a sample contained 1 gram of C-14, the time required to decay to only 0.0625 g would be:

a. 11.46×10^3 years

b. 5.73×10^3 years

c. 22.92×10^3 years

d. none of the above

10.79 The time required for $\frac{1}{2}$ of the atoms in a sample of a radioactive element to disintegrate is known as the element's:

a. decay period

b. lifetime

c. radioactive period

d. half-life

Chemistry for Thought

10.80 One (unrealized) goal of ancient alchemists was to change one element into another (such as lead to gold). Do such changes occur naturally? Explain your reasoning.

10.81 Refer to Figure 10.2. Some isotopes used as positron emitters in PET scans are fluorine-18, oxygen-15, and nitrogen-13. What element results in each case when a positron is emitted?

10.82 Consider the concept of half-life and decide if, in principle, a radioactive isotope ever completely disappears by radioactive decay. Explain your reasoning.

10.83 Do you think Earth is more or less radioactive than it was when it first formed? Explain your answer.

10.84 Nuclear wastes typically have to be stored for at least 20 half-lives before they are considered safe. This can be a time of hundreds or thousands of years for some isotopes. With that in mind, would you consider sending such wastes into outer space a responsible solution to the nuclear waste disposal problem? Explain your answer.

10.85 Read Chemistry Around Us 10.2 and write an equation to represent the radioactive decay of radon-222. Then, write an equation to represent the decay of the daughter produced by the radon decay. The daughter decays by alpha emission. Then, write an equation for the decay of the daughter produced by this second reaction, which decays by beta emission. In each of the three reactions, assume that only one particle in addition to the daughter is produced. What element is radon converted into by this series of three decays?

10.86 Uranium-238 is the most abundant naturally occurring isotope of uranium. It undergoes radioactive decay to form other isotopes. In the first three steps of this decay, uranium-238 is converted to thorium-234, which is converted to protactinium-234, which is converted to uranium-234. Assume that only one particle in addition to the daughter is produced in each step, and use equations for the decay processes to determine the type of radiation emitted during each step.

APPENDIX A

The International System of Measurements

The International System of Units (SI units) was established in 1960 by the International Bureau of Weights and Measures. The system was established in an attempt to streamline the metric system, which included certain traditional units that had historical origins but that were not logically related to other metric units. The International System established fundamental units to represent seven basic physical quantities. These quantities and the fundamental units used to express them are given in Table A.1.

TABLE A.1 Fundamental SI Units

Physical Quantity	SI Unit	Abbreviation
Length	meter	m
Mass	kilogram	kg
Temperature	kelvin	K
Amount of substance	mole	mol
Electrical current	ampere	A
Time	second	s
Luminous intensity	candela	cd

All other SI units are derived from the seven fundamental units. Prefixes are used to indicate multiples or fractions of the fundamental units (Table 1.2). In some cases, it has been found convenient to give derived units specific names. For example, the derived unit for force is $kg \, m/s^2$, which has been given the name newton (abbreviated N) in honor of Sir Isaac Newton (1642–1726). Some examples of derived units are given in Table A.2.

TABLE A.2 Examples of Derived SI Units

Physical Quantity	Definition in Fundamental Units	Specific Name	Abbreviation
Volume	m^3 (see NOTE below)	—	—
Force	$kg \, m/s^2$	newton	N
Energy	$kg \, m^2/s^2 = N \, m$	joule	J
Power	$kg \, m^2/s^3 = J/s$	watt	W
Pressure	$kg/m \, s^2 = N/m^2$	pascal	Pa
Electrical charge	$A \, s$	coulomb	C
Electrical potential	$kg \, m^2/A \, s^3 = W/A$	volt	V
Electrical resistance	$kg \, m^2/A^2 \, s^3 = V/A$	Ohm	Ω
Frequency	$1/s$	hertz	Hz

NOTE: The liter (L), a popular volume unit of the metric system, has been redefined in terms of SI units as $1 \, dm^3$ or $1/1000 \, m^3$. It and its fractions (mL, µL, etc.) are still widely used to express volume.

Answers to
Even-Numbered End-of-Chapter Exercises

CHAPTER 1

1.2 Mass is a measurement of the amount of matter in an object. Weight is a measurement of the gravitational force acting on an object.

1.4 **a.** The distance the ball could be thrown on the moon would be greater. Although the throwing force would be the same, the smaller gravitational force pulling the ball downward would allow the ball to travel farther before hitting the surface of the moon.

b. The rolling distance would be the same in both locations because gravity does not influence rolling motion on a smooth surface.

1.6 The gravitational force is greater the closer an object is to the center of Earth. Thus, a person would weigh more at the poles (closer to the center) and less at the equator.

1.8 **a.** Physical: Both pieces of the stick are still made of the same substance.

b. Chemical: The smoke is a new substance.

c. Physical: Each smaller piece is still the substance salt.

d. Chemical: The change in color indicates a change in the substances present.

1.10 **a.** Physical: The state of a substance (solid, liquid, or gas) can be changed without changing the identity of the substance.

b. Chemical: The word *reacts* indicates a chemical change.

c. Physical: Freezing is a physical change, so the temperature at which it occurs is a physical property.

d. Chemical: Because rusting is a chemical change (rust is a different substance from the metal), the ability to rust or not to rust is a chemical property.

e. Physical: Color is a physical property because it can be observed without trying to change the substance into another substance.

1.12 **a.** Yes: The change in melting point indicates a different substance has been formed. Also, the evolved gas is a different substance.

b. No: It must be different because its melting point is different from that of succinic acid.

c. The succinic acid molecules must be larger because the atoms of the succinic acid were divided between the molecules of the new solid and the molecules of the evolved gas.

d. Heteroatomic: Because the succinic acid changed into two different substances, molecules of succinic acid must contain at least two different kinds of atoms.

1.14 Heteroatomic: Because only a single substance is formed, its molecules must contain at least one atom from each of the two reacting substances.

1.16 Heteroatomic: The molecules of the products hydrogen and oxygen contain hydrogen atoms and oxygen atoms, respectively. The two different kinds of atoms must have come from the water.

1.18 **a.** A is a compound because compounds are made up of heteroatomic molecules.

b. D is an element because elements are made up of homoatomic molecules.

c. E is a compound because it can be changed into the simpler substances G and J. The simpler substances G and J cannot be classified without conducting further tests on them to see if they can be changed into still other simpler substances.

1.20 **a.** R cannot be classified as an element or a compound. Even though none of the tests changed it into a simpler substance, all possible tests were not done on it, and some test might exist that would change it.

b. T can be classified as a compound because it was changed into simpler substances.

c. The remaining solid cannot be classified because no tests were done to see if it could be changed into simpler substances.

1.22 **a.** homogeneous

b. homogeneous

c. heterogeneous

d. heterogeneous

e. homogeneous

f. heterogeneous

g. homogeneous

1.24 **a.** pure substance

b. solution

e. solution

g. solution

1.26 In a modern society, quantities must be expressed with precision. Goods, services, time, and so on must be clearly expressed as quantities in order for trade, purchases, salaries, and the like to be carried out or determined.

1.28 The units are given in parentheses following the measuring device: clock (hours, minutes, and seconds), bathroom scale (pounds and ounces), ruler (inches or centimeters), kitchen measuring cup (cups, ounces, or milliliters), cooking thermometer

(Fahrenheit or Celsius degrees), fever thermometer (Fahrenheit or Celsius degrees), kitchen measuring spoons (tablespoons, or teaspoons), automobile speedometer (miles per hour or kilometers per hour), directional compass (degrees), tire pressure gauge (pounds per square inch), protractor (degrees).

1.30 **a.** metric

b. metric

c. nonmetric

d. metric

e. nonmetric

f. metric

1.32 **c.** meters

e. liters

1.34 **a.** 1 million microliters

b. 75 thousand watts

c. 15 million hertz

d. 200×10^{-12} meters

1.36 0.240 L, 240 cm^3

1.38 8.8 lb

1.40 **a.** About 3 in.

b. 65 K

c. About 2 kg

1.42 **a.** 0.001 kg, or 1 g

b. About 68 fl oz

c. 333 mg, or about 300 mg

1.44 97.0°F in the morning and 99.0°F at bedtime

1.46 **a.** Incorrect: No zero should precede the first number in the nonexponential number. 2.7×10^{-3} is correct.

b. Correct

c. Incorrect: The decimal is not in the standard position in the nonexponential number. 7.19×10^{-5} is correct.

d. Incorrect: No nonexponential number is given. 1×10^3 is correct.

e. Incorrect: The decimal is not in the standard position in the nonexponential number. 4.05×10^{-4} is correct.

f. Incorrect: No exponent is used, and the decimal is not in the standard position in the nonexponential number. 1.19×10^{-1} is correct.

1.48 **a.** 1.4×10^4

b. 3.65×10^2

c. 2.04×10^{-3}

d. 4.618×10^2

e. 1.00×10^{-3}

f. 9.11×10^2

1.50 1.86×10^5 mi/s; 1.100×10^9 km/h

1.52 0.000 000 000 000 000 000 000 105

1.54 **a.** 9.0×10^{-5}

b. 1.4×10^7

c. 7.6×10^{-2}

d. 1.9×10^1

e. 2.6×10^{12}

1.56 **a.** $(1.44 \times 10^2)(8.76 \times 10^{-2}) = 1.26 \times 10^1$

b. $(7.51 \times 10^2)(1.06 \times 10^2) = 7.96 \times 10^4$

c. $(4.22 \times 10^{-2})(1.19 \times 10^{-3}) = 5.02 \times 10^{-5}$

d. $(1.28000 \times 10^5)(3.16 \times 10^{-5}) = 4.04 \times 10^0$

1.58 **a.** 2.6×10^{-5}

b. 2.2×10^2

c. 6.4×10^{-4}

d. 7.25×10^{-4}

e. 3.1×10^{-2}

1.60 **a.** 1.7×10^{-1}

b. 1.0×10^{-9}

c. 2.6×10^6

d. 2.3×10^0

e. 1.5×10^7

1.62 **a.** 0.01 cm

b. 0.01 mm

c. 0.1°

d. 0.1 lb/in.^{2}

1.64 **a.** 6.00 mL

b. 37.0°C

c. 9.00 s

d. 15.5°

1.66 **a.** Measured: 5.06 lb; exact: 16 potatoes; 0.316 lb/potato

b. Measured: the individual percentages; exact: the 5 players; 71.34%/player. Note that the sum of the five percentages has four significant figures, so the calculated average must have four.

1.68 **a.** 3

b. 3

c. 4

d. 2

e. 4

f. 5

1.70 **a.** 5.2

b. 0.104

c. 0.518

d. 1.0×10^2

e. 2.52×10^{-18}

1.72 **a.** 6.2

b. 230

c. 0.589

d. 0.58

e. 27.75

f. 21.64

1.74 **a.** 4.60×10^{-3}

  b. 2.208

  c. 2.65

  d. -13

  e. 3

  f. 0.81

1.76 **a.** Black rectangle: Area = 124.8 cm^2; perimeter = 44.80 cm

  Red rectangle: Area = 48.9 cm^2; perimeter = 45.24 cm

  Green rectangle: Area = 8.11 cm^2; perimeter = 11.46 cm

  Orange rectangle: Area = 9.0 cm^2; perimeter = 27.80 cm

  b. Black rectangle: Area = 0.01248 m^2; perimeter = 0.4480 m

  Red rectangle: Area = 0.00489 m^2; perimeter = 0.4524 m

  Green rectangle: Area = 0.000811 m^2; perimeter = 0.1146 m

  Orange rectangle: Area = 0.00090 m^2; perimeter = 0.2780 m

  c. No

1.78 **a.** $\dfrac{0.015 \text{ grain}}{1 \text{ mg}}$

  b. $\dfrac{0.0338 \text{ fl oz}}{1 \text{ mL}}$

  c. $\dfrac{1 \text{ L}}{1.057 \text{ qt}}$

  d. $\dfrac{1 \text{ m}}{1.094 \text{ yd}}$

1.80 The factor is $\dfrac{1 \text{ km}}{0.621 \text{ mi}}$. The race is about 42 km long.

1.82 1.06 cups or 1 cup

1.84 39.6 lb; your baggage is not overweight.

1.86 1.31 g/L

1.88 $25.73/$467.80 \times 100 = 5.500\%$

1.90 71%

1.92 IgG = 75.09%; IgA = 16.23%; IgM = 7.58%; IgD = 1.10%; IgE = 0.008%

1.94 **a.** 0.792 g/mL

  b. 1.03 g/mL

  c. 1.98 g/L

  d. 8.90 g/cm^3

1.96 Volume = 63.0 cm^3; density = 11.4 g/cm^3

1.98 380 mL

1.100 **a.** 4.5×10^6 mm

  b. 6.0×10^3 g

  c. 9.86×10^{12} km

  d. 1.91×10^2 mg

  e. 5.0×10^{-6} mg

1.102 54 days

1.104 9.5×10^2 mg

1.106 b and d

1.108 b

1.110 d

1.112 d

1.114 b

1.116 d

1.118 c

1.120 b

1.122 d

1.124 c

1.126 d

CHAPTER 2

2.2 **a.**

  b.

  c.

  d.

2.4 **a.** H_2O

  b. H_2O_2

  c. H_2SO_4

  d. C_2H_6O

2.6 **a.** One carbon (C), four hydrogen (H)

  b. One hydrogen (H), one chlorine (Cl), four oxygen (O)

  c. One carbon (C), five hydrogen (H), one nitrogen (N)

  d. Three carbon (C), eight hydrogen (H)

2.8 **a.** The two hydrogen atoms should be denoted by a subscript: H_2S

  b. The letter L of chlorine should be lowercase: $HClO_2$

  c. The coefficient 2 should not be used to indicate the number of atoms in the molecule. Subscripts should be used instead: H_2N_4

  d. The numbers in the formula should be written as subscripts: C_2H_6

2.10 **a.** Charge = 4+; mass = 9 u

  b. Charge = 9+; mass = 19 u

  c. Charge = 20+; mass = 43 u

  d. Charge = 47+; mass = 107 u

2.12 **a.** 4

  b. 9

  c. 20

  d. 47

2.14 **a.** 19 electrons, 19 protons

  b. 48 electrons, 48 protons

  c. 51 electrons, 51 protons

2.16 **a.** 16 protons, 18 neutrons, 16 electrons

   b. 40 protons, 51 neutrons, 40 electrons

   c. 54 protons, 77 neutrons, 54 electrons

2.18 **a.** $^{28}_{14}\text{Si}$

   b. $^{40}_{18}\text{Ar}$

   c. $^{88}_{38}\text{Sr}$

2.20 **a.** Atomic number = 4; mass number = 9; $^{9}_{4}\text{Be}$

   b. Atomic number = 9; mass number = 19; $^{19}_{9}\text{F}$

   c. Atomic number = 20; mass number = 43; $^{43}_{20}\text{Ca}$

   d. Atomic number = 47; mass number = 107; $^{107}_{47}\text{Ag}$

2.22 **a.** $^{37}_{17}\text{Cl}$

   b. $^{65}_{29}\text{Cu}$

   c. $^{66}_{30}\text{Zn}$

2.24 Atomic weights rounded to nearest whole number give He = 4.0 u and C = 12 u. Three He total 12 u, which will balance one C, which also totals 12 u.

2.26 Ca, calcium

2.28 N, nitrogen

2.30 **a.** 80.06 u

   b. 92.03 u

   c. 98.06 u

   d. 28.02 u

   e. 44.03 u

2.32 Ethylene, C_2H_4

2.34 $y = 3$

2.36 **a.** 14 neutrons

   b. 27.0 u

2.38 Calculated value = 10.81 u

   Periodic table value = 10.81 u

2.40 Calculated value = 63.55 u

   Periodic table value = 63.55 u

2.42 1.90 g F

2.44 **a.** 1 mol Si atoms = 6.02×10^{23} Si atoms

   6.02×10^{23} Si atoms = 28.1 g Si

   1 mol Si atoms = 28.1 g Si

   b. 1 mol Ca atoms = 6.02×10^{23} Ca atoms

   6.02×10^{23} Ca atoms = 40.1 g Ca

   1 mol Ca atoms = 40.1 g Ca

   c. 1 mol Ar atoms = 6.02×10^{23} Ar atoms

   6.02×10^{23} Ar atoms = 40.0 g Ar

   1 mol Ar atoms = 40.0 g Ar

2.46 **a.** 35.1 g Si

   b. 6.66×10^{-23} g Ca

   c. 3.09×10^{23} Ar atoms

2.48 Molecular weight BF_3 = 67.8 u

   Molecular weight H_2S = 34.1 u

   0.68 g BF_3

2.50 **a.** 1. $2\,C_6H_6$ molecules contain 12 C atoms and 12 H atoms.

   2. $10\,C_6H_6$ molecules contain 60 C atoms and 60 H atoms.

   3. $100\,C_6H_6$ molecules contain 600 C atoms and 600 H atoms.

   4. $6.02 \times 10^{23}\,C_6H_6$ molecules contain 36.12×10^{23} C atoms and 36.12×10^{23} H atoms.

   5. 1 mol of C_6H_6 molecules contains 6 mol of C atoms and 6 mol of H atoms.

   6. 78.1 g of C_6H_6 contains 72.0 g of C and 6.1 g of H.

   b. 1. $2\,NO_2$ molecules contain 2 N atoms and 4 O atoms.

   2. $10\,NO_2$ molecules contain 10 N atoms and 20 O atoms.

   3. $100\,NO_2$ molecules contain 100 N atoms and 200 O atoms.

   4. $6.02 \times 10^{23}\,NO_2$ molecules contain 6.02×10^{23} N atoms and 12.04×10^{23} O atoms.

   5. 1 mol of NO_2 molecules contains 1 mol of N atoms and 2 mol of O atoms.

   6. 46.0 g of NO_2 contains 14.0 g of N and 32.0 g of O.

   c. 1. 2 HCl molecules contain 2 H atoms and 2 Cl atoms.

   2. 10 HCl molecules contain 10 H atoms and 10 Cl atoms.

   3. 100 HCl molecules contain 100 H atoms and 100 Cl atoms.

   4. 6.02×10^{23} HCl molecules contain 6.02×10^{23} H atoms and 6.02×10^{23} Cl atoms.

   5. 1 mol of HCl molecules contains 1 mol of H atoms and 1 mol of Cl atoms.

   6. 36.5 g of HCl contains 1.01 g of H and 35.5 g of Cl.

2.52 **a.** 4.5 mol H atoms

   b. 6.00×10^{23} O atoms

   c. 97.3% Cl

2.54 34.6 g C_2H_6O

2.56 In CH_4, mass % H = 25.3%.

   In C_2H_6, mass % H = 20.1%.

2.58 Statements 4–6 are:

   4. $6.02 \times 10^{23}\,H_3PO_4$ molecules contain 18.06×10^{23} H atoms, 6.02×10^{23} P atoms, and 24.08×10^{23} O atoms.

   5. 1 mol of H_3PO_4 molecules contains 3 mol of H atoms, 1 mol of P atoms, and 4 mol of O atoms.

   6. 98.0 g of H_3PO_4 contains 3.03 g of H, 31.0 g of P, and 64.0 g of O.

   a. 1.45 g of H

   b. 5.00 mol of O atoms

   c. 8.42×10^{21} P atoms

2.60 Magnetite = 72.36% Fe; hematite = 69.94% Fe. Magnetite has the higher % Fe by mass.

2.62 U-238 would have a greater density because each U-238 nucleus would contain 3 more neutrons than a U-235 nucleus and thus have a greater mass. However, the size (volume) of each atom (measured from the nucleus to the outer electrons) would be the same, so one cm^3 of each isotope would contain the same number of atoms.

2.64 2.32×10^{-23} g

2.66 The density would increase tremendously because many more of the smaller atoms (measured from the nucleus to the outer electrons) would fit into a cm³. Each of the smaller atoms would have the same mass as the larger atoms, so the density (mass/volume) would be much greater.

2.68 b

2.70 a

2.72 b

2.74 b

2.76 a

2.78 d

2.80 b

2.82 c

2.84 a

2.86 b

CHAPTER 3

3.2   **a.** Group VA (15), period 5

    b. Group IIB (12), period 5

    c. Group IIIA (13), period 4

    d. Group IVB (4), period 5

3.4   **a.** Kr, krypton

    b. Sn, tin

    c. Mo, molybdenum

    d. Nd, neodymium

3.6   **a.** 4

    b. 10

    c. 18

3.8   **a.** period

    b. period

    c. group

    d. group

3.10 protons

3.12 **a.** 2

    b. 6

    c. 8

3.14 Four orbitals; one $2s$, and three $2p$.

3.16 Seven orbitals in the subshell can contain a maximum of 14 electrons.

3.18 **a.** 8

    b. 5

    c. 4

    d. 4

3.20 Cesium, Cs, has chemical properties most like sodium, Na. Both elements have one valence-shell electron.

3.22 An ore deposit that contained copper might also be expected to contain gold and silver because all three elements belong to group IB (11) of the periodic table and thus have similar chemical properties.

3.24 **a.** $1s^2 2s^2 2p^6 3s^2 3p^6 4s^2 3d^{10} 4p^6 5s^1$; one electron is unpaired.

    b. $1s^2 2s^2 2p^6 3s^2 3p^2$; two electrons are unpaired.

    c. $1s^2 2s^2 2p^6 3s^2 3p^6 4s^2 3d^2$; two electrons are unpaired.

    d. $1s^2 2s^2 2p^6 3s^2 3p^6$; no electrons are unpaired.

3.26 **a.** $1s^2 2s^2 2p^6 3s^2 3p^6 4s^1$; 7 s electrons.

    b. $1s^2 2s^2 2p^6 3s^2 3p^1$; 1 unpaired electron.

    c. $1s^2 2s^2 2p^6 3s^2$; 4 subshells are filled.

3.28 **a.** C, carbon

    b. Na, sodium

    c. Si, silicon, and S, sulfur

    d. Nb, niobium

    e. V, vanadium, and Co, cobalt

3.30 **a.** $[Ar]4s^2 3d^{10} 4p^3$

    b. $[Ar]4s^2 3d^5$

    c. $[Ne]3s^2 3p^2$

    d. $[Kr]5s^2 4d^{10} 5p^5$

3.32 Na: $[Ne]3s^1$

    Mg: $[Ne]3s^2$

    Al: $[Ne]3s^2 3p^1$

    Si: $[Ne]3s^2 3p^2$

    P: $[Ne]3s^2 3p^3$

    S: $[Ne]3s^2 3p^4$

    Cl: $[Ne]3s^2 3p^5$

    Ar: $[Ne]3s^2 3p^6$

3.34 **a.** p area

    b. d area

    c. f area

    d. s area

3.36 **a.** transition

    b. representative

    c. inner-transition

    d. noble gas

    e. representative

3.38 **a.** nonmetal

    b. metal

    c. metalloid

    d. metalloid

    e. metal

3.40 **a.** K

    b. Bi

    c. Sr

    d. Sn

3.42 **a.** Ga

    b. Sb

c. C

d. Te

3.44 a. Si

b. Ba

c. Br

d. Ca

3.46 The chemical properties would be the same because chemical properties depend on the number of valence electrons in the atom. The number of valence electrons in isotopes of the same element is identical.

3.48 8.0 mg

3.50 The element has an atomic weight of 40.0, and so is calcium (Ca). It is a representative element and will conduct electricity because it is a metal.

3.52 d

3.54 d

3.56 a

3.58 b

3.60 b

3.62 b

3.64 d

3.66 d

3.68 a

3.70 b

CHAPTER 4

4.2 a. $\ddot{\text{:}}\overset{\cdot}{\text{I}}\cdot$

b. $\text{Sr}\cdot$

c. $\cdot\dot{\text{Sn}}\cdot$

d. $\overset{\cdot\cdot}{\text{:}\dot{\text{S}}}\cdot$

4.4 a. $[\text{Kr}]\,5s^2 4d^{10} 5p^2$

b. $[\text{Xe}]\,6s^1$

c. $[\text{Kr}]\,5s^2 4d^{10} 5p^1$

d. $[\text{Kr}]\,5s^2$

4.6 a. $\cdot\dot{\text{Sn}}\cdot$

b. $\text{Cs}\cdot$

c. $\dot{\text{In}}\cdot$

d. $\dot{\text{Sr}}\cdot$

4.8 a. $\dot{\text{E}}\cdot$

b. $\cdot\dot{\text{E}}\text{:}$

4.10 a. Add 4 or lose 14

b. Add 31 or lose 1

c. Add 5 or lose 13

d. Add 16 or lose 2

4.12 a. Lose 1: $\text{Cs} \rightarrow \text{Cs}^+ + e^-$

b. Add 2: $\text{O} + 2e^- \rightarrow \text{O}^{2-}$

c. Add 3: $\text{N} + 3e^- \rightarrow \text{N}^{3-}$

d. Add 1: $\text{I} + e^- \rightarrow \text{I}^-$

4.14 a. P^{3-}

b. Sr^{2+}

c. I^-

4.16 a. sulfur, S

b. aluminum, Al

c. sodium, Na

d. chlorine, Cl

4.18 a. argon, Ar

b. argon, Ar

c. krypton, Kr

d. xenon, Xe

4.20 a. $\text{Ca} \rightarrow \text{Ca}^{2+} + 2e^-; \text{Cl} + e^- \rightarrow \text{Cl}^-$
Formula is CaCl_2.

b. $\text{Li} \rightarrow \text{Li}^+ + e^-; \text{Br} + e^- \rightarrow \text{Br}^-$
Formula is LiBr.

c. $\text{Mg} \rightarrow \text{Mg}^{2+} + 2e^-; \text{S} + 2e^- \rightarrow \text{S}^{2-}$
Formula is MgS.

4.22 a. BaSe

b. Ba_3P_2

c. BaI_2

d. Ba_3As_2

4.24 a. binary

b. binary

c. not binary

d. not binary

e. binary

4.26 a. lithium ion

b. magnesium ion

c. barium ion

d. cesium ion

4.28 a. bromide ion

b. oxide ion

c. phosphide ion

d. telluride ion

4.30 a. strontium sulfide

b. calcium fluoride

c. barium chloride

d. lithium oxide

e. magnesium oxide

4.32 a. lead (II) oxide and lead (IV) oxide

b. copper (I) chloride and copper (II) chloride

c. gold (I) sulfide and gold (III) sulfide

d. cobalt (II) oxide and cobalt (III) oxide

4.34 a. plumbous oxide and plumbic oxide

b. cuprous chloride and cupric chloride

c. aurous sulfide and auric sulfide

d. cobaltous oxide and cobaltic oxide

4.36   **a.** PbO_2

      b. $CoCl_2$

      c. CuS

      d. NiN

      e. Pt_3P_2

4.38   **a.** 102.9 u

      b. 78.1 u

      c. 159.2 u

      d. 34.8 u

4.40   **a.** Na^+ and Br^-

      b. Ca^{2+} and F^-

      c. Cu^+ and S^{2-}

      d. Li^+ and N^{3-}

4.42   **a.** 22.99 g of Na^+ and 79.90 g of Br^-

      b. 40.08 g of Ca^{2+} and 38.00 g of F^-

      c. 127.10 g of Cu^+ and 32.06 g of S^{2-}

      d. 20.82 g of Li^+ and 14.01 g of N^{3-}

4.44   **a.** 6.02×10^{23} Na^+ ions and 6.02×10^{23} Br^- ions

      b. 6.02×10^{23} Ca^{2+} ions and 12.04×10^{23} F^- ions

      c. 12.04×10^{23} Cu^+ ions and 6.02×10^{23} S^{2-} ions

      d. 18.06×10^{23} Li^+ ions and 6.02×10^{23} N^{3-} ions

4.46

4.48

      a.

      b. $:\ddot{O}::C::\ddot{O}:$ or $O=C=O$

      c. $H:\ddot{Se}:H$ or $H-Se-H$

      d.

4.50   **a.**

      b.

      c.

4.52   **a.** $H:\ddot{S}:H$ Bent

      b. $:\ddot{Cl}:\ddot{P}:\ddot{Cl}:$ Triangular pyramid with P at the top

      c. $:\ddot{F}:\ddot{O}:\ddot{F}:$ Bent

      d.   Tetrahedral with Sn in center

4.54   **a.** $\left[:\ddot{O}:\ddot{N}::\ddot{O}:\right]^-$ Bent

      b.   Triangular pyramid with Cl at the top

      c.   Flat triangle with C in the center

      d.   Triangular pyramid with O at the top

4.56   **a.**

      b.

      c.   Nonpolarized

4.58   **a.** polar covalent

      b. polar covalent

      c. nonpolar covalent

      d. polar covalent

      e. ionic

4.60   **a.** polar

      b. polar

      c. nonpolar

4.62   **a.** polar covalent

      b. polar covalent

      c. nonpolar covalent

4.64   **a.** Polar

      b. Polar

      c. Nonpolar

4.66 **a.** phosphorus trichloride

b. dinitrogen pentoxide

c. carbon tetrachloride

d. boron trifluoride

e. carbon disulfide

4.68 **a.** ClO_2

b. N_2O

c. SO_2

d. CCl_4

4.70 **a.** $Ca(NO_2)_2$, calcium nitrite

b. $Mg(ClO)_2$, magnesium hypochlorite

c. $Cs_2Cr_2O_7$, cesium dichromate

d. K_2SO_3, potassium sulfite

4.72 **a.** $Ba(OH)_2$

b. $MgSO_3$

c. $CaCO_3$

d. $(NH_4)_2SO_4$

e. $LiHCO_3$

4.74 **a.** M_2SO_3

b. $MC_2H_3O_2$

c. MCr_2O_7

d. MPO_4

e. $M(NO_3)_3$

4.76 The stronger the attractive forces between molecules, the higher the melting and boiling points. Thus, we can conclude that the forces attracting dimethyl ether molecules to one another are weaker than the forces attracting ethyl alcohol molecules to one another. Because both compounds are covalently bonded, it is likely that the attractive forces are polar forces, and we might conclude that ethyl alcohol is more polar than dimethyl ether.

4.78 The predominant forces that exist between molecules of noble gases are dispersion forces because noble gas molecules are nonpolar. Dispersion forces increase with increasing mass of the particles involved, so the boiling points of the noble gases would increase in the order of He, Ne, Ar, Kr, Xe, and Rn.

4.80 The relatively high melting point indicates that it is unlikely that weak dispersion forces are the predominant forces present. Dipolar forces are also quite weak and therefore are also unlikely to be the forces present.

4.82 NH_3

$$H-N-H\cdots N-H$$

with H above and below each N (structure showing NH_3 hydrogen bonding)

4.84 8.06 g

4.86 N_2O, covalent, dinitrogen monoxide

4.88 a

4.90 a

4.92 c

4.94 b

4.96 b

4.98 c

4.100 c

4.102 a

4.104 a

4.106 b

CHAPTER 5

5.2

	Reactants	Products
a.	H_2, Cl_2	HCl
b.	$KClO_3$	KCl, O_2
c.	magnesium oxide, carbon	magnesium, carbon monoxide
d.	ethane, oxygen	carbon dioxide, water

5.4 **a.** Not consistent: The number of O atoms is not the same on both sides.

b. Consistent

c. Not consistent: The masses of reactants and products are not equal.

d. Consistent

5.6

	Left side	Right side	Classification
a.	1Ag, 1Cu, 2N, 6O	1Ag, 1Cu, 1N, 3O	Unbalanced
b.	4N, 8O	4N, 8O	Balanced
c.	1Mg, 2O	2Mg, 2O	Unbalanced
d.	4H, 1S, 6O, 1Ca	4H, 1S, 6O, 1Ca	Balanced

5.8 **a.** $2KClO_3(s) \rightarrow 2KCl(s) + 3O_2(g)$

b. $2C_2H_6(g) + 7O_2(g) \rightarrow 4CO_2(g) + 6H_2O(\ell)$

c. $2N_2(g) + 5O_2(g) \rightarrow 2N_2O_5(g)$

d. $MgCl_2(s) + H_2O(g) \rightarrow MgO(s) + 2HCl(g)$

e. $CaH_2(s) + 2H_2O(\ell) \rightarrow Ca(OH)_2(s) + 2H_2(g)$

f. $2Al(s) + Fe_2O_3(s) \rightarrow Al_2O_3(s) + 2Fe(s)$

g. $2Al(s) + 3Br_2(g) \rightarrow 2AlBr_3(s)$

h. $Hg_2(NO_3)_2(aq) + 2NaCl(aq) \rightarrow$
$$Hg_2Cl_2(s) + 2NaNO_3(aq)$$

5.10 **a.** $+1$

b. $+3$

c. 0

d. $+1$

e. $+3$

f. $+5$

5.12 **a.** Cr $(+6)$

b. S $(+2)$

c. N $(+5)$

d. P $(+5)$

e. Cl $(+7)$

f. Cl $(+3)$

5.14 **a.** oxidized

b. neither

c. reduced

d. oxidized

e. oxidized

5.16 **a.** $2Cu(s) + O_2(g) \rightarrow 2CuO(s)$
  $\quad\quad 0 \quad\quad\ 0 \quad\quad\quad +2\ -2$

  Oxidizing agent: O_2

  Reducing agent: Cu

b. $Cl_2(g) + 2KI(aq) \rightarrow 2KCl(aq) + I_2(aq)$
  $\quad 0 \quad\quad +1\ -1 \quad\quad +1\ -1 \quad\quad\ 0$

  Oxidizing agent: Cl_2

  Reducing agent: KI (I)

c. $3MnO_2(s) + 4Al(s) \rightarrow 2Al_2O_3(s) + 3Mn(s)$
  $\quad\ +4\ -2 \quad\ 0 \quad\quad +3\ -2 \quad\quad 0$

  Oxidizing agent: MnO_2 (Mn)

  Reducing agent: Al

d. $2H^+(aq) + 3SO_3^{2-}(aq) + 2NO_3^-(aq) \rightarrow$
  $\quad +1 \quad\quad\ +4\ -2 \quad\quad$
  $\quad\quad\quad\quad\quad 2NO(g) + H_2O(\ell) + 3SO_4^{2-}(aq)$
  $\quad\quad\quad\quad\quad +2\ -2 \quad +1\ -2 \quad\ +6\ -2$

  Oxidizing agent: NO_3^- (N)

  Reducing agent: SO_3^{2-} (S)

e. $Mg(s) + 2HCl(aq) \rightarrow MgCl_2(aq) + H_2(g)$
  $\quad 0 \quad\quad +1\ -1 \quad\quad +2\ -1 \quad\quad 0$

  Oxidizing agent: HCl (H^+)

  Reducing agent: Mg

f. $4NO_2(g) + O_2(g) \rightarrow 2N_2O_5(g)$
  $\quad +4\ -2 \quad\ 0 \quad\quad +5\ -2$

  Oxidizing agent: O_2

  Reducing agent: NO_2 (N)

5.18 Oxidizing agent: NaOH (H)

  Reducing agent: Al

5.20 **a.** Nonredox, decomposition

b. Redox, single replacement

c. Nonredox, double replacement

d. Nonredox, combination

e. Redox, combination

f. Redox, combination

5.22 Nonredox, decomposition

5.24 Redox

5.26 Redox

5.28 **a.** Li^+, NO_3^-

b. Na^+, HPO_4^{2-}

c. Ca^{2+}, ClO_3^-

d. K^+, OH^-

e. Mg^{2+}, Br^-

f. NH_4^+, SO_4^{2-}

5.30 **a.** Total ionic: $SO_2(aq) + H_2O(\ell) \rightarrow 2H^+(aq) + SO_3^{2-}(aq)$

  Spectator ions: none

  Net ionic: $SO_2(aq) + H_2O(\ell) \rightarrow 2H^+(aq) + SO_3^{2-}(aq)$

b. Total ionic: $Cu^{2+}(aq) + SO_4^{2-}(aq) + Zn(s) \rightarrow$
  $\quad\quad\quad\quad\quad\quad Cu(s) + Zn^{2+}(aq) + SO_4^{2-}(aq)$

  Spectator ions: SO_4^{2-}

  Net ionic: $Cu^{2+}(aq) + Zn(s) \rightarrow Cu(s) + Zn^{2+}(aq)$

c. Total ionic:
  $2K^+(aq) + 2Br^-(aq) + 4H^+(aq) + 2SO_4^{2-}(aq) \rightarrow$
  $Br_2(aq) + SO_2(aq) + 2K^+(aq) + SO_4^{2-}(aq) + 2H_2O(\ell)$

  Spectator ions: K^+, one SO_4^{2-}

  Net ionic: $2Br^-(aq) + 4H^+(aq) + SO_4^{2-}(aq) \rightarrow$
  $\quad\quad\quad\quad\quad Br_2(aq) + SO_2(aq) + 2H_2O(\ell)$

d. Total ionic:
  $Ag^+(aq) + NO_3^-(aq) + Na^+(aq) + OH^-(aq) \rightarrow$
  $\quad\quad\quad\quad\quad AgOH(s) + Na^+(aq) + NO_3^-(aq)$

  Spectator ions: Na^+, NO_3^-

  Net ionic: $Ag^+(aq) + OH^-(aq) \rightarrow AgOH(s)$

e. Total ionic: $BaCO_3(s) + 2H^+(aq) + 2NO_3^-(aq) \rightarrow$
  $\quad\quad\quad Ba^{2+}(aq) + 2NO_3^-(aq) + CO_2(g) + H_2O(\ell)$

  Spectator ions: NO_3^-

  Net ionic:
  $BaCO_3(s) + 2H^+(aq) \rightarrow Ba^{2+}(aq) + CO_2(g) + H_2O(\ell)$

f. Total ionic:
  $N_2O_5(aq) + H_2O(\ell) \rightarrow 2H^+(aq) + 2NO_3^-(aq)$

  Spectator ions: none

  Net ionic: $N_2O_5(aq) + H_2O(\ell) \rightarrow 2H^+(aq) + 2NO_3^-(aq)$

5.32 **a.** $H^+(aq) + NO_3^-(aq) + K^+(aq) + OH^-(aq) \rightarrow$
  $\quad\quad\quad\quad\quad K^+(aq) + NO_3^-(aq) + H_2O(\ell)$

  Spectator ions: K^+, NO_3^-

  Net ionic: $H^+(aq) + OH^-(aq) \rightarrow H_2O(\ell)$

b. $3H^+(aq) + PO_4^{3-}(aq) + 3NH_4^+(aq) + 3OH^-(aq) \rightarrow$
  $\quad\quad\quad\quad\quad 3NH_4^+(aq) + PO_4^{3-}(aq) + 3H_2O(\ell)$

  Spectator ions: NH_4^+, PO_4^{3-}

  Net ionic: $3H^+(aq) + 3OH^-(aq) \rightarrow 3H_2O(\ell)$

c. $H^+(aq) + I^-(aq) + Na^+(aq) + OH^-(aq) \rightarrow$
  $\quad\quad\quad\quad\quad Na^+(aq) + I^-(aq) + H_2O(\ell)$

  Spectator ions: Na^+, I^-

  Net ionic: $H^+(aq) + OH^-(aq) \rightarrow H_2O(\ell)$
  In every net ionic reaction, H^+ reacts with OH^- to form H_2O

5.34 Exothermic: Heat is a product of the reaction.

5.36 The insulation around the ice slows its rate of melting by preventing it from absorbing heat. However, this prevents the ice from performing its cooling function in the cooler. For the cooler to work properly, heat must move from the warmer food to the cooler ice, where it is absorbed as the ice melts.

5.38 **a.** $S(s) + O_2(g) \rightarrow SO_2(g)$

    2. $1 \text{ mol S} + 1 \text{ mol O}_2 \rightarrow 1 \text{ mol SO}_2$

    3. 6.02×10^{23} S atoms $+ 6.02 \times 10^{23}$ O$_2$ molecules $\rightarrow$
6.02×10^{23} SO$_2$ molecules

    4. $32.06 \text{ g S} + 32.00 \text{ g O}_2 \rightarrow 64.06 \text{ g SO}_2$

  b. $Sr(s) + 2H_2O(\ell) \rightarrow Sr(OH)_2(s) + H_2(g)$

    2. $1 \text{ mol Sr} + 2 \text{ mol H}_2O \rightarrow 1 \text{ mol Sr(OH)}_2 + 1 \text{ mol H}_2$

    3. 6.02×10^{23} Sr atoms $+ 12.04 \times 10^{23}$ H$_2$O
molecules $\rightarrow 6.02 \times 10^{23}$ Sr(OH)$_2$ molecules
$+ 6.02 \times 10^{23}$ H$_2$ molecules

    4. $87.62 \text{ g Sr} + 36.03 \text{ g H}_2O \rightarrow 121.64 \text{ g Sr(OH)}_2 +$
2.02 g H_2

  c. $2H_2S(g) + 3O_2(g) \rightarrow 2H_2O(g) + 2SO_2(g)$

    2. $2 \text{ mol H}_2S + 3 \text{ mol O}_2 \rightarrow 2 \text{ mol H}_2O + 2 \text{ mol SO2}$

    3. 12.04×10^{23} H$_2$S molecules $+ 18.06 \times 10^{23}$
O$_2$ molecules $\rightarrow 12.04 \times 10^{23}$ H$_2$O molecules $+$
12.04×10^{23} SO$_2$ molecules

    4. $68.15 \text{ g H}_2S + 96.00 \text{ g O}_2 \rightarrow$
$36.03 \text{ g H}_2O + 128.12 \text{ g SO}_2$

  d. $4NH_3(g) + 5O_2(g) \rightarrow 4NO(g) + 6H_2O(g)$

    2. $4 \text{ mol NH}_3 + 5 \text{ mol O}_2 \rightarrow 4 \text{ mol NO} + 6 \text{ mol H}_2O$

    3. 24.08×10^{23} NH$_3$ molecules $+ 30.10 \times 10^{23}$ O$_2$
molecules $\rightarrow 24.08 \times 10^{23}$ NO molecules $+$
36.12×10^{23} H$_2$O molecules

    4. $68.14 \text{ g NH}_3 + 160.00 \text{ g O}_2 \rightarrow$
$120.04 \text{ g NO} + 108.10 \text{ g H}_2O$

  e. $CaO(s) + 3C(s) \rightarrow CaC_2(s) + CO(g)$

    2. $1 \text{ mol CaO} + 3 \text{ mol C} \rightarrow 1 \text{ mol CaC}_2 + 1 \text{ mol CO}$

    3. 6.02×10^{23} CaO molecules $+ 18.06 \times 10^{23}$ C
atoms $\rightarrow 6.02 \times 10^{23}CaC_2$ molecules $+$
6.02×10^{23} CO molecules

    4. $56.08 \text{ g CaO} + 36.03 \text{ g C} \rightarrow$
$64.10 \text{ g CaC}_2 + 28.01 \text{ g CO}$

5.40 $2SO_2(g) + O_2(g) \rightarrow 2SO_3(g)$

    2. $2 \text{ mol SO}_2 + 1 \text{ mol O}_2 \rightarrow 2 \text{ mol SO}_3$

    3. 12.0×10^{23} SO$_2$ molecules $+ 6.02 \times 10^{23}$ O$_2$
molecules $\rightarrow 12.0 \times 10^{23}$ SO$_3$ molecules

    4. $128 \text{ g SO}_2 + 32.0 \text{ g O}_2 \rightarrow 160 \text{ g SO}_3$

Factors:

$$\frac{12.0 \times 10^{23} \text{ SO}_2 \text{ molecules}}{6.02 \times 10^{23} \text{ O}_2 \text{ molecules}}; \quad \frac{12.0 \times 10^{23} \text{ SO}_3 \text{ molecules}}{12.0 \times 10^{23} \text{ SO}_2 \text{ molecules}};$$

$$\frac{32.0 \text{ g O}_2}{128 \text{ g SO}_2}; \quad \frac{128 \text{ g SO}_2}{160 \text{ g SO}_3}; \quad \frac{32.0 \text{ g O}_2}{160 \text{ g SO}_3}; \quad \frac{128 \text{ g SO}_2}{32.0 \text{ g O}_2};$$

$$\frac{2 \text{ mol SO}_2}{1 \text{ mol O}_2}; \quad \frac{2 \text{ mol SO}_3}{2 \text{ mol SO}_2}; \quad \frac{1 \text{ mol O}_2}{2 \text{ mol SO}_3}; \quad \frac{1 \text{ mol O}_2}{2 \text{ mol SO}_2}; \quad \frac{2 \text{ mol SO}_3}{32.0 \text{ g O}_2};$$

$$\frac{128 \text{ g SO}_2}{2 \text{ mol SO}_2}; \quad \frac{2 \text{ mol SO}_2}{32.0 \text{ g O}_2}; \quad \frac{2 \text{ mol SO}_3}{1 \text{ mol O}_2}; \quad \frac{160 \text{ g SO}_3}{32.0 \text{ g O}_2}; \quad \frac{32.0 \text{ g O}_2}{1 \text{ mol O}_2}$$

This list does not include all possible factors.

5.42 892 g

5.44 445 g

5.46 495 g

5.48 1.02×10^3 g

5.50 256 g

5.52 **a.** O$_2$ is the limiting reactant.

  b. 71.9 g NO$_2$

5.54 144 g

5.56 72.96%

5.58 72.5%

5.60 89.0%

5.62 $BaCl_2(aq) + Na_2SO_4(aq) \rightarrow 2NaCl(aq) + BaSO_4(s)$

  Net ionic: $Ba^{2+}(aq) + SO_4^{2-}(aq) \rightarrow BaSO_4(s)$

5.64 7.502 g

5.66 c

5.68 c

5.70 b

5.72 b

5.74 c

5.76 a

5.78 b

5.80 a

5.82 d

5.84 b

5.86 c

5.88 d

CHAPTER 6

6.2 **a.** 121 mL

  b. 158 mL

  c. 67.9 mL

6.4 2 mL

6.6 **a.** The density will decrease because density is equal to mass/volume, and the volume of carbon dioxide will increase on heating while the mass remains unchanged.

  b. 1.67 g/L

6.8 As the ball travels upward and slows, more and more of its kinetic energy is converted into potential energy. At the maximum height, when the ball stops before starting to fall, all the energy is in the form of potential energy. As the ball falls faster and faster, more and more of the potential energy is converted back into kinetic energy. If there were no energy losses from air resistance, the ball would gain back as much kinetic energy as it had when it was initially thrown.

6.10 Each type of molecule has the same kinetic energy of 3.18×10^{10} u cm^2/s^2.

6.12 **a.** Particles (molecules) in a liquid move freely and therefore assume the container shape, but the cohesive forces between the particles prevent them from separating completely from each other and filling the container.

 b. There is very little space between the particles (molecules) of liquids and solids. As a result, increased pressure cannot push the particles closer together and will have little influence on their volumes.

 c. Gas molecules move around essentially unrestricted, and on average, equal numbers collide with each of the walls of a container in a given amount of time. The collisions cause the pressure on the walls, so the pressure on all the walls is the same.

6.14 Heat must be added to a substance to convert it from the solid to the liquid state, and then from the liquid to the gaseous state. If the added heat causes no temperature difference between the two states, as when a solid melts to a liquid at the melting point, then the added heat increases the potential energy of the particles of the substance. Thus, at the melting point, the particles of both solid and liquid forms of a substance will have the same kinetic energy because they are at the same temperature. However, particles of the liquid will have more potential energy, corresponding to the energy that had to be added to change the solid to a liquid. Similar arguments apply for a comparison of the liquid and gaseous states of a substance.

6.16 **a.** gaseous

 b. solid

 c. gaseous

 d. gaseous

6.18 **a.** 0.957 atm

 b. 727 torr

 c. 14.1 psi

 d. 0.966 bar

6.20 **a.** 14.3 atm

 b. 14.4 bar

 c. 1.09×10^4 mmHg

 d. 427 in. Hg

6.22 **a.** 337 K

 b. $-259°C$

 c. 4 K

6.24 Column A: $P_f = 1.50$ atm; Column B: $V_f = 4.49$ L; Column C: $T_f = 221$ K

6.26 147 mL

6.28 2.73 atm

6.30 28.1 L

6.32 684 mL

6.34 4.6 L

6.36 2.5 L

6.38 100 atm

6.40 1.4×10^4 ft^3

6.42 842 mL

6.44 5.00 g/L

6.46 **a.** 0.0130 mol

 b. 7.65 atm

 c. 8.35 L

6.48 3.19 atm

6.50 26.3 g

6.52 2.7 mol

6.54 46.0 g/mol

6.56 44.1 g/mol; the gas is most likely CO_2

6.58 370 torr

6.60 O_2 mass/H_2 mass = 16 according to Graham's law. From the periodic table, O_2 mass/H_2 mass = 32 u/2 u = 16. The agreement is very good.

6.62 The helium-filled balloon would appear to go flat first because helium molecules diffuse through the rubber balloon faster than do nitrogen molecules. The rates of diffusion according to Graham's law would be He rate/N_2 rate = $\sqrt{28.0 \text{ u}/4.00 \text{ u}}$ = 2.6. Thus, He would be expected to escape through the balloon 2.6 times as fast as nitrogen.

6.64 **a.** exothermic

 b. exothermic

 c. endothermic

6.66 A change in state refers to a process in which a substance changes from one of the three states of matter—solid, liquid, or gas—to another of the three states.

6.68 Liquid methylene chloride evaporates very rapidly, and when it evaporates, it absorbs heat. Thus, the liquid acted to cool and temporarily anesthetize any tissue on which it was sprayed.

6.70 The boiling points of the two liquids are different. Thus, a measurement of the temperature of each boiling liquid would differentiate between them. The ethylene glycol would have the higher boiling point of the two.

6.72 The temperature of boiling water is influenced by the atmospheric pressure on the water. On top of Mount Everest the atmospheric pressure is quite low, so water boils at a temperature significantly lower than the normal boiling point of 100°C. The temperature of the burning campfire would not be influenced by the altitude. Thus, the potato would cook faster if it were put into the campfire.

6.74 Since sand does not sublime, heat the mixture and lead the hot iodine vapor out of the hot chamber and to a cool surface. The pure iodine vapor will condense to a solid on the cool surface, and can be scraped off and collected.

6.76 **a.** 3.6×10^2 cal

   b. 7×10^3 cal

   c. 2.4×10^3 cal

6.78 **a.** 4.07×10^7 cal

   b. 7.07×10^7 cal

   c. 5.71×10^7 cal

6.80 7.72×10^4 cal

6.82 1.78×10^{-3} g/mL

6.84 1.16 L The reaction is a redox combination reaction.

6.86 The air of the atmosphere contains about 21 volume percent oxygen at any altitude. However, the pressure of the atmosphere at high altitudes is less than the pressure at lower altitudes. If the air at a higher altitude is at the same temperature as at a lower altitude, and since $P = nRT/V$, the same volume (V) of a lung full of air at a higher altitude and lower pressure (P) contains fewer moles of gas (n), and so fewer moles of oxygen. So, from the point of view of the number of molecules of oxygen, the higher altitude air has less oxygen, but from a percentage point of view it still contains 21% oxygen.

6.88 a

6.90 c

6.92 d

6.94 d

6.96 a

6.98 c

6.100 a

6.102 c

6.104 a

6.106 a

6.108 d

6.110 a

6.112 d

6.114 d

6.116 b

CHAPTER 7

7.2 **a.** Solvent: water; solute: sodium hypochlorite

   b. Solvent: isopropyl alcohol; solute: water

   c. Solvent: water; solute: hydrogen peroxide

   d. Solvent: SD alcohol; solutes: water, glycerin, fragrance, menthol, benzophenone, coloring

7.4 **a.** Not a solution because it is not homogeneous. It is cloudy rather than clear.

   b. Solution; solvent: water; solutes: all dissolved components.

   c. Not a solution because it is not homogeneous. Bits of pulp can be seen.

   d. Solution if prepared carefully with a tea bag or good strainer; solvent: water; solutes: all dissolved components.

   e. Not a solution because it is not homogeneous.

7.6 **a.** The oil and vinegar are immiscible.

   b. The alcohol is soluble in the water.

   c. The tar is soluble in the chloroform.

7.8 **a.** unsaturated

   b. supersaturated

   c. saturated

7.10 The solution could be carefully cooled to a temperature lower than room temperature. Another method would be to allow some of the solvent to evaporate.

7.12 **a.** soluble

   b. insoluble

   c. slightly soluble

   d. very soluble

   e. slightly soluble

7.14 Add pure water to the mixture. The $CaCl_2$ is soluble and will dissolve in the water. The liquid that contains the dissolved $CaCl_2$ can then be poured off carefully, leaving behind the insoluble solid $CaCO_3$.

7.16 **a.** Soluble in benzene

   b. Soluble in benzene

   c. Soluble in water

   d. Soluble in benzene

7.18 The structure indicates that Freon-114 is a nonpolar material. Grease is also nonpolar, as indicated by its insolubility in water. Thus, grease would dissolve readily in Freon-114.

7.20 **a.** 0.364 M

   b. 0.860 M

   c. 1.75 M

7.22 **a.** 0.500 M

   b. 0.200 M

   c. 0.26 M

7.24 **a.** 0.376 mol

   b. 0.0750 mol

   c. 571 mL

7.26 **a.** 0.584 g

   b. 0.157 L

   c. 1.18×10^3 mL

7.28 **a.** 5.0%

   b. 5.0%

   c. 5.0%

7.30 **a.** 6.98%

   b. 5.08%

   c. 13.0%

   d. 51.3%

7.32 **a.** 12.5%

   b. 13.4%

   c. 6.24%

7.34 **a.** 8.00%

    b. 8.00%

    c. 2.3%

    d. 34.2%

7.36 0.025 L or 25 mL

7.38 **a.** 5.00%

    b. 5.00%

    c. 8.77%

7.40 About 40%

7.42 **a.** Put 0.0300 mol (4.26 g) of solid Na_2SO_4 into a 200 mL volumetric flask and add enough pure water to fill the flask to the mark after the solid dissolves.

    b. Put 0.0625 mol (11.8 g) of solid $Zn(NO_3)_2$ into a 250 mL volumetric flask and add enough pure water to fill the flask to the mark after the solid dissolves.

    c. Mix together 3.38 g of solid NaCl and 147 g (147 mL) of pure water. Allow the NaCl to dissolve.

    d. Put 0.94 g of solid KCl into a 125 mL volumetric flask and add enough pure water to fill the flask to the mark after the solid dissolves.

7.44 %(w/w) = 25.0; %(w/v) = 23.8

7.46 **a.** 32.3 g

    b. 0.700 mol

    c. 31.3 mL

    d. 2.10 g

7.48 **a.** Add 1.67 L of 18.0 M H_2SO_4 solution to enough water to give 5.00 L of solution. The 18.0 M acid should be added to water, and not the water to the acid. One way to make the desired solution is to add the 1.67 L of 18.0 M acid to about 3 L of pure water. Stir the resulting solution, then add enough additional water to bring the total volume to 5.00 L.

    b. Put 41.7 mL of 3.00 M $CaCl_2$ solution into a 250 mL volumetric flask and add enough pure water to fill the flask to the mark when the contents are well mixed.

    c. Put 30.0 mL of a 10.0% solution into a calibrated volumetric container (such as a volumetric flask or large volumetric cylinder) and add enough pure water to bring the total volume of well-mixed solution to 200 mL.

    d. Put 100 mL of a 50.0% solution into a calibrated volumetric container and add enough pure water to bring the total volume of well-mixed solution to 500 mL.

7.50 **a.** 0.0113 M

    b. 0.100 M

    c. 0.0375 M

    d. 0.0265 M

7.52 16.6 g

7.54 20.0 mL

7.56 50.7 mL

7.58 27.8 mL

7.60 1.3 g

7.62 Because ice cream is essentially a solution of substances dissolved in water, the freezing point of the solution is lower than that of pure water. The temperature of a mixture of ice and pure water is the same as the freezing point of pure water. Thus, a mixture of ice and water has a temperature higher than the freezing point of the ice cream solution. A mixture of ice, salt, and water has a temperature (freezing point) lower than that of pure water. If enough salt is added to the ice and water mixture, the temperature can be made lower than the freezing point of the ice cream solution, and it will freeze.

7.64 **a.** Boiling point = 101.6°C; freezing point = −5.58°C

    b. Boiling point = 100.78°C; freezing point = −2.79°C

    c. Boiling point = 102.3°C; freezing point = −8.37°C

    d. Boiling point = 103.1°C; freezing point = −11.2°C

7.66 **a.** Boiling point = 100.39°C; freezing point = −1.40°C

    b. Boiling point = 100.23°C; freezing point = −0.837°C

    c. Boiling point = 102.51°C; freezing point = −8.98°C

7.68 **a.** Osmolarity = 0.20

    b. Osmolarity = 0.60

    c. Osmolarity = 3.18

7.70 $\pi = 4.65 \times 10^3$ torr = 4.65×10^3 mmHg = 6.12 atm

7.72 $\pi = 1.12 \times 10^4$ torr = 1.12×10^4 mmHg = 14.7 atm

7.74 $\pi = 5.88 \times 10^4$ torr = 5.88×10^4 mmHg = 77.4 atm

7.76 $\pi = 3.5 \times 10^3$ torr = 3.5×10^3 mmHg = 4.6 atm

7.78 $\pi = 9.95 \times 10^3$ torr = 9.95×10^3 mmHg = 13.1 atm

7.80 Water will flow from the 5.00% solution into the 10.0% solution. The 10.0% solution will become diluted as osmosis takes place.

7.82 Much of the soiling material to be removed by washing consists of fats and oils, which are nonpolar and will not dissolve in water. The soaps or detergents act as emulsifying agents, which allow the nonpolar materials to be suspended in water and removed from the material being washed.

7.84 **a.** Na^+ and Cl^-

    b. urea molecules

    c. K^+, Cl^- and glucose molecules

7.86 1.07 g/mL

7.88 10.2 mL

7.90 a

7.92 c

7.94 b

7.96 c

7.98 b

7.100 b

7.102 b

7.104 b

7.106 a

7.108 d

7.110 a

7.112 d

7.114 c

CHAPTER 8

8.2 **a.** Nonspontaneous: The rocket engines of the shuttle must operate continuously to push the shuttle into an orbit.

b. Spontaneous: Once ignited, the fuel continues to burn without additional input of energy.

c. Nonspontaneous: Heat must be supplied continuously to keep the water boiling.

d. Nonspontaneous: The only way to raise the temperature of a substance is to supply heat.

e. Nonspontaneous: To make your room orderly, someone must provide some work to move the clothing, etc., around. The work done represents an input of energy to the room.

8.4 **a.** Exergonic: The person doing the pushing is giving up energy to the automobile.

b. Endergonic: The ice must gain energy in the form of heat in order to melt.

c. Exergonic: The surroundings lose heat to the ice in order for the ice to melt.

d. Exergonic: In order to condense, the steam must give up some energy in the form of heat.

e. Endergonic: The heat given up by the steam as it condenses is accepted by the surroundings.

8.6 **a.** Both the energy and the entropy of the water decrease when water freezes. The process is spontaneous if the temperature of the surroundings is lower than 0°C.

b. The energy of the water decreases while the entropy stays essentially constant. The process is spontaneous.

c. The energy of the perfume increases slightly as energy is absorbed during evaporation. The entropy of the perfume increases a great deal as the molecules evaporate from the liquid and disperse throughout the room. The process is spontaneous.

8.8 **a.** Entropy is highest at the end of the play because the players are in a more mixed-up state as compared to just before the play began.

b. The answer is the 10% copper/gold alloy or the 2% alloy, depending on the atoms discussed. A lower concentration of copper atoms allows more disorder among the copper atoms, but more order among the gold atoms. A higher concentration of copper allows less disorder for the copper atoms, but increases the disorder of the gold atoms.

c. Entropy is highest with the purse on the ground and the contents scattered.

d. The mixture characteristic of coins in a piggy bank would have the highest entropy of the choices.

e. The mixture of loose pearls in a box would have the highest entropy of the choices.

8.10 The answers are subjective and depend on the meaning given to the terms *very slow, slow,* and *fast.*

a. slow

b. very slow

c. slow

d. fast

e. very slow

8.12 **a.** Any of the following could be measured at different times: the weight of the block, the weight or volume of the water produced by the melting block, and the physical dimensions of the melting block.

b. Either of the following could be measured at different times: the distance a weighted object penetrates into a sample, and the force required to break a specific-sized piece.

c. Either of the following could be measured at different times: the length of the unburned candle and the weight of the candle.

8.14 **a.** Rate = 0.0330 M/min

b. Rate = 0.0194 M/min

8.16 Rate = 3.96×10^{-3} M/min

8.18 Rate = 1.51×10^{-6} M/sec

8.20 **a.** Assumption #1 is being violated. According to assumption #1, decreasing the concentrations of reactants would decrease the number of collisions and so decrease the rate at which the molecules would react.

b. Assumption #2 is being violated. According to assumption #2, molecules must collide with a certain minimum energy in order to react. At a lower temperature, fewer molecules would have the necessary minimum energy, so the reaction should go slower, not faster.

c. Assumption #1 is being violated. According to assumption #1, increasing the concentration of either of the reactants should increase the number of collisions and so increase the rate of the reaction.

8.22 An increase in temperature increases the speeds of the molecules involved in a reaction. The increased speed acts to increase the reaction rate in two ways. First, it causes more collisions to occur between molecules in a given amount of time. Second, it causes the energy of the colliding molecules to be greater so that more of the collisions will involve molecules that have at least the minimum energy needed for a reaction to take place.

8.24 **a.**

Reaction progress

b.

8.26 a.

AT ROOM TEMP.

b.

AT 150° C

Both reactions are exothermic and the products have 500 kJ/mol less energy than the reactants, so both reactions give up 500 kJ/mol. The reaction that takes place at room temperature has a smaller activation energy than the reaction that must be heated to 150°C.

8.28 a. Rapid. The oppositely charged ions in solution will be attracted to each other and undergo many collisions per second.

b. Won't react. The reactant particles in solution both have a negative electrical charge and will repel each other. Collisions are highly unlikely.

c. Rapid. The reacting particles are in the gaseous state where they move about freely, making effective collisions likely.

d. Slow to rapid, depending on the state of division of the solid. If large lumps of CaO were exposed to gaseous CO_2 the reaction would be slow, since only the CaO on the surface of the lumps would be available. If the CaO were divided into dust-size articles and blown into gaseous CO_2 the reaction would proceed more rapidly.

8.30 Heat the reaction mixture, add a catalyst, or increase the concentration of one or more of the reactants.

8.32 About 7.5 minutes

8.34 Catalysts provide reaction pathways with lower activation energy. They may do this by reacting to form an intermediate structure that yields products when it breaks up or by providing a surface on which the reactants react.

8.36 a. Equilibrium would be indicated when the intensity of the violet color of the reaction mixture remained constant with time.

b. Equilibrium would be indicated when the amount of undissolved sugar remained constant with time.

c. Equilibrium would be indicated when either the intensity of the red-brown color of the reaction mixture remained constant with time or the total pressure of the reaction mixture remained constant with time.

8.38 The concentrations of H_2 and Br_2 would decrease, and the concentration of HBr would increase. The intensity of the red-brown color of the mixture would stop decreasing and would become constant when equilibrium was established.

8.40 a. $K = \dfrac{[CO_2]^2}{[CO]^2[O_2]}$

b. $K = \dfrac{[NO_2]^2}{[N_2O_4]}$

c. $K = \dfrac{[CO_2]^4[H_2O]^6}{[C_2H_6]^2[O_2]^7}$

d. $K = \dfrac{[NO]^2[Cl_2]}{[NOCl]^2}$

e. $K = \dfrac{[O_2][ClO_2]^4}{[Cl_2O_5]^2}$

8.42 a. $K = \dfrac{[Fe(CN)_6^{3-}]}{[Fe^{3+}][CN^-]^6}$

b. $K = \dfrac{[Ag(NH_3)_2^+]}{[Ag^+][NH_3]^2}$

c. $K = \dfrac{[AuCl_4^-]}{[Au^{3+}][Cl^-]^4}$

8.44 a. $CH_4 + 2O_2 \rightleftarrows CO_2 + 2H_2O$

b. $3H_2 + CO \rightleftarrows CH_4 + H_2O$

c. $2O_3 \rightleftarrows 3O_2$

d. $4NO_2 + 6H_2O \rightleftarrows 4NH_3 + 7O_2$

8.46 $K = 0.47$

8.48 $K = 0.099$

8.50 a. The concentration of products will be larger than the concentration of reactants.

b. The concentration of products will be larger than the concentration of reactants.

c. The concentration of products will be smaller than the concentration of reactants.

d. The concentration of products will be smaller than the concentration of reactants.

8.52 a. Shift left

b. Shift right

c. Shift left

8.54 a. Shift right: The color will become less blue and more purple.

b. Shift right: The amount of solid $PbCl_2$ will increase.

c. Shift right: The intensity of the violet color of the mixture will decrease.

d. Shift right: The intensity of the violet color of the mixture will decrease.

e. Shift left: The intensity of the brown color of the mixture will increase.

8.56 a. The concentrations of NO_2 and O_2 will increase and the concentration of N_2O will decrease.

b. The concentrations of SO_2 and O_2 will increase and the concentration of SO_3 will decrease.

c. The concentration of CO_2 will increase and the concentrations of CO and O_2 will decrease.

8.58 a. Shift right

b. Shift left

c. Shift right

d. Shift left

e. No shift will occur

f. Shift left

8.60 Apply pressure to the balloon and decrease its volume. This would increase the concentrations of the gaseous reactants A and B inside the balloon and speed up the reaction.

8.62 a. Criterion 2

b. Criterion 3

c. Criterion 2

d. Criterion 3

e. Criterion 3

8.64 a

8.66 a

8.68 c

8.70 b

8.72 a

8.74 a

8.76 a

8.78 a

8.80 a

8.82 c

8.84 d

CHAPTER 9

9.2 a. $HBrO_2(aq) \rightarrow H^+(aq) + BrO_2^-(aq)$

b. $HS^-(aq) \rightarrow H^+(aq) + S^{2-}(aq)$

c. $HBr(aq) \rightarrow H^+(aq) + Br^-(aq)$

d. $HC_2H_3O_2(aq) \rightarrow H^+(aq) + C_2H_3O_2^-(aq)$

9.4 a. Arrhenius base: $LiOH(aq) \rightarrow Li^+(aq) + OH^-(aq)$

b. Not an Arrhenius base

c. Arrhenius base: $Sr(OH)_2(aq) \rightarrow Sr^{2+}(aq) + 2OH^-(aq)$

d. Not an Arrhenius base

9.6 a. Acids: $HC_2O_4^-$ and H_3O^+
Bases: H_2O and $C_2O_4^{2-}$

b. Acids: HNO_2 and H_3O^+
Bases: H_2O and NO_2^-

c. Acids: H_2O and HPO_4^{2-}
Bases: PO_4^{3-} and OH^-

d. Acids: H_2SO_3 and H_3O^+
Bases: H_2O and HSO_3^-

e. Acids: H_2O and HF
Bases: F^- and OH^-

9.8

	Acid	Conjugate base	Base	Conjugate acid
a.	$HC_2O_4^-$	$C_2O_4^{2-}$	H_2O	H_3O^+
b.	HNO_2	NO_2^-	H_2O	H_3O^+
c.	H_2O	OH^-	PO_4^{3-}	HPO_4^{2-}
d.	H_2SO_3	HSO_3^-	H_2O	H_3O^+
e.	H_2O	OH^-	F^-	HF

9.10 a. $HF + H_2O \rightleftarrows H_3O^+ + F^-$

b. $HClO_3 + H_2O \rightleftarrows H_3O^+ + ClO_3^-$

c. $HClO + H_2O \rightleftarrows H_3O^+ + ClO^-$

d. $HS^- + H_2O \rightleftarrows H_3O^+ + S^{2-}$

9.12 a. SO_4^{2-}

b. CH_3NH_2

c. ClO_4^-

d. NH_3

e. Cl^-

9.14 a. H_2CO_3

b. HS^-

c. H_2S

d. $H_2C_2O_4$

e. $H_2N_2O_2$

9.16 The substance added to complete each equation appears in blue.

a. $H_2AsO_4^-(aq) + NH_3(aq) \rightarrow NH_4^+(aq) + HAsO_4^{2-}(aq)$

b. $C_6H_5NH_2(aq) + H_2O(\ell) \rightarrow C_6H_5NH_3^+(aq) + OH^-(aq)$

c. $S^{2-}(aq) + H_2O(\ell) \rightarrow HS^-(aq) + OH^-(aq)$

d. $(CH_3)_2NH(aq) + HBr(aq) \rightarrow (CH_3)_2NH_2^+(aq) + Br^-(aq)$

e. $H_2PO_4^-(aq) + CH_3NH_2(aq) \rightarrow$
$HPO_4^{2-}(aq) + CH_3NH_3^+(aq)$

9.18 a. $HS^-(aq) + NH_3(aq) \rightarrow S^{2-}(aq) + NH_4^+(aq)$

b. $H_2O(\ell) + ClO_3^-(aq) \rightarrow OH^-(aq) + HClO_3(aq)$

c. $H_2O(\ell) + NH_2^-(aq) \rightarrow OH^-(aq) + NH_3(aq)$

d. $HBO_3^{2-}(aq) + H_2O(\ell) \rightarrow BO_3^{3-}(aq) + H_3O^+(aq)$

e. $HNO_2(aq) + NH_3(aq) \rightarrow NO_2^-(aq) + NH_4^+(aq)$

9.20 hydrocyanic acid

9.22 a. hydrotelluric acid

b. hypochlorous acid

c. sulfurous acid

d. nitrous acid

9.24 succinic acid

9.26 H_2CO_3

9.28 **a.** 2.3×10^{-13} M

　b. 7.7×10^{-11} M

　c. 1.1×10^{-12} M

　d. 1.3×10^{-5} M

　e. 3.0×10^{-13} M

9.30 **a.** 1.4×10^{-10} M

　b. 1.4×10^{-13} M

　c. 2.0×10^{-15} M

　d. 5.9×10^{-12} M

　e. 1.1×10^{-6} M

9.32 In Exercise 9.28: (a) acidic, (b) acidic, (c) acidic, (d) basic, (e) acidic

　In Exercise 9.30: (a) basic, (b) basic, (c) basic, (d) basic, (e) acidic

9.34 **a.** acidic

　b. basic

　c. acidic

　d. basic

　e. acidic

9.36 **a.** 8.39, basic

　b. 10.97, basic

　c. 7.50, basic

　d. 1.64, acidic

　e. 4.71, acidic

9.38 **a.** 2.66, acidic

　b. 11.41, basic

　c. 5.12, acidic

　d. 10.40, basic

　e. 4.94, acidic

9.40 **a.** 5.4×10^{-10} M

　b. 2.8×10^{-3} M

　c. 3.8×10^{-6} M

9.42 **a.** $[H^+] = 1.1 \times 10^{-4}$ M; $[OH^-] = 8.9 \times 10^{-11}$ M

　b. $[H^+] = 1.0 \times 10^{-4}$ M; $[OH^-] = 1.0 \times 10^{-10}$ M

　c. $[H^+] = 1.4 \times 10^{-12}$ M; $[OH^-] = 7.2 \times 10^{-3}$ M

9.44 **a.** 8.9×10^{-9} M, basic

　b. 1.2×10^{-4} M, acidic

　c. 4.2×10^{-8} M, basic

　d. 4.0×10^{-8} M, basic

　e. 5.9×10^{-7} M, acidic

9.46 **a.** 1.2×10^{-3} M, acidic

　b. 7.8×10^{-5} M, acidic

　c. 4.8×10^{-3} M, acidic

　d. 8.5×10^{-4} M, acidic

9.48 **a.** Add 1.0 L of dilute (6 M) HNO_3 to 1.0 L of pure water and mix well.

　b. Add 50 mL of concentrated (15 M) NH_3 to 450 mL of pure water and mix well.

　c. Add 83 mL of concentrated (12 M) HCl to 4.9 L of pure water and mix well.

9.50 **a.** $H_2SO_4(aq) + 2H_2O(\ell) \rightarrow 2H_3O^+(aq) + SO_4{}^{2-}(aq)$

　b. $H_2SO_4(aq) + CaO(s) \rightarrow CaSO_4(aq) + H_2O(\ell)$

　c. $H_2SO_4(aq) + Mg(OH)_2(s) \rightarrow MgSO_4(aq) + 2H_2O(\ell)$

　d. $H_2SO_4(aq) + CuCO_3(s) \rightarrow CuSO_4(aq) + CO_2(g) + H_2O(\ell)$

　e. $H_2SO_4(aq) + 2KHCO_3(s) \rightarrow K_2SO_4(aq) + 2CO_2(g) + 2H_2O(\ell)$

　f. $H_2SO_4(aq) + Mg(s) \rightarrow MgSO_4(aq) + H_2(g)$

9.52 **a.** Total ionic: $2H^+(aq) + SO_4{}^{2-}(aq) + 2H_2O(\ell) \rightarrow 2H_3O^+(aq) + SO_4{}^{2-}(aq)$

　Net ionic: $2H^+(aq) + 2H_2O(\ell) \rightarrow 2H_3O^+(aq)$

　Simplified net ionic: $H^+(aq) + H_2O(\ell) \rightarrow H_3O^+(aq)$

　b. Total ionic: $2H^+(aq) + SO_4{}^{2-}(aq) + CaO(s) \rightarrow Ca^{2+}(aq) + SO_4{}^{2-}(aq) + H_2O(\ell)$

　Net ionic: $2H^+(aq) + CaO(s) \rightarrow Ca^{2+}(aq) + H_2O(\ell)$

　c. Total ionic: $2H^+(aq) + SO_4{}^{2-}(aq) + Mg(OH)_2(s) \rightarrow Mg^{2+}(aq) + SO_4{}^{2-}(aq) + 2H_2O(\ell)$

　Net ionic: $2H^+(aq) + Mg(OH)_2(s) \rightarrow Mg^{2+}(aq) + 2H_2O(\ell)$

　d. Total ionic: $2H^+(aq) + SO_4{}^{2-}(aq) + CuCO_3(s) \rightarrow Cu^{2+}(aq) + SO_4{}^{2-}(aq) + CO_2(g) + H_2O(\ell)$

　Net ionic: $2H^+(aq) + CuCO_3(s) \rightarrow Cu^{2+}(aq) + CO_2(g) + H_2O(\ell)$

　e. Total ionic: $2H^+(aq) + SO_4{}^{2-}(aq) + 2KHCO_3(s) \rightarrow 2K^+(aq) + SO_4{}^{2-}(aq) + 2CO_2(g) + 2H_2O(\ell)$

　Net ionic: $2H^+(aq) + 2KHCO_3(s) \rightarrow 2K^+(aq) + 2CO_2(g) + 2H_2O(\ell)$

　Simplified net ionic: $H^+(aq) + KHCO_3(s) \rightarrow K^+(aq) + CO_2(g) + H_2O(\ell)$

　f. Total ionic: $2H^+(aq) + SO_4{}^{2-}(aq) + Mg(s) \rightarrow Mg^{2+}(aq) + SO_4{}^{2-}(aq) + H_2(g)$

　Net ionic: $2H^+(aq) + Mg(s) \rightarrow Mg^{2+}(aq) + H_2(g)$

9.54 $2HCl(aq) + MgO(s) \rightarrow MgCl_2(aq) + H_2O(\ell)$

　$2HCl(aq) + Mg(OH)_2(s) \rightarrow MgCl_2(aq) + 2H_2O(\ell)$

　$2HCl(aq) + MgCO_3(s) \rightarrow MgCl_2(aq) + H_2O(\ell) + CO_2(g)$

　$2HCl(aq) + Mg(HCO_3)_2(s) \rightarrow MgCl_2(aq) + 2H_2O(\ell) + 2CO_2(g)$

　$2HCl(aq) + Mg(s) \rightarrow MgCl_2(aq) + H_2(g)$

9.56 **a.** Molecular: $Sn(s) + H_2SO_3(aq) \rightarrow SnSO_3(aq) + H_2(g)$

　Total ionic: $Sn(s) + 2H^+(aq) + SO_3{}^{2-}(aq) \rightarrow Sn^{2+}(aq) + SO_3{}^{2-}(aq) + H_2(g)$

　Net ionic: $Sn(s) + 2H^+(aq) \rightarrow Sn^{2+}(aq) + H_2(g)$

b. Molecular: $3Mg(s) + 2H_3PO_4(aq) \rightarrow Mg_3(PO_4)_2(s) + 3H_2(g)$

Total ionic: $3Mg(s) + 6H^+(aq) + 2PO_4^{3-}(aq) \rightarrow Mg_3(PO_4)_2(s) + 3H_2(g)$

Net ionic: $3Mg(s) + 6H^+(aq) + 2PO_4^{3-}(aq) \rightarrow Mg_3(PO_4)_2(s) + 3H_2(g)$

c. Molecular: $Ca(s) + 2HBr(aq) \rightarrow CaBr_2(aq) + H_2(g)$

Total ionic: $Ca(s) + 2H^+(aq) + 2Br^-(aq) \rightarrow Ca^{2+}(aq) + 2Br^-(aq) + H_2(g)$

Net ionic: $Ca(s) + 2H^+(aq) \rightarrow Ca^{2+}(aq) + H_2(g)$

9.58 a. Molecular: $RbOH(aq) + HBr(aq) \rightarrow RbBr(aq) + H_2O(\ell)$

Total ionic:

$Rb^+(aq) + OH^-(aq) + H^+(aq) + Br^-(aq) \rightarrow Rb^+(aq) + Br^-(aq) + H_2O(\ell)$

Net ionic: $OH^-(aq) + H^+(aq) \rightarrow H_2O(\ell)$

b. Molecular: $2RbOH(aq) + H_2SO_3(aq) \rightarrow Rb_2SO_3(aq) + 2H_2O(\ell)$

Total ionic:

$2Rb^+(aq) + 2OH^-(aq) + 2H^+(aq) + SO_3^{2-}(aq) \rightarrow 2Rb^+(aq) + SO_3^{2-}(aq) + 2H_2O(\ell)$

Net ionic: $2OH^-(aq) + 2H^+(aq) \rightarrow 2H_2O(\ell)$

Simplified net ionic: $OH^-(aq) + H^+(aq) \rightarrow H_2O(\ell)$

c. Molecular:

$2RbOH(aq) + H_3PO_3(aq) \rightarrow Rb_2HPO_3(aq) + 2H_2O(\ell)$

Total ionic:

$2Rb^+(aq) + 2OH^-(aq) + 2H^+(aq) + HPO_3^{2-}(aq) \rightarrow 2Rb^+(aq) + HPO_3^{2-}(aq) + 2H_2O(\ell)$

Net ionic: $2OH^-(aq) + 2H^+(aq) \rightarrow 2H_2O(\ell)$

Simplified net ionic: $OH^-(aq) + H^+(aq) \rightarrow H_2O(\ell)$

9.60 a. Molecular: $2KOH(aq) + H_3PO_4(aq) \rightarrow K_2HPO_4(aq) + 2H_2O(\ell)$

Total ionic:

$2K^+(aq) + 2OH^-(aq) + 2H^+(aq) + HPO_4^{2-}(aq) \rightarrow 2K^+(aq) + HPO_4^{2-}(aq) + 2H_2O(\ell)$

Net ionic: $2OH^-(aq) + 2H^+(aq) \rightarrow 2H_2O(\ell)$

Simplified net ionic: $OH^-(aq) + H^+(aq) \rightarrow H_2O(\ell)$

b. Molecular: $3KOH(aq) + H_3PO_4 \rightarrow K_3PO_4(aq) + 3H_2O(\ell)$

Total ionic:

$3K^+(aq) + 3OH^-(aq) + 3H^+(aq) + PO_4^{3-}(aq) \rightarrow 3K^+(aq) + PO_4^{3-}(aq) + 3H_2O(\ell)$

Net ionic: $3OH^-(aq) + 3H^+(aq) \rightarrow 3H_2O(\ell)$

Simplified net ionic: $OH^-(aq) + H^+(aq) \rightarrow H_2O(\ell)$

c. Molecular: $KOH(aq) + H_2C_2O_4(aq) \rightarrow KHC_2O_4(aq) + H_2O(\ell)$

Total ionic:

$K^+(aq) + OH^-(aq) + H^+(aq) + HC_2O_4^-(aq) \rightarrow K^+(aq) + HC_2O_4^-(aq) + H_2O(\ell)$

Net ionic: $OH^-(aq) + H^+(aq) \rightarrow H_2O(\ell)$

9.62 a. Cation: Cu^{2+}; anion: Cl^-

b. Cation: NH_4^+; anion: SO_4^{2-}

c. Cation: Li^+; anion: PO_4^{3-}

d. Cation: Mg^{2+}; anion: CO_3^{2-}

e. Cation: Ca^{2+}; anion: $C_2H_3O_2^-$

f. Cation: K^+; anion: NO_3^-

9.64 a. Acid: HCl; base: $Cu(OH)_2$

b. Acid: H_2SO_4; base: NH_3 or NH_4OH

c. Acid: H_3PO_4; base: $LiOH$

d. Acid: H_2CO_3; base: $Mg(OH)_2$

e. Acid: $HC_2H_3O_2$; base: $Ca(OH)_2$

f. Acid: HNO_3; base: KOH

9.66 a. 1.3×10^2 g; the product would be $MgSO_4$

b. 1.8×10^2 g; the product would be $Na_2B_4O_7$

9.68 a. Acid: HNO_3; solid: $MgCO_3$

b. Acid: HCl; solid: CaO

c. Acid: H_2SO_4; solid $RbHCO_3$

9.70 a. $2HNO_3(aq) + MgCO_3(s) \rightarrow Mg(NO_3)_2(aq) + H_2O(\ell) + CO_2(g)$

b. $2HCl(aq) + CaO(s) \rightarrow CaCl_2(aq) + H_2O(\ell)$

c. $H_2SO_4(aq) + 2RbHCO_3(s) \rightarrow Rb_2SO_4(aq) + 2H_2O(\ell) + 2CO_2(g)$

9.72 a. 1/2 mol

b. 1/2 mol

c. 1/3 mol

9.74 a. 0.44 eq; 4.4×10^2 meq

b. 0.45 eq; 4.5×10^2 meq

c. 6.24×10^{-2} eq; 62.4 meq

9.76 a. 3.50×10^{-2} eq; 35.0 meq

b. 4.01×10^{-2} eq; 40.1 meq

c. 1.35×10^{-1} eq; 135 meq

d. 4.17×10^{-2} eq; 41.7 meq

9.78 9.98×10^{-3} mol, 1.74 g

9.80 B, A, C, D

9.82 a. Acid order is B, A, C, D.

b. Conjugate base order is D, C, A, B.

9.84 a. $HNO_2(aq) \rightleftarrows H^+(aq) + NO_2^-(aq)$

$$K_a = \frac{[H^+][NO_2^-]}{[HNO_2]}$$

b. $HCO_3^-(aq) \rightleftarrows H^+(aq) + CO_3^{2-}(aq)$

$$K_a = \frac{[H^+][CO_3^{2-}]}{[HCO_3^{2-}]}$$

c. $H_2PO_4^-(aq) \rightleftarrows H^+(aq) + HPO_4^{2-}(aq)$

$$K_a = \frac{[H^+][HPO_4^{2-}]}{[H_2PO_4^-]}$$

d. $HS^-(aq) \rightleftarrows H^+(aq) + S^{2-}(aq)$

$$K_a = \frac{[H^+][S^{2-}]}{[HS^-]}$$

e. $HBrO(aq) \rightleftarrows H^+(aq) + BrO^-(aq)$

$$K_a = \frac{[H^+][BrO^-]}{[HBrO]}$$

9.86 According to the definitions given in the chapter, the person should be given the 20% acetic acid solution. If the other solution were the one desired, the term *dilute* should have been used rather than *weak*.

9.88 Titrations are done in order to analyze acids or bases. The amount of acid or base contained in a titrated sample can be determined from the volume of titrant added and the concentration of the titrant.

9.90 a. The endpoint and equivalence point will be the same if the chosen indicator changes color at the same pH as the pH of the salt solution produced by the reaction of the acid and base involved in the titration.

b. The endpoint and equivalence point will not be the same if the chosen indicator changes color at a pH that is different from the pH of the salt solution produced by the reaction of the acid and base involved in the titration.

9.92 a. 0.100 mol

b. 0.225 mol

9.94 a. $2NaOH(aq) + H_2CrO_4(aq) \rightarrow Na_2CrO_4(aq) + 2H_2O(\ell)$

b. $NaOH(aq) + HClO_3(aq) \rightarrow NaClO_3(aq) + H_2O(\ell)$

c. $3NaOH(aq) + H_3AsO_3(aq) \rightarrow Na_3AsO_3(aq) + 3H_2O(\ell)$

9.96 a. $2HCl(aq) + Sr(OH)_2(aq) \rightarrow SrCl_2(aq) + 2H_2O(\ell)$

b. $3HCl(aq) + Ni(OH)_3(aq) \rightarrow NiCl_3(aq) + 3H_2O(\ell)$

c. $2HCl(aq) + Zn(OH)_2(aq) \rightarrow ZnCl_2(aq) + 2H_2O(\ell)$

9.98 0.1660 M

9.100 a. 33.3 mL

b. 41.7 mL

c. 100 mL

d. 120 mL

e. 85.0 mL

f. 83.3 mL

9.102 a. 3.50 M

b. 0.891 M

c. 12.4 M

9.104 122.0 g/mol

9.106 The ions produced when Na_2CO_3 dissolves in water are Na^+ and CO_3^{2-}. The Na^+ ion is the conjugate acid of the strong base

NaOH and therefore is a very weak Brønsted acid. The CO_3^{2-} ion is the conjugate base of the weak acid HCO_3^- and therefore will act as a base in a hydrolysis reaction with water and will contribute to the pH of the solution. The reaction is

$$CO_3^{2-}(aq) + H_2O(\ell) \rightarrow HCO_3^-(aq) + OH^-(aq)$$

Thus, we see that the hydrolysis reaction contributes OH^- ions to the solution, causing it to have a pH greater than 7.

9.108 a. pH greater than 7. The ions produced when the salt dissolves in water are Na^+ and OCl^-. The Na^+ is the conjugate acid of a strong base (NaOH). It will be a weak acid and will not undergo a significant hydrolysis reaction. The OCl^- ion is the conjugate base of a weak acid (HOCl) and will therefore undergo a significant hydrolysis reaction that will contribute to the pH of the solution. The reaction is:

$$OCl^-(aq) + H_2O(\ell) \rightarrow HOCl(aq) + OH^-$$

The OH^- ions produced by this reaction will cause the pH of the solution to be greater than 7.

b. pH greater than 7. The explanation is the same as in part a. The hydrolysis reaction of the CHO_2^- ion is:

$$CHO_2^-(aq) + H_2O(\ell) \rightarrow HCHO_2(aq) + OH^-(aq)$$

c. pH equal to 7. The ions produced when the salt dissolves in water are K^+ and NO_3^-. The K^+ is the conjugate acid of the strong base KOH, and the NO_3^- is the conjugate base of the strong acid HNO_3. Since both ions come from strong sources, they are both weak in their respective behaviors as a conjugate acid or base. As a result, neither of them will undergo a significant hydrolysis reaction that will influence the solution pH.

d. pH greater than 7. The explanation is the same as in part a. The hydrolysis reaction of the PO_4^{3-} ion is:

$$PO_4^{3-}(aq) + H_2O(\ell) \rightarrow HPO_4^{2-}(aq) + OH^-(aq)$$

9.110 Different salts will hydrolyze to different extents in water solution and produce solutions of different pH. In titrations, the pH at the equivalence point is the pH characteristic of a solution of the salt produced by the reaction of the acid and base involved in the titration. Different acids and bases will produce different salts and possibly different solution pH values at the equivalence point. Thus, an indicator must be chosen that will change color at a pH as near as possible to the pH of the salt solution resulting from the titration.

9.112 The buffering anions present in the solution of the two salts are HPO_4^{2-} and $H_2PO_4^-$. When acid is added to the solution, the buffering reaction (like Equation 9.48) is:

$$HPO_4^{2-}(aq) + H^+(aq) \rightleftarrows H_2PO_4^-(aq)$$

When base is added, the buffering reaction (like Equation 9.49) is:

$$H_2PO_4^-(aq) + OH^-(aq) \rightleftarrows HPO_4^{2-}(aq) + H_2O(\ell)$$

9.114 $HCO_3^-(aq) + H^+(aq) \rightleftarrows H_2CO_3(aq)$

9.116 a. pH = 3.85

b. pH = 3.85

c. The difference is that the buffer of part b would have a greater buffer capacity; that is, it could buffer against larger added amounts of acid or base.

9.118 a. pH = 4.54

b. pH = 7.81

c. pH = 6.12

9.120 $[HPO_4^{2-}]/[H_2PO_4^-] = 2.8$

9.122 $Cl^-(aq) + H_3O^+(aq) \rightarrow HCl(aq) + H_2O(\ell)$

$$\left[:\ddot{\underset{..}{Cl}}:\right]^- + \left[H:\ddot{\underset{H}{O}}:H\right]^+ \rightarrow H:\ddot{\underset{..}{Cl}}: + H:\ddot{\underset{H}{O}}:$$

$\quad$ base $\qquad$ acid $\qquad$ acid $\qquad$ base

A base donates a pair of electrons to form a covalent bond, while an acid accepts a pair.

9.124 $pH = 6.63$

9.126 a

9.128 c

9.130 c

9.132 c

9.134 c

9.136 a

9.138 d

9.140 d

9.142 b

9.144 c

9.146 c

CHAPTER 10

10.2 **a.** Beta, gamma, and positron

$\quad$ **b.** Alpha and positron

$\quad$ **c.** Gamma and neutron

10.4 **a.** A beta particle is identical with an electron.

$\quad$ **b.** An alpha particle is made up of two protons and two neutrons.

$\quad$ **c.** A positron is a positively charged electron.

10.6 **a.** The mass number is reduced by 4, and the atomic number is reduced by 2.

$\quad$ **b.** The mass number is unchanged, and the atomic number is increased by 1.

$\quad$ **c.** The mass number is unchanged, and the atomic number is reduced by 1.

$\quad$ **d.** The mass number and atomic number are both unchanged.

$\quad$ **e.** The mass number is unchanged, and the atomic number is reduced by 1.

10.8 **a.** $^{27}_{13}Al$

$\quad$ **b.** $^{90}_{38}Sr$

$\quad$ **c.** $^{76}_{36}Kr$

10.10 The quantity added to complete each equation appears in blue.

$\quad$ **a.** $^{234}_{90}Th \rightarrow {}^{0}_{-1}\beta + {}^{234}_{91}Pa$

$\quad$ **b.** $^{35}_{16}S \rightarrow {}^{35}_{17}Cl + {}^{0}_{-1}\beta$

$\quad$ **c.** $^{248}_{100}Fm \rightarrow {}^{4}_{2}\alpha + {}^{244}_{98}Cf$

$\quad$ **d.** $^{226}_{88}Ra \rightarrow {}^{222}_{86}Rn + {}^{4}_{2}\alpha$

$\quad$ **e.** $^{113}_{47}Ag \rightarrow {}^{113}_{48}Cd + {}^{0}_{-1}\beta$

$\quad$ **f.** $^{41}_{18}Ar + {}^{0}_{-1}e \rightarrow {}^{41}_{17}Cl$

10.12 **a.** $^{157}_{63}Eu \rightarrow {}^{0}_{-1}\beta + {}^{157}_{64}Gd$

$\quad$ **b.** $^{190}_{78}Pt \rightarrow {}^{4}_{2}\alpha + {}^{186}_{76}Os$

$\quad$ **c.** $^{138}_{62}Sm + {}^{0}_{-1}e \rightarrow {}^{138}_{61}Pm$

$\quad$ **d.** $^{188}_{80}Hg \rightarrow {}^{188}_{79}Au + {}^{0}_{1}\beta$

$\quad$ **e.** $^{234}_{90}Th \rightarrow {}^{0}_{-1}\beta + {}^{234}_{91}Pa$

$\quad$ **f.** $^{218}_{85}At \rightarrow {}^{4}_{2}\alpha + {}^{214}_{83}Bi$

10.14 The half-life of a dozen cookies would be the time it takes for the number to be reduced to 6. The half-life of a $500 checking account would be the time it takes to reduce the amount in the account to $250.

10.16 0.56 ng remains

10.18 1.68×10^4 years old

10.20 2.24×10^4 years old

10.22 Long-term exposure to low-level radiation is more likely to cause changes to genetic material in cells. Such changes might result in genetic mutations, cancer, or other serious consequences. Short-term exposure to intense radiation destroys tissue rapidly and causes radiation sickness, which can lead to death, depending on the dose of radiation received.

10.24 A physical unit of radiation, such as a curie, indicates the activity of a source of radiation in terms of the number of nuclear decays that occur per minute. A biological unit of radiation, such as a rem, indicates the damage caused by radiation in living tissue.

10.26 3.2 R of X-rays

10.28 1.7×10^5 disintegrations/s

10.30 Hot spots are concentrations of radioisotope in tissue. Cold spots are areas of tissue that reject or keep out radioactive tracers.

10.32 $^{51}_{24}Cr + {}^{0}_{-1}e \rightarrow {}^{51}_{23}V$. The daughter is vanadium-51.

10.34 Use a radioactive isotope of oxygen to make water by burning hydrogen in it. Mix the radioactive water with hydrogen peroxide. Add the catalyst and collect the oxygen gas given off. If the gas is radioactive, it came from the water; if it is not radioactive, it came from the hydrogen peroxide.

10.36 Use a Geiger-Müller tube to measure the radioactivity of a gallon of water that contains the radioisotope. Pour that water into the pool and allow time for it to become completely and uniformly mixed with the water in the pool. Then, remove a gallon of the pool water and measure its radioactivity. For example, if the pool water has a radioactivity one ten-thousandth that of the original tracer-containing water, it means the original gallon of water has been diluted by a factor of 10,000, and therefore the pool contains 10,000 gal of water. Thus, the pool's volume is equal to the factor by which the radioactivity has been reduced.

10.38 $^{24}_{12}Mg + {}^{4}_{2}\alpha \rightarrow {}^{1}_{0}n + {}^{27}_{14}Si$

10.40 $^{238}_{92}U + {}^{1}_{0}n \rightarrow {}^{239}_{94}Pu + 2\ {}^{0}_{-1}\beta$

10.42 A moderator slows neutrons so that their chances of being captured and interacting with a target nucleus are greater.

10.44 $^{66}_{30}Zn + {}^{1}_{1}p \rightarrow {}^{67}_{31}Ga$

10.46 $2\,{}^{1}_{1}H \rightarrow {}^{2}_{1}H + {}^{0}_{1}\beta$

10.48 Nuclear fission is a process in which heavy nuclei split into smaller nuclei when they are hit by a neutron. Energy is released by the process. An example is:

$$^{235}_{92}U + ^{1}_{0}n \rightarrow ^{139}_{54}Xe + ^{95}_{38}Sr + 2^{1}_{0}n$$

When nuclear fusion occurs, small nuclei fuse together and form heavier nuclei. Energy is released by the process. An example is: $^{1}_{1}H + ^{2}_{1}H \rightarrow ^{3}_{2}He + \gamma$

10.50 $^{238}_{94}Pu \rightarrow ^{4}_{2}\alpha + ^{234}_{92}U$

10.52 The positive nuclei of the reacting hydrogen atoms naturally repel each other. In order to overcome this repulsion and get close enough together to fuse, the nuclei must have very high speeds (kinetic energy). The necessary high kinetic energy can only be obtained by the nuclei at very high temperatures. Remember, temperature is a measurement of the average kinetic energy of particles in matter.

10.54 3.5×10^{20} counts/minute $= 1.6 \times 10^{8}$ Ci

10.56 a

10.58 a

10.60 a

10.62 d

10.64 d

10.66 a

10.68 d

10.70 a

10.72 b

10.74 b

10.76 a

10.78 c

CHAPTER 11

11.2 Hair, skin, clothing, paper, plastic, carpet

11.4 They all contain carbon.

11.6 Covalent

11.8 **a.** Organic; most organic compounds are flammable.

b. Inorganic; a high-melting solid is typical of ionic compounds.

c. Organic; many organic compounds are insoluble in water.

d. Organic; most organic compounds can undergo combustion.

e. Organic; a low-melting solid is characteristic of covalent materials, which are usually organic.

11.10 Organic compounds do not readily accept electrons to transport them, thus making the flow of electrical current impossible.

11.12 The ability of carbon to bond to itself repeatedly; isomerism.

11.14 The s orbital of the hydrogen overlaps with the sp^3 orbital of carbon.

11.16 A p orbital and an sp^3 orbital both have a two-lobed shape. A p orbital has equal-sized lobes, whereas an sp^3 orbital has different-sized lobes.

11.18 a.

b.

c.

d.

11.20 b, d, e

11.22 a. Incorrect; H cannot have two bonds.

b. Correct

c. Incorrect; the carbon with the C $=$ O bond has five bonds.

d. Incorrect; both carbons of the C $=$ C double bond have three bonds.

e. Correct

11.24

a. $CH_3CH_2CH_2CH{=}CH{-}\overset{\overset{O}{\|}}{C}{-}CH_2CH_3$

b. $CH_3CH_2{-}\overset{\overset{O}{\|}}{C}{-}NH_2$

11.26 a.

b.

11.28 a. branched

b. normal

c. normal

d. normal

e. branched

f. branched

11.30 a. same

b. same

c. isomers

d. same

11.32 a. six

b. six

c. eight

11.34 a. 3-methylpentane

b. 2-methylpropane

c. 3,6-diethyl-4-methylnonane

d. 4,7-diethyl-5-methyldecane

e. 3-ethyl-5-methylnonane

Solutions to Learning Checks

CHAPTER 1

1.1 A change is chemical if new substances are formed and physical if no new substances are formed.

a. Chemical: The changes in taste and odor indicate that new substances form.

b. Physical: The handkerchief doesn't change, and the evaporated water is still water.

c. Chemical: The changes in color and taste indicate that new substances form.

d. Chemical: The gases and smoke released indicate that new substances form.

e. Physical: The air is still air, as indicated by appearance, odor, and so on.

f. Physical: On being condensed, the steam forms water, which is the same substance present before the change.

1.2 a. Triatomic and heteroatomic: The three atoms of two kinds (oxygen and hydrogen) make it both triatomic (three atoms) and heteroatomic (two or more kinds of atoms).

b. Triatomic and homoatomic: The three atoms make it triatomic, but all three atoms are the same (oxygen).

c. Polyatomic and heteroatomic: The total of five atoms makes it polyatomic, and the two different kinds of atoms (carbon and hydrogen) make it heteroatomic.

1.3 Molecules of elements can be diatomic, triatomic, or polyatomic, but all atoms will be identical and will be represented by the same color. Molecules of compounds must be diatomic, triatomic, or polyatomic (more than one atom), and the atoms will not all be identical.

a. Compound: polyatomic and heteroatomic

b. Element: diatomic and homoatomic

c. Element: polyatomic and homoatomic

d. Compound: polyatomic and heteroatomic

1.4 The product would have to be a compound because it would be polyatomic (more than one atom because two substances were used) and heteroatomic (different atoms from the two different substances).

1.5 Substitute the given radius and π values into the equation:

$$A = \pi r^2 = (3.14)(3.5 \text{ cm})^2 = (3.14)(12.25 \text{ cm}^2) = 38 \text{ cm}^2$$

1.6 The volume of a rectangular object is equal to the product of the three sides.

$$V = (30.0 \text{ cm})(20.0 \text{ cm})(15.0 \text{ cm})$$
$$= 9000 \text{ cm}^3, \text{ or } (9.00 \times 10^3 \text{ cm}^3)$$

$1 \text{ cm}^3 = 1 \text{ mL}$, so the volume is 9.00×10^3 mL. A liter is equal to 1000 mL, so 9000 mL = 9 L. The correct answers are 9.00×10^3 mL and 9.00 L.

1.7 Because 1 kg = 1000 g,

$$0.819 \text{ kg} \times \frac{1000 \text{ g}}{1 \text{ kg}} = 819 \text{ g}$$

1.8 Use Equation 1.4 and the fact that 77°F = 25°C, as shown in Example 1.7.

$$K = °C + 273 = 25°C + 273 = 298 \text{ K}$$

1.9 a. Scientific notation is used. In nonscientific notation, the number is written 588.

b. Scientific notation is not used. In scientific notation, the number is written 4.39×10^{-4}.

c. Scientific notation is used. In nonscientific notation, the number is written 0.0003915.

d. Scientific notation is not used. In scientific notation, the number is written 9.870×10^3.

e. Scientific notation is not used. In scientific notation, the number is written 3.677×10^1.

f. Scientific notation is not used. In scientific notation, the number is written 1.02×10^{-1}.

1.10 The two important things to note are the placement of the decimal in the standard position and the correct value of the exponent.

a. Written incorrectly: The decimal is not in standard position. Move the decimal 1 place to the left and increase the exponent by 1 to account for moving the decimal: 6.25×10^5.

b. Written incorrectly: The decimal is not in the standard position, and no exponent is used. Move the decimal 3 places to the right and use −3 for the exponent to account for moving the decimal: 9.8×10^{-3}.

c. Written incorrectly: The decimal is not in the standard position. Move the decimal 3 places to the right and reduce the exponent by 3 to account for moving the decimal: 4.1×10^{-6}.

d. Written correctly.

1.11 Perform multiplications and divisions separately for the nonexponential and exponential terms, then combine the results into the final answer.

a. $(2.4 \times 10^3)(1.5 \times 10^4) = (2.4 \times 1.5)(10^3 \times 10^4) = (3.6)(10^{3+4}) = 3.6 \times 10^7$

b. $(3.5 \times 10^2)(2.0 \times 10^{-3}) = (3.5 \times 2.0)(10^2 \times 10^{-3}) = (7.0)(10^{2+(-3)}) = 7.0 \times 10^{-1}$

c. $\dfrac{6.3 \times 10^5}{2.1 \times 10^3} = \left(\dfrac{6.3}{2.1}\right)\left(\dfrac{10^5}{10^3}\right) = (3.0)(10^{5-3})$

$$= 3.0 \times 10^2$$

d. $\dfrac{4.4 \times 10^{-2}}{8.8 \times 10^{-3}} = \left(\dfrac{4.4}{8.8}\right)\left(\dfrac{10^{-2}}{10^{-3}}\right) = (.50)(10^{-2-(-3)})$

$$= .50 \times 10^1 = 5.0 \times 10^0, \text{ or } 5.0$$

1.12 The primary challenge is to interpret correctly the significance of zeros. Leading zeros to the left of nonzero numbers are not significant; all other zeros, including trailing zeros, are significant.

    a. 3, trailing zero is significant.

    b. 4

    c. 3, leading zeros are not significant.

    d. 2

    e. 1, leading zeros are not significant.

    f. 4, trailing zero is significant.

1.13 **a.** 3 significant figures; original decimal position is 2 places to the right of standard position, so exponent is +2. 1.01×10^2 m.

    b. 4 significant figures; original decimal position is 3 places to the right of standard position, so exponent is +3. 1.200×10^3 g.

    c. 3 significant figures; original decimal position is 3 places to the left of standard position, so exponent is −3. 2.30×10^{-3} kg.

    d. 4 significant figures; original decimal position is 3 places to the right of standard position, so exponent is +3. 1.296×10^{3}°C.

    e. 4 significant figures; original decimal position is 1 place to the right of standard position, so exponent is +1. 2.165×10^1 mL.

    f. 2 significant figures; original decimal position is 2 places to the left of standard position, so exponent is −2. 1.5×10^{-2} km.

1.14 **a.** Answer will be rounded to 2 significant figures to match 0.0019.

$$(0.0019)(21.39) = 0.04064 = 4.1 \times 10^{-2}$$

    b. Answer will be rounded to 2 significant figures to match 4.1.

$$\dfrac{8.321}{4.1} = 2.0295 = 2.0$$

    c. Answer will be rounded to 3 significant figures to match 0.0911 and 3.22.

$$\dfrac{(0.0911)(3.22)}{(1.379)} = 0.21272 = 0.213, \text{ or } 2.13 \times 10^{-1}$$

1.15 In each case, the sum or difference is rounded to have the same number of places to the right of the decimal as the least number of places in the terms added or subtracted.

    a. Answer will be rounded to have one place to the right of the decimal to match 3.2.

$$8.01 + 3.2 = 11.21 = 11.2$$

    b. Answer will be rounded to have no places to the right of the decimal to match 3000.

$$3000 + 20.3 + 0.009 = 3020.309 = 3020$$

    c. Answer will be rounded to have two places to the right of the decimal to match both 4.33 and 3.12.

$$4.33 - 3.12 = 1.21$$

    d. Answer will be rounded to have two places to the right of the decimal to match 2.42.

$$6.023 - 2.42 = 3.603 = 3.60$$

1.16 Two factors result from the relationship 1 g = 1000 mg. The factors are:

$$\dfrac{1 \text{ g}}{1000 \text{ mg}} \quad \text{and} \quad \dfrac{1000 \text{ mg}}{1 \text{ g}}$$

The first factor is used because it cancels the mg unit and generates the g unit:

Step 1. 1.1 mg

Step 2. 1.1 mg   =   g

Step 3. $1.1 \text{ mg} \times \dfrac{1 \text{ g}}{1000 \text{ mg}}$ =   g

Step 4. $\dfrac{(1.1)(1 \text{ g})}{(1000)} = 0.0011 \text{ g} = 1.1 \times 10^{-3} \text{ g}$

The number of significant figures matches the number in 1.1 because the 1 and 1000 are exact numbers.

1.17 Two factors, 1 km/1000 m and 0.621 mi/1 km, are used to convert meters to miles, and two factors, 60 s/1 min and 60 min/1 h, are used to convert seconds in the denominator to hours.

$$10.0 \dfrac{\text{m}}{\text{s}} \times \dfrac{1 \text{ km}}{1000 \text{ m}} \times \dfrac{0.621 \text{ mi}}{1 \text{ km}} \times \dfrac{60 \text{ s}}{1 \text{ min}}$$

$$\times \dfrac{60 \text{ min}}{1 \text{ h}} = 22.356 \text{ mi/h}$$

Rounding to 3 significant figures gives 22.4 mi/h.

1.18 **a.** $\% = \dfrac{\text{part}}{\text{total}} \times 100; \% = \dfrac{\$988}{\$1200} \times 100 = 82.3\%$

    b. $\% = \dfrac{\text{part}}{\text{total}} \times 100$

$$90.4\% = \dfrac{\text{part}}{83} \times 100$$

$$\dfrac{(90.4\%)(83)}{100} = \text{part}$$

$$= \text{no. voting to not take final}$$

$$= 75.0$$

Number voting to take exam = 83 − 75 = 8.

1.19 The value of the density gives two factors that may be used to solve problems of this type:

$$\dfrac{2.7 \text{ g}}{1.0 \text{ cm}^3} \quad \text{and} \quad \dfrac{1.0 \text{ cm}^3}{2.7 \text{ g}}$$

    a. The sample volume is 60.0 cm³, and we wish to use a factor that will convert this to grams. The first factor given above will work.

$$60.0 \text{ cm}^3 \times \dfrac{2.7 \text{ g}}{1.0 \text{ cm}^3} = 162 \text{ g (calculator answer)}$$

$$= 1.6 \times 10^2 \text{ g (properly rounded answer)}$$

b. The sample mass is 98.5 g, and we wish to convert this to cm^3. The second factor given above will work.

$$98.5 \text{ g} \times \frac{1.0 \text{ cm}^3}{2.7 \text{ g}} = 36.48 \text{ cm}^3 \text{ (calculator answer)}$$

$$= 3.6 \times 10^1 \text{ cm}^3 \text{ (properly rounded answer)}$$

1.20 a. The sample mass is equal to the difference between the mass of the container with the sample inside and the mass of the empty container:

$$m = 64.93 \text{ g} - 51.22 \text{ g} = 13.71 \text{ g}$$

The density of the sample is equal to the sample mass divided by the sample volume:

$$d = \frac{m}{V} = \frac{13.71 \text{ g}}{10.00 \text{ mL}}$$

$$= 1.371 \text{ g/mL (properly rounded answer)}$$

b. The identity can be determined by calculating the density of the sample from the data and comparing the density with the known densities of the two possible metals. The volume of the sample is equal to the difference between the cylinder readings with the sample present and with the sample absent:

$$V = 25.2 \text{ mL} - 21.2 \text{ mL} = 4.0 \text{ mL}$$

According to Table 1.3, 1 mL = 1 cm^3, so the sample volume is equal to 4.0 cm^3. The density of the sample is equal to the sample mass divided by the sample volume:

$$d = \frac{m}{V} = \frac{35.66 \text{ g}}{4.0 \text{ cm}^3}$$

$$= 8.9 \text{ g/cm}^3 \text{ (properly rounded answer)}$$

A comparison with the known densities of nickel and chromium allows the metal to be identified as nickel.

CHAPTER 2

2.1 The number of atoms in each molecule is represented by a subscript in the formula.

a. H_3PO_4 **b.** SO_3 **c.** $C_6H_{12}O_6$

2.2 a. The atomic number, Z, equals the number of protons: $Z = 4$. The mass number, A, equals the sum of the number of protons and the number of neutrons: $A = 4 + 5 = 9$. According to the periodic table, the element with an atomic number of 4 is beryllium, with the symbol Be. The isotope symbol is $^9_4$Be.

b. According to the periodic table, chlorine has the symbol Cl and an atomic number, Z, of 17. The mass number, 37, is equal to the sum of the number of protons and neutrons: $A = \#p + \#n$. Therefore, the number of neutrons is equal to the mass number, A, minus the number of protons: $\#n = A - 17 = 37 - 17 = 20$.

c. According to the periodic table, the element with the symbol Si is silicon, which has an atomic number of 14. Therefore, the atom contains 14 protons. Since $A = \#p + \#n$, we see that $\#n = A - \#p = 28 - 14 = 14$.

2.3 a. Nitrogen atoms have a periodic table weight of 14.01 u, which rounds to 14 u. Iron atoms have a periodic table weight of 55.85 u, which rounds to 56 u. Thus, four nitrogen atoms (4×14 u = 56 u) will balance one iron atom.

b. Carbon atoms have a periodic table weight of 12.01 u, which rounds to 12 u. Helium atoms have a periodic table weight of 4.00 u, which rounds to 4 u. Thus, two carbon atoms (2×12 u = 24 u) will balance six helium atoms (6×4 u = 24 u).

c. Calcium atoms have a periodic table weight of 40.08 u, which rounds to 40 u. Argon atoms have a periodic table weight of 39.95 u, which rounds to 40 u. Thus, two calcium atoms (2×40 u = 80 u) will balance two argon atoms (2×40 u = 80 u).

2.4 Molecular weights are obtained by adding the atomic weights of the atoms in molecules. We have used four significant figures in atomic weights.

a. H_2SO_4; molecular weight = 2(at. wt. H) + (at. wt. S) + 4(at. wt. O)

$$MW = 2(1.008 \text{ u}) + 32.06 \text{ u} + 4(16.00 \text{ u})$$

$$= 2.016 \text{ u} + 32.06 \text{ u} + 64.00 \text{ u}$$

$$= 98.08 \text{ u}$$

b. C_3H_8O; molecular weight = 3(at. wt. C) + 8(at. wt. H) + (at. wt. O)

$$MW = 3(12.01 \text{ u}) + 8(1.008 \text{ u}) + 16.00 \text{ u}$$

$$= 36.03 \text{ u} + 8.064 \text{ u} + 16.00 \text{ u}$$

$$= 60.09 \text{ u}$$

2.5 a. Because naturally occuring fluorine consists of only a single isotope, the equation for calculating atomic weight becomes:

$$\text{At. wt.} = \frac{(\% \text{ fluorine-19})(\text{mass fluorine-19})}{100}$$

$$= \frac{(100)(19.00 \text{ u})}{100} = 19.00 \text{ u}$$

The calculated value is the same as the periodic table value.

b.

$$\text{At. wt.} = \frac{(\% \text{ Mg-24})(\text{mass Mg-24})}{100}$$

$$+ \frac{(\% \text{ Mg-25})(\text{mass Mg-25}) + (\% \text{ Mg-26})(\text{mass Mg-26})}{100}$$

$$= \frac{(78.70)(23.99 \text{ u}) + (10.13)(24.99 \text{ u})}{100}$$

$$+ \frac{(11.17)(25.98 \text{ u})}{100}$$

$$= 24.31 \text{ u}$$

Once again, the calculated value is the same as the periodic table value.

2.6 The mass ratio is obtained by dividing the mass of the magnesium sample by the mass of the carbon sample:

$$\frac{\text{Mg mass}}{\text{C mass}} = \frac{13.66 \text{ g}}{6.748 \text{ g}} = 2.024$$

The number of atoms in each sample is calculated using the factors obtained earlier from the mass of one atom of each element:

$$13.66 \text{ g Mg} \times \frac{1 \text{ Mg atom}}{4.037 \times 10^{-23} \text{ g Mg}}$$

$$= 3.384 \times 10^{23} \text{ Mg atoms}$$

$$6.748 \text{ g } \cancel{C} \times \frac{1 \text{ C atom}}{1.994 \times 10^{-23} \text{ g } \cancel{C}} = 3.384 \times 10^{23} \text{ C atoms}$$

2.7 The known quantity is one O atom, and the unit of the unknown is grams of O. The factor comes from the relationship 6.02×10^{23} O atoms = 16.00 g O. This relationship provides two factors:

$$\frac{6.02 \times 10^{23} \text{ O atoms}}{16.00 \text{ g O}} \quad \text{and} \quad \frac{16.00 \text{ g O}}{6.02 \times 10^{23} \text{ O atoms}}$$

The second factor is the one used:

$$(1 \text{ } \cancel{\text{O atom}})\left(\frac{16.00 \text{ g O}}{6.02 \times 10^{23} \text{ } \cancel{\text{O atoms}}}\right) = 2.66 \times 10^{-23} \text{ g O}$$

The ratio of this mass to the mass of a carbon atom is:

$$\frac{2.66 \times 10^{-23} \text{ g O}}{1.994 \times 10^{-23} \text{ g C}} = 1.33 \text{ (properly rounded answer)}$$

The ratio of the atomic weights of oxygen and carbon is:

$$\frac{16.00 \text{ } \cancel{u}}{12.01 \text{ } \cancel{u}} = 1.33 \text{ (rounded to 3 significant figures)}$$

The two ratios are the same to 3 significant figures.

2.8 MW = 1(At. wt. C) + 1(At. wt. O) = 1(12.0 u) + 1(16.0 u) = 28.0 u

Application of the mole concept to CO molecules gives the following relationships:

$$1 \text{ mol CO molecules} = 6.02 \times 10^{23} \text{ CO molecules}$$
$$= 28.0 \text{ g CO}$$

The calculation of the mass of one CO molecule is done as follows: The known quantity is 1 CO molecule, and the unit of the unknown is g CO. The factor comes from the relationship 6.02×10^{23} CO molecules = 28.0 g CO.

$$(1 \text{ } \cancel{\text{CO molecule}})\left(\frac{28.0 \text{ g CO}}{6.02 \times 10^{23} \text{ } \cancel{\text{CO molecules}}}\right)$$
$$= 4.65 \times 10^{-23} \text{ g CO}$$

Comparing the mass of a CO molecule to the mass of a CO_2 molecule gives:

$$\frac{\text{mass of 1 CO molecule}}{\text{mass of 1 CO}_2 \text{ molecule}} = \frac{4.65 \times 10^{-23} \text{ g}}{7.31 \times 10^{-23} \text{ g}} = 0.636$$

Comparing the molecular weight of CO to the molecular weight of CO_2 gives:

$$\frac{\text{MW CO}}{\text{MW CO}_2} = \frac{28.0 \text{ u}}{44.0 \text{ u}} = 0.636$$

It is seen that the ratio of the actual masses of the molecules is identical to the ratio of their relative masses (their molecular weights).

2.9 The relationships between moles of atoms in 1 mol of molecules is given by the subscripts of the atoms. Thus, 1 mol of glucose molecules would contain 6 mol of C atoms, 12 mol of H atoms, and 6 mol of O atoms. One-half mol of glucose molecules would contain half as many moles of each atom, or 3 mol of C atoms, 6 mol of H atoms, and 3 mol of O atoms.

2.10 $\% = \dfrac{\text{part}}{\text{total}} \times 100$

In each compound, the part will be the mass of carbon associated with some mass of compound (the total). The mass relationships

are easily obtained by assuming a sample size equal to 1.00 mol of each compound.

CO_2: 1.00 mol CO_2 molecules = 1.00 mol C atoms + 2.00 mol O atoms or, using atomic weights,

$$44.01 \text{ g CO}_2 = 12.01 \text{ g C} + 32.00 \text{ g O}$$

$$\% \text{ C} = \frac{(12.01 \text{ g C}) \times 100}{(44.01 \text{ g CO}_2)} = 27.29\% \text{ C}$$

CO: 1.00 mol CO molecules = 1.00 mol C atoms + 1.00 mol O atoms

$$28.01 \text{ g CO} = 12.01 \text{ g C} + 16.00 \text{ g O}$$

$$\% \text{ C} = \frac{(12.01 \text{ g C})}{(28.01 \text{ g CO})} \times 100 = 42.88\% \text{ C}$$

CHAPTER 3

3.1 Each element is found at the intersection of the given group (vertical columns) and period (horizontal rows).

   a. Ge   **b.** Re

3.2 **a.** Period 1 has 2 elements, H and He.

   b. Group IIB(12) has 4 elements, Zn, Cd, Hg, and element 112.

3.3 **a.** The maximum number of electrons that can be found in any orbital, including a $4p$ orbital, is 2.

   b. All d subshells, including the $5d$, contain 5 orbitals with a capacity of 2 electrons each. Thus, the $5d$ subshell can contain a maximum of 10 electrons.

   c. Shell number 1 contains only a single s orbital, the $1s$ orbital, and so can contain a maximum of 2 electrons.

3.4 The valence shell is the last shell with electrons. According to Table 3.2, group VA(15) elements all have 5 electrons in the valence shell. Group VIA(16) elements have 6, group VIIA(17) have 7, and the noble gases VIIIA(18) have 8.

3.5 **a.** Strontium (Sr) is in group IIA(2) and so contains 2 electrons in its valence shell.

   b. Once again, the roman numeral of the group heading gives the number of valence-shell electrons. The element is Ge, and because it is in group IVA(14), it has 4 electrons in the valence shell.

   c. The fifteenth element in period 4 is As, which is in group VA(15) and therefore has 5 electrons in the valence shell.

3.6 The subshell filling order is obtained from Figure 3.7 or Figure 3.8. All subshells except the last one are filled to their capacities as follows: s subshells—filled with 2, p subshells—filled with 6, d subshells—filled with 10, and f subshells—filled with 14. The last subshell has only enough electrons added to bring the total number of electrons to the desired value.

   a. Element 9 requires 9 electrons: $1s^2 2s^2 2p^5$; 1 unpaired

   b. Mg is element 12 and so needs 12 electrons:

$$1s^2 2s^2 2p^6 3s^2; \text{ 0 unpaired}$$

   c. The element in group VIA(16) and period 3 is S, which is element 16. Therefore, it needs 16 electrons:

$$1s^2 2s^2 2p^6 3s^2 3p^4; \text{ 2 unpaired}$$

   d. An atom with 23 protons needs 23 electrons:

$$1s^2 2s^2 2p^6 3s^2 3p^6 4s^2 3d^3; \text{ 3 unpaired}$$

3.7 **a.** From Learning Check 3.6, the electronic configuration of element number 9 is $1s^22s^22p^5$. The $1s^2$ electrons represent the configuration of helium. So, we can write the configuration as $[He]2s^22p^5$.

  b. The configuration of Mg is $1s^22s^22p^63s^2$. The first 10 electrons represent the configuration of neon: $[Ne]3s^2$.

  c. The configuration of S is $1s^22s^22p^63s^23p^4$. Once again, the first 10 electrons represent the configuration of neon: $[Ne]3s^23p^4$.

  d. The 23 electrons are represented by $1s^22s^22p^63s^23p^64s^2\,3d^3$. The first 18 electrons represent the configuration of argon: $[Ar]4s^23d^3$.

3.8 **a.** On the basis of the location of each element in Figure 3.9, the distinguishing electrons are of the following types: element 38(Sr): s; element 47(Ag): d; element 50(Sn): p; element 86(Rn): p.

  b. Their classifications based on Figure 3.10 are element 38(Sr): representative element; element 47(Ag): transition element; element 50(Sn): representative element; element 86(Rn): noble gas.

3.9 The periodic table and Figure 3.12 are used.

  a. Xe: nonmetal

  b. As: metalloid

  c. Hg: metal

  d. Ba: metal

  e. Th: metal

3.10 **a.** The slowest reaction involves Li, and the fastest involves K.

   Li
   Na
   K

  b. The rate of the reaction increases coming down the group, and according to Table 3.3, the ionization energy decreases coming down the group.

  c. The decrease in ionization energy coming down the group means less energy is required to remove electrons from atoms of the elements located farther down the group. The reactions in Figure 3.17 involve the removal of an electron from atoms of the metals, so the easier it is to remove electrons (less energy required), the faster the reaction should go. This trend in speed is what is observed.

CHAPTER 4

4.1 The subshell filling order is obtained from Figure 3.7 or Figure 3.8.

   F; 9 electrons: $1s^22s^22p^5$

   K; 19 electrons: $1s^22s^22p^63s^23p^64s^1$

   The configuration of F can be changed to that of He ($1s^2$) by giving up 7 electrons or to that of Ne ($1s^22s^22p^6$) by accepting 1 electron. The configuration of K can be changed to that of Ar ($1s^22s^22p^63s^23p^6$) by giving up 1 electron or to that of Kr ($1s^22s^22p^63s^23p^64s^23d^{10}4p^6$) by accepting 17 electrons.

4.2 **a.** Element number 9 is fluorine, which is in group VIIA(17) and so has 7 valence-shell electrons. The Lewis structure is $:\overset{..}{\underset{..}{F}}\cdot$.

  b. Magnesium is in group IIA(2) and so has 2 valence-shell electrons. The Lewis structure is $\overset{.}{Mg}\cdot$.

  c. Sulfur is in group VIA(16) and so has 6 valence-shell electrons. The Lewis structure is $:\overset{.}{\underset{.}{S}}\cdot$.

  d. Krypton is a noble gas and so has 8 valence-shell electrons. The Lewis structure is $:\overset{..}{\underset{..}{Kr}}:$.

4.3 **a.** Lithium contains 3 electrons, with one of them classified as a valence-shell electron: $[He]2s^1$ and Li$\cdot$.

  b. Bromine contains 35 electrons, with 7 of them classified as valence-shell electrons: $[Ar]4s^23d^{10}4p^5$ and $:\overset{..}{\underset{..}{Br}}\cdot$.

  c. Strontium contains 38 electrons, with 2 classified as valence-shell electrons: $[Kr]5s^2$ and $\overset{.}{Sr}\cdot$.

  d. Sulfur contains 16 electrons, with 6 classified as valence-shell electrons: $[Ne]3s^23p^4$ and $:\overset{..}{\underset{.}{S}}\cdot$.

4.4 The change that will actually take place is the one that involves the fewest electrons. Thus, F will gain 1 electron rather than give up 7, and K will give up 1 electron rather than gain 17:

$$F + 1e^- \rightarrow F^-$$

$$K \rightarrow K^+ + 1e^-$$

4.5 **a.** Element number 34 is a nonmetal. It is in group VIA(16) and will, when treated using the rules for nonmetals, accept $8 - 6$ or 2 electrons during ionic bond formation:

$$Se + 2e^- \rightarrow Se^{2-}$$

  b. Rb, a metal in group IA(1), will lose 1 electron during ionic bond formation:

$$Rb \rightarrow Rb^+ + 1e^-$$

  c. Element 18 is a nonmetallic noble gas. It will not react to form ionic bonds.

  d. In, a metal in group IIIA(13), will lose 3 electrons during ionic bond formation:

$$In \rightarrow In^{3+} + 3e^-$$

4.6 **a.** Mg, a metal in group IIA(2), will lose 2 electrons:

$$Mg \rightarrow Mg^{2+} + 2e^-$$

   O, a nonmetal in group VIA(16), will gain 2 electrons:

$$O + 2e^- \rightarrow O^{2-}$$

   The positive and negative ions are combined in the lowest numbers possible to give the compound formula. The combining requirement is that the total number of positive charges from the positive ion must equal the total number of negative charges from the negative ion. In the case of Mg^{2+} and O^{2-}, one of each ion satisfies the requirement, and the formula is MgO.

  b. K, a group IA(1) metal, will lose 1 electron:

$$K \rightarrow K^+ + 1e^-$$

   S, a group VIA(16) nonmetal, will gain 2 electrons:

$$S + 2e^- \rightarrow S^{2-}$$

   In the formula, $2K^+$ will be needed for each S^{2-}. Thus, the compound formula is K_2S.

  c. Ca, a group IIA(2) metal, will lose 2 electrons:

$$Ca \rightarrow Ca^{2+} + 2e^-$$

   Br, a group VIIA(17) nonmetal, will gain 1 electron:

$$Br + 1e^- \rightarrow Br^-$$

   In the formula, $2Br^-$ will combine with each Ca^{2+}: $CaBr_2$

4.7   **a.** magnesium oxide

     b. potassium sulfide

     c. calcium bromide

4.8   **a.** Charge balance requires that $2Br^-$ combine with each Co^{2+}: $CoBr_2$. Names are cobalt(II) bromide and cobaltous bromide.

     b. Charge balance requires that $3Br^-$ combine with each Co^{3+}: $CoBr_3$. Names are cobalt(III) bromide and cobaltic bromide.

4.9   **a.** For H_2S, the molecular weight is the sum of the atomic weights of the atoms in the formula:

$$MW = (2)(\text{at. wt. H}) + (1)(\text{at. wt. S})$$
$$= (2)(1.01\text{ u}) + (1)(32.1\text{ u})$$
$$MW = 34.1\text{ u}$$

For CaO, the formula weight is also equal to the sum of the atomic weights of the atoms in the formula:

$$FW = (1)(\text{at. wt. Ca}) + (1)(\text{at. wt. O})$$
$$= (1)(40.1\text{ u}) + (1)(16.0\text{ u})$$
$$FW = 56.1\text{ u}$$

     b. According to Section 2.6, 1.00 mol of a molecular compound has a mass in grams equal to the molecular weight of the compound. Thus, 1.00 mol H_2S = 34.1 g H_2S. For ionic compounds, 1.00 mol of compound has a mass in grams equal to the formula weight of the compound. Thus, 1.00 mol CaO = 56.1 g CaO.

     c. In Section 2.6, we learned that 1.00 mol of a molecular compound contains Avogadro's number of molecules. Thus, 1.00 mol H_2S = 6.02×10^{23} molecules of H_2S. In the case of ionic compounds, 1.00 mol of compound contains Avogadro's number of formula units. That is, 1.00 mol of calcium oxide contains Avogadro's number of CaO units, where each unit represents one Ca^{2+} ion and one O^{2-} ion. Thus:

1.00 mol CaO = 6.02×10^{23} CaO units   or

1.00 mol CaO = 6.02×10^{23} Ca^{2+} ions

$$+ \; 6.02 \times 10^{23}\; O^{2-}\text{ ions}$$

4.10   **a.** N is in group VA(15) and therefore has 5 valence-shell electrons. Two N atoms will provide 10 total electrons to satisfy the octets. At least 2 electrons are needed to bond the atoms. This leaves only 8 electrons to satisfy the remaining octet requirements of both atoms. This deficiency of electrons means some multiple bonds will be needed between the atoms:

$$:\!\dot{\text{N}}\cdot + \cdot\dot{\text{N}}\!: \longrightarrow \;:\!\text{N}\!:::\!\text{N}\!: \quad \text{or} \quad \text{N}\!\equiv\!\text{N}$$

     b. Br is in group VIIA(17) and therefore has 7 valence-shell electrons. Two atoms will provide 14 electrons, and the bonding is readily shown:

$$:\!\ddot{\text{Br}}\cdot + \cdot\ddot{\text{Br}}\!: \longrightarrow \;:\!\ddot{\text{Br}}\!:\!\ddot{\text{Br}}\!: \quad \text{or} \quad \text{Br}\!-\!\text{Br}$$

4.11   **a.** Each C contributes 4 valence-shell electrons, and each H contributes 1. The total available is 8:

$$\text{H}\!:\!\overset{\textstyle\text{H}}{\underset{\textstyle\ddot{\text{H}}}{\text{C}}}\!:\!\text{H} \quad \text{or} \quad \text{H}\!-\!\overset{\textstyle\text{H}}{\underset{\textstyle\text{H}}{\overset{\textstyle|}{\underset{\textstyle|}{\text{C}}}}}\!-\!\text{H}$$

     b. Each H contributes 1 electron, the C contributes 4, and the O contributes 6. The total of 12 electrons cannot satisfy all octets unless two pair are shared between the C and O atoms:

$$\text{H}\!:\!\overset{}{\underset{\textstyle\ddot{\text{H}}}{\text{C}}}\!:\!:\!\ddot{\text{O}}\!: \quad \text{or} \quad \text{H}\!-\!\overset{}{\underset{\textstyle\text{H}}{\overset{\textstyle|}{\underset{\textstyle|}{\text{C}}}}}\!=\!\text{O}$$

     c. Each H contributes 1 electron, each O contributes 6, and the N contributes 5. The 24 total electrons can satisfy all octets only if two pair are shared between the N and one of the O atoms:

$$\text{H}\!:\!\ddot{\text{O}}\!:\!\overset{}{\underset{\textstyle:\ddot{\text{O}}:}{\text{N}}}\!:\!:\!\ddot{\text{O}}\!: \quad \text{or} \quad \text{H}\!-\!\text{O}\!-\!\overset{}{\underset{\textstyle\text{O}}{\overset{\textstyle}{\underset{\textstyle|}{\text{N}}}}}\!=\!\text{O}$$

4.12   **a.** The P provides 5 electrons, each O provides 6, and the 3− charge indicates an additional 3 electrons. The total number of electrons is therefore 32, which is enough to satisfy all octets without any multiple bonds:

$$\left[\begin{array}{c} :\!\ddot{\text{O}}\!: \\ :\!\ddot{\text{O}}\!:\!\text{P}\!:\!\ddot{\text{O}}\!: \\ :\!\ddot{\text{O}}\!: \end{array}\right]^{3-}$$

     b. The S provides 6 electrons, each O provides 6, and the 2− charge indicates an additional 2 electrons. The total number of electrons is therefore 26, which is enough to satisfy all octets without any multiple bonds:

$$\left[\begin{array}{c} :\!\ddot{\text{O}}\!: \\ :\!\ddot{\text{O}}\!:\!\ddot{\text{S}}\!:\!\ddot{\text{O}}\!: \end{array}\right]^{2-}$$

     c. The N provides 5 electrons, each H provides 1, and the + charge indicates that 1 must be subtracted. The total number of electrons is therefore 8, which is enough to satisfy the octet of N, and each H requirement of 2 electrons without any multiple bonds:

$$\left[\begin{array}{c} \text{H} \\ \text{H}\!:\!\ddot{\text{N}}\!:\!\text{H} \\ \ddot{\text{H}} \end{array}\right]^{+}$$

4.13   **a.** The 3 electron pairs around the B atom will arrange themselves in a flat triangle around the B. Therefore, the molecule will be flat (planar) with the F atoms forming a triangle around the B.

     b. The Lewis structure is:

$$\begin{array}{c} :\!\ddot{\text{Cl}}\!: \\ :\!\ddot{\text{Cl}}\!:\!\ddot{\text{C}}\!:\!\ddot{\text{Cl}}\!: \\ :\!\ddot{\text{Cl}}\!: \end{array}$$

The 4 electron pairs around the C atom will arrange themselves in a tetrahedral orientation. Thus, the molecule will be tetrahedral with C in the center and a Cl atom at each of the corners of the tetrahedron around the C.

     c. The Lewis structure is:

$$:\!\ddot{\text{O}}\!:\!\ddot{\text{S}}\!:\!:\!\ddot{\text{O}}\!:$$

The 3 electron pairs around the S will occupy the corners of a triangle around the S. This will give rise to a bent molecule

$$O-\overset{\cdot\cdot}{\underset{\diagdown}{S}}$$
$$O$$

d. The Lewis structure is:

$$:\overset{\cdot\cdot}{S}::C::\overset{\cdot\cdot}{S}:$$

Each double bond on the central C atom is treated like a single pair of electrons when VSEPR theory is used. Thus, the double bonds are treated like 2 pairs and will arrange themselves to be on opposite sides of the C atom. A linear molecule results: S=C=S.

4.14 a. The Lewis structure from Learning Check 4.12 is:

$$\left[\begin{array}{c}:\overset{\cdot\cdot}{O}:\\:\overset{\cdot\cdot}{O}:\overset{\cdot\cdot}{P}:\overset{\cdot\cdot}{O}:\\:\overset{\cdot\cdot}{O}:\end{array}\right]^{3-}$$

We see that phosphorus is the central atom, and it has four electron pairs around it. The four pairs will be located at the corners of a tetrahedron with the P in the middle. The four O atoms will occupy the corners of the tetrahedron, giving the same shape as that of the SO_4^{2-} ion of Example 4.14a.

b. The Lewis structure from Learning Check 4.12 is:

$$\left[\begin{array}{c}:\overset{\cdot\cdot}{O}:\\:\overset{\cdot\cdot}{O}:\overset{\cdot\cdot}{S}:\overset{\cdot\cdot}{O}:\end{array}\right]^{2-}$$

We see that sulfur is the central atom, and it has four electron pairs around it. The four pairs will be located at the corners of a tetrahedron with the S in the middle. The shape of the ion is determined by the location of the oxygen atoms and the S atom. The O atoms will occupy three of the corners of the tetrahedron, and one pair of electrons will occupy the fourth. Thus, the shape of the ion will be a triangular-based pyramid with the O atoms at the three corners of the base and the S atom at the top.

c. The Lewis structure from Learning Check 4.12 is:

$$\left[\begin{array}{c}H\\H:\overset{\cdot\cdot}{N}:H\\\overset{\cdot\cdot}{H}\end{array}\right]^{+}$$

We see that nitrogen is the central atom, and it has four electron pairs around it. The four pairs will be located at the corners of a tetrahedron with the N in the middle. The four H atoms will thus be located at the corners of the tetrahedron with the N in the middle, giving a tetrahedral-shaped ion like the SO_4^{2-} ion of Example 4.14a.

4.15 a. No polarization because bound atoms are identical.

b. Br is located higher in the group and so is more electronegative than I:

$$\overset{\delta+}{I}-\overset{\delta-}{Br}$$
$$\longrightarrow$$

c. Br is farther toward the right in the periodic table and so is more electronegative than H:

$$\overset{\delta+}{H}-\overset{\delta-}{Br}$$
$$\longrightarrow$$

4.16 The values of ΔEN are used.

a. ΔEN = 4.0 − 0.8 = 3.2. Bond is classified as ionic.

b. ΔEN = 3.5 − 3.0 = 0.5. Bond is classified as polar covalent.

c. ΔEN = 3.0 − 1.5 = 1.5. Bond is classified as polar covalent.

d. ΔEN = 3.5 − 0.8 = 2.7. Bond is classified as ionic.

4.17 a. sulfur trioxide

b. boron trifluoride

c. disulfur heptoxide

d. carbon tetrachloride

4.18 a. Ca is a group IIA(2) metal and so forms Ca^{2+} ions. Formula: $CaHPO_4$; name: calcium hydrogen phosphate.

b. Mg is a group IIA(2) metal and so forms Mg^{2+} ions. Formula: $Mg_3(PO_4)_2$; name: magnesium phosphate.

c. K is a group IA(1) metal and so forms K^+ ions. Formula: $KMnO_4$; name: potassium permanganate.

d. The ammonium ion, NH_4^+, acts as the positive ion in this compound. Formula: $(NH_4)_2Cr_2O_7$; name: ammonium dichromate.

4.19 Assuming that dispersion forces are significant in establishing the melting and boiling points of these elements leads to the conclusion that, in each pair, the element with the higher atomic weight would have the higher melting and boiling points.

a. Se   **b.** Sb   **c.** Ne

CHAPTER 5

5.1 Both subscripts and coefficients are used.

N_2 : 2 atoms—on the basis of subscript.

$3H_2$: 6 atoms—on the basis of subscript and coefficient.

$2NH_3$: 2 N atoms (on the basis of coefficient) and 6 H atoms (on the basis of subscript and coefficient).

5.2 $2SO_2 + O_2 \rightarrow 2SO_3$

5.3 a. Oxygen is −2 (Rule 5).

Sulfur is +6 (Rule 6).

b. Calcium is +2 (Rule 3).

Oxygen is −2 (Rule 5).

Chlorine is +5 (Rule 6).

c. Oxygen is −2 (Rule 5).

Chlorine is +7 (Rule 7).

5.4 a. Zn = 0 (Rule 1), H^+ = +1 (Rule 2), Zn^{2+} = +2 (Rule 2), H_2 = 0 (Rule 1).

Zn is oxidized, so it is the reducing agent.

H^+ is reduced, so it is the oxidizing agent.

b. In KI: K = +1 (Rule 3), I = −1 (Rule 6).

Cl_2 = 0 (Rule 1).

In KCl: K = +1 (Rule 3), Cl = −1 (Rule 6).

$I_2 = 0$ (Rule 1).

   v I in KI is oxidized, so KI is the reducing agent.

     Cl_2 is reduced, so Cl_2 is the oxidizing agent.

  c. In IO_3^-: $O = -2$ (Rule 5), $I = +5$ (Rule 7).

    In HSO_3^-: $H = +1$ (Rule 4), $O = -2$ (Rule 5), $S = +4$ (Rule 7).

    $I^- = -1$ (Rule 2).

    In HSO_4^-: $H = +1$ (Rule 4), $O = -2$ (Rule 5), $S = +6$ (Rule 7).

    S in HSO_3^- is oxidized, so HSO_3^- is the reducing agent.

    I in IO_3^- is reduced, so IO_3^- is the oxidizing agent.

5.5 a. The O.N. of H changes from $+1$ to 0, and the O.N. of I changes from -1 to 0. Reaction is redox. Because one substance changes into two substances, the reaction is a decomposition.

  b. The O.N. of H does not change. The O.N. of O changes from -1 (in a peroxide) to -2 (in water) and 0 (in O_2). This is an example in which the same element is both oxidized and reduced. The reaction is redox. Because one substance changes into two substances, the reaction is a decomposition.

  c. No oxidation numbers change. The reaction is nonredox. This is a double-replacement (metathesis) reaction.

  d. The O.N. of P changes from 0 to $+5$, and the O.N. of O changes from 0 to -2. The reaction is redox. Because two substances combine to form a single substance, the reaction is a combination.

  e. The O.N. of Na does not change. The O.N. of I changes from -1 to 0, and the O.N. of Cl changes from 0 to -1. The reaction is redox. Because the Cl simply replaces the I in the compound, this is a single-replacement reaction.

5.6 a. Total ionic: $2Na^+(aq) + 2I^-(aq) + Cl_2(aq) \rightarrow 2Na^+(aq) + 2Cl^-(aq) + I_2(aq)$

    Na^+ is a spectator ion.

    Net ionic: $2I^-(aq) + Cl_2(aq) \rightarrow 2Cl^-(aq) + I_2(aq)$

  b. Total ionic: $Ca^{2+}(aq) + 2Cl^-(aq) + 2Na^+(aq) + CO_3^{2-}(aq) \rightarrow 2Na^+(aq) + 2Cl^-(aq) + CaCO_3(s)$

    Na^+ and Cl^- are spectator ions.

    Net ionic: $Ca^{2+}(aq) + CO_3^{2-}(aq) \rightarrow CaCO_3(s)$

  c. Total ionic: $Ba^{2+}(aq) + 2OH^-(aq) + 2H^+(aq) + SO_4^{2-}(aq) \rightarrow 2H_2O(\ell) + BaSO_4(s)$

    There are no spectator ions, so the net ionic reaction is the same as the total ionic reaction.

5.7 a. 2. $1 \text{ mol } N_2(g) + 3 \text{ mol } H_2(g) \longrightarrow 2 \text{ mol } NH_3(g)$

    3. $6.02 \times 10^{23} \text{ } N_2 \text{ molecules} + 3(6.02 \times 10^{23}) \text{ } H_2 \text{ molecules} \longrightarrow 2(6.02 \times 10^{23}) \text{ } NH_3 \text{ molecules}$

    4. $28.0 \text{ g } N_2 + 3(2.00)\text{g } H_2 \longrightarrow 2(17.0)\text{g } NH_3$

  b. **Step 1.** $2.11 \text{ mol } H_2$

    Step 2. $2.11 \text{ mol } H_2 = \text{ mol } NH_3$

    Step 3. $2.11 \text{ mol } H_2 \times \dfrac{2 \text{ mol } NH_3}{3 \text{ mol } H_2} = \text{ mol } NH_3$

Step 4. $2.11 \times \dfrac{2 \text{ mol } NH_3}{3} = 1.406 \text{ mol } NH_3 = 1.41 \text{ mol } NH_3$

The answer was rounded to 3 significant figures to match the three in 2.11. The 2 and 3 are exact counting numbers.

  c. **Step 1.** $9.47 \text{ g } H_2$

    Step 2. $9.47 \text{ g } H_2 \qquad\qquad = \text{ g } N_2$

    Step 3. $9.47 \text{ g } H_2 \times \dfrac{28.0 \text{ g } N_2}{6.00 \text{ g } H_2} = \text{ g } N_2$

    Step 4. $9.47 \times \dfrac{28.0 \text{ g } N_2}{6.00} = 44.19 \text{ g } N_2 = 44.2 \text{ g } N_2$

The answer was rounded to 3 significant figures to match the three in 9.47, 28.0 g N_2 and 6.00.

5.8 The mass of NH_3 possible from each reactant is calculated. The reactant giving the smallest mass of NH_3 is the limiting reactant, and the mass of NH_3 calculated for the limiting reactant is the amount that will be produced:

$$2.00 \text{ mol } H_2 \times \frac{34.0 \text{ g } NH_3}{3 \text{ mol } H_2} = 22.7 \text{ g } NH_3$$

$$15.5 \text{ g } N_2 \times \frac{34.0 \text{ g } NH_3}{28.0 \text{ g } N_2} = 18.8 \text{ g } NH_3$$

The results show the limiting reactant to be the 15.5 g of N_2, and the amount of NH_3 produced will be 18.8 g.

5.9 a. $\% \text{ yield} = \dfrac{\text{actual yield}}{\text{theoretical yield}} \times 100$

    $= \dfrac{17.43 \text{ g}}{21.34 \text{ g}} \times 100 = 81.68\%$

  b. The theoretical yield must be calculated.

$$510 \text{ g } CaCO_3 \times \frac{56.08 \text{ g } CaO}{100.1 \text{ g } CaCO_3} = 285.722 \text{ g } CaO = 286 \text{ g } CaO$$

The % yield is then calculated using this theoretical yield:

$$\% \text{ yield} = \frac{\text{actual yield}}{\text{theoretical yield}} \times 100$$

$$= \frac{235 \text{ g}}{286 \text{ g}} \times 100 = 82.2\%$$

CHAPTER 6

6.1 a. $d = \dfrac{\text{mass}}{\text{volume}}$

    copper: $d = \dfrac{114.2 \text{ g}}{12.8 \text{ mL}} = 8.92 \text{ g/mL}$

    glycerin: $d = \dfrac{63.0 \text{ g}}{50.0 \text{ mL}} = 1.26 \text{ g/mL}$

    helium: $d = \dfrac{0.286 \text{ g}}{1500 \text{ mL}} = 1.91 \times 10^{-4} \text{ g/mL}$

  b. Copper: Doubling the pressure will not influence the volume, mass, or density of a solid.

    Glycerin: Doubling the pressure will not significantly influence the volume and will not influence the mass or density of a liquid.

Helium: The increased pressure will cause the volume to decrease. The mass will not be changed, so the density will increase.

6.2 $d = \dfrac{m}{V}$, so $m = Vd$

$m = (1200 \text{ mL})(1.18 \times 10^{-3} \text{ g/mL})$

$= 1.42 \text{ g}$

6.3 $\text{KE} = \frac{1}{2}mv^2 = \frac{1}{2}(3.00 \text{ g})(10.0 \text{ cm/s})^2 = 1.50 \times 10^2 \text{ g cm}^2/\text{s}^2$

$\text{KE} = \frac{1}{2}mv^2 = \frac{1}{2}(3.00 \text{ g})(20.0 \text{ cm/s})^2 = 6.00 \times 10^2 \text{ g cm}^2/\text{s}^2$

6.4 The factors used come from the information in Table 6.3.

   a. $670 \text{ torr} \times \dfrac{1 \text{ atm}}{760 \text{ torr}} = 0.882 \text{ atm}$

   b. $670 \text{ torr} \times \dfrac{14.7 \text{ psi}}{760 \text{ torr}} = 13.0 \text{ psi}$

6.5   **a.** $\text{K} = {}^\circ\text{C} + 273 = 27 + 273 = 300 \text{ K}$

   b. $\text{K} = {}^\circ\text{C} + 273 = 0 + 273 = 273 \text{ K}$

   c. ${}^\circ\text{C} = \text{K} - 273 = 0 - 273 = -273{}^\circ\text{C}$

   d. ${}^\circ\text{C} = \text{K} - 273 = 100 - 273 = -173{}^\circ\text{C}$

6.6   **a.** Equation 6.8 is used. Temperatures must be converted to kelvins:

$P_i = 1.90 \text{ atm} \qquad P_f = 1.00 \text{ atm}$

$V_i = 10.0 \text{ L} \qquad V_f = ?$

$T_i = 30{}^\circ\text{C} = 303 \text{ K} \quad T_f = -10.2{}^\circ\text{C} = 262.8 \text{ K}$

$$\frac{P_i V_i}{T_i} = \frac{P_f V_f}{T_f}$$

$$\frac{(1.90 \text{ atm})(10.0 \text{ L})}{303 \text{ K}} = \frac{(1.00 \text{ atm})(V_f)}{262.8 \text{ K}}$$

Solve for V_f:

$$V_f = \frac{(1.90 \text{ atm})(10.0 \text{ L})(262.8 \text{ K})}{(1.00 \text{ atm})(303 \text{ K})} = 16.5 \text{ L}$$

   b. $P_i = 800 \text{ torr} \qquad P_f = 900 \text{ torr}$

$V_i = 500 \text{ mL} \qquad V_f = 250 \text{ mL}$

$T_i = 300 \text{ K} \qquad T_f = ?$

$$\frac{P_i V_i}{T_i} = \frac{P_f V_f}{T_f}$$

$$\frac{(800 \text{ torr})(500 \text{ mL})}{300 \text{ K}} = \frac{(900 \text{ torr})(250 \text{ mL})}{T_f}$$

Invert both sides and solve for T_f:

$$T_f = \frac{(300 \text{ K})(900 \text{ torr})(250 \text{ mL})}{(800 \text{ torr})(500 \text{ mL})} = 169 \text{ K}$$

$${}^\circ\text{C} = \text{K} - 273 = 169 \text{ K} - 273 = -104{}^\circ\text{C}$$

6.7 To use the ideal gas law, all quantities must have units to match those of R. The only unit that needs to be changed is the temperature:

$$\text{K} = {}^\circ\text{C} + 273 = 30{}^\circ\text{C} + 273 = 303 \text{ K}$$

$$PV = nRT$$

Solve for P:

$$P = \frac{nRT}{V}$$

$$= \frac{(2.15 \text{ mol})\left(0.0821 \dfrac{\text{L atm}}{\text{mol K}}\right)(303 \text{ K})}{(12.6 \text{ L})}$$

$$= 4.24 \text{ atm}$$

6.8 The ideal gas law in the form $PV = mRT/MW$ is solved for MW:

$$\text{MW} = \frac{mRT}{PV}$$

The temperature must be converted to kelvins:

$$\text{K} = 27{}^\circ\text{C} + 273 = 300 \text{ K}$$

$$\text{MW} = \frac{(3.35 \text{ g})\left(0.0821 \dfrac{\text{L atm}}{\text{mol K}}\right)(300 \text{ K})}{(1.21 \text{ atm})(2.00 \text{ L})}$$

$$= 34.1 \text{ g/mol}$$

This matches the molecular weight of H_2S.

6.9 $P_{\text{total}} = P_{\text{He}} + P_{N_2} + P_{O_2}$

The partial pressures of N_2 and O_2 are changed to torr before adding. The factors used come from information in Table 6.3.

$P_{N_2}: 0.200 \text{ atm} \times \dfrac{760 \text{ torr}}{1 \text{ atm}} = 152 \text{ torr}$

$P_{O_2}: 7.35 \text{ psi} \times \dfrac{760 \text{ torr}}{14.7 \text{ psi}} = 380 \text{ torr}$

$P_{\text{total}} = P_{\text{He}} + P_{N_2} + P_{O_2} = 310 \text{ torr} + 152 \text{ torr} + 380 \text{ torr}$

$= 842 \text{ torr}$

6.10 The lighter (least massive) helium molecules will diffuse faster than the neon molecules. The relationship between the diffusion rates is given by Graham's law.

$$\frac{\text{Rate of He}}{\text{Rate of Ne}} = \sqrt{\frac{\text{mass of Ne}}{\text{mass of He}}} = \sqrt{\frac{20.18}{4.003}}$$

$$= \sqrt{5.041} = 2.25$$

Thus, rate of He = (2.25) rate of Ne. So, helium is seen to diffuse at a rate more than twice that of neon.

6.11   **a.** Evaporation is endothermic.

   b. Freezing is exothermic.

   c. Melting is endothermic.

6.12 The lower the forces between particles, the higher will be the vapor pressure.

   a. Both substances form hydrogen bonds, but dispersion forces are stronger in propyl alcohol. Thus, methyl alcohol has the higher vapor pressure.

   b. The only forces acting between the molecules in each case are dispersion forces. These are stronger between the heavier nitrogen molecules, so helium has the higher vapor pressure.

   c. Molecules of HF form strong hydrogen bonds with one another, whereas the only forces between neon molecules are weak dispersion forces. Thus, neon has the higher vapor pressure.

6.13 Heat absorbed = (mass)(specific heat)(temp. change). The specific heat of helium is obtained from Table 6.8.

$$\text{Heat absorbed} = (1000 \text{ g})(1.25 \text{ cal/g °C})(700°C - 25°C)$$
$$= 8.44 \times 10^5 \text{ cal}$$

6.14 Heat absorbed = (mass)(heat of vaporization). The heat of vaporization for water was given in the text.

$$\text{Heat absorbed} = (5000 \text{ g})(540 \text{ cal/g}) = 2.70 \times 10^6 \text{ cal}$$

CHAPTER 7

7.1 **a.** Solvent is gold; solutes are copper, zinc, and nickel.

b. Solvent is nitrogen; solutes are oxygen, argon, and the other gases.

7.2 At 80°C, a saturated solution of KNO_3 contains about 181 g of solute/100 g of H_2O, and a saturated solution of NaBr contains about 118 g of solute per 100 g H_2O. At 50°C both solutions are still saturated, but the KNO_3 solution contains only about 102 g of solute per 100 g H_2O, while the NaBr solution contains about 117 g of solute per 100 g H_2O. Thus, the solubility of KNO_3 is seen to be much more temperature dependent than the solubility of NaBr.

7.3 **a.** The solubility of nitrates and lack of solubility of sulfates and carbonates of barium makes barium nitrate, $Ba(NO_3)_2$, the salt to use. More of it would dissolve and would release more Ba^{2+} ions.

b. Like dissolves like, so oil will dissolve best in oily solvents such as light mineral oil or gasoline. Light mineral oil is the solvent used because it leaves a residue that is similar to the natural oily secretions found on the bird feathers. Gasoline would be hazardous to use (flammable) and leaves no such useful residue.

7.4 **a.** The data may be substituted directly into Equation 7.5:

$$M = \frac{1.25 \text{ mol solute}}{2.50 \text{ L solution}} = 0.500 \frac{\text{mol solute}}{\text{L solution}}$$

The solution is 0.500 molar, or 0.500 M.

b. In this problem, the volume of solution must be converted to liters before the data are substituted into Equation 7.5:

$$(225 \text{ mL solution})\left(\frac{1 \text{ L}}{1000 \text{ mL}}\right) = 0.225 \text{ L solution}$$

$$M = \frac{0.486 \text{ mol solute}}{0.225 \text{ L solution}} = 2.16 \frac{\text{mol solute}}{\text{L solution}}$$

The solution is 2.16 molar, or 2.16 M.

c. In this problem, the volume of solution must be converted to liters and the mass of solute must be converted to moles before the data can be substituted into Equation 7.5:

$$(100 \text{ mL solution})\left(\frac{1 \text{ L}}{1000 \text{ mL}}\right) = 0.100 \text{ L solution}$$

$$(2.60 \text{ g NaCl})\left(\frac{1 \text{ mol NaCl}}{58.44 \text{ g NaCl}}\right) = 0.0445 \text{ mol NaCl}$$

In the last calculation, the factor $\frac{1 \text{ mol NaCl}}{58.44 \text{ g NaCl}}$ comes from the calculated formula weight of 58.44 u for NaCl. The data are now substituted into Equation 7.5:

$$M = \frac{0.0445 \text{ mol NaCl}}{0.100 \text{ L solution}} = 0.445 \frac{\text{mol NaCl}}{\text{L solution}}$$

The solution is 0.445 molar, or 0.445 M.

7.5 **a.** To calculate %(w/w), the mass of solute contained in a specific mass of solution is needed. The mass of solute is 0.900 g, and the mass of solution is equal to the solvent mass (100 g) plus the solute mass (0.900 g):

$$\%(w/w) = \frac{\text{solute mass}}{\text{solution mass}} \times 100 = \frac{0.900 \text{ g}}{100.9 \text{ g}} \times 100$$
$$= 0.892 \%(w/w)$$

To calculate %(w/v), the number of grams of solute must be known along with the number of milliliters of solution. The mass of solute is 0.900 g, and the solution volume is 100 mL:

$$\%(w/v) = \frac{\text{g of solute}}{\text{mL of solution}} \times 100 = \frac{0.900 \text{ g}}{100 \text{ mL}} \times 100$$
$$= 0.900 \%(w/v)$$

b. The given quantity is 30 mL of beverage, and the desired quantity is milliliters of pure alcohol. The %(v/v) provides the necessary factor:

$$30 \text{ mL beverage} \times \frac{45 \text{ mL alcohol}}{100 \text{ mL beverage}} = 14 \text{ mL alcohol}$$

7.6 **a.** $M = \dfrac{\text{mol of solute}}{\text{liters of solution}}$

$$\text{mol of solute} = (M)(\text{liters of solution})$$
$$= (1.00 \text{ M})(0.500 \text{ L})$$
$$= 0.500 \text{ mol of solute}$$

The solute is $MgCl_2$, which has a formula weight of 95.2 u. The formula weight provides the factor in the following calculation:

$$0.500 \text{ mol MgCl}_2 \times \frac{95.2 \text{ g MgCl}_2}{1 \text{ mol MgCl}_2} = 47.6 \text{ g MgCl}_2$$

The solution is prepared by putting 47.6 g of $MgCl_2$ into a 500-mL flask and adding pure water to the mark, making certain all the solid solute dissolves and the resulting solution is well mixed.

b. $\%(w/v) = \dfrac{\text{g of solute}}{\text{mL of solution}} \times 100$

$$\text{g of solute} = \frac{\%(w/v)(\text{mL of solution})}{100}$$
$$= \frac{(12.0\%)(100 \text{ mL})}{100} = 12.0 \text{ g}$$

The solution is prepared by putting 12.0 g of $MgCl_2$ into a 100-mL flask and adding pure water to the mark, making certain all the solid solute dissolves and the resulting solution is well mixed.

c. $\%(v/v) = \dfrac{\text{solute volume}}{\text{solution volume}} \times 100$

$$\text{solute volume} = \frac{\%(v/v)(\text{solution volume})}{100}$$
$$= \frac{(20.0\%)(1.00 \text{ L})}{100} = 0.200 \text{ L}$$
$$= 200 \text{ mL}$$

The solution is prepared by putting 200 mL of ethylene glycol into a 1.00-L flask and adding pure water to the mark, making certain the two liquids are completely mixed to form the final solution.

7.7 Equation 7.9 is used to calculate the volume of 6.00 M NaOH solution needed.

$$(C_c)(V_c) = (C_d)(V_d)$$

$$(6.00 \text{ M})(V_c) = (0.250 \text{ M})(500 \text{ mL})$$

$$V_c = \frac{(0.250 \text{ M})(500 \text{ mL})}{(6.00 \text{ M})} = 20.8 \text{ mL}$$

The solution is prepared by putting 20.8 mL of 6.00 M NaOH solution into a 500-mL flask, adding pure water to the mark, and making certain the resulting solution is well mixed.

7.8 **a.** Because $CaCl_2$ is a strong electrolyte, it dissociates completely:

$$CaCl_2 \rightarrow Ca^{2+} + 2Cl^-. \text{ Thus, } n = 3.$$

$$\Delta T_b = nK_bM = (3)(0.52°C/M)(0.100 \text{ M}) = 0.16°C$$

Because the boiling point of a solution is higher than the boiling point of the pure solvent, ΔT_b is added to the boiling point of the pure solvent.

Solution B.P. = 100.00°C + 0.16°C = 100.16°C

$$\Delta T_f = nK_fM = (3)(1.86°C/M)(0.100 \text{ M}) = 0.558°C$$

Because the freezing point of a solution is lower than the freezing point of the pure solvent, ΔT_f is subtracted from the freezing point of the solvent.

Solution F.P. = 0.00°C − 0.558°C = −0.558°C = −0.56°C

$$\pi = nMRT$$

$$= (3)(0.100 \text{ mol/L})(62.4 \text{ L torr/K mol})(300 \text{ K})$$

$$= 5.62 \times 10^3 \text{ torr} = 7.39 \text{ atm}$$

b. Because ethylene glycol does not dissociate in solution, $n = 1$.

$$\Delta T_b = nK_bM = (1)(0.52°C/M)(0.100 \text{ M}) = 0.052°C$$

Solution B.P. = 100.00°C + 0.052°C = 100.05°C

$$\Delta T_f = nK_fM = (1)(1.86°C/M)(0.100 \text{ M}) = 0.186°C$$

Solution F.P. = 0.00°C − 0.186°C = −0.186°C

$$= −0.19°C$$

$$\pi = nMRT$$

$$= (1)(0.100 \text{ mol/L})(62.4 \text{ L torr/K mol})(300 \text{ K})$$

$$= 1.87 \times 10^3 \text{ torr} = 2.46 \text{ atm}$$

CHAPTER 8

8.1 Because only Ce^{4+} and Fe^{2+} are mixed, the initial concentration of Ce^{3+} is 0.

$$\text{Rate} = \frac{C_t - C_0}{\Delta t} = \frac{1.50 \times 10^{-5} \text{ mol/L} - 0 \text{ mol/L}}{75.0 \text{ s}}$$

$$= 2.00 \times 10^{-7} \frac{\text{mol/L}}{\text{s}}$$

8.2 **a.** $K = \dfrac{[NO_2]^2}{[N_2O_4]}$

b. $K = \dfrac{[N_2O][NO_2]}{[NO]^3}$

8.3 **a.** $K = \dfrac{[IBr]^2}{[Br_2][I_2]}$

$$= \frac{(1.96 \times 10^{-2})^2}{(1.50 \times 10^{-1})(5.00 \times 10^{-2})}$$

$$= 5.12 \times 10^{-2}$$

The small value of K indicates the equilibrium position is toward the left.

b. $K = \dfrac{[NH_3]^2}{[N_2][H_2]^3}$

$$= \frac{(9.00)^2}{(9.23 \times 10^{-3})(2.77 \times 10^{-2})^3}$$

$$= 4.13 \times 10^8$$

The large K value indicates the equilibrium position is toward the right.

8.4 **a.** $\text{Heat} + NH_4NO_3(s) \rightleftarrows NH_4^+(aq) + NO_3^-(aq)$

The equilibrium will shift to the right when heat is added in an attempt to use up the added heat. This shift corresponds to dissolving more of the solid at the higher temperature. The solubility is higher at a higher temperature.

b. $2SO_2(g) + O_2(g) \rightleftarrows 2SO_3(g)$

If O_2 were removed from an equilibrium mixture, the equilibrium would shift left in an attempt to replenish the removed O_2. This shift would reduce the amount of SO_3 present at equilibrium, which means the concentration of SO_3 would be lower than it was in the initial equilibrium.

CHAPTER 9

9.1 **a.** $HC_2H_3O_2 + H_2O \rightleftharpoons H_3O^+ + C_2H_3O_2^-$
 acid   base   acid   base
 conjugate pair
 conjugate pair

b. $NO_2^- + H_2O \rightleftharpoons OH^- + HNO_2$
 base   acid   base   acid
 conjugate pair
 conjugate pair

c. $HS^- + H_2O \rightleftharpoons H_3O^+ + S^{2-}$
 acid   base   acid   base
 conjugate pair
 conjugate pair

9.2 **a.** The anhydrous compound is called hydrogen iodide. The name of the water solution is obtained by dropping *hydrogen* from the anhydrous compound name and adding the prefix *hydro-* to the stem *iod*. The *-ide* suffix on the *iod* stem is replaced by the suffix *-ic* to give the name *hydroiodic*. The word *acid* is added, giving the final name *hydroiodic acid* for the water solution.

b. The anhydrous compound is called hydrogen bromide. The name of the water solution is obtained the same way it was for HI(aq) in part a, using the stem *brom* in place of *iod*. The resulting name for the water solution is *hydrobromic acid*.

9.3 The removal of the two H^+ ions leaves behind the CO_3^{2-} polyatomic ion. From Table 4.6, this ion is the carbonate ion. According to rules 1-4, the *-ate* suffix is replaced by the *-ic* suffix to give the name *carbonic acid*.

9.4 **a.** Solution is basic because $[OH^-]$ is greater than 1.0×10^{-7} mol/L:

$$[H_3O^+] = \frac{1.0 \times 10^{-14}(\text{mol/L})^2}{[OH^-]} = \frac{1.0 \times 10^{-14}(\text{mol/L})^2}{1.0 \times 10^{-5} \text{ mol/L}}$$

$$= 1.0 \times 10^{-9} \text{ mol/L}$$

b. Solution is basic because $[H_3O^+]$ is less than 1.0×10^{-7} mol/L:

$$[OH^-] = \frac{1.0 \times 10^{-14}(\text{mol/L})^2}{[H_3O^+]} = \frac{1.0 \times 10^{-14}(\text{mol/L})^2}{1.0 \times 10^{-9} \text{ mol/L}}$$

$$= 1.0 \times 10^{-5} \text{ mol/L}$$

This solution has the same $[H_3O^+]$ and $[OH^-]$ as the solution in part a.

c. Solution is acidic because $[H_3O^+]$ is greater than 1.0×10^{-7} mol/L:

$$[OH^-] = \frac{1.0 \times 10^{-14}(\text{mol/L})^2}{[H_3O^+]} = \frac{1.0 \times 10^{-14}(\text{mol/L})^2}{1.0 \times 10^{-2} \text{ mol/L}}$$

$$= 1.0 \times 10^{-12} \text{ mol/L}$$

9.5 **a.** $[H^+] = 1 \times 10^{-14}$ mol/L. pH is the negative of the exponent used to express $[H^+]$, so pH $= -(-14) = 14.0$.

b. $[OH^-] = 1.0$ mol/L. The value of $[H^+]$ is calculated:

$$[H^+] = \frac{1 \times 10^{-14}(\text{mol/L})^2}{[OH^-]} = \frac{1 \times 10^{-14}(\text{mol/L})^2}{1.0 \text{ mol/L}}$$

$$= 1 \times 10^{-14} \text{ mol/L}$$

pH is the negative of the exponent used to express $[H^+]$, so pH $= -(-14) = 14.0$.

c. $[OH^-] = 1 \times 10^{-8}$ mol/L. The value of $[H^+]$ is calculated:

$$[H^+] = \frac{1 \times 10^{-14}(\text{mol/L})^2}{[OH^-]} = \frac{1 \times 10^{-14}(\text{mol/L})^2}{1 \times 10^{-8} \text{ mol/L}}$$

$$= 1 \times 10^{-6} \text{ mol/L}$$

pH is the negative of the exponent used to express $[H^+]$, so pH $= -(-6) = 6.0$.

9.6 **a.** pH $= 10.0$, and

$$[H^+] = 1 \times 10^{-pH} = 1 \times 10^{-10} \text{ mol/L}$$

$$[OH^-] = \frac{K_w}{[H^+]} = \frac{1 \times 10^{-14} (\text{mol/L})^2}{1 \times 10^{-10} \text{ mol/L}} = 1 \times 10^{-4} \text{ mol/L}$$

b. pH $= 4.0$, and $[H^+] = 1 \times 10^{-pH} = 1 \times 10^{-4}$

$$[OH^-] = \frac{K_w}{[H^+]} = \frac{1 \times 10^{-14} (\text{mol/L})^2}{1 \times 10^{-4} \text{ mol/L}} = 1 \times 10^{-10} \text{ mol/L}$$

c. pH $= 5.0$, and $[H^+] = 1 \times 10^{-pH} = 1 \times 10^{-5}$

$$[OH^-] = \frac{K_w}{[H^+]} = \frac{1 \times 10^{-14} (\text{mol/L})^2}{1 \times 10^{-5} \text{ mol/L}} = 1 \times 10^{-9} \text{ mol/L}$$

9.7 **a.** The calculator answer is 4.3768. The correctly rounded value is 4.38.

b. The calculator answer is 8.0915. The correctly rounded value is 8.09.

9.8 **a.** The calculator answer is 0.0017783, or 1.7783×10^{-3}. The correctly rounded value should have two significant figures to match the two figures to the right of the decimal in the pH value. The answer is 1.8×10^{-3} mol/L.

b. The calculator answer is 4.6774×10^{-9}. The correctly rounded answer is 4.7×10^{-9} mol/L.

9.9 According to Table 9.4, concentrated NH_3 stock solution is 15 M. Equation 7.9 is used to determine the amount of this solution needed.

$$C_c V_c = C_d V_d$$

$$(15 \text{ M})(V_c) = (3.0 \text{ M})(500 \text{ mL})$$

$$V_c = \frac{(3.0 \text{ M})(500 \text{ mL})}{(15 \text{ M})}$$

$$= 100 \text{ mL}$$

Because the resulting solution molarity is given using only two significant figures, the volumes can be measured using graduated cylinders. Measure 100 mL of stock (15 M) aqueous ammonia, put it into a container and add 400 mL of distilled water. Mix the resulting solution well.

9.10 $2HCl(aq) + Sr(HCO_3)_2(s) \rightarrow$
$$SrCl_2(aq) + 2CO_2(g) + 2H_2O(\ell)$$

Total Ionic: $2H^+(aq) + 2Cl^-(aq) + Sr(HCO_3)_2(s) \rightarrow$
$$Sr^{2+}(aq) + 2Cl^-(aq) + 2CO_2(g) + 2H_2O(\ell)$$
Cl^- is a spectator ion.

Net Ionic: $2H^+(aq) + Sr(HCO_3)_2(s) \rightarrow$
$$Sr^{2+}(aq) + 2CO_2(g) + 2H_2O(\ell)$$

9.11 **a.** Molecular: $Ca(s) + 2H_2O(\ell) \rightarrow H_2(g) + Ca(OH)_2(s)$

Total Ionic: $Ca(s) + 2H_2O(\ell) \rightarrow H_2(g) + Ca(OH)_2(s)$

Net Ionic: $Ca(s) + 2H_2O(\ell) \rightarrow H_2(g) + Ca(OH)_2(s)$

All three equations are the same because none of the reactants or products form ions in solution.

b. Molecular: $Mg(s) + H_2SO_4(aq) \rightarrow H_2(g) + MgSO_4(aq)$

Total Ionic: $Mg(s) + 2H^+(aq) + SO_4^{2-}(aq) \rightarrow$
$$H_2(g) + Mg^{2+}(aq) + SO_4^{2-}(aq)$$
The SO_4^{2-} is a spectator ion.

Net Ionic: $Mg(s) + 2H^+(aq) \rightarrow H_2(g) + Mg^{2+}(aq)$

9.12 **a.** Molecular: $HNO_3(aq) + NaOH(aq) \rightarrow$
$$H_2O(\ell) + NaNO_3(aq)$$

Total Ionic: $H^+(aq) + NO_3^-(aq) + Na^+(aq) + OH^-(aq) \rightarrow$
$$H_2O(\ell) + Na^+(aq) + NO_3^-(aq)$$
Na^+ and NO_3^- are spectator ions.

Net Ionic: $H^+(aq) + OH^-(aq) \rightarrow H_2O(\ell)$

b. Molecular: $H_2SO_4(aq) + 2KOH(aq) \rightarrow$
$$2H_2O(\ell) + K_2SO_4(aq)$$

Total Ionic:
$2H^+(aq) + SO_4^{2-}(aq) + 2K^+(aq) + 2OH^-(aq) \rightarrow$
$$2H_2O(\ell) + 2K^+(aq) + SO_4^{2-}(aq)$$

K^+ and SO_4^{2-} are spectator ions.

Net Ionic: $2H^+(aq) + 2OH^-(aq) \rightarrow 2H_2O(\ell)$

9.13 $6HCl(aq) + 2Al(s) \rightarrow 2AlCl_3(aq) + 3H_2(g)$

$6HCl(aq) + Al_2O_3(s) \rightarrow 2AlCl_3(aq) + 3H_2O(\ell)$

$3HCl(aq) + Al(OH)_3(s) \rightarrow AlCl_3(aq) + 3H_2O(\ell)$

$6HCl(aq) + Al_2(CO_3)_3(s) \rightarrow$
$$2AlCl_3(aq) + 3H_2O(\ell) + 3CO_2(g)$$

$3HCl(aq) + Al(HCO_3)_3(s) \rightarrow$
$$AlCl_3(aq) + 3H_2O(\ell) + 3CO_2(g)$$

9.14 a. The dissociation reaction is $NaCl(aq) \rightarrow Na^+(aq) + Cl^-(aq)$. Thus, 1 mol of NaCl produces 1 mol of Na^+ or 1 mol of positive charges. Thus, 1 mol NaCl = 1 eq NaCl. Therefore,

$$0.10 \text{ mol NaCl} \times \frac{1 \text{ eq NaCl}}{1 \text{ mol NaCl}} = 0.10 \text{ eq NaCl}$$

Also, because 1 eq = 1000 meq,

$$0.10 \text{ eq} \times \frac{1000 \text{ meq}}{1 \text{ eq}} = 1.0 \times 10^2 \text{ meq}$$

b. The dissociation reaction is $Mg(NO_3)_2(aq) \rightarrow Mg^{2+}(aq) + 2NO_3^-(aq)$. Thus, 1 mol of $Mg(NO_3)_2$ produces 1 mol of Mg^{2+} or 2 mol of positive charges. Thus, 1 mol $Mg(NO_3)_2$ = 2 eq $Mg(NO_3)_2$. Therefore,

$$0.10 \text{ mol Mg(NO}_3)_2 \times \frac{2 \text{ eq Mg(NO}_3)_2}{1 \text{ mol Mg(NO}_3)_2}$$
$$= 0.20 \text{ eq Mg(NO}_3)_2$$

Also, because 1 eq = 1000 meq,

$$0.20 \text{ eq} \times \frac{1000 \text{ meq}}{1 \text{ eq}} = 2.0 \times 10^2 \text{ meq}$$

9.15 The Cl^- ion has a single charge, so we may write:

$$1.00 \text{ mol NaCl} = 1.00 \text{ mol Cl}^- = 1.00 \text{ eq Cl}^- = 1.00 \text{ eq NaCl}$$

As in Example 9.11, the pattern for the calculation is liters solution $A \rightarrow$ eq A, and the pathway is liters Cl^- solution $\rightarrow$ eq Cl^-. The conversion factor comes from the concentration of 0.103 eq/L given in the problem:

$$(0.250 \text{ L Cl}^- \text{ solution})\left(\frac{0.103 \text{ eq Cl}^-}{1.00 \text{ L solution}}\right) = 0.0258 \text{ eq Cl}^-$$

The equivalents of Cl^- are converted to the quantities asked for by using factors from the relationships given above and the formula weight for NaCl, 58.44 u:

$$(0.0258 \text{ eq Cl}^-)\left(\frac{1.00 \text{ eq NaCl}}{1.00 \text{ eq Cl}^-}\right) = 0.0258 \text{ eq NaCl}$$

$$(0.0258 \text{ eq NaCl})\left(\frac{1.00 \text{ mol NaCl}}{1.00 \text{ eq NaCl}}\right) = 0.0258 \text{ mol NaCl}$$

$$(0.0258 \text{ mol NaCl})\left(\frac{58.44 \text{ g NaCl}}{1.00 \text{ mol NaCl}}\right) = 1.51 \text{ g NaCl}$$

9.16 a. $HPO_4^{2-} \rightleftarrows H^+ + PO_4^{3-}$

$$K_a = \frac{[H^+][PO_4^{3-}]}{[HPO_4^{2-}]}$$

b. $HNO_2 \rightleftarrows H^+ + NO_2^-$

$$K_a = \frac{[H^+][NO_2^-]}{[HNO_2]}$$

c. $HF \rightleftarrows H^+ + F^-$

$$K_a = \frac{[H^+][F^-]}{[HF]}$$

9.17 a. In a series of similar oxyacids, acid strength increases with increases in the number of oxygens: $HClO_3$, $HClO_2$, $HClO$.

b. The anion from the weakest acid will be the strongest base. HNO_3 is a stronger acid than HNO_2, so NO_2^- is a stronger base than NO_3^-.

c. The anion of an acid behaves as a weaker acid than the acid from which it came. The anion actually behaves as a base in this case, so $HC_2H_3O_2$ is a stronger acid than $C_2H_3O_2^-$.

9.18 The pattern is liters solution $A \rightarrow$ mol B, and the pathway is liters NaOH solution $\rightarrow$ mol NaOH $\rightarrow$ mol H_3PO_4. In combined form, the steps in the factor-unit method are:

Step 1. 0.0141 L NaOH solution

Step 2. 0.0141 L NaOH solution = mol H_3PO_4

Step 3. $0.0141 \text{ L NaOH solution} \times \dfrac{0.250 \text{ mol NaOH}}{1 \text{ L NaOH solution}}$

$$\times \frac{1 \text{ mol H}_3PO_4}{3 \text{ mol NaOH}} = \text{mol H}_3PO_4$$

Step 4. $(0.0141)\left(\dfrac{0.250}{1}\right)\left(\dfrac{1 \text{ mol H}_3PO_4}{3}\right)$

$$= 0.00118 \text{ mol H}_3PO_4$$

The molarity of the H_3PO_4 solution is calculated:

$$M = \frac{\text{mol H}_3PO_4}{\text{L H}_3PO_4 \text{ solution}}$$
$$= \frac{0.00118 \text{ mol H}_3PO_4}{0.0250 \text{ L H}_3PO_4 \text{ solution}}$$
$$= \frac{0.0472 \text{ mol H}_3PO_4}{\text{L H}_3PO_4 \text{ solution}} = 0.0472 \text{ M}$$

9.19 a. The ions resulting from dissociation are Na^+ and NO_3^-. Na^+ is the cation of a strong base, and NO_3^- is the anion of a strong acid. Neither ion will hydrolyze significantly in water, so the solution pH will be about 7.

b. The ions resulting from dissociation are Na^+ and NO_2^-. Na^+ is the cation of a strong base and will not hydrolyze significantly. NO_2^- is the anion of a weak acid and will behave as a base in the hydrolysis reaction. The solution will therefore be basic, and the pH will be higher than 7.

c. The ions resulting from dissociation are K^+ and BO_3^{3-}. K^+ is the cation of a strong base and will not hydrolyze significantly. BO_3^{3-} is the anion of a weak acid and will behave as a base in the hydrolysis reaction. The solution will therefore be basic, and the pH will be higher than 7.

9.20 a. Because the acid (HCOOH) and conjugate base ($HCOO^-$) concentrations are the same, the pH will equal pK_a for the acid. This is shown by the following calculation:

$$pH = pK_a + \log \frac{[B^-]}{[HB]}$$
$$= 3.74 + \log \frac{[HCOO^-]}{[HCOOH]}$$
$$= 3.74 + \log \frac{0.22 \text{ mol/L}}{0.22 \text{ mol/L}}$$
$$= 3.74 + \log 1$$
$$= 3.74 + 0 = 3.74$$

b. The acid is H_2SO_3, and the conjugate base is HSO_3^- (produced when the $NaHSO_3$ salt dissociates).

$$pH = pK_a + \log \frac{[B^-]}{[HB]}$$
$$= 1.82 + \log \frac{(0.10 \text{ mol/L})}{(0.25 \text{ mol/L})}$$
$$= 1.82 + \log 0.40$$
$$= 1.82 - 0.40$$
$$= 1.42$$

c. The acid is H_2CO_3 and the conjugate base is HCO_3^-. The desired pH is 7.40. Substitution into Equation 9.54 gives:

$$pH = pK_a + \log \frac{[HCO_3^-]}{[H_2CO_3]}$$

$$7.40 = 6.37 + \log \frac{[HCO_3^-]}{[H_2CO_3]}$$

$$7.40 - 6.37 = \log \frac{[HCO_3^-]}{[H_2CO_3]}$$

$$1.03 = \log \frac{[HCO_3^-]}{[H_2CO_3]}$$

We now know the log of the desired ratio. Use Step 3 in Table 9.3 to get the antilog of 1.03. The correctly rounded value is 11. Thus, the concentration of HCO_3^- must be 11 times the concentration of H_2CO_3 in the blood to maintain the desired pH.

CHAPTER 10

10.1 a. $^{131}_{53}I$ **b.** $^{0}_{-1}\beta$ **c.** $^{4}_{4}X$ **d.** $^{0}_{-4}X$

10.2 $^{234}_{90}Th \rightarrow {}^{0}_{-1}\beta + {}^{234}_{91}Pa$

In general: $^{A}_{Z}X \rightarrow {}^{0}_{-1}\beta + {}_{Z+1}^{A}Y$

When a β particle is emitted, the daughter has the same mass number and an atomic number higher by 1 than that of the decaying nucleus.

10.3 $^{A}_{Z}X \rightarrow {}^{0}_{1}\beta + {}_{Z-1}^{A}Y$

The atomic number of the daughter is 1 less than that of the decaying nucleus.

10.4 $^{A}_{Z}X + {}^{0}_{-1}e \rightarrow {}_{Z-1}^{A}Y$

The atomic number of the daughter is 1 less than that of the decaying nucleus.

10.5 a. $^{50}_{25}Mn \rightarrow {}^{0}_{1}\beta + {}^{50}_{24}Cr$

b. $^{54}_{25}Mn + {}^{0}_{-1}e \rightarrow {}^{54}_{24}Cr$

c. $^{56}_{25}Mn \rightarrow {}^{0}_{-1}\beta + {}^{56}_{26}Fe$

d. $^{224}_{88}Ra \rightarrow {}^{4}_{2}\alpha + {}^{220}_{86}Rn$

10.6 During the first half-life, the original 200.00 mg will be reduced to 100.00 mg. This will be reduced to 50.00 mg during the second half-life and to 25.00 mg during the third. Thus, 3 half-lives or 3(7.7 min) = 23.1 min will be required.

10.7 The elapsed time of 79.8 hours is 6 half-lives (79.8/13.3 = 6). The fraction remaining is $1/2 \times 1/2 \times 1/2 \times 1/2 \times 1/2 \times 1/2 = 1/64$ of the original diagnostic dose.

10.8 Equation 10.2 is used: $\dfrac{I_x}{I_y} = \dfrac{d_y^2}{d_x^2}$

$$\frac{10.0 \text{ units}}{I_5} = \frac{(5 \text{ ft})^2}{(25 \text{ ft})^2}$$

$$I_5 = \frac{(10.0 \text{ units})(25 \text{ ft})^2}{(5 \text{ ft})^2} = 250 \text{ units}$$

10.9 $^{55}_{25}Mn + {}^{1}_{1}p \rightarrow {}^{1}_{0}n + {}^{55}_{26}Fe$

The product is iron-55.

CHAPTER 11

11.1 Look for the presence of carbon atoms.

    a. inorganic   **d.** inorganic

    b. organic   **e.** organic

    c. organic   **f.** inorganic

11.2 Refer to Table 11.1 for help.

    a. organic

    b. organic

    c. inorganic

11.3 Compound (a) is a structural isomer because it has the same molecular formula but a different structural formula.

11.4 a. $CH_3 - \overset{\overset{\displaystyle OH}{|}}{CH} - CH_2 - CH_2 - CH_3$

    b. $CH_3 - \overset{\overset{\displaystyle}{|}}{\underset{\underset{\displaystyle CH_3}{|}}{CH}} - CH_2 - \overset{\overset{\displaystyle O}{||}}{C} - OH$

11.5 a. The number of hydrogen atoms is twice the carbon atoms plus two: C_8H_{18}.

    b. $CH_3-CH_2-CH_2-CH_2-CH_2-CH_2-CH_2-CH_3$

11.6 a. Same molecule: In both molecules, the five carbons are bonded in a continuous chain.

    b. Same molecule: In both molecules, there is a continuous chain of four carbons with a branch at position 2.

    c. Structural isomers: Both molecules have a continuous chain of five carbons, but the branch is located at different positions.

11.7 a. a butane

    b. a pentane

    c. a butane

11.8 a.

absolute specificity The characteristic of an enzyme that it acts on one and only one substrate.

absolute zero The temperature at which all motion stops; a value of 0 on the Kelvin scale.

acetal A compound that contains the $-\overset{\displaystyle \text{OR}}{\underset{\displaystyle \text{OR}}{\overset{|}{\underset{|}{\text{C}}}}}-\text{H}$ arrangement of atoms.

acetone breath A condition in which acetone can be detected in the breath.

acid dissociation constant The ratio of product molarities (H^+ and acid anion) multiplied together and divided by the molarity of undissociated acid. All molarities are measured at equilibrium.

acidic solution A solution in which the concentration of H_3O^+ is greater than the concentration of OH^-; also, a solution in which pH is less than 7.

acidosis A low blood pH.

activation energy Energy needed to start some spontaneous processes. Once started, the processes continue without further stimulus or energy from an outside source.

activator A substance that binds to an allosteric enzyme and increases its activity.

active site The location on an enzyme where a substrate is bound and catalysis occurs.

activity series A tabular representation of the tendencies of metals to react with H^+.

acute radiation syndrome A condition following short-term exposure to intense radiation.

addition polymer A polymer formed by the linking together of many alkene molecules through addition reactions.

addition reaction A reaction in which a compound adds to a multiple bond.

adipose tissue A kind of connective tissue where triglycerides are stored.

aerobic In the presence of oxygen.

alcohol A compound in which an $-OH$ group is connected to an aliphatic carbon atom.

alcoholic fermentation The conversion of glucose to ethanol.

aldehyde A compound that contains the $-\overset{\displaystyle \text{O}}{\overset{||}{\text{C}}}-\text{H}$ group; the general formula is $R-\overset{\displaystyle \text{O}}{\overset{||}{\text{C}}}-\text{H}$.

aliphatic compound Any organic compound that is not aromatic.

alkaline solution See **basic solution**.

alkaloids A class of nitrogen-containing organic compounds obtained from plants.

alkalosis An abnormally high blood pH.

alkane A hydrocarbon that contains only single bonds.

alkene A hydrocarbon that contains one or more double bonds.

alkoxy group The $-O-R$ functional group.

alkyl group A group differing by one hydrogen from an alkane.

alkyl halide See **haloalkane**.

alkyne A hydrocarbon that contains one or more triple bonds.

allosteric enzyme An enzyme with quaternary structure whose activity is changed by the binding of modulators.

alpha-amino acid An organic compound containing both an amino group and a carboxylate group, with the amino group attached to the carbon next to the carboxylate group.

α-helix The helical structure in proteins that is maintained by hydrogen bonds.

alpha particle The particle that makes up alpha rays; identical to the helium nucleus and composed of two protons and two neutrons.

amide An organic compound having the functional group $-\overset{\displaystyle \text{O}}{\overset{||}{\text{C}}}-\overset{\displaystyle }{\underset{|}{\text{N}}}-$.

amide linkage The carbonyl carbon-nitrogen single bond of the amide group.

amine An organic compound derived by replacing one or more of the hydrogen atoms of ammonia with alkyl or aromatic groups, as in RNH_2, R_2NH, and R_3N.

amino acid pool The total supply of amino acids in the body.

amino acid residue An amino acid that is a part of a peptide, polypeptide, or protein chain.

amphetamines A class of drugs structurally similar to epinephrine, used to stimulate the central nervous system.

anabolism All reactions involved in the synthesis of biomolecules.

anaerobic In the absence of oxygen.

anion A negatively charged ion.

anomeric carbon An acetal, ketal, hemiacetal, or hemiketal carbon atom giving rise to two stereoisomers.

anomers Stereoisomers that differ in the three-dimensional arrangement of groups at the carbon of an acetal, ketal, hemiacetal, or hemiketal group.

antibiotic A substance produced by one microorganism that kills or inhibits the growth of other microorganisms.

antibody A substance that helps protect the body from invasion by foreign materials known as antigens.

anticodon A three-base sequence in tRNA that is complementary to one of the codons in mRNA.

antioxidant A substance that prevents another substance from being oxidized.

apoenzyme A catalytically inactive protein formed by removal of the cofactor from an active enzyme.

aromatic hydrocarbon Any organic compound that contains the characteristic benzene ring or similar feature.

Arrhenius acid Any substance that provides H^+ ions when dissolved in water.

Arrhenius base Any substance that provides OH^- ions when dissolved in water.

atom The limit of chemical subdivision for matter.

atomic mass unit (u) A unit used to express the relative masses of atoms. One u is equal to $\frac{1}{12}$ the mass of an atom of carbon-12.

atomic number of an atom A number equal to the number of protons in the nucleus of an atom. Symbolically it is represented by Z.

atomic orbital A volume of space around atomic nuclei in which electrons of the same energy move. Groups of orbitals with the same n values form subshells.

atomic weight The mass of an average atom of an element expressed in atomic mass units.

Avogadro's law Equal volumes of gases measured at the same temperature and pressure contain equal numbers of molecules.

balanced equation An equation in which the number of atoms of each element in the reactants is the same as the number of atoms of that same element in the products.

basic solution A solution in which the concentration of OH^- is greater than the concentration of H_3O^+; also, a solution in which pH is greater than 7; also called alkaline solution.

basic unit of measurement A specific unit from which other units for the same quantity are obtained by multiplication or division.

becquerel A physical unit of radiation measurement corresponding to one nuclear disintegration per second.

Benedict's reagent A mild oxidizing solution containing Cu^{2+} ions used to test for the presence of aldehydes.

β-oxidation process A pathway in which fatty acids are broken down into molecules of acetyl CoA.

beta particle The particle that makes up beta rays; identical to an electron but produced in the nucleus when a neutron is changed into a proton and an electron.

β-pleated sheet A secondary protein structure in which protein chains are aligned side by side in a sheetlike array held together by hydrogen bonds.

binary compound A compound made up of two different elements.

biochemistry A study of the compounds and processes associated with living organisms.

biological unit of radiation A radiation measurement unit that indicates the damage caused by radiation in living tissue.

biomolecule A general term referring to an organic compound essential to life.

blood sugar level The amount of glucose present in blood, normally expressed as milligrams per 100 mL of blood.

boiling point The temperature at which the vapor pressure of a liquid is equal to the prevailing atmospheric pressure.

bond polarization A result of the attraction of shared electrons to the more electronegative atom of a bonded pair of atoms.

Boyle's law A gas law that describes the pressure and volume behavior of a gas sample kept at constant temperature. Mathematically, it is $PV = k$.

branched alkane An alkane in which at least one carbon atom is not part of a continuous chain.

branching chain reaction See **expanding chain reaction**.

breeder reaction A nuclear reaction in which isotopes that will not undergo spontaneous fission are changed into isotopes that will.

Brønsted acid Any hydrogen-containing substance that is capable of donating a proton (H^+) to another substance.

Brønsted base Any substance capable of accepting a proton from another substance.

buffer A solution with the ability to resist changing pH when acids (H^+) or bases (OH^-) are added.

buffer capacity The amount of acid (H^+) or base (OH^-) that can be absorbed by a buffer without causing a significant change in pH.

carbaminohemoglobin Hemoglobin combined with carbon dioxide.

carbocation An ion of the form $-\overset{+}{\underset{|}{C}}-$.

carbohydrate A polyhydroxy aldehyde, ketone, or substance that yields such compounds on hydrolysis.

carbonyl group The $-\overset{\overset{\displaystyle O}{\|}}{C}-$ group.

carboxyl group The $-\overset{\overset{\displaystyle O}{\|}}{C}-OH$ group.

carboxylate ion The $R-\overset{\overset{\displaystyle O}{\|}}{C}-O^-$ ion that results from the dissociation of a carboxylic acid.

carboxylic acid An organic compound that contains the $-\overset{\overset{\displaystyle O}{\|}}{C}-OH$ functional group.

carboxylic acid anhydride An organic compound that contains the $-\overset{\overset{\displaystyle O}{\|}}{C}-O-\overset{\overset{\displaystyle O}{\|}}{C}-$ functional group.

carboxylic acid chloride An organic compound that contains the $-\overset{\overset{\displaystyle O}{\|}}{C}-Cl$ functional group.

carboxylic ester A compound with the $-\overset{\overset{\displaystyle O}{\|}}{C}-OR$ functional group.

catabolism All reactions involved in the breakdown of biomolecules.

catalyst A substance that changes (usually increases) reaction rates without being used up in the reaction.

cation A positively charged ion.

cellular respiration The entire process involved in the use of oxygen by cells.

central dogma of molecular biology The well-established process by which genetic information stored in DNA molecules is expressed in the structure of synthesized proteins.

cephalin A phosphoglyceride containing ethanolamine or serine.

chain reaction A nuclear reaction in which the products of one reaction cause a repeat of the reaction to take place.

Charles's law A gas law that describes the temperature and volume behavior of a gas sample kept at constant pressure. Mathematically, it is $V/T = k'$.

chemical change A change matter undergoes that involves changes in composition.

chemical property A property matter demonstrates when attempts are made to change it into new substances.

chemiosmotic hypothesis The postulate that a proton flow across the inner mitochondrial membrane during operation of the electron transport chain provides energy for ATP synthesis.

chiral A descriptive term for compounds or objects that cannot be superimposed on their mirror image.

chiral carbon A carbon atom with four different groups attached.

chloride shift The movement of chloride ions from one fluid to another to maintain charge neutrality.

chromosome A tightly packaged bundle of DNA and protein that is involved in cell division.

chylomicron A lipoprotein aggregate found in the lymph and the bloodstream.

cis- On the same side (as applied to geometric isomers).

citric acid cycle A series of reactions in which acetyl CoA is oxidized to carbon dioxide and reduced forms of coenzymes FAD and NAD$^+$ are produced.

codon A sequence of three nucleotide bases that represents a code word on mRNA molecules.

coenzyme An organic molecule required by an enzyme for catalytic activity.

cofactor A nonprotein molecule or ion required by an enzyme for catalytic activity.

cohesive force The attractive force between particles; it is associated with potential energy.

cold spot Tissue from which a radioactive tracer is excluded or rejected.

colligative property A solution property that depends only on the concentration of solute particles in solution.

colloid A homogeneous mixture of two or more substances in which the dispersed substances are present as larger particles than are found in solutions.

combination reaction A chemical reaction in which two or more substances react to form a single substance.

combined gas law A gas law that describes the pressure, volume, and temperature behavior of a gas sample. Mathematically, it is $PV/T = k''$.

common catabolic pathway The reactions of the citric acid cycle plus those of the electron transport chain and oxidative phosphorylation.

competitive inhibitor An inhibitor that competes with substrate for binding at the active site of an enzyme.

complementary DNA strands Two strands of DNA in a double-helical form such that adenine and guanine of one strand are matched and hydrogen bonded to thymine and cytosine, respectively, of the second strand.

complete protein Protein in food that contains all essential amino acids in the proportions needed by the body.

complex carbohydrates The polysaccharides amylose and amylopectin, collectively called starch.

complex lipid An ester-containing lipid with more than two types of components: an alcohol, fatty acids, plus other components.

compound A pure substance consisting of two or more kinds of atoms in the form of heteroatomic molecules or individual atoms.

compound formula A symbol for the molecule of a compound, consisting of the symbols of the atoms found in the molecule. Atoms present in numbers higher than 1 have the number indicated by a subscript.

compressibility The change in volume of a sample resulting from a pressure change acting on the sample.

concentration The amount of solute contained in a specific amount of solution.

condensation An exothermic process in which a gas or vapor is changed to a liquid or solid.

condensation polymerization The process by which monomers combine with the simultaneous elimination of a small molecule.

condensed structural formula A structural molecular formula showing the general arrangement of atoms but without showing all the covalent bonds.

conformations The different arrangements of atoms in space achieved by rotation about single bonds.

conjugate acid–base pair A Brønsted acid and its conjugate base.

conjugate base The species remaining when a Brønsted acid donates a proton.

conjugated protein A protein made up of amino acid residues and other organic or inorganic components.

copolymer An addition polymer formed by the reaction of two different monomers.

Cori cycle The process in which glucose is converted to lactate in muscle tissue, lactate is reconverted to glucose in the liver, and glucose is returned to the muscle.

covalent bond The attractive force that results between two atoms that are both attracted to a shared pair of electrons.

critical mass The minimum amount of fissionable material needed to sustain a critical chain reaction at a constant rate.

critical reaction A constant-rate chain reaction.

crystal lattice A rigid three-dimensional arrangement of particles.

C-terminal residue An amino acid on the end of a chain that has an unreacted or free carboxylate group.

curie A physical unit of radiation measurement corresponding to 3.7×10^{10} nuclear disintegrations per second.

cycloalkane An alkane in which carbon atoms form a ring.

cyclotron A cyclic particle accelerator that works by changing electrical polarities as charged particles cross a gap. The particles are kept in a spiral path by a strong magnetic field.

cytochrome An iron-containing enzyme located in the electron transport chain.

daily reference values (DRV) A set of standards for nutrients and food components (such as fat and fiber) that have important relationships with health; used on food labels as part of the Daily Values.

daily values (DV) Reference values developed by the FDA specifically for use on food labels. The Daily Values represent two sets of standards: Reference Daily Intakes (RDI) and Daily Reference Values (DRV).

Dalton's law of partial pressures The total pressure exerted by a mixture of gases is equal to the sum of the partial pressures of the gases in the mixture.

daughter nuclei The new nuclei produced when unstable nuclei undergo radioactive decay.

decomposition A change in chemical composition that can result from heating.

decomposition reaction A chemical reaction in which a single substance reacts to form two or more simpler substances.

dehydration reaction A reaction in which water is chemically removed from a compound.

denaturation The process by which a protein loses its characteristic native structure and function.

density The number given when the mass of a sample of a substance is divided by the volume of the same sample.

deoxyhemoglobin (hemoglobin) Nonoxygenated hemoglobin.

deoxyribonucleic acid (DNA) A nucleic acid found primarily in the nuclei of cells.

derived unit of measurement A unit obtained by multiplication or division of one or more basic units.

dextrorotatory Describes substances that rotate plane-polarized light to the right.

dialysis A process in which solvent molecules, other small molecules, and hydrated ions pass from a solution through a membrane.

dialyzing membrane A semipermeable membrane with pores large enough to allow solvent molecules, other small molecules, and hydrated ions to pass through.

diatomic molecule A molecule that contains two atoms.

diffusion A process that causes substances to spontaneously intermingle when they are brought together.

dimer Two identical molecules bonded together.

dipeptide A compound formed when two amino acids are bonded by an amide linkage.

dipolar force The attractive force that exists between the positive end of one polar molecule and the negative end of another.

diprotic acid An acid that gives up two protons (H^+) per molecule when dissolved.

disaccharide A carbohydrate formed by the combination of two monosaccharides.

dispersed phase The substance present in a colloidal dispersion in amounts less than the amount of dispersing medium.

dispersing medium The substance present in a collodial dispersion in the largest amount.

dispersion force A very weak attraction force acting between the particles of all matter; results from momentary nonsymmetric electron distributions in molecules or atoms.

disruptive force The force resulting from particle motion; associated with kinetic energy.

dissolving A term used to describe the process of solution formation when the solvent and solutes form a homogeneous mixture.

distinguishing electron The last or highest-energy electron found in an element.

disulfide A compound containing an —S—S— group.

disulfide bridge A bond produced by the oxidation of —SH groups on two cysteine residues. The bond loops or holds two peptide chains together.

double bond The bond resulting from the sharing of two pairs of electrons.

double-replacement reaction A chemical reaction in which two compounds react and exchange partners to form two new compounds.

effective collision A collision that causes a reaction to occur between colliding molecules.

effusion A process in which a gas escapes from a container through a small hole.

electrolyte A solute that when dissolved in water forms a solution that conducts electricity.

electron capture A mode of radioactive decay for some unstable nuclei in which an electron from outside the nucleus is drawn into the nucleus, where it combines with a proton to form a neutron.

electron transport chain A series of reactions in which protons and electrons from the oxidation of foods are used to reduce molecular oxygen to water.

electronegativity The tendency of an atom to attract shared electrons of a covalent bond.

electronic configurations The detailed arrangements of electrons indicated by a specific notation, $1s^2 2s^2 2p^4$, and so on.

element A pure substance consisting of only one kind of atom in the form of homoatomic molecules or individual atoms.

elemental symbol A symbol assigned to an element that is based on the name of the element and consists of one capital letter or a capital letter followed by a lowercase letter.

elimination reaction A reaction in which two or more covalent bonds are broken and a new multiple bond is formed.

emulsifying agent (stabilizing agent) A substance that when added to colloids prevents them from coalescing and settling.

enantiomers Stereoisomers that are mirror images.

endergonic process A process that gains or accepts energy as it takes place.

endothermic reaction A reaction that absorbs heat.

endpoint of a titration The point at which the titration is stopped on the basis of an indicator color change or pH meter reading.

entropy A measurement or indication of the disorder or randomness of a system; the more disorderly a system, the higher its entropy.

enzyme A protein molecule that catalyzes chemical reactions.

enzyme activity The rate at which an enzyme catalyzes a reaction.

enzyme induction The synthesis of an enzyme in response to a cellular need.

enzyme inhibitor A substance that decreases the activity of an enzyme.

enzyme international unit (IU) A quantity of enzyme that catalyzes the conversion of 1 μmol of substrate per minute under specified conditions.

equilibrium concentration The unchanging concentration of reactants and products in a reaction system that is in a state of equilibrium.

equilibrium constant A constant that relates the equilibrium concentrations of products to those of reactants, each being raised to exponents obtained from stoichiometric coefficients.

equilibrium expression An equation relating the equilibrium constant and reactant and product concentrations.

equivalence point of a titration The point at which the unknown solution has exactly reacted with the known solution. Neither is in excess.

equivalent of salt The amount that will produce 1 mol of positive charges on dissolving and dissociating into ions.

essential amino acid An amino acid that cannot be synthesized within the body at a rate adequate to meet metabolic needs.

essential fatty acid A fatty acid needed by the body that cannot be synthesized within the body.

ester A compound in which the —OH of an acid is replaced by an —OR.

ester linkage The carbonyl carbon–oxygen single bond of the ester group.

esterification The process of forming an ester.

ether A compound that contains a $-\overset{|}{\underset{|}{C}}-O-\overset{|}{\underset{|}{C}}-$ functional group.

eukaryotic cell A cell containing membrane-enclosed organelles, particularly a nucleus.

evaporation An endothermic process in which a liquid is changed to a gas; also called vaporization.

exact numbers Numbers that have no uncertainty; numbers from defined relationships, counting numbers, and numbers that are part of reduced simple fractions.

exergonic process A process that gives up energy as it takes place.

exon A segment of a eukaryotic DNA molecule that is coded for amino acids.

exothermic reaction A reaction that liberates heat.

expanded structural formula A structural molecular formula showing all the covalent bonds.

expanding chain reaction A reaction in which the products of one reaction cause more than one more reaction to occur; also called branching chain reaction.

extracellular fluid Body fluid located outside cells.

extremozyme A nickname for certain enzymes isolated from microorganisms that thrive in extreme environments.

factors used in the factor-unit method Fractions obtained from fixed relationships between quantities.

family of the periodic table See **Group of the periodic table**.

fat A triglyceride that is a solid at room temperature.

fat mobilization The hydrolysis of stored triglycerides followed by the entry of fatty acids and glycerol into the bloodstream.

fatty acid A long-chain carboxylic acid found in fats.

feedback inhibition A process in which the end product of a sequence of enzyme-catalyzed reactions inhibits an earlier step in the process.

fermentation A reaction of sugars, starch, or cellulose to produce ethanol and carbon dioxide.

fibrous protein A protein made up of long rod-shaped or stringlike molecules that intertwine to form fibers.

first ionization energy The energy required to remove the first electron from a neutral atom.

Fischer projection A method of depicting three-dimensional shapes for chiral molecules.

fluid-mosaic model A model of membrane structure in which proteins are embedded in a flexible lipid bilayer.

formula weight The sum of the atomic weights of the atoms shown in the formula of an ionic compound.

functional group A unique reactive combination of atoms that differentiates molecules of organic compounds of one class from those of another.

furanose ring A five-membered sugar ring system containing an oxygen atom.

gamma ray A high-energy ray that is like an X ray but with a higher energy.

gas law A mathematical relationship that describes the behavior of gases as they are mixed, subjected to pressure or temperature changes, or allowed to diffuse.

Geiger-Müller tube A radiation-detection device operating on the principle that ions form when radiation passes through a tube filled with low-pressure gas.

gene An individual section of a chromosomal DNA molecule that is the fundamental unit of heredity.

genome A summation of all the genetic material (chromosomes) of a cell; a person's genetic blueprint.

geometric isomers Molecules that differ in the three-dimensional arrangements of their atoms in space and not in the order of linkage of atoms.

globular protein A spherical protein that usually forms stable suspensions in water or dissolves in water.

glucogenic amino acid An amino acid whose carbon skeleton can be converted metabolically to an intermediate used in the synthesis of glucose.

gluconeogenesis The synthesis of glucose from noncarbohydrate molecules.

glucosuria A condition in which elevated blood sugar levels result in the excretion of glucose in the urine.

glycogenesis The synthesis of glycogen from glucose.

glycogenolysis The breakdown of glycogen to glucose.

glycolipid A complex lipid containing sphingosine, a fatty acid, and a carbohydrate.

glycolysis A series of reactions by which glucose is oxidized to pyruvate.

glycoside Another name for a carbohydrate containing an acetal or ketal group.

glycosidic linkage The carbon-oxygen-carbon linkage that joins the components of a glycoside to the ring.

Graham's law A mathematical expression that relates rates of effusion or diffusion of two gases to the masses of the molecules of the two gases.

gray A biological unit of radiation measurement that corresponds to the transfer of 1 J of energy to 1 kg of tissue.

group of the periodic table A vertical column of elements that have similar chemical properties; also called family of the periodic table.

half-life The time required for one-half the unstable nuclei in a sample to undergo radioactive decay.

haloalkane A derivative of an alkane in which one or more hydrogens are replaced by halogens; also called alkyl halide.

Haworth structure A method of depicting three-dimensional carbohydrate structures.

heat of fusion The amount of heat energy required to melt exactly 1 g of a solid substance at constant temperature.

heat of vaporization The amount of heat energy required to vaporize exactly 1 g of a liquid substance at constant temperature.

hemiacetal A compound that contains the functional group

$$-\overset{\displaystyle OH}{\underset{\displaystyle |}{\overset{\displaystyle |}{C}}}-H.$$

OR

hemiketal A compound that contains the $-\overset{\displaystyle OH}{\underset{\displaystyle |}{\overset{\displaystyle |}{C}}}-\overset{}{\underset{\displaystyle |}{\overset{\displaystyle |}{C}}}-\overset{}{\underset{\displaystyle |}{\overset{\displaystyle |}{C}}}-$ group.

OR

Henderson-Hasselbalch equation A relationship between the pH of a buffer, pK_a, and the concentrations of acid and salt in the buffer.

heteroatomic molecule A molecule that contains two or more kinds of atoms.

heterocyclic ring A ring in which one or more atoms is an atom other than carbon.

heterogeneous matter Matter with properties that are not the same throughout the sample.

heterogeneous nuclear RNA (hnRNA) RNA produced when both introns and exons of eukaryotic cellular DNA are transcribed.

heterogeneous catalyst A catalytic substance normally used in the form of a solid with a large surface area on which reactions take place; also called surface catalyst.

high-energy compound A substance that on hydrolysis liberates a great amount of free energy.

homeostasis A condition of chemical balance in the body.

homoatomic molecule A molecule that contains only one kind of atom.

homogeneous catalyst A catalytic substance that is distributed uniformly throughout the reaction mixture.

homogeneous matter Matter that has the same properties throughout the sample.

homologous series Compounds of the same functional class that differ by a —CH_2— group.

hormone A chemical messenger secreted by specific glands and carried by the blood to a target tissue, where it triggers a particular response.

hot spot Tissue in which a radioactive tracer concentrates.

Hund's rule A statement of the behavior of electrons when they occupy orbitals: Electrons will not join other electrons in an orbital if an empty orbital of the same energy is available for occupancy.

hybrid orbital An orbital produced from the combination of two or more nonequivalent orbitals of an atom.

hydrate A salt that contains specific numbers of water molecules as part of the solid crystalline structure.

hydrated ion An ion in solution that is surrounded by water molecules.

hydration The addition of water to a multiple bond.

hydrocarbon An organic compound that contains only carbon and hydrogen.

hydrogen bonding The result of attractive dipolar forces between molecules in which hydrogen atoms are covalently bonded to very electronegative elements (O, N, or F).

hydrogenation A reaction in which the addition of hydrogen takes place.

hydrolysis Bond breakage by reaction with water.

hydrolysis reaction Any reaction with water; for salts, a reaction of the acidic cation and/or basic anion of the salt with water.

hydrophobic Characterizing molecules or parts of molecules that repel (are insoluble in) water.

hydroxy group The —OH functional group.

hyperglycemia A higher-than-normal blood sugar level.

hyperventilation Rapid, deep breathing.

hypoglycemia A lower-than-normal blood sugar level.

hypoventilation Slow, shallow breathing.

ideal gas law A gas law that relates the pressure, volume, temperature, and number of moles in a gas sample. The equation is $PV = nRT$.

immiscible A term used to describe liquids that are insoluble in each other.

inborn error of metabolism A disease in which a genetic change causes a deficiency of a particular protein, often an enzyme.

induced-fit theory The theory of enzyme action proposing that the conformation of an enzyme changes to accommodate an incoming substrate.

inhibitor A substance that decreases reaction rates.

inner-transition element An element in which the distinguishing electron is found in an *f subshell*.

inorganic chemistry The study of the elements and all noncarbon compounds.

insoluble substance A substance that does not dissolve to a significant extent in the solvent.

internal energy The energy associated with vibrations within molecules.

International Unit (IU) A measure of vitamin activity, determined by biological methods.

interstitial fluid The fluid surrounding individual tissue cells.

intracellular fluid Body fluid located inside cells.

intron A segment of a eukaryotic DNA molecule that carries no codes for amino acids.

inverse square law of radiation A mathematical way of saying that the intensity of radiation is inversely proportional to the square of the distance from the source of the radiation.

invert sugar A mixture of equal amounts of glucose and fructose.

ionic bond The attractive force that holds together ions of opposite charge.

ion product of water The equilibrium constant for the dissociation of pure water into H_3O^+ and OH^-.

isoelectric point The characteristic solution pH at which an amino acid has a net charge of zero.

isoelectronic A term meaning "same electronic," used to describe atoms or ions that have identical electronic configurations.

isoenzyme A slightly different form of the same enzyme produced by different tissues.

isomerism A property in which two or more compounds have the same molecular formula but different arrangements of atoms.

isotopes Atoms having the same atomic number but different mass numbers; that is, they are atoms of the same element that contain different numbers of neutrons in their nuclei.

ketal A compound that contains the
$$-\overset{|}{\underset{|}{C}}-\overset{OR}{\underset{OR}{\overset{|}{\underset{|}{C}}}}-\overset{|}{\underset{|}{C}}-\text{ group.}$$

ketoacidosis A low blood pH due to elevated levels of ketone bodies.

ketogenic amino acid An amino acid whose carbon skeleton can be converted metabolically to acetyl CoA or acetoacetyl CoA.

ketone A compound that contains the
$$-\overset{|}{\underset{|}{C}}-\overset{O}{\overset{\parallel}{C}}-\overset{|}{\underset{|}{C}}-\text{ group;}$$
the general formula is $R-\overset{O}{\overset{\parallel}{C}}-R'$.

ketone bodies Three compounds—acetoacetate, β-hydroxybutyrate, and acetone—formed from acetyl CoA.

ketonemia An elevated level of ketone bodies in the blood.

ketonuria The presence of ketone bodies in the urine.

ketosis A condition in which ketonemia, ketonuria, and acetone breath exist together.

kinetic energy The energy a particle has as a result of its motion. Mathematically, it is $KE = \frac{1}{2}mv^2$.

lactate fermentation The production of lactate from glucose.

lattice site The individual location occupied by a particle in a crystal lattice.

law of conservation of matter Atoms are neither created nor destroyed in chemical reactions.

Le Châtelier's principle The position of equilibrium will shift to compensate for changes made that upset a system at equilibrium.

lecithin A phosphoglyceride containing choline.

levorotatory Describes substances that rotate plane-polarized light to the left.

Lewis structure A representation of an atom or ion in which the elemental symbol represents the atomic nucleus and all but the valence-shell electrons. The valence-shell electrons are represented by dots arranged around the elemental symbol.

limiting reactant The reactant present in a reaction in the least amount based on its reaction coefficients and molecular weight. The limiting reactant determines the maximum amount of product that can be formed.

limiting-reactant principle The maximum amount of product possible from a reaction is determined by the amount of reactant present in the least amount, based on its reaction coefficient and molecular weight.

linear accelerator A particle accelerator that works by changing electrical polarities as charged particles cross gaps between segments of a long tube.

lipid A biological compound that is soluble only in nonpolar solvents.

lipid bilayer A structure found in membranes, consisting of two sheets of lipid molecules arranged so that the hydrophobic portions are facing each other.

lock-and-key theory The theory of enzyme specificity proposing that a substrate has a shape fitting that of the enzyme's active site, as a key fits a lock.

macronutrient A substance needed by the body in relatively large amounts.

major mineral A mineral found in the body in quantities greater than 5 g.

Markovnikov's rule In the addition of H—X to an alkene, the hydrogen becomes attached to the carbon atom that is already bonded to more hydrogens.

mass A measurement of the amount of matter in an object.

mass number of an atom A number equal to the sum of the number of protons and neutrons in the nucleus of an atom. Symbolically, it is represented by A.

matter Anything that has mass and occupies space.

melting point The temperature at which a solid changes to a liquid; the solid and liquid have the same vapor pressure.

messenger RNA (mRNA) RNA that carries genetic information from the DNA in the cell nucleus to the site of protein synthesis in the cytoplasm.

metabolic acidosis A condition of acidosis resulting from causes other than hypoventilation.

metabolic alkalosis A condition of alkalosis resulting from causes other than hyperventilation.

metabolic pathway A sequence of reactions used to produce one product or accomplish one process.

metabolism The sum of all reactions occurring in an organism.

metal An element found in the left two-thirds of the periodic table. Most have the following properties: high thermal and electrical conductivities, high malleability and ductility, and a metallic luster.

metallic bond An attractive force responsible for holding solid metals together, originating from the attraction between positively charged atomic kernels that occupy the lattice sites and mobile electrons that move freely through the lattice.

metalloids Elements that form a narrow diagonal band in the periodic table between the metals and nonmetals; they have properties somewhat between those of metals and nonmetals.

micelle A spherical cluster of molecules in which the polar portions of the molecules are on the surface and the nonpolar portions are located in the interior.

micronutrient A substance needed by the body only in small amounts.

mineral A metal or nonmetal used in the body in the form of ions or compounds.

mitochondrion A cellular organelle where reactions of the common catabolic pathway occur.

mixture A physical blend of matter that can be physically separated into two or more components.

moderator A material capable of slowing down neutrons that pass through it.

modulator A substance that binds to an enzyme at a location other than the active site and alters the catalytic activity.

molarity A solution concentration that is expressed in terms of the number of moles of solute contained in 1 L of solution.

mole The number of particles (atoms or molecules) contained in a sample of element or compound with a mass in grams equal to the atomic or molecular weight, respectively.

molecular equation An equation written with each compound represented by its un-ionized formula.

molecular weight The relative mass of a molecule expressed in atomic mass units and calculated by adding the atomic weights of the atoms in the molecule.

molecule The smallest particle of a pure substance that has the properties of that substance.

monomer The starting material that becomes the repeating unit of polymers.

monoprotic acid An acid that gives up only one proton (H^+) per molecule when dissolved.

monosaccharide A simple carbohydrate most commonly consisting of three to six carbon atoms.

mutagen A chemical that induces mutations by reacting with DNA.

mutation Any change resulting in an incorrect base sequence on DNA.

native state The natural three-dimensional conformation of a functional protein.

net ionic equation An equation that contains only un-ionized or insoluble materials and ions that undergo changes as the reaction proceeds. All spectator ions are eliminated.

network solid A solid in which the lattice sites are occupied by atoms that are covalently bonded to each other.

neurotransmitter A substance that acts as a chemical bridge in nerve impulse transmission between nerve cells.

neutral A term used to describe any water solution in which the concentrations of H_3O^+ and OH^- are equal; also, a water solution with pH = 7.

neutralization reaction A reaction in which an acid and base react completely, leaving a solution that contains only a salt and water.

noble gas configuration An electronic configuration in which the last eight electrons occupy and fill the *s* and *p* subshells of the highest-occupied shell.

noncompetitive inhibitor An inhibitor that binds to the enzyme at a location other than the active site.

nonelectrolyte A solute that when dissolved in water forms a solution that does not conduct electricity.

nonessential amino acid An amino acid that can be synthesized within the body in adequate amounts.

nonmetal An element found in the right one-third of the periodic table. Most occur as brittle, powdery solids or gases and have properties generally opposite those of metals.

nonpolar covalent bond A covalent bond in which the bonding pair of electrons is shared equally by the bonded atoms.

nonpolar molecule A molecule that contains no polarized bonds, or a molecule containing polarized bonds in which the resulting charges are distributed symmetrically throughout the molecule.

normal alkane Any alkane in which all the carbon atoms are aligned in a continuous chain.

normal boiling point The temperature at which the vapor pressure of a liquid is equal to one standard atmosphere (760 torr); also called standard boiling point.

N-terminal residue An amino acid on the end of a chain that has an unreacted or free amino group.

nuclear fission A process in which large nuclei split into smaller, approximately equal-sized nuclei when bombarded by neutrons.

nuclear fusion A process in which small nuclei combine or fuse to form larger nuclei.

nucleic acid A biomolecule involved in the transfer of genetic information from existing cells to new cells.

nucleic acid backbone The sugar-phosphate chain that is common to all nucleic acids.

nucleotide The repeating structural unit or monomer of polymeric nucleic acids.

nucleus The central core of atoms that contains protons, neutrons, and most of the mass of atoms.

nutrition An applied science that studies food, water, and other nutrients and the ways living organisms use them.

octet rule A rule for predicting electron behavior in reacting atoms: Atoms will gain or lose sufficient electrons to achieve an outer electron arrangement identical to that of a noble gas. This arrangement usually consists of eight electrons in the valence shell.

oil A triglyceride that is a liquid at room temperature.

Okazaki fragment A DNA fragment produced during replication as a result of strand growth in a direction away from the replication fork.

optically active molecule A molecule that rotates the plane of polarized light.

optimum pH The pH at which enzyme activity is highest.

optimum temperature The temperature at which enzyme activity is highest.

organelle A membrane-enclosed structure in the interior of a cell.

organic chemistry The study of carbon-containing compounds.

organic compound A compound that contains the element carbon.

osmolarity The product of n and M in the equation $\pi = nMRT$.

osmosis The process in which solvent flows through a semipermeable membrane into a solution.

osmotic pressure The hydrostatic pressure required to prevent the net flow of solvent through a semipermeable membrane into a solution.

oxidation Originally, a process involving a reaction with oxygen. Today it means a number of things, including a process in which electrons are given up, hydrogen is lost, or an oxidation number increases.

oxidation number A positive or negative number assigned to the elements in chemical formulas according to a specific set of rules; also called oxidation state.

oxidation state See **oxidation number**.

oxidative deamination An oxidation process resulting in the removal of an amino group.

oxidative phosphorylation A process coupled with the electron transport chain whereby ADP is converted to ATP.

oxidizing agent The substance that contains an element that is reduced during a chemical reaction.

oxyhemoglobin Hemoglobin combined with oxygen.

partial pressure The pressure an individual gas of a mixture would exert if it were in the container alone at the same temperature as the mixture.

Pauli exclusion principle A statement of the behavior of electrons when they occupy orbitals: Only electrons spinning in opposite directions can simultaneously occupy the same orbital.

peptide An amino acid polymer of short chain length.

peptide linkage (peptide bond) The amide linkage between amino acids that results when the amino group of one acid reacts with the carboxylate group of another.

percent A solution concentration that expresses the amount of solute in 100 parts of solution.

percentage yield The percentage of the theoretical amount of product actually produced by a reaction.

periodic law A statement about the behavior of the elements when they are arranged in a specific order. In its present form, it is stated as follows: Elements with similar chemical properties occur at regular (periodic) intervals when the elements are arranged in order of increasing atomic numbers.

period of the periodic table A horizontal row of elements.

pH The negative logarithm of the molar concentration of H^+ (H_3O^+) in a solution.

phenol A compound in which an —OH group is connected to a benzene ring. The parent compound is also called phenol.

phenyl group A benzene ring with one hydrogen absent, C_6H_5—.

phosphoglyceride A complex lipid containing glycerol, fatty acids, phosphoric acid, and an alcohol component.

phospholipid A phosphorus-containing lipid.

phosphoric anhydride A compound that contains the

$$-O-\overset{\overset{\displaystyle O}{\|}}{\underset{\underset{\displaystyle O^-}{|}}{P}}-O-\overset{\overset{\displaystyle O}{\|}}{\underset{\underset{\displaystyle O^-}{|}}{P}}-O-\text{ group.}$$

physical change A change matter undergoes without changing composition.

physical property A property of matter that can be observed or measured without trying to change the composition of the matter being studied.

physical unit of radiation A radiation measurement unit that indicates the activity of the source of the radiation; for example, the number of nuclear decays per minute.

pK_a The negative logarithm of K_a.

plasma The liquid portion of whole blood.

plasmid Circular, double-stranded DNA found in the cytoplasm of bacterial cells.

polar covalent bond A covalent bond that shows bond polarization; that is, the bonding electrons are shared unequally.

polar molecule A molecule that contains polarized bonds and in which the resulting charges are distributed asymmetrically throughout the molecule.

polyatomic ion A covalently bonded group of atoms that carries a net electrical charge.

polyatomic molecule A molecule that contains three or more atoms.

polycyclic aromatic compound A derivative of benzene in which carbon atoms are shared between two or more benzene rings.

polyfunctional compound A compound with two or more functional groups.

polymer A very large molecule made up of repeating units.

polymerization A reaction that produces a polymer.

polypeptide An amino acid polymer of intermediate chain length containing up to 50 amino acid residues.

polyribosome A complex of mRNA and several ribosomes; also called polysome.

polysaccharide A carbohydrate formed by the combination of many monosaccharide units.

polysome See **polyribosome**.

polyunsaturated A term usually applied to molecules with several double bonds.

position of equilibrium An indication of the relative amounts of reactants and products present at equilibrium.

positron A positively charged electron.

potential energy The energy a particle has as a result of attractive or repulsive forces acting on it.

pressure A force per unit area of surface on which the force acts. In measurements and calculations involving gases, it is often expressed in units related to measurements of atmospheric pressure.

primary alcohol An alcohol in which the carbon bearing the —OH group is attached to one other carbon atom.

primary amine An amine having one alkyl or aromatic group bonded to nitrogen, as in $R—NH_2$.

primary protein structure The linear sequence of amino acid residues in a protein chain.

product of a reaction A substance produced as a result of a chemical reaction; written on the right side of the equation representing the reaction.

prokaryotic cell A simple unicellular organism that contains no nucleus and no membrane-enclosed organelles.

prostaglandin A substance derived from unsaturated fatty acids, with hormonelike effects on a number of body tissues.

prosthetic group The non–amino acid parts of conjugated proteins.

protein An amino acid polymer made up of more than 50 amino acids.

protein turnover The continuing process in which body proteins are hydrolyzed and resynthesized.

pure substance Matter that has a constant composition and fixed properties.

pyranose ring A six-membered sugar ring system containing an oxygen atom.

quaternary ammonium salt An ionic compound containing a positively charged ion in which four alkyl or aromatic groups are bonded to nitrogen, as in R_4N^+.

quaternary protein structure The arrangement of subunits that form a larger protein.

rad A biological unit of radiation measurement that corresponds to the transfer of 1×10^{-2} J or 2.4×10^{-3} cal of energy to 1 kg of tissue.

radical An electron-deficient particle that is very reactive; also called free radical.

radioactive dating A process for determining the age of artifacts and rocks, based on the amount and half-life of radioisotopes contained in the object.

radioactive decay A process in which an unstable nucleus changes energy states and in the process emits radiation.

radioactive nuclei Nuclei that undergo spontaneous changes and emit energy in the form of radiation.

radioisotope An isotope of an element that emits nuclear radiation.

reactant of a reaction A substance that undergoes chemical change during a reaction; written on the left side of the equation representing the reaction.

reaction mechanism A detailed explanation of how a reaction actually takes place.

reaction rate The speed at which a chemical reaction takes place.

recombinant DNA DNA of an organism that contains genetic material from another organism.

reducing agent The substance that contains an element that is oxidized during a chemical reaction.

reducing sugar A sugar that can be oxidized by Cu^{2+} in solution.

reduction Originally, a process in which oxygen was lost. Today it means a number of things, including a process in which electrons are gained, hydrogen is accepted, or an oxidation number decreases.

reference daily intakes (RDI) A set of standards for protein, vitamins, and minerals used on food labels as part of the Daily Values.

relative specificity The characteristic of an enzyme that it acts on several structurally related substances.

rem A biological unit of radiation measurement that corresponds to the health effect produced by one roentgen of gamma rays or X rays regardless of the type of radiation involved.

renal threshold The blood glucose level at which glucose begins to be excreted in the urine.

replication The process by which an exact copy of a DNA molecule is produced.

replication fork A point where the double helix of a DNA molecule unwinds during replication.

representative element An element in which the distinguishing electron is found in an s or a p subshell.

respiratory acidosis A condition of acidosis caused by hypoventilation.

respiratory alkalosis A condition of alkalosis caused by hyperventilation.

restriction enzyme A protective enzyme found in some bacteria that catalyzes the cleaving of all but a few specific types of DNA.

retrovirus A virus in which RNA directs the synthesis of DNA.

ribonucleic acid (RNA) A nucleic acid found mainly in the cytoplasm of cells.

ribosomal RNA (rRNA) RNA that constitutes about 65% of the material in ribosomes, the sites of protein synthesis.

ribosome A subcellular particle that serves as the site of protein synthesis in all organisms.

roentgen A biological unit of radiation measurement used with X rays and gamma rays; the quantity of radiation that generates 2.1×10^9 ion pairs per cubic centimeter of dry air or 1.8×10^{12} ion pairs per gram of tissue.

salt A solid crystalline ionic compound at room temperature that contains the cation of a base and the anion of an acid.

saponification The basic cleavage of an ester linkage.

satiety A state of satisfaction or fullness.

saturated hydrocarbon Another name for an alkane.

saturated solution A solution that contains the maximum amount possible of dissolved solute in a stable situation under the prevailing conditions of temperature and pressure.

scientific models Explanations for observed behavior in nature.

scientific notation A way of representing numbers consisting of a product between a nonexponential number and 10 raised to a whole-number exponent that may be positive or negative.

scintillation counter A radiation-detection device that works on the principle that phosphors give off light when struck by radiation.

secondary alcohol An alcohol in which the carbon bearing the —OH group is attached to two other carbon atoms.

secondary amine An amine having two alkyl or aromatic groups bonded to nitrogen, as in R_2NH.

secondary protein structure The arrangement of protein chains into patterns as a result of hydrogen bonds between amide groups of amino acid residues in the chain. The common secondary structures are the α-helix and the β-pleated sheet.

semiconservative replication A replication process that produces DNA molecules containing one strand from the parent and a new strand that is complementary to the strand from the parent.

shape Shape depends on the physical state of matter.

shell A location and energy of electrons around a nucleus that is designated by a value for n, where $n = 1, 2, 3$, etc.

side reaction A reaction that does not give the desired product of a reaction.

significant figures The numbers in a measurement that represent the certainty of the measurement, plus one number representing an estimate.

simple carbohydrates Monosaccharides and disaccharides, commonly called sugars.

simple ion An atom that has acquired a net positive or negative charge by losing or gaining electrons.

simple lipid An ester-containing lipid with just two types of components: an alcohol and one or more fatty acids.

simple protein A protein made up entirely of amino acid residues.

single-replacement reaction A chemical reaction in which an element reacts with a compound and displaces another element from the compound.

soap A salt of a fatty acid often used as a cleaning agent.

solubility The maximum amount of solute that can be dissolved in a specific amount of solvent under specific conditions of temperature and pressure.

soluble substance A substance that dissolves to a significant extent in the solvent.

solute One or more substances present in a solution in amounts less than that of the solvent.

solution A homogeneous mixture of two or more pure substances.

solvent The substance present in a solution in the largest amount.

specific heat The amount of heat energy required to raise the temperature of exactly 1 g of a substance by exactly 1°C.

spectator ion An ion in a total ionic reaction that is not changed as the reaction proceeds. It appears in identical forms on the left and right sides of the equation.

sphingolipid A complex lipid containing the aminoalcohol sphingosine.

spontaneous process A process that takes place naturally with no apparent cause or stimulus.

stable substance A substance that does not undergo spontaneous changes under the surrounding conditions.

standard atmosphere The pressure needed to support a 760-mm column of mercury in a barometer tube.

standard boiling point See **normal boiling point**.

standard conditions A set of specific temperature and pressure values used for gas measurements.

standard position for a decimal In scientific notation, the position to the right of the first nonzero digit in the nonexponential number.

state of equilibrium A condition in a reaction system in which the rates of the forward and reverse reactions are equal.

stem cell A cell with the ability to divide for indefinite periods in culture and to give rise to specialized cells.

stereochemical specificity An enzyme's ability to distinguish between stereoisomers.

stereoisomers Compounds with the same structural formula but different spatial arrangements of atoms.

steroid A compound containing four rings fused in a particular pattern.

stoichiometry The study of mass relationships in chemical reactions.

strong acid or base An acid or base that dissociates (ionizes) essentially completely when dissolved to form a solution.

structural isomers Compounds that have the same molecular formula but in which the atoms bond in different patterns.

sublimation The endothermic process in which a solid is changed directly to a gas without first becoming a liquid.

subshell A component of a shell that is designated by a letter from the group $s, p, d,$ and f.

substrate The substance that undergoes a chemical change catalyzed by an enzyme.

subunit A polypeptide chain having primary, secondary, and tertiary structural features that is a part of a larger protein.

sulfhydryl group The —SH functional group.

supercritical mass The minimum amount of fissionable material that must be present to cause a branching chain reaction to occur.

supercritical reaction A branching chain reaction.

supersaturated solution An unstable solution that contains an amount of solute greater than the solute solubility under the prevailing conditions of temperature and pressure.

surface catalyst See **heterogeneous catalyst**.

tertiary alcohol An alcohol in which the carbon bearing the —OH group is attached to three other carbon atoms.

tertiary amine An amine having three alkyl or aromatic groups bonded to nitrogen, as in R_3N.

tertiary protein structure A specific three-dimensional shape of a protein resulting from interactions between R groups of the amino acid residues in the protein.

thermal expansion A change in volume of a sample resulting from a change in temperature of the sample.

thermonuclear reaction A nuclear fusion reaction that requires a very high temperature to start.

thioester bond The carbon–sulfur bond of the
$$\overset{\text{O}}{\underset{}{\overset{\|}{-\text{C}}}}-\text{S}-$$
functional group.

thiol A compound containing an —SH group.

titration An analytical procedure in which one solution (often a base) of known concentration is slowly added to a measured volume of an unknown solution (often an acid). The volume of the added solution is measured with a buret.

Tollens' reagent A mild oxidizing solution containing silver ions used to test for aldehydes.

torr The pressure needed to support a 1-mm column of mercury in a barometer tube.

total ionic equation An equation written with all soluble ionic substances represented by the ions they form in solution.

trace mineral A mineral found in the body in quantities smaller than 5 g.

tracer A radioisotope used medically because its progress through the body or localization in specific organs can be followed.

trans- On opposite sides (as applied to geometric isomers).

transaminase An enzyme that catalyzes the transfer of an amino group.

transamination The enzyme-catalyzed transfer of an amino group to a keto acid.

transcription The transfer of genetic information from a DNA molecule to a molecule of messenger RNA.

transfer RNA (tRNA) RNA that delivers individual amino acid molecules to the site of protein synthesis.

transition element An element in which the distinguishing electron is found in a *d* subshell.

translation The conversion of the code carried by messenger RNA into an amino acid sequence of a protein.

transuranium element A synthetic element with atomic number greater than that of uranium.

triatomic molecule A molecule that contains three atoms.

triglyceride (triacylglycerol) A triester of glycerol in which all three alcohol groups are esterified.

triple bond The bond resulting from the sharing of three pairs of electrons.

triprotic acid An acid that gives up three protons (H^+) per molecule when dissolved.

turnover number The number of molecules of substrate acted on by one molecule of enzyme per minute.

Tyndall effect A property of colloids in which the path of a beam of light through the colloid is visible because the light is scattered.

universal gas constant The constant that relates pressure, volume, temperature, and the number of moles of gas in the ideal gas law.

unsaturated hydrocarbon A hydrocarbon containing one or more multiple bonds.

urea cycle A metabolic pathway in which ammonium ions are converted to urea.

valence shell The outermost (highest-energy) shell that contains electrons.

vaporization See **evaporation**.

vapor pressure The pressure exerted by vapor that is in equilibrium with its liquid.

vector A carrier of foreign DNA into a cell.

virus An infectious particle composed of only proteins and DNA or RNA.

vitamin An organic nutrient that the body cannot produce in the small amounts needed for good health.

vitamin international unit A measure of vitamin activity, determined by biological methods.

volume/volume percent A concentration that expresses the volume of liquid solute contained in 100 volumes of solution.

VSEPR theory A theory based on the mutual repulsion of electron pairs; used to predict molecular shapes.

water of hydration Water retained as a part of the solid crystalline structure of some salts.

wax An ester of a long-chain fatty acid and a long-chain alcohol.

weak (or moderately weak) acid or base An acid or base that dissociates (ionizes) less than completely when dissolved to form a solution.

weight A measurement of the gravitational force acting on an object.

weight/volume percent A concentration that expresses the grams of solute contained in 100 mL of solution.

weight/weight percent A concentration that expresses the mass of solute contained in 100 mass units of solution.

zwitterion A dipolar ion that carries both a positive and a negative charge as a result of an internal acid-base reaction in an amino acid molecule.

zymogen or proenzyme The inactive precursor of an enzyme.

INDEX

Heartburn, 316
Heat
    absorption of, 223
    addition of, 199
    endothermic process, 223–224
    exothermic process, 223–224
    fusion, 201
    release of, 223
    vaporization, 201
Heating, 131
Heavy metal ions, 455
Helicobacter pylori, 315
Helium, 194
Heme groups, 805
Hemiacetals
    conversion, 477–479
    definition of, 477
    formation of, 480–481, 570–571
Hemiketals, 478–479, 570–571
Hemochromatosis, 90
Hemoglobin
    amino acid residues in, 638
    molecular weight of, 633–634
    physiological importance, 805
    proton release, 808
    quaternary structure, 644
    structure, 644
Hemophilia, 705
Hemoproteins, 637
Henderson-Hasselbalch equation, 313
Heptane, 375
Heroin, 545
Heteroatomic molecules, 9, 12
Heterocyclic compounds, 682
Heterocyclic ring, 452
Heterogeneous catalyst, 264
Heterogeneous matter, 11
Heterogeneous nuclear RNA, 698
Hexane, 196, 375, 387
Hexokinase, 634, 751
High-density lipoproteins, 615, 777
High-energy compound, 731
High-fiber foods, 34
High-fructose corn syrup, 584–585
High-protein diet, 629–630
Histidine, 625, 787
Homoatomic molecules, 9, 11
Homocystinuria, 658
Homogeneous catalyst, 264
Homogeneous matter, 10
Homologous series, 386
Hormones. *See also specific hormone*
    protein, 633
    steroid, 612–614
Hospitals, 260
Hot spot, 341
"Huffing," 186
Human growth hormone, 634, 638
Humans
    basic needs of, 362
    fatty acid synthesis by, 597
    folic acid requirement for, 669
    insulin produced by. *See* Insulin
    short-term whole-body radiation exposure
        effects on, 336
Humectants, 386
Hummingbirds, 579
Hund's rule, 81

Hybrid orbital, 364
Hydrated ions, 221
Hydrates, 297–298
Hydration, 297, 411, 779
Hydrobromic acid, 302
Hydrocarbons, 369, 389
Hydrochloric acid, 294, 297, 302
Hydrogen
    addition of, 476–477
    atoms, 726
    burning of, 150
    conversion of, 351
    oxidation number of, 150, 295
    oxygen binding to, 303
Hydrogenation, 408–409, 601–602
Hydrogen bond(s), 643
    covalent, 285
    formation of, 472, 642
    representation of, 134
    R-group interactions, 641–643
Hydrogen-bonded acids, 501
Hydrogen bonding, 134, 438–439, 687
Hydrogen-boron bond, 127
Hydrogen chloride, 124–125, 286
Hydrogen gas, 295, 445
Hydrogen halides, 409
Hydrogen ions
    concentration of, 291
    excretion, 814
    molar concentration of, 289
    pH and, 290
Hydrogen peroxide, 49, 155
Hydrolysis, 599, 645, 657
Hydrolysis reactions
    description of, 513
    of salts, 310–311
Hydronium ions
    equilibrium concentrations, 302
    formation, 284
Hydrophilic molecule, 604
Hydrophobic, 387
Hydrophobic forces, 643
Hydrophobic interactions, 642–643
Hydrophobic molecule, 604
Hydrostatic pressure, 238
Hydroxide, 290
Hydroxy-bearing carbon atom, 437
Hydroxyl group, 434
Hypercholesterolemia, 675
Hyperglycemia, 748
Hypertension, 109
Hyperventilation, 814
Hypoglycemia, 748
Hypothermia, 19, 265
Hypoventilation, 814

I

Ibuprofen, 538
Ice, 134
Icebergs, 135
Ice storms, 388
Ideal gas law, 191
Immiscible, 217
Inborn errors of metabolism, 658
Indicator, of organic compound, 308
Indicator papers, 308
Induced-fit theory, 661
Induced nuclear reactions, 344–347

Inhalant abuse, 186
Inhibitor, 264
Initiation complex, 702
Inner-transition elements, 87
Inorganic acids, 515–517
Inorganic chemistry, 362
Inorganic compounds, 363
Inorganic phosphate, 731
Insoluble substance, 217
Insulin
    amino acid sequences for, 631
    biochemical balance, 765
    molecular weight of, 634
    primary structure, 638
Interionic attractions, 220
Intermolecular hydrogen bonds, 546
Internal energy, 259–260
International Union of Pure and Applied
    Chemistry (IUPAC) nomenclature system
    alcohols, 434
    alkenes, 400
    amides, 545
    aromatic compounds, 419
    carboxylic acids, 498
    description of, 376
    esters, 510
    ethers, 452
    rules, 381
International unit, 662
Interparticle forces, 131–137
Interstitial fluid, 803–804
Interstitial spaces, 809
Intracellular fluid, 803–804
Intramolecular reaction, 480
Introns, 698
Inverse square law of radiation, 337
Invert sugar, 571, 580
Iodine, 91, 726–727
Iodine-123, 341
Iodine-131, 341
Ion(s). *See also specific ions*
    carboxylate, 502
    formation of, 242
    rules, 150–151
    spectator, 159
Ionic attractions, 643
Ionic bond, 106–107
Ionic bonding, 106–108, 661
Ionic compounds, 108–110
    dissolution, 220
    formulas, 112–113
    names, 110–111
    naming, 129
    smallest unit, 112–113
    solubilities, 222
Ionic equations, 157, 159
Ionic shapes, 120–121
Ionic solute, 220
Ionizable hydrogens, 303
Ionization energy, 93
Ion product, 288
Iron, 90, 726–727
Irradiated food, 350
Irreversible inhibition, of enzymes, 664–666
Isobutane, 375
Isocitrate dehydrogenase, 756
Isoelectric point, 626
Isoelectronic, 107

oxygenation, 808
oxygen in
    combination, 641
    exchange, 808
    hemoglobin reaction, 806
    transfusion of, 817
Redox reactions, 161
Reducing agent, 152, 628
Reducing sugars, 573
Reduction, 150
Reduction-oxidation reactions, 149–154
Reference daily intake, 718
Relative masses, 54
Relative specificity, 656
Rem, 338–339
Renal threshold, 748
Replacement reactions, 155–157
Replication, 689–691
Replication fork, 690
Representative elements, 87
Reproductive rights, 710
Repulsion, 181
Resiniferatoxin, 483
Respiratory acidosis, 816
Respiratory alkalosis, 816
Restriction enzymes, 706
Reverse activation energies, 271
Reversible inhibition, of enzymes,
    668–670
R groups
    extension of, 639
    interactions of, 641–643
    polarity of, 624
    tertiary protein structure, 642
Rheumatoid arthritis, 612
Rhubarb, 497
Riboflavin, 424, 659
Ribonucleic acid, 576, 692–696
    cellular distribution of, 694
    characteristics of, 692
    definition of, 682
    distribution, 694
    messenger. See Messenger RNA
    polymerase, 697
    ribosomal, 694
    structure of, 692–693
    sugar component, 682
    synthesis of, 697–698
Ribose, 560, 576, 682
Ribosomal RNA, 694
Ribosomal subunit, 702
Ribosome, 694
Ribosome shift, 704
Ribulose, 569
Ring structure, 381
RNA. See Ribonucleic acid
Roentgen, 338
Roentgen, Wilhelm, 339
rRNA. See Ribosomal RNA
R side chains, 624
RTX. See Resiniferatoxin
Rubbing alcohol, 446
Rubidium, 104
Rutherford, Ernest, 75, 347

S

Saccharin, 574
Salt bridges, 642

Salts
    amine, 532–535
    carboxylic acid, 504–505
    definition of, 297
    description of, 283, 297–300
    equivalent of, 299
    hydrolysis reactions of, 310–312
    ions, 299
    potassium, 600
    preparation, 297–298
    quaternary ammonium, 534
    sodium, 600
    table. See Sodium chloride
Saltwater solution, 220
Saponifiable lipids, 594
Saponification, 511, 513–514, 600
Satiety, 721
Saturated fatty acids, 598, 720
Saturated hydrocarbons, 369
Saturated solution, 218
Scanning tunneling microscope, 51
Schrödinger, Ernest, 77
Scientific measurements, 14
Scientific models, 8
Scientific notation, 19–21
Scintillation counter, 340
Scintillations, 340
Sebum, 602
Secondary alcohol, 437, 443
Secondary amine
    cation, 534
    definition of, 527
    hydrogen bonding of, 531
    usage, 535
Secondary protein structures, 638–639
Selenium, 727
Self-ionization, 287–289
Semiconservative replication, 689
Semipermeable membrane, 238
Serine
    phosphate group attachment, 604
    products derived from, 787
    structural formula of, 625
Serotonin, 540–541
Serum albumin, 780
Sex hormones, 613–614
Shared electron pairs, 114
Shells, 77–78
    electron-filling order, 81–82
    electron occupancy, 79
    model, 79–81
    relative energies, 81
    valence, 80
Short-term intense radiation, 336
Short-term whole-body radiation exposure, 337
Sickle-cell anemia, 641
Sickle-cell disease, 638, 641, 705
Sickle-cell trait, 641
Side reactions, 165
Sigma bond, 415
Significant figures, 22–27, 36
Silicon, 727
Silicon dioxide, 133
Silk, 539
Silver chloride, 262
Simple carbohydrates, 719
Simple ion, 106, 150
Simple lipid, 594

Simple protein, 637
Simplified carbon cycle, 728
Single-replacement reaction, 155
SI units, 14
Skin moisturizers, 386
Skunks, 453
Sleep, 796
Small numbers, 19–22, 36
Smoking
    cessation of, 422, 517
    peptic ulcer disease and, 316
"Sniffing," 186
Soaps
    bile salts and, comparison between, 611
    cleaning action, 242
    creation of, 600
    products, usage, 505
    triglycerides in, 600
Soddy, Frederick, 347
Sodium, 727
Sodium acetate, 311
Sodium benzoate, 505
Sodium carbonate decahydrate, 298
Sodium chloride (table salt)
    crystal lattice, 112
    dissolving, 238, 311
    solution, 159
Sodium glycocholate, 611
Sodium hydroxide, 296, 312
Sodium phosphate dodecahydrate, 298
Sodium salts, 600
Sodium stearate, 505
Sodium sulfate decahydrate, 298
Sodium tetraborate decahydrate, 298
Sodium thiosulfate pentahydrate, 298
Solid oxygen, 135
Solids
    heating of, 199–200
    kinetic molecular view of, 182
    melting of, 135
    properties of, 179, 182
    shape of, 178–179
    vapor pressures, 199
    volume of, 178–179
Solid state, 181–182
Solid water, 135
Solubility, 217–220
Soluble substance, 217
Solute
    definition of, 216, 239
    dispersed phase, 239
    dissociation effects, 237
    dissolution rate for, 223
    excess of, 219
    moles, 225
    solubilities, 218–219, 223
Solute ions, 221
Solution-forming process, 221
Solutions, 215, 220–224
    analysis, 307
    boiling points of, 235
    colligative properties of, 235
    concentrations of, 224–228
    definition of, 11
    dilution of, 231
    dissolved solute, 239
    electrolyte, 234
    freezing points of, 235

Features within the Text